STRUCTURES TECHNOLOGY

Historical Perspec
and
Evolution

STRUCTURES TECHNOLOGY

Historical Perspective
and
Evolution

Ahmed K. Noor, Editor
University of Virginia
Charlottesville, Virginia
NASA Langley Research Center
Hampton, Virginia

AIAA

Published by
American Institute of
Aeronautics and Astronautics, Inc.
Reston, Virginia

1998

Publishers since 1930

American Institute of Aeronautics and Astronautics, Inc., Reston, Virginia

Library of Congress Cataloging-in-Publication Data

Noor, Ahmed K., editor, 1938–
Structures technology : historical perspective and evolution / by Ahmed K. Noor, editor.
p. cm.
Includes bibliographical references and index.
ISBN 1-56347-116-7
1. Airframes—Design and construction. 2. Space vehicles—Design and construction. 3. Structural engineering. I. Noor, Ahmed Khairy, 1938– .

TL671.6.S78 1998
629.1′2—DC21 98-5292
CIP

Table of Contents

Preface

Structures technology encompasses a wide range of component technologies from materials development to analysis, design, testing, production, and maintenance. Materials and structures have been largely responsible for major performance improvements in many aerospace systems. The maturation of computational structures technology and the development of advanced composite materials, witnessed during the past 30 years, have improved structural performance, reduced operational risk, and shortened development time. In addition to being enabling technologies for future aeronautical and space systems, materials and structures continue to be key elements in determining the reliability, performance, testability, and cost effectivenesss of these systems.

Over the past 40 years, several articles in *Aerospace America* and its predecessors have addressed the state of the component structures technologies and their future directions. These articles were authored by leading experts in the field, some of whom are no longer with us today. The present monograph is an attempt to rescue the insights, concerns, and dreams of these authors.

The first paper in the monograph, which is reprinted from the November 1997 issue of *Aerospace America*, deals with new structures for future aerospace systems and provides a contrast between our current thinking and past technology plans. The remainder of the mongraph is divided into two major sections. The first presents a technology overview, and the second examines special aircraft and spacecraft structures. The first section consists of 18 articles and is divided into five subsections covering materials and structures, general structures technology, aircraft structures, space structures, and structural dynamics technology. The second section consists of 31 articles and is divided into five subsections covering structural configurations, thermal-protection systems, subsonic aircraft, supersonic and hypersonic vehicles, and structures for space systems.

The editor expresses his sincere thanks to the individuals who helped in this volume: John Calderone and the staff of AIAA's Books program for their cooperation in publishing this volume, Derek Wilson for improving many of the figures, and Mary Fox for her secretarial support. It is my earnest hope that this effort will serve the profession well by: 1) putting in perspective the thinking process that went into planning the structures technology programs over the past 40 years and 2) showing how close the predictions of eminent scientists and engineers were to reality and the reasons why they were not close in some cases.

Ahmed K. Noor
January 1998
Hampton, Virginia

Introduction

New Structures for New Aerospace Systems

Ahmed K. Noor
University of Virginia's Center for Advanced Computational Technology, NASA Langley Research Center

Samuel L. Venneri
NASA Headquarters

Donald B. Paul
Air Force Research Lab, Wright–Patterson Air Force Base

James C. I. Chang
Air Force Office of Scientific Research

Advances in computational structures technology and composite materials will enable breakthroughs in vehicle performance while reducing cost and schedule. The maturation of this technology and the development of advanced composite materials have improved structural performance, reduced operational risk, and shortened development time in numerous aerospace systems. These systems will face additional challenges in the future.

For aircraft, these include affordability, safety, and environmental compatibility. For military aircraft, emphasis will change from best performance to low cost at acceptable performance. For space systems, new challenges are a result of a shift in strategy from long-term, complex, expensive missions to those that are small, inexpensive, and fast.

Smart Materials and Structures

Smart structures sense external stimuli, process the sensed information, and respond with active control. Response can consist of deforming or deflecting the structure or communicating the information to another control center. Smart materials deform or deflect the structure by changing their physical properties when subjected to electric, magnetic, or thermal loads. An extension of this is the intelligent, self-healing vehicle, whose built-in redundancy and onboard self-inspection detect damage and respond with autonomous adjustments and repair.

The active elements in smart structures can be embedded in or attached to the structure. Typical sensors include fiber optics, piezoelectric ceramics, and polymers. Embedded sensors can be either discrete or distributed to provide built-in structural quality assessment capabilities, both during material processing

Aerospace America, Vol. 35, No. 11, Nov. 1997, pp. 26–31.

and vehicle operation. Sensors can also be used for monitoring in-service or environmental loading and for shape sensing. Typical smart structure actuators include shape memory alloys (SMAs), piezoelectric and electrostrictive ceramics, microelectromechanical devices, magnetostrictive materials, and electro- and magneto-rheological fluids and elastomers.

The first applications of smart materials and structural concepts will be in rotorcraft blades, aircraft wings, air inlets, engine nozzles, large deployable precision space systems, and robust microspacecraft. Expected benefits include enhanced handling qualities (by changing control surface shape to manipulate lift or reduce drag, producing twist in aircraft wings or helicopter rotor blades, or affecting flow conditions over the lifting surface); vibration suppression (including flutter and buffet control); alleviation of noise and vibration; and monitoring of vehicle health. For space systems, smart structures provide a robust design approach for meeting precision requirements. They can significantly reduce cost and schedule by decreasing the requirements on analysis, development testing, hardware processes testing, and quality control.

Shape adaptive structures and aerodynamic load control are being studied by DOD and NASA. Design concepts include airfoil warping, camber shaping/control surface deformation, and variable stiffness structures. The goal of such research is to enhance flight vehicle performance (while reducing weight and the need for discrete, external control surfaces). Some adaptive structures concepts twist the airfoil, vary its camber, and deform leading- and trailing-edge control surfaces through SMA actuation to enhance maneuvering and lift. Such techniques can increase aircraft survivability and reduce drag.

Active aerodynamic load control can be achieved via self-straining actuators (SMAs or piezoelectric devices) embedded within the structure. The actuators expand or contract on command. This changes the shape of the active airfoil element, which in turn changes the aerodynamic load on the lifting surface. Future piezoelectric materials will have to withstand harsh environments and must be expansive enough to deform large aerodynamic surfaces. Anisotropic actuators will be needed to control bending and torsional response independently.

One near term use for smart structures is monitoring of vehicle health. An onboard distributed fiber optic network, connecting sensors to processors, can be used for this purpose. One type of sensor being developed will measure the "sounds" of crack growth—transducers emit acoustic signals throughout the structure and measure changes in the structural response, indicating crack initiation or growth at remote sites. Other sensors detect and measure separation of composite material layers. Each processor receives signals and analyzes an array of sensors to determine if and where damage has occurred.

Two areas of special interest are reducing the oscillations of primary structures due to unsteady external forces and reducing the transmission of acoustic energy through the structure. Active control systems that use piezoelectric actuators are being developed. Actuators will be attached to the skin and substructures of a craft. Alleviation of these dynamic loads will increase structure life and reduce maintenance time and costs.

Some elements of these technologies are already being demonstrated; however, several technical issues must be addressed before the technology's full potential can be realized. These involve structure fabrication methods, reliable actuator material, lightweight structural materials capable of physical and virtual shape changes, and reconfigurable adaptive control systems.

Multifunctional Structures

Multifunctional structures (MFS), in addition to supporting loads, use sensors to detect and evaluate loads or failure and to interact with the surrounding electromagnetic environment. MFS represents a new manufacturing and integration technology by which communications and electronics equipment are integrated into conformal load-bearing structures. The technology is enabled by advances in large-scale integrated electronics packaging, lightweight composite structures, and high-conductivity materials. In MFS, electronic assemblies (multichip modules), miniature sensors, and actuators are embedded into load-carrying structures, along with associated cabling for power and data transmission (Fig. 1). This level of integration effectively eliminates traditional boards and boxes, large connectors, bulky cables, and thermal base plates, yielding major weight, volume, and cost savings.

Current MFS research for aircraft addresses antenna/airframe proliferation, integration, and maintenance. Present flight vehicle designs have almost 100 antennas occupying roughly 60 apertures. To accommodate electromagnetic windows, these antennas require local reinforcement of the airframe structure, increasing structural weight and cost. Externally mounted antennas degrade the aircraft's aerodynamic performance and require extensive maintenance. Integrating antennas and other electronics into load-bearing skin structures will yield lower cost, lighter airframes, increased antenna performance, and lower manufacturing and maintenance costs.

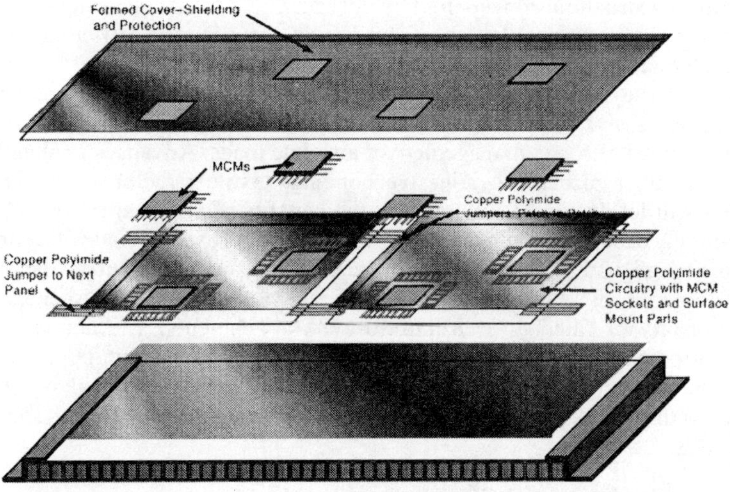

Fig. 1 Multifunctional structure panel with integral electronic, structural, and thermal control.

MFS technology offers significant savings in the mass and volume of spacecraft by eliminating electrical chassis and cabling and placing most sensors and batteries on the bus structure, which also provides structural and thermal control. The technology offers several benefits to future spacecraft. Its inherent modularity supports low-cost mass production and assembly, significantly reducing life cycle costs. It eliminates cables and connectors, minimizing the ''touch labor'' needed during final spacecraft integration, whose robustness and reliability it increases. It enables reworkable ''plug-and-play'' spacecraft. And it provides integrated thermal control, flex interconnect architecture for power distribution, and data storage and electronics in load-bearing structure.

Affordable Composite Structures

For many future aircraft, the use of composite primary structure could have more impact on affordability than any other technology area. The cost of manufacturing composite structure has proved to be the biggest obstacle to its widespread use. This is because of design and manufacturing approaches that use composite materials in the conventional ''metals fashion'' of assembling large numbers of mechanically fastened parts.

Affordable composite structures can be achieved by proper material selection, changing load paths, using robust low-cost manufacturing and joining/assembly techniques, and developing approaches for subsystem integration. A fully coordinated design approach involving larger, integrated components to maximize producibility, quality, and design efficiency is needed to fully exploit the weight and cost benefits of composites. This will require composites to be considered as early as possible in the design process so that load paths are defined that offer manufacturability and do not penalize the composite structure's efficiency.

Low-cost composite manufacturing processes include tow placement, resin transfer molding, resin film infusion, pultrusion, and nonautoclave processing. A promising structural concept for low-cost automated manufacturing is advanced grid stiffened structures, which evolved from early isogrid stiffening concepts and feature a lattice of rigid, interconnected ribs.

Because the properties of composite materials are directionally dependent, they enable a structure's strength and stiffness to tailored in directions that allow the most efficient management of airframe loads. Advances in manufacturing using fiber placement, adhesive bonding, textile structures, and low-cost tooling will let designers fully exploit the benefits of composites.

Naturally occurring composite structural members have evolved with extremely complex load paths for system performance. This is prevalent in the skelctal formation in birds' wings, the damage-tolerant structure of beetle shells, and the directionality of fibers in tree limb-to-trunk attachments. The ability to tailor material properties will allow structural designs to achieve similar complexity. These technologies will be transferred into the automotive, railroad, marine, and infrastructure sectors to produce low-cost, vehicle load-bearing frames and bridge structures.

Extreme Environment Structures

Hot structures are an enabling technology for airframes and engines operating in the high-speed flight regime required for future transpacific and transatmospheric vehicles (TPV and TAV), as well as for space-transportation systems. Airframes will need new lightweight structural concepts that can accept high temperatures (400–1500°F) and high acoustic content (noise levels up to 170 dB). This creates an entirely new environment within which large areas of the vehicle will now be exposed, simultaneously, to extreme thermal and acoustic load levels.

State-of-the-art concepts for such vehicles can easily exceed twice the weight of structures designed for nonextreme environments. Design life requirements of future systems also far exceed those of current vehicles.

For future launch vehicles, advanced materials and structural concepts will be needed for primary structure, leading edges/nosecaps, cryotanks, and thermal-protection systems (TPS). For the primary structure, candidate materials are high-temperature polymeric-matrix and advanced metal-matrix and ceramic-matrix composites. Reliable bonded and bolted joint concepts are needed for these materials. Refractory composites, active cooling, and reusable ablators are being considered for the leading edges/nosecaps. Composite and metallic sandwich construction are candidates for the cryotanks. The tanks must be integrated with the vehicle and with a global health monitoring system during design. Candidate materials for TPS include ultrahigh-temperature ceramic composites and long-life, low-cost carbon-carbon and CMCs. Use of refractory composite hot structures in the primary structure could eliminate the requirement for TPS.

Flexible Load-Bearing Structures

Future flexible load-bearing structures include inflatable deployable aperture structures for antennas, radars, solar sails, and reflectors; flexible wall multilayer structures for lunar and Mars habitats; and novel flexible load-bearing concepts for aircraft structures.

Inflatable deployable structures offer low launch volume and mass. Following the successful flight of the inflatable antenna experiment in May 1996, NASA and DOD undertook a space inflatable technology program. It addresses concepts and component technologies for long-life space missions of the 25-m-class apertures supported by large rigidizable structures. Studies focus on thin-film membrane materials, fabrication techniques, and rigidization methods; inflation systems; and the interfaces between their structure and the rest of the spacecraft. Rigidization methods include gel impregnation, cold rigidization, UV curing, yielding of aluminum laminate, and foam injection. One inflation concept is an onboard gas generation system that uses chemical reactions between liquids or liquids and solids.

Inflatable structures are candidates for the Mars transit vehicle and for the habitats on the lunar/Martian surfaces (Fig. 2). The habitats will encounter large surface temperature gradients, radiation from solar flares and galactic cosmic rays, and micromechanical impacts. The primary force controlling the structural design is the difference between internal and external pressures (external pressure is nearly zero on the Moon and less than 1% of the internal pressure on Mars), and the structure behaves as a pressure vessel. Inflatable structures have low mass and a small ratio of stowable size to deployed volume for economical transportation from Earth. These structures can be prefabricated and tested on Earth, then deployed on site through controlled internal pressurization.

Fig. 2 Flexible wall multilayer structure for lunar and Mars habitat.

Among the novel flexible load-bearing concepts considered for aircraft structures are the expandable fuel cell (EFC) and the compliant trailing edge. Located on the external surface of the vehicle, EFCs are conformal to its outer moldline when they are empty and are inflated when filled with fuel. They significantly increase aircraft range.

The compliant trailing edge integrates structures and control technologies into a continuous trailing-edge surface that fully complies with requirements for aerodynamic performance, flight control, and structural stiffness. The integrated flexible structures provide more-efficient control surfaces (reduced maneuver drag) than do conventional designs.

Practical use of these concepts will require development of technologies to ensure their affordability, durability, and supportability in the operational environment.

Computational Methods and Simulation-Based Design

High-fidelity finite element models are routinely used to predict the loads and responses of aerospace vehicles. Advances are still needed in several areas of computational technology, including computational development of new materials and processes, accurate prediction of damage initiation and propagation and of the safe life of the vehicle, and intelligent simulation-based design. The latter refers to the seamless process of simulating the entire life cycle of the aerospace system before physical prototyping.

Computationally based material development involves hierarchical, physics-based models spanning several length scales, time scales, and disciplines. These range from first-principle quantum mechanics and atomistic models to phenomenological models to process modeling and engineering design. Work will also focus on the ability to model and assess flaws and damage in materials under realistic service conditions, especially in regions of high stress gradients such as joints and interfaces. Eventually, this work will enable materials at the atomic/molecular level to be multifunctional via the specification of particular properties.

Modeling and simulation of a system's entire life cycle demands a high level of confidence and functionality. Among the weak links in the process is the reliable prediction of structural failure modes, ultimate strength, residual strength, and fatigue life. Simulation tools for damage initiation and propagation are needed, as are methodologies for accurate prediction of safe operating cycles for airframes and propulsion systems and of the useful life of space systems. Accomplishing this task will require an understanding of the physical phenomena associated with damage and failure; development of a framework for modeling material degradation and structural damage; hierarchical multiscale computational strategies; novel and accelerated test methods, measurement techniques, and scaling laws; and validation and verification methods.

Intelligent simulation-based design (ISBD) refers to simulation of the entire life cycle of the aerospace system. This is a seamless process performed in a distributed synthetic environment linking design and manufacturing teams, facilities, and resources. The conceptual, preliminary, and detailed design phases will merge into a single continuous design process with progressively refined models. Computational tools include high-fidelity, rapid modeling facilities and physics-based deterministic, nondeterministic, and qualitative simulation tools for structures, aerodynamics, controls, thermal management, power, propulsion, and optics. They also include tools for mission design, cost estimating, product assurance, safety analysis, risk management, virtual manufacturing, qualification testing, and life cycle optimization. Realization of ISBD requires integration and deployment of new technologies, including high-capacity computing, communications, and networking; synthetic/immersive environments; product data management systems; CAD/CAM/CAE systems; computational intelligence and its soft computing tools; knowledge-based engineering; and virtual manufacturing.

An extension of ISBD is the intelligent synthesis environment being developed by NASA. This radically different approach for aerospace system development and mission synthesis provides a holistic view of the development process (including design, manufacturing, certification, operations, and training). It addresses the entire mission and life cycle of the system. Synthesis begins before the requirements are firmed up, much less analyzed. Systems and missions will be rapidly configured and assessed for scientific payoff and system performance. The environment is expected to enhance significantly the creativity and productivity of the engineering and science teams and of the system's affordability and performance.

Future Directions

Not every area of structures technology is expected to yield breakthroughs in the near future. However, steady progress in all the component technologies will improve the structural efficiency of aerospace systems. Demands for efficiency and multifunctionality will drive structures technology to develop new capabilities in which material selection and structural forms are highly complex and integrated.

Future structural research issues involve integrating existing and new materials into functional systems with high quality and low-cost features. Future efforts will address advanced load path management and other technology challenges to enable more-affordable, lighter, stronger, stiffer, safer, and more-durable vehicles for different flight regimes and for planetary atmospheric entry and flights throughout the solar system. These activities will dispel the notion that high performance can be achieved only at high cost. Today, structures technology is derived from mission and vehicle requirements. A cultural shift is needed to effect a change to missions and vehicles that are enabled by innovative structures technology.

Technology Overview

Materials and Structures

Structures and Materials Impasse?

William A. Mrazek
NASA Marshall Space Flight Center

No experienced automotive engineer would start the design of a chassis without knowing whether to design for a one-quarter-ton truck, a convertible, or a sports car. However, designers of the large launch vehicle of the Advanced Saturn class worked for months without such basic definitions and with but one requirement—to meet a firm schedule for a first-launch date. Never before has the designer faced so many critical decisions with such a minimum of available data from which to make assumptions.

With the advent of the space age and with the President's commitment more than a year ago to land an American on the moon and return him alive in this decade, new dimensions and new definitions were coined by the materials and structures men.

A new word, ''manrating,'' for launch vehicles appeared. The definition could be simplified by using the expression, ''safe as a rocking chair.''

But we have already, of course, launched men on existing vehicles, despite the fact that these vehicles, the Redstone and the Atlas, were never built with manned flight in mind. How was the demonstrated flight safety achieved with these vehicles? Entering into redesign for incorporating manrating criteria was not possible in the available time. The only remaining and acceptable approach was a repeated testing of all components, to assure a mission success. This testing procedure was backed up by painstaking preparations, selections, tagging of components, and supervision of the manufacturing of all items going into these vehicles. Even after all these preparations, still-unknown environments, possibly aggravating load factors, could have spelled loss. Then, too, statistics could have caught up with the man in the capsule and forced him to an early abort.

The aircraft industry uses regulations or manuals defining safety aspects, safety factors, and fatigue requirements for structures and materials in airplane construction. There are no such regulations or manuals to govern the design and construction of the large launch vehicles we are now developing to carry multiple-man crews on training flights around the globe and finally on to lunar exploration. From now on, manrating manned launch vehicles will be deeply embedded in every designer's consciousness.

Design approaches in structures, down to the smallest details, are looked upon with the knowledge that the life of men could be endangered by a slight miscalculation. Detailed analysis, more thorough and minute than ever before, considering the most remote cross-feed influences, is being done.

Astronautics, Vol. 8, No. 1, Jan. 1963, pp. 36–39.

Test programs and methods must be more comprehensive and more exhaustive to eliminate all doubts and to establish, under the most adverse load conditions, clear safety margins.

Design must allow manufacturing processes compatible with the state of the art in tooling, with joining methods and, most important, with complete inspection feasibility to establish the quality specified. Design must also establish that the designer and analyst have assumed the right factors for calculations.

At the start of this era of unprecedented large launch vehicles, of extreme exploitation of all known technologies, the designer finds himself more than ever before "boxed-in" between almost solid walls which spell MANRATING/RELIABILITY/PERFORMANCE/STATE OF TECHNOLOGY/SCHEDULE/ and COST.

The conservative aspects of manrating can only be partially expressed in terms of reliability.

Structural design approaches applicable to manrating should avoid a sudden catastrophic failure even with the low percentage of possible losses guaranteed by high reliability.

Two structural design approaches are generally feasible: pressure-stabilized units for prelaunch handling and self-supporting, buckling-resistant, unpressurized, stiff structures.

Considering the complexity of a launch operation—including cost of launch equipment refurbishing time, cost of static-test facilities, and last, but not least, tooling and handling costs—it seemed only reasonable to prefer large launch vehicle structures of self-supporting design over those which need internal pressures to prestress the external load-carrying walls of the containers for the support of their own weight and useful loads above them. In addition to the material to be used, launch-crew and flight-crew safety on the pad is one factor in the choice of the design approach. Table 1 compares structural weights for different design approaches and materials.

A design in steel would result in lower weights compared to one in aluminum if the designs were based on prestressing (pressurizing) the container; this design is ruled out for reasons just given. Aluminum yields lower weights using shell or skin/stringer designs. Additionally, for pad safety, any substantially rigid design will give the flight crew a few more precious seconds to abandon ship (to abort) if commanded or advised to do so by an "emergency detection system."

Table 1 Steel vs aluminum structural weights

	Design determined by:			
	Elastic buckling		Inelastic buckling	
	Shell	Plate	Skin/stringer	Tension
$\frac{W_S}{W_A}$	1.73	2.10	$2.00 \rightarrow 1.00$	0.75
Structural component	← Use aluminum → Unpressurized shrouds, adapters (tanks)			← Use steel → Pressure maintained in tanks

This system would sense decay or loss of pressure in vital parts of the launch-vehicle system and either indicate an unsafe condition or trigger the escape system. In a pressure-stabilized, thin-shell steel design, primary structural failure need not occur for a rapid collapse and disintegration of the shell to come about. Powerplant and accessory failures can trigger conditions for which the shell is simply not designed nor strong enough to withstand.

The question might arise: Is this decision right? Is it an unnecessary sacrifice of performance, using aluminum alloys in conjunction with a self-supporting structure?

If all stages of the C-5 Advanced Saturn were built as a pressure-stabilized steel structure, for example, this would bring approximately 25% weight saving, as indicated in Table 1. The performance calculations show that, for such a C-5, every tenth complete vehicle launched would be a bonus, considering the tradeoff factor for payload. Cost of special tooling and handling equipment, on the other hand, would probably consume or surpass the cost of manufacture of two complete launch vehicles and thus would not allow a "cash-in" on a lighter structure before the 30th flight. This also assumes that no other losses occur due to the lesser ruggedness of a pressurized design. There is no obvious practical payoff in such an approach to structural design.

Performance, cost, state of the art, and schedule have intricate interwoven relationships. All these factors have to be carefully weighed before design decisions are fully justifiable.

But let us shift our attention mainly to materials—alloys for containers which will carry a variety of propellants, from kerosene at room temperature to liquid oxygen and liquid hydrogen at 20 K.

Ease of manufacture (mainly joining), reproducible machine welds, ease of hand-weld repairs of unavoidable weld flaws, ease of hand-welding of pipe connections, filler necks, measuring bosses, etc.—these factors determine the preference for an alloy with good weld properties. The past has seen cooperation with the aluminum industry to explore fully the use of 5000 series alloys and to push toward better properties of these alloys. Table 2 outlines the introduction and the merits of these alloys in launch vehicles.

Even for the C-5 Saturn, 5456 aluminum alloy was originally contemplated, but more thorough metallurgical research disclosed incompatible shortcomings in the form of embrittlement along the weld-bead boundary in weldments above 0.5 in. thickness. It was found that, due to nonideal cooling rates in the thicker welded plates, a second phase (beta) characteristically forms at crown laps and along the line of fusion in inert-arc welds. This beta phase (aluminum-magnesium) was associated with premature weld fracture and accelerated corrosion. It was not practical to eliminate this second phase in weldments by controlled cooling. Although additional research could undoubtedly solve this problem, time was not available for it. So the alloy was not selected for the C-5 first stage.

After careful consideration of all the 2000 series alloys, the 2219 alloy, being simpler in chemistry and having been developed with welding in mind, was selected as the most appropriate replacement for 5456. The parent material and weldments in thicknesses greater than 1/2 in. have been investigated fully at temperatures down to 20 K.

Table 2 Use of aluminum alloys

Year	Vehicle	Type STR weld thickness	Alloy	Ult tens str / Weld str at room temp
1951–55	Redstone	doublers $< \frac{1}{8}''$	5052-N22	31,000 psi / 26–31,000 psi
1954–60	Jupiter ‘C’	butt welds $< \frac{1}{4}''$	5036-N34	44,000 psi / 37–39,000 psi
1955–60	Jupiter	butt welds milling $< \frac{5}{8}''$	early 5086-N34 / later 5456-N34	— / —
1955–60	Thor	butt welds $< \frac{3}{8}''$	2014-T6	70,000 psi / ?–60,000 psi
1957–61	Titan	butt welds $< \frac{3}{8}''$	2014-T6	70,000 psi / ?–60,000 psi
1959–62	Saturn C-1	butt welds $> \frac{3}{8}''$	5456-N34	53,000 psi / 47–49,000 psi
1962	Saturn C-5	butt welds $\gtrsim 2''$	2219-T87	63,000 psi / 37–42,000 psi

Reproducible machine welds with a repeatable ultimate strength for multiple passes of 37,000–42,000 psi are presently the basis for the container designs of the large C-5 vehicle. Proof of the reproducible weldability of this material is given by the fact that several (12 total) weldments of 5 in. thickness are required to manufacture Y-shaped rings used in connecting the shells to the bulkhead. In the process of machining after welding, only approximately 12% of the total weld cross section remains, representing a Y-shaped area. Biaxial yield strength in this welded Y-shaped ring is well above 23,000 psi without cracks or rejectable porosity.

The C-5 upper stages—the S-II built by NAA Space and Information Systems Division and the S-IV and S-IVB built by Douglas Aircraft Company—are constructed with 2014 aluminum alloy, which has acceptable properties in weld thicknesses of less than 3/8 in.

Other aluminum alloys generally used in the launch vehicle, mainly in unpressurized areas, belong to the 6000 and 7000 series. They are preferred in built-up structures of riveted and screwed design for highly loaded and stressed structural assemblies.

Machined wrought aluminum parts or high-strength forgings are dispersed over the whole launch vehicle in almost all interstages.

It has to be stated that the final choice of 2219 came almost by chance—as a last resort for solution of an advanced Saturn design. We have presently exhausted the known potential of commercial high-strength weldable aluminum alloys. It should be stated that much more research on aluminum alloy is necessary to replenish our resources and knowledge. The negligible level of support in materials research is almost lethal. We need research on high-strength alloys with good weld properties, as well as research on high-grade aluminum casting material with improved weldability and satisfactory cryogenic behavior. Never before have the designers and material specialists found so little support having such impact on efforts to widen their knowledge in the materials field.

But it is believed that aluminum alloy will stay with us even for larger future launch and space vehicles.

This discussion would not be complete without mentioning titanium. A thorough investigation was made and test series completed to explore the usefulness of titanium and its commercial alloys for oxidizer (liquid oxygen) tanks. It was found, as expected, that a small energy release by impact or puncture of liquid oxygen-filled test tanks was sufficient to initiate a spontaneous oxidation resulting in an explosion-like destruction and burning of the titanium vessel.

This was enough evidence for ruling out titanium alloys for use in oxidizer containers. The small weight savings possible by application of titanium to the fuel container alone, in addition to the more complicated and costly manufacturing methods of titanium tanks, caused abandonment of titanium at the present time as a material for use in large launch vehicle tanks. It is under consideration for use in thrust-distribution and other secondary structures.

Other important materials widely used in large launch vehicles are plastic foams for insulation, reinforced plastics for load-carrying shells and skins, and honeycomb filler materials.

Adhesives play a most important role in bonding insulation material to metals, plastic-sheet sealers to insulation, and metal to glass-fiber-reinforced honeycomb. Operating temperatures cover a wide range from 20 to well above 550 K.

In the growing field of plastics and adhesives, the research effort is on a starvation level. Many promising approaches will have to be curtailed in the near future, leaving the designers and materials men without the needed advancements in the state of the art.

Figure 1 shows the design approaches to containers for large vehicles as used during the last 10 years. It summarizes in a simplified fashion the evolution of certain principles used. The lower row is the integral skin-stringer ring-frame design generally used at present, with the variation of diagonally arranged integral reinforcements (waffle pattern) used in some designs. All of these designs became feasible only because of advances made in the production of reasonably large plates of aluminum alloy in thicknesses up to and greater than 2 in. High performance and economical design are achieved with milling of patterns from flat plates, which are afterwards shaped into cylindrical portions of the containers and welded together. Large plates may be chemically milled to required patterns and thicknesses, sized to accurate shapes, and finally assembled by welding, in an orange-peel fashion, to form spherical or ellipsoidal bulkheads.

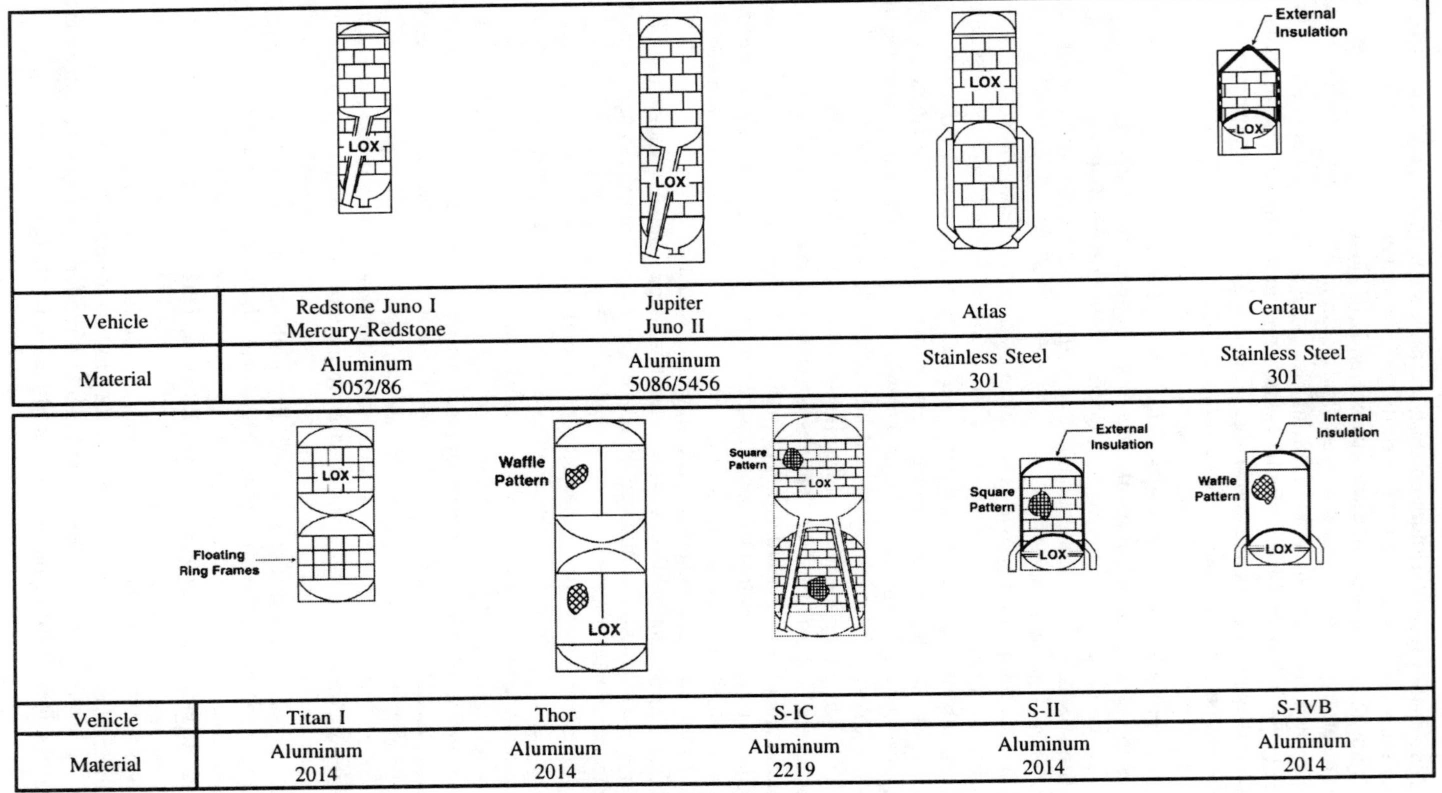

Fig. 1 Designs for propellant tanks.

The choice of two separated bulkheads or an integrally stiffened insulated bulkhead, as in the S-IVB and S-II stages, depends much upon the individual stage-optimization studies. Length and weight of the stage are traded off in these two stages against slightly more complicated tooling and manufacture of honeycomb-type common bulkheads separating both propellants. These bulkheads are designed against maximum buckling loads.

Unpressurized sections of the launch vehicle, starting from the lower level throughout the top of the second stage, are of stringer-ring-frame-shell designs with variations in the chosen profiles, depending on the load condition in each particular station. These particular designs for the first two stages were favored because of their flexibility and ease of manufacture without the use of expensive tooling.

In the third stage, the S-IVB, more sophisticated designs are employed, requiring more effort in tooling and manufacturing. This results in an appreciable return in weight savings, reflected ultimately in payload capability.

Decision-making requires a certain time, especially when concerned with the choice of mode in tackling the technologically most difficult task of this decade—that of landing a man on the moon and returning him safely to Earth. The decision had to include a flexible design, which laid out the basic design to a most probable load condition and which made certain that all variations occurring in the criteria and specifications could be taken up by slight alterations in the sheet and plate thicknesses, or variations in mill pattern and weld islands, without stepping out of the state of the art in welding, and particularly without upsetting the tooling design. It was recognized that it is most important that early tool concept and design has to start simultaneously with the airframe design to conserve lead-time.

Now, a half year after the decisions were made as to how to reach the moon, and after a full year of layout work and detailed structural design, we find that every single choice of design was correctly made. No changes causing a delay have been necessary. Tooling production is on schedule, and soon the first hardware will come from the production lines to undergo strict structural tests to prove correctness of the analysis and to confirm safety factors and design margins.

It can be stated that with this design we were able to achieve and retain the initial lead-time advantage. We dipped deep into the barrel of our present knowledge to fill all requests of the engineers. I am somewhat fearful that the next time we may find the barrel empty unless we replenish it by recognizing the urgency of research in all areas of large launch vehicle structures and the materials used in them.

Structures and Materials (State of the Art—1960)

George Gerard
New York University

The profound influence of spaceflight and hypersonic re-entry upon structures, materials, and design concepts has resulted in a notable record of progress in these areas during the past year. This review of some of the highlights in the areas pertinent to structural design necessarily reflects a broad spectrum of activity. In the interests of solving the extremal problem of maximum information in minimum space, only a brief treatment of the individual areas can be presented.

Thermal-Protection Systems

One of the most outstanding examples of the significant rewards to be gained from an integrated approach to structures-materials systems is the success achieved in the area of thermal protection. Recognition that thermal-protection techniques cover a broad spectrum of varied application from short-time ballistic re-entry to long-time hypersonic atmospheric flight has provided intelligent guidance for major structures-materials research and development efforts.

Among the noteworthy highlights in this area during the past year are broad analyses of the potentials of thermal-protection systems, the emergence of ablation as an applied science as evidenced by the relatively large number of basic papers in this area, and the development of practical radiation-cooled structures.

In the high-heat-flux drag re-entry area, developments have proceeded at a rapid pace, with the result that there are now a rather wide variety of materials available ranging from castable plastics to reinforced ceramics. The selection of materials for a particular application can now be based on efficiency rather than feasibility. Of considerable interest, too, are the development of low-heat-flux ablation-insulation coatings. Their continued development offers an interesting potential of substantial significance.

A major development during the past year involves the availability of pyrolytic graphite as a high-strength, ultrahigh-temperature material in various sizes and shapes. The material could have a major impact on both short- and long-time thermal-protection-system designs both from the standpoint of feasibility and efficiency. The use of such materials for major structural components, however, will involve special design techniques to cope with the many problems associated

Astronautics, Vol. 5, No. 11, Nov. 1960, pp. 47, 48, 64.

with brittle materials. There is some research effort in this area at present, and a more significant effort would seem to be warranted.

The development of refractory materials for hot structures as well as radiation-shielded structures constitutes a major portion of our national research and development efforts in airframe materials. Important progress has been made in the development of columbium- and vanadium-base alloys as well in tungsten- and molybdenum-base alloys. Concurrent development of coatings for the refractory metals is proceeding apace with the alloys.

An interesting analysis of the structural behavior of refractory materials revealed that metallics and nonmetallics are quite similar in this respect: Significant reductions in strength occur at about 0.6 of the absolute melting point. Thus, radical departures in the structural behavior of nonmetallic refractories from that associated with metals are not to be expected.

Shell Structures

The acute importance of high-strength/weight-ratio shell structure in missile and space vehicle booster has resulted in renewed research and development efforts in this field. Results become available from research on the instability of pressure-stabilized shells under compression and bending, and other research programs into the instability behavior of stiffened and unstiffened shells got well under way during the past year. Because of the greater relative importance of tension structures in solid-propellant engines, it is notable that progress was evident this year in the development of high-strength/weight-ratio pressure vessels.

After the initial disappointments with high-strength-steel pressure vessels, sufficient progress has been made in recognition of the importance of careful design and fabrication as well as screening of materials to permit the successful construction of steel pressure vessels above the 200,000-psi strength level. Particularly dramatic was the successful fabrication of all-beta-titanium alloy 10-in.-diam welded pressure vessels that consistently resisted hoop stresses over 200,000 psi before bursting.

Although developments in the pressure-vessel field have been most encouraging, research directed toward even higher strength levels is continuing. Considerable progress has been made in understanding the interrelations among pressure-vessel and material strengths as affected by ductility and stress concentrations. Further development of high-strength materials of improved ductility as well as more-advanced fabrication concepts will ultimately lead to significant improvements in this area.

Concurrent with these developments with homogeneous materials are the significant results beginning to come to the fore in the field of filamentary materials. Significant progress has been reported during the past year in the development of inherently high-strength glass, beryllium, steel, titanium, and refractory wire filaments. Basic understanding of the use of such filaments in combination with a ductile matrix to form shell structures is progressing at an encouraging rate as a result of an expanding research effort. Finally, important results have been reported from research on optimum configurations for shells of filamentary materials.

Because of the large size of solid-propellant rockets, research on the structural integrity of the propellant grain has assumed considerable importance within the past year or two. As a consequence, the behavior of viscoelastic solids has been under intense investigation. It is anticipated that the coming year will see a rapid expansion of research effort in this important problem area.

Manned Space Vehicle Concepts

Although current satellites and space probes are relatively free of major structural and material problems, active consideration of manned space vehicles has raised questions concerning the space environment and its long-time effect upon structures and materials. At the same time, the acute need for minimum-weight structures has led to radical vehicle configuration proposals as well as high imaginative use of materials in erectable space structures. A particular noteworthy achievement during the past year in this regard is the success of the Echo satellite.

In the environmental area, the determination that the solar-flare radiation hazard may be of primary importance for manned space vehicle structures has far-reaching implications. Current estimates of the severity of this hazard indicate that spaceflight beyond the proximity of the Earth may not be possible unless new shielding techniques are developed. Recognition of this problem has already resulted in the interesting proposal of concentrated shielding "storm cellars."

Another potentially significant item during the past year was the development of high-strength lead alloy to shield structures. Strengthened by dispersion hardening, this alloy has over five times the strength of commercial wrought lead and 1000 times the creep strength at 300°F.

A rather incompletely understood environmental factor is concerned with the influence of meteoroid impact upon space vehicle design. Although progress in assessing the meteoroid environment in space has been disappointing, the considerable effort of hypervelocity-impact research is beginning to pay dividends. Particularly noteworthy are experimental results obtained during the past year that indicate the effectiveness of the Whipple meteoroid-bumper concept. Much remains to be accomplished, however, particularly in the difficult area of equipment development for hypervelocity testing.

The important design concepts and engineering criteria underlying long-lifetime space vehicles of significant size have only recently achieved a first-round level of appreciation. It is particularly notable, however, that progress in assessing the structural significance of meteoroid and radiation environment in space has been rather disappointing, particularly in view of the profound influences these factors will have on fundamental design concepts. These environmental factors can only be evaluated in space, and consequently satellite and space probe experiments devoted specifically to meteoroid and radiation effects are urgently needed by those involved in structural design of space vehicles.

What may be lacking in our understanding of the space environment is compensated by some of the intriguing structures-materials concepts that are now emerging for coping with the special requirements of space vehicles. The enormous potentialities of tailoring materials to structural requirements offer wide freedom in design, seemingly to be limited only by lack of imagination. The success of ablative materials in severe thermal environments and the potential of filamentary materials and inflatable structures clearly portend a new era of significant gains in structures-materials design.

Bibliography

Butz, J. S., Jr., "Space Stations Require Radical Design, Structures," *Aviation Week,* Aug. 1, 1960.

Coyne, J. E., and Sharp, W. H., "Titanium for Rocket Motor Cases," *Metal Progress,* March 1960.

Dow, N. F., "The Ionizing Radiation in Space-Structural Implications," *Aero/Space Engineering,* May 1960.

Espey, G. B., "Sheet Alloys Graded by Sharp-Notch Sensitivity," *Metal Progress,* Aug. 1960.

Glaser, P. E., "Thermal Protection of Space Vehicles," *Astronautics,* April 1960.

"Important Research Problems in Advanced Flight Structure Design 1960," NASA TN D-518.

Jaffe, R. I., and Maykuth, D. J., "Refractory Materials," *Aero/Space Engineering,* June, July 1960.

Olshaker, A. E., "An Experimental Investigation in Lead of the Whipple Meteor Bumper," M.I.T. thesis, 1960.

Pellini, W. S., and Harris, W. J., "Flight in the Thermosphere," *Metal Progress,* March, April, May, and June 1960.

Williams, M. L., Gerard, C., and Hoffman, H., "Selected Areas of Structural Research in Rocket Vehicles," *XI International Astronautical Congress, Stockholm,* Vol. 1, 1960, pp. 146–166.

Wood, R. M., and Tagliani, R. J., "Heat Protection by Ablation," *Aero/Space Engineering,* July 1960.

Yaffee, M., "Ablation Wins Missile Performance Gain," *Aviation Week,* July 18, 1960.

Following Technical Reports of Defense Metals Information Center, Battelle Memorial Institute: "Design Information for 5Cr Alloy Steels for Aircraft and Missiles," Rept. 116; "Current Tests for Evaluating Fracture Toughness of Sheet Metals at High Strength Levels," Rept. 124; "Physical and Mechanical Properties of Columbium and Columbium-Base Alloys," Rept. 125; "Design Information on Nickel-Base Alloys for Aircraft and Missiles," Rept. 132.

New Era Dawns for Flight Materials and Structures

George Gerard
New York University

The so-called flight corridor, now virtually a cliche, dramatically illustrates altitude-vs-velocity conditions for hypervelocity flight. This flight corridor as a thermal environment directs our attention to inadequately explored problems of materials and structures. Figure 1 depicts the flight corridor and other flight regions, such as that typical for the long-range ballistic missile, in terms of altitude vs velocity and heating.

As indicated in the figures, re-entry vehicles undergo intense transient heating. Heating rates as high as 500 Btu/ft^2 per s have been stated for presently designed ballistic missile nose cones. Orbital re-entry bodies, particularly those that are manned, will be subject to lower heating rates because of the necessity for limiting decelerations. The total time a vehicle descending from orbit undergoes heating, however, will be considerably increased.

These vehicles, such as the boost-glide aircraft or glider released from a satellite, will be characterized by steady-state heating. Steady-state heating rates are an order of magnitude less than for re-entry vehicles, but the long periods of heating for them pose materials problems that, as of now, appear considerably more serious than the nose cone problem. The current material solutions for nose cones are based on transient heating conditions. These solutions do not have application in steady-state thermal environments.

Steady-State Heating Affects Structure

There are two distinct aspects of the boost-glide class of vehicle which present materials problems. Foremost is the leading-edge structure, which may be subject to stagnation temperatures on the order of 3000°F. Fortunately, this structure is lightly loaded, which means materials may be selected primarily on the basis of suitable thermal and oxidation characteristics. However, the long heating times involved at 3000°F make this a particularly difficult problem to solve in terms of a reasonably light structure.

The second aspect of boost-glide vehicles centers about the load-carrying structure, which is generally sufficiently removed from the stagnation regions so that maximum equilibrium temperatures are of the order of 2000°F. Because of the low wing loadings associated with boost-glide vehicles, some enthusiasts hold hopes for the use of heat-sustaining materials in load-carrying structures. Others are exploring the use of thermal-protection techniques in which the load-

Astronautics, Vol. 5, No. 8, Aug. 1958, pp. 20, 21, 46.

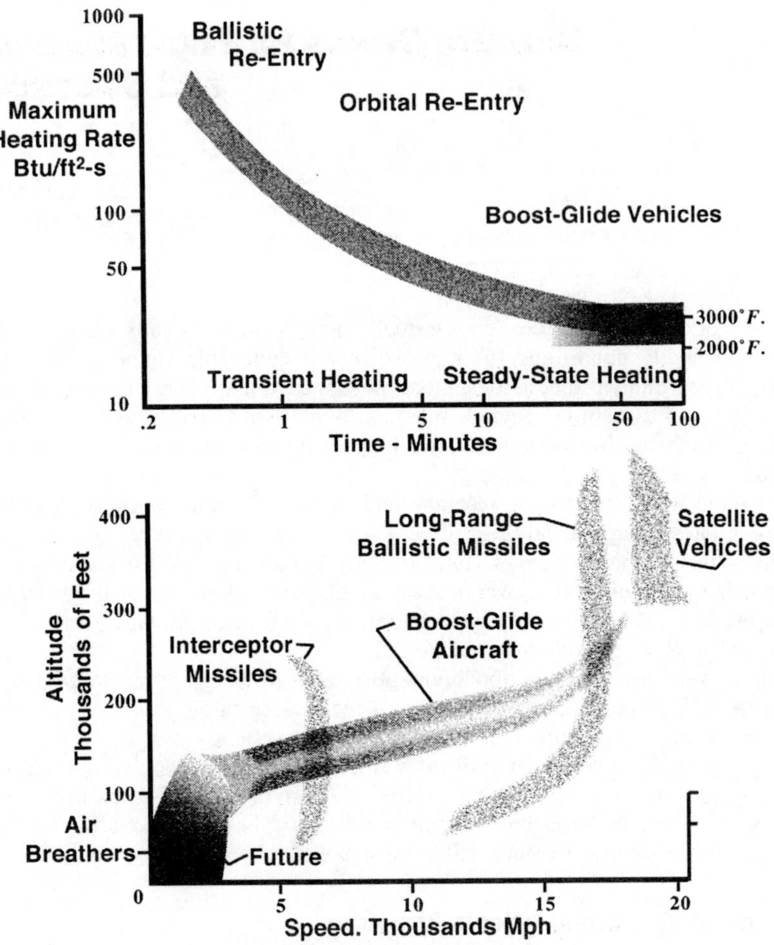

Fig. 1 Velocity and heating in hypervelocity flight.

carrying structure, shielded from the surface, operates at a considerably lower temperature.

Both approaches have merit, and signify our entry into a new era in material-structural development in which the ingenuity of the structures engineer in exploiting the desirable characteristics of materials through materials design is a necessity. A complete recognition of the importance of materials design holds great promise of achieving significant developments in lightweight, elevated-temperature structures.

In somewhat broad perspective, we have just left a long era in which the problem of selecting materials for the primary structure of aircraft was virtually nonexistent—that is, aluminum alloys were employed almost exclusively. Now the thermal environments associated with hypersonic weapon systems have opened a new era in which the problems of material development, selection, and fabrication have added a new dimension to the complexity of these advanced systems.

In the realm of large heat-flux transient heating, heat sink and ablation materials have been successfully employed for ballistic re-entry bodies. Considerable research is being directed toward the development of more efficient *heat-absorbing* materials—beryllium, graphite, reinforced plastics, and ceramics. These developments for ballistic re-entry bodies will provide a broad background for orbital re-entry vehicles.

Likewise, the development of graphite, ceramics, and coated molybdenum as *heat-sustaining* materials holds considerable promise for the steady-state heating environment of leading-edge structures of boost-glide vehicles. Another approach may be the development of slowly ablating materials for use on leading-edge structures. In all cases, serious consideration should be given to any significant weight advantage accruing through use of limited lifetime, replaceable leading edges.

Structures governed primarily by thermal rather than load-carrying requirements have material requirements which are stringent in terms of heat-absorption or heat-sustaining characteristics and relatively minor in terms of mechanical strength.

But, when viewing the primary load-carrying structure of boost-glide vehicles, material requirements include both major strength and thermal characteristics. Thus, the problem is considerably different than for the structures discussed previously, and in all probability the really successful solutions will involve extensive material-structural design.

Before considering the various approaches to the design of a 2000°F load-carrying structure, let us consider the roles performed by the outer structure of an airframe. In general, the covering provides both contouring and load-carrying functions. Thus, minimum weight considerations for compression surfaces often lead to optimum stress levels considerably below the yield strength. In hypersonic vehicles, the skin provides a thermal-shielding function, as well as contouring and load-carrying functions.

The most direct approach to the design of primary structure is the use of heat-sustaining materials. In this case, the three functions are satisfied by the temperature-resistant skin. Thermal shielding is provided by a high-emissivity coating, which permits radiation of a major portion of the heat, thus effectively cooling the structure by "radiation cooling."

Because of the greatly deteriorated weight-strength characteristics of materials in the 1200–2000°F range, the use of the heat-sustaining approach necessitates a low load-carrying function for the skin. To some extent, this is already true, since glide-type vehicles necessarily operate at low wing loadings. In addition, by use of an internal supporting structure, the surface load may be reduced further.

Thus, the strength requirements for the heat-resistant structure can be reduced by proper design to the point where the skin effectively provides only contouring and thermal-shielding functions. Contouring requirements provide certain minimum stiffness criteria to avoid panel flutter. This approach, in effect, results in material requirements similar to those for the leading-edge structure.

Thermal Shielding Helps

An alternate approach to the design of primary structure for boost-glide vehicles is to employ thermal-protection techniques for the load-carrying structure. In this manner thermal-shielding and contouring functions are separated from the load-carrying function.

It is apparent from weight-strength considerations that 1000–1200°F is the current upper limit of equilibrium temperatures that can be met with the use of heat-sustaining materials in relatively efficient load-carrying structures. To go beyond 1200°F with current materials introduces a prohibitive weight penalty as compared to a 7075-T6 aluminum alloy primary structure at room temperature. Beyond 1000°F, and in all probability at considerably lower equilibrium temperatures, thermal protection of the load-carrying structure by various material-design techniques appears to offer an attractive design approach.

Thermal protection of primary airframe structures appears very efficient in terms of complete system weight. Operation of the load-carrying structure at relatively low temperatures permits the use of low-density alloys; avoids a host of materials problems associated with, for example, high-temperature seals, lubricants, and elastomers; and does not require special cooling for the pilot, fuel, and equipment.

A promising development for hypersonic gliders consists of an external high-temperature radiation shield which radiates a major portion of the heat flux absorbed from the boundary layer. A moderate thickness of fibrous insulation is placed adjacent to this radiation shield. The small portion of the heat flux transmitted through this low-density insulation is removed by a cooling system attached to the primary structure. The amount of coolant required to keep the interior below 200°F is sufficiently small so that low-density aluminum or magnesium alloys can be used for the load-carrying structure.

With this approach, it appears feasible to construct a relatively efficient composite structure which can be subjected to external equilibrium temperatures up to 2000°F. The major materials requirements involved in this thermal-protection technique up to 2000°F, and beyond, are concerned with the radiation shield and are of a similar nature to the leading-edge problem.

It is the intent of this rather brief article to sketch an introductory background to the relation between the hypersonic environment and the material-design aspects of hypersonic vehicles. It is evident that we have entered an era which places tremendous emphasis on materials-design technology.

Development of new forms of construction and fabrication is essential for the effective utilization of materials. Development of materials themselves must be guided by more precise definitions of the environment and design conditions. This realization is slowly taking place, with the result that a two-way bridge of understanding and sharing of problems is developing between structures and materials-oriented engineers.

Structures Technology—General

Developing Structures Technology for the Day After Tomorrow

George W. Brooks
NASA Langley Research Center

In the years to come an aircraft's structure will play an increasing role in making it economically and ecologically competitive in the world market, and progress in performance will impose new standards of excellence on the structure beyond the usual ones of structural weight fraction and useful life. The structure will also be judged by its ability to withstand high temperatures and impinging unsteady flows and to accommodate highly interactive control systems for improving ride quality. To prepare for these new demands on the structure, NASA has set these goals for its airframe structures program:

- Automatic analysis and design.
- Build confidence in advanced composites.
- Improve the technology base for future supersonic and hypersonic vehicles.
- Validate concepts for active controls.
- Develop methods for predicting aircraft loads and aeroelasticity.
- Generate design methods for assuring structural integrity.

In recent years three important factors have emerged which support the development of advanced automated analysis and design. The first is the need for more-sophisticated analysis and design procedures to optimize the vehicle as a coordinated structural system to gain higher structural efficiency and vehicle performance. The second is the need to lower the cost of analyzing and designing high-performance aircraft by reducing both the timespan required to complete the design and the man-hours that go into it. The third is the new options the high-speed digital computer has offered the designer for analyzing integrated structures and rapidly iterating subsystem tradeoffs.

Table 1 shows representative examples of computer programs for analysis and design under development or in use by the Langley Research Center. Other significant programs with related objectives are being developed by industry, by the Department of Defense, and by other NASA centers.

Analysis programs vary from rather simple special-purpose programs, such as BUCLASP for analyzing panels, to highly sophisticated general-purpose programs such as NASTRAN for analyzing complete complex vehicle structures. NASTRAN finds wide use in the public sector as well as in the aircraft industry, and NASA will continue to update it to satisfy industry needs. It seems clear that the efficient analysis of structures requires a full spectrum of analytical tools of which these programs are representative. A major driver is the cost of the

Astronautics & Aeronautics, Vol. 11, No. 7, July 1973, pp. 56–66.

Table 1 Acronyms for automated analysis and design programs

ASOP	Automated Structural Optimization Program
BOSOR	Buckling of Shells of Revolution
BUCLASP	Buckling of Laminated Stiffened Plates
CASE	Optimization Program for Stiffened Cylindrical Shell Structures Developed at Case Western Reserve University
COHEN	Comprehensive Shell of Revolution Analysis Programs Developed by G. A. Cohen
DAWNS	Design of Aircraft Wing Structures
FADES	Fuselage Analysis and Design Synthesis
IPAD	Integrated Programs for Aerospace-Vehicle Design
NASTRAN	NASA Structural Analysis Program
ODIN	Optimal Design Integration Procedure
SALORS	Structural Analysis of Layered Orthotropic Ring-Stiffened Shells of Revolution
SAVES	Sizing of Aerospace Vehicle Structures
SECAR	Structurally Efficient Cones with Arbitrary Rings
SNAP	Structural Network Analysis Program
STAGS	Structural Analysis of General Shells
SWIFT	Strength of Wings Including Flutter

computer time which dictates the use of the simplest analytical program that will produce the desired result.

In general, the analysis programs tell the designer the structure's responses—such as stresses and mode shapes—to a given set of constraints such as loads, boundary conditions, and geometry. The designer needs, in addition, programs which aid him in defining the most advantageous structural concepts or configurations for given constraints and forcing functions. The spectrum of such programs also varies from simple programs for panel structures such as CASE to very sophisticated programs such as IPAD, which encompasses the complete aircraft design procedure.

The IPAD idea arose when a few years ago NASA felt the need to help provide the technology to improve future aircraft design. The concept (Fig. 1) makes use

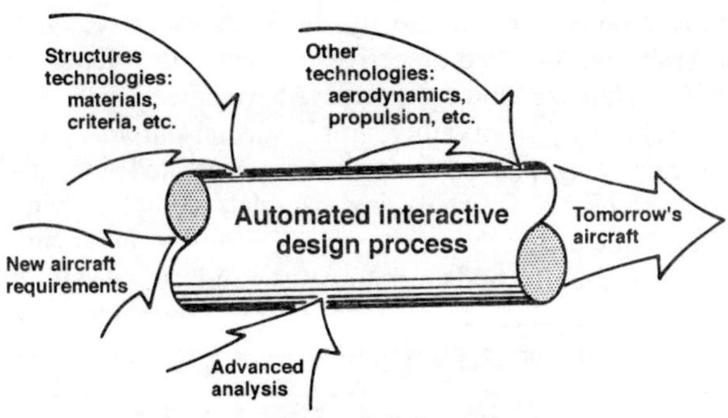

Fig. 1 Future automated aircraft design.

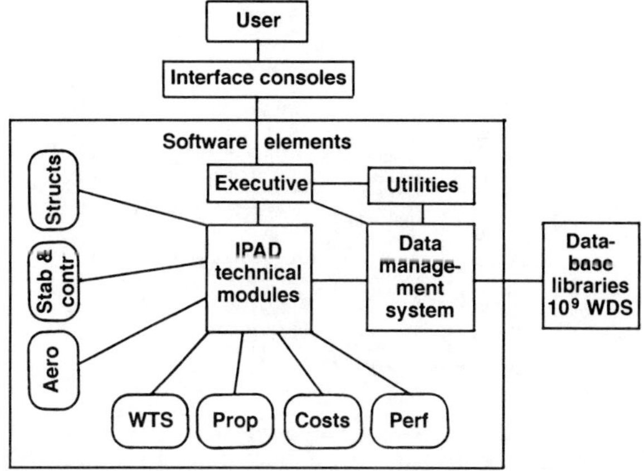

Fig. 2 Integrated Program for Aerospace-Vehicle Design (IPAD).

of advanced high-speed computers and analytical programs to integrate structures technologies with other technologies, such as aerodynamics and propulsion, in the overall design. NASA aims to develop an automated, interactive system for designing an optimal aircraft to satisfy a given set of requirements. Two feasibility studies in progress seek to identify desirable options, and the aircraft industry is working closely with NASA to make sure the system fulfills its needs.

In the evolving IPAD system (Fig. 2) the software component contains four major elements: an Executive, a Database Manager, Utilities, and Operational Modules.

The Executive program provides the interface for the user, translates engineering commands into computer language, and controls the computations and data flow.

The Database Manager, under the control of the Executive, builds and updates the database and provides security against systems failures and unauthorized access.

The Utility programs provide data display, man/machine interfaces, and software conversion routines.

The Operational Modules carry out computations in specific engineering design disciplines and contain disciplinary analysis programs such as NASTRAN. Together with the database library, they represent the depository of the users' proprictary technology.

The concept of IPAD is not new. Many design programs superficially resemble it, and some such as ODIN and the Pilot IPAD are already being used for preliminary design. What is new is the attempt to provide wide versatility to the user, to store and use all data needed in the design cycle from preliminary to final design, to include technical modules representative of current knowledge, and to provide a system adaptable to a broad spectrum of aircraft.

During the past decade substantial progress has been made in the development of boron and graphite filamentary composite materials. Great credit is due the

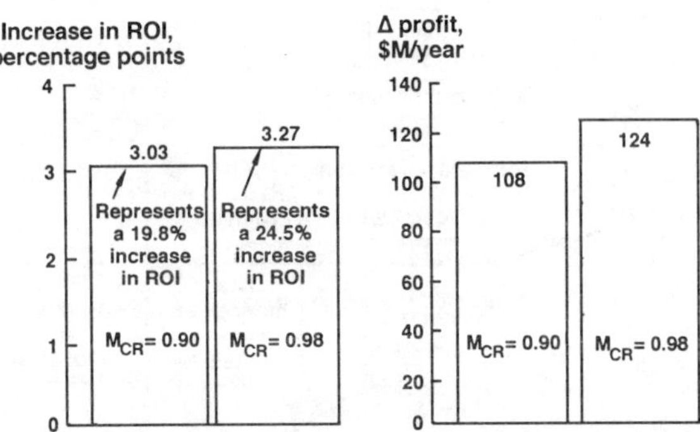

Fig. 3 Economic gains from composites. For a fleet of 280 Mach-0.90 transports compared to metal aircraft containing the same technology.

U.S. Air Force Materials Laboratory for its strong leadership and substantial contributions. But there is general agreement, as expressed during the Composites Recast[1] exercise, that much remains to be done before realizing the potential advantages of these materials in increased aircraft performance, payloads, and profits. NASA shares the belief that the promise of advanced composite structures merits strong support of a broad base of technology development and demonstration. Advanced composites technology comprises a key element in our long-range plan.

NASA has set as the major goal in composite airframe structures helping provide the technical credibility which will enable the industry to use composite materials in the primary structural components of transport aircraft with confidence. The industry needs confidence to use composites on aircraft that accumulate upward of 15 years and 60,000 h of flight service.

NASA plans to continue systems studies as the technology develops to assure identifying the potential advantages and highest payoff options of composites in civil applications. As an example of possible advantages, the recent Advanced Transport Technology[2] study showed that judiciously applying composites to parts making up about 45% of the structural weight of a reconfigured transport could reduce structural weight by approximately 26%. Composites could increase return on investment (ROI) by 19.8% (Fig. 3) for a fleet of 280 aircraft operating at a cruise Mach number of 0.90, compared to the metal baseline aircraft containing the same technology. This would raise profits by $108 million per year. The advantages would be even greater for aircraft cruising at a Mach number of 0.98. ROI would rise 24.5% and yearly profits, $124 million.

The foregoing results relate to three-engine, statistically stable aircraft employing supercritical aerodynamics and satisfying FAR 36—10 EPNdB noise requirements. The basis of comparison included a basic ROI of 12.23% for M = 0.9 and 13.3% for M = 0.98, a range of 300 n mile, and a payload of 40,000 lb.

The first step in building confidence in advanced composite structures is to enlarge the technology base. NASA has mapped this program for doing so:

• Focus on improving and using a few select materials with the emphasis on graphite.

• Establish a database for fatigue, fracture, and sound absorption of composites.

• Develop failure analyses for bonds, joints, and fiber-matrix interfaces.

• Develop good practices for composite design to emphasize unique material properties and update current computer programs to accept anisotropic materials.

• Promote development of NDE methodology for lab and field use.

• Enhance fabrication by building and testing realistic structural components for static and fatigue loads.

We emphasize graphite both because of the material's high promise for low cost and the flexibility it offers for control of structural properties.

The second and equally important step is to expand user confidence through developing, manufacturing, testing, and demonstrating in flight realistic structural components. Focused activities (Fig. 4) include secondary and primary structures for CTOL vehicles, stiffness controlled structures for STOL vehicles, and high-temperature composite structures to improve the payload fractions, essential ingredients of Advanced Supersonic Technology (AST).

NASA has reached its first milestone (Fig. 5). Minor secondary structures such as spoilers are now flying. If resources permit, flight tests of reinforced primary structures on two C-130 aircraft should follow in 1974, and a year later major secondary structures should go into service. By 1978 we hope that flight tests of all-composite primary structures can begin. It is expected that serious consideration can be given to making production commitments of advanced composites for civil applications three to five years later. Promised payoffs can then become reality.

A host of major structural problems must be overcome before the technology base suffices for designing efficient supersonic and hypersonic vehicles. NASA has work underway on three: transonic flutter, supersonic transport structural concepts, and cooling for hypersonic structures.

Designing transport aircraft to fly through the transonic region with adequate flutter margins presents a formidable task. Figure 6 illustrates the difficulty of

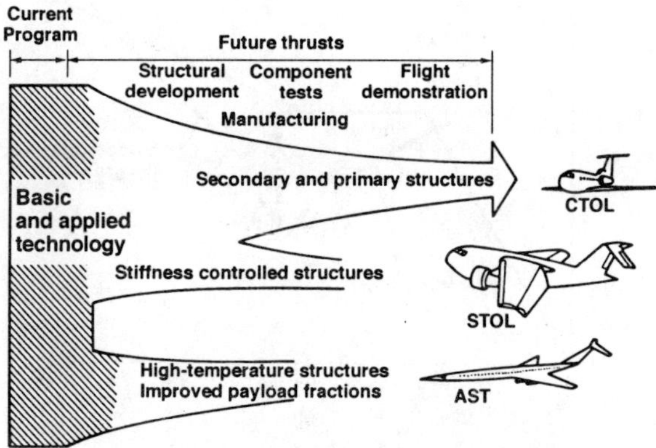

Fig. 4 NASA programs for composites for airframes.

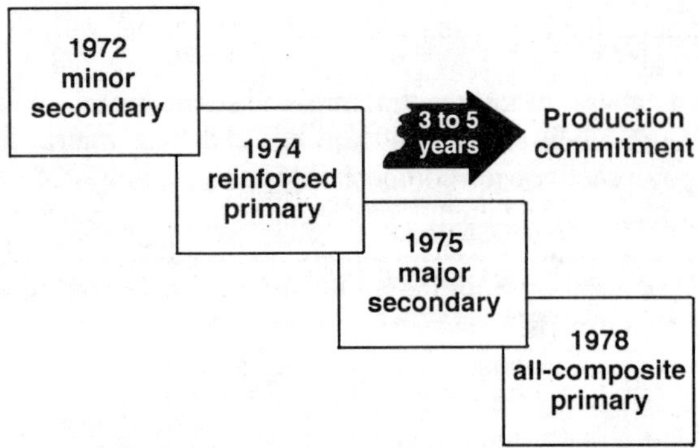

Fig. 5 NASA milestones for civil applications of composites.

providing adequate stiffness and mass balance to avoid flutter for a realistic supersonic transport configuration. Both subsonic and supersonic theories predict unsatisfactory flutter margins. Even if the designer succeeds in modifying the aircraft structure to move the flutter boundary above the operating boundary in these regions, he faces two difficult problems. First, the trends of the theory show that the minimum flutter speed occurs in the transonic region. The trends are supported by data gathered from tests of several configurations in transonic wind tunnels. Second, no aerodynamic theories or basic aerodynamic data exist for predicting flutter in this flight regime. The only reliable information for the designer is that which he gathers from tests of dynamic models of his design in a transonic wind tunnel. This information comes too late to properly influence selection of the basic configuration without expensive redesign.

The inability to predict transonic flutter arises from the difficulty of the problem and past allocation of resources to other pressing needs. NASA future plans call

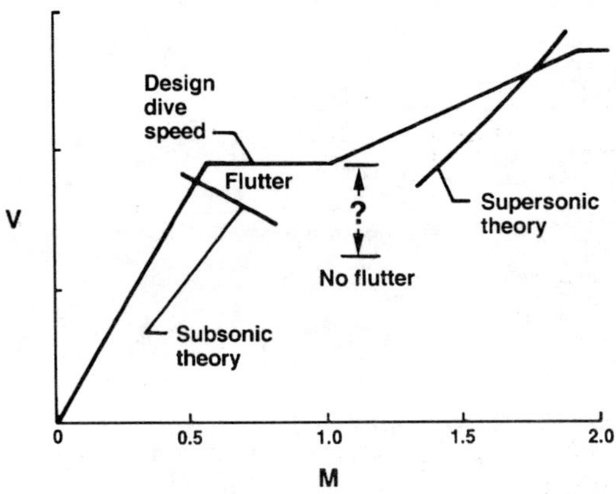

Fig. 6 Transonic flutter gap.

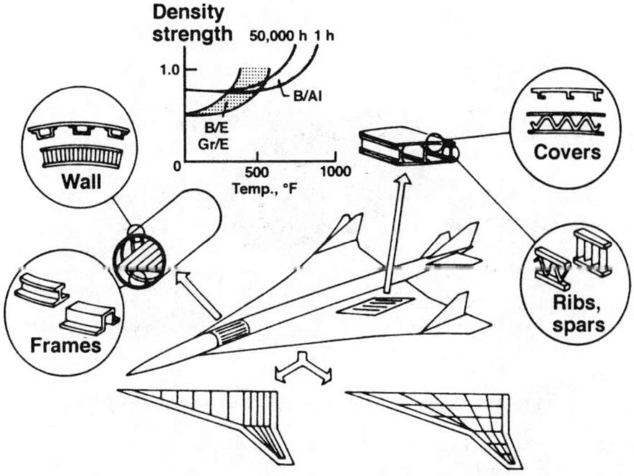

Fig. 7 Structural concepts for advanced supersonic transport.

for increasing emphasis in the area of aeroelasticity for transonic flight. Promising success with steady transonic calculations have encouraged us to undertake four exploratory theoretical studies of unsteady transonic flow. We will then select the best approach for further development into a working transonic flutter theory. Simultaneously we will measure detailed pressures on oscillating airfoils, which will guide development of the advanced theories and provide information for verification of new analytical approaches. Additional programs in planning include the study of pressures on oscillating three-dimensional supersonic airfoils and a study of the effect of lift on flutter. We feel these efforts, although probably not adequate, will make substantial progress toward understanding transonic flutter.

NASA has recently launched a comprehensive program for advanced supersonic aircraft structures technology. A major portion of this program is directed toward developing improved structural materials and concepts (Fig. 7).

As its principal objectives the effort will assess the relative merits of various materials and concepts, identify preferable methods for construction, and make realistic weight estimates for wing and fuselage designs. The studies will define tradeoffs in weight and materials selection for critical variables such as temperature (Mach No.) and engine location and delve into loads, fatigue, panel flutter, fabrication, sealing, and cost.

The number of potential uses of composite materials for reducing structural weight and improving payload fractions is high. But as mentioned previously, developing the processability and integrity of the high-temperature composite matrices, such as the polyimides and aluminum, to the stage of readiness which justifies their use for long-duration exposure and realistic service life will take considerable effort. We believe this program responds to long-range need. No doubt we will reshape it as we learn.

Some studies[3,4] have indicated that the use of hydrogen fuel, coupled with new air-breathing propulsion and aerodynamic and structural concepts, offers major advances for high-speed aeronautical transportation. Transports capable of 6000–

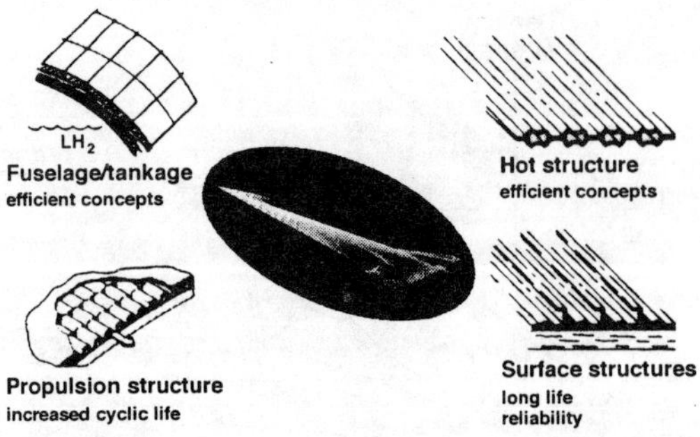

Fig. 8 Hypersonic structures technology.

8000-mile range at 6–10 times the speed of sound and new military systems are among the important possibilities. Hydrogen fuel appears essential. Not only does it provide propulsion efficiency and cool the structure, but also it reduces fossil-fuel depletion and environmental pollution.

This new class of vehicle will need the synergistic coordination of aircraft aerodynamics, propulsion, and structural disciplines. Because of the major impact of aerodynamic heating and carrying liquid hydrogen onboard, radically new structures with integral thermal control systems must evolve (Fig. 8).

R&D for hypersonic vehicles becomes highly involved because of the interactions among these various systems and requires test facilities which will permit realistic simulation of these interactions. For the past two decades NASA has been building such facilities, the use of which has brought us to the threshold of major gains in hypersonic technology.

Although numerous concepts have been offered as potential solutions for cooling hypersonic vehicle structures, the best solution has yet to be defined. One of the more promising concepts would use the liquid-hydrogen fuel to cool the airframe (Fig. 9). Hydrogen must also cool the propulsion system.

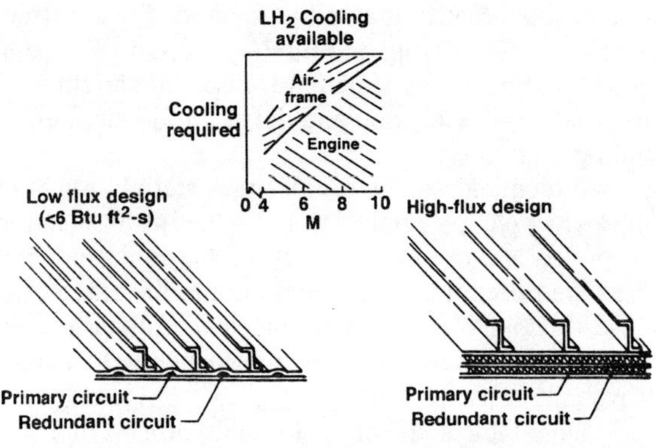

Fig. 9 Actively cooled surface structures.

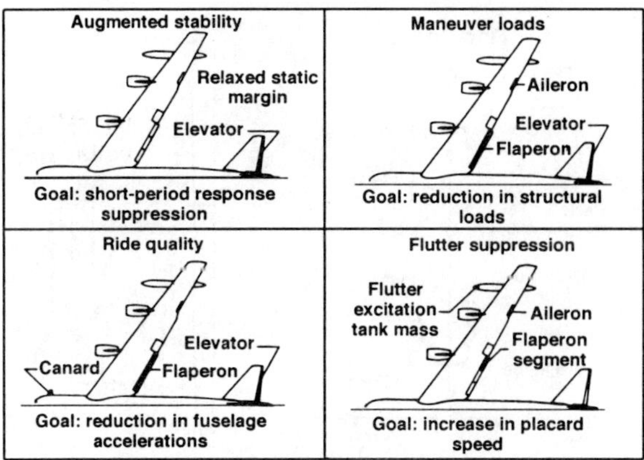

Fig. 10 Aims for active controls.

Our studies predict that the hydrogen can cool both the engine and the airframe up to about Mach 7.0. Above this Mach number the surface temperature must be allowed to rise, or the heat input to the structure must be reduced with external insulation as is being considered for the Space Shuttle.

In all of our work in this area, we strive for maximum simplicity, light weight, high reliability, and long life. In the case of actively cooled structures, the problem reduces to that of integrating the coolant passages into the structure without increasing the vehicle's weight or reducing its reliability.

The potential of active controls for improving the structural and aerodynamic performance of aircraft is well recognized.[5,6] The current thrust is toward establishing strong R&D programs for transforming promising control concepts into flight hardware with predictable performance and high reliability. NASA expects to work closely with industry, DOD, and DOT in developing this technology over the next several years. Its program will coordinate work on flight requirements, advanced fly-by-wire control concepts, onboard computer logic and hardware, aerodynamics, structural systems, and other disciplines. Advanced control concepts may require putting greater faith in electronic systems, as is done in spaceflight, if inherent stability and control requirements are relaxed to gain the full potential payoff of active control systems.

The U.S. Air Force Flight Dynamics Laboratory is sponsoring Boeing flight tests of a B-52 equipped with a variety of automated control surfaces to confirm predicted improvements in augmented stability, maneuver load control, ride quality control, and flutter suppression. NASA will participate by conducting tests of a one-thirtieth-scale aeroelastic model of the control-equipped airplane in the 16-ft Transonic Dynamics Tunnel. Flight validation of the wind-tunnel test data would establish such model testing as a relatively inexpensive, timely means of verifying the effectiveness of advanced control systems for future airplanes.

The program will test active controls for four different purposes (Fig. 10) but with this common rationale: In present aircraft extreme conditions, seldom met, often set structural design specifications. Using active controls to avoid the

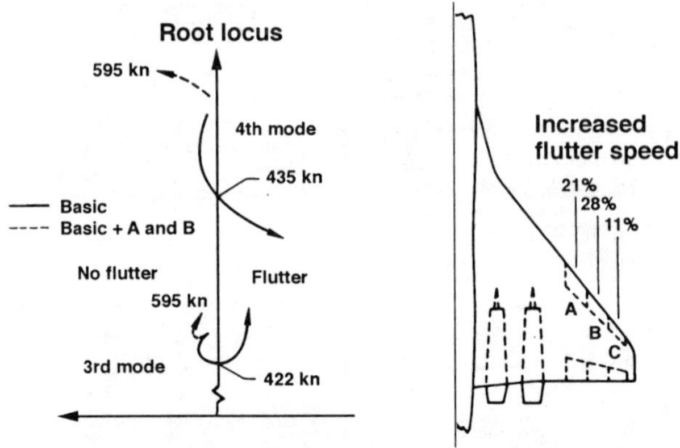

Fig. 11 Analyzed flutter suppression systems.

extremes can save substantial weight by reducing the range of loadings on portions of the structure.

For example, empennages are sized to maintain inherent static stability when the aircraft center of gravity has moved as far to the rear as it will ever go. Sizing the tail only for the usual movement of the center of gravity and letting the controls handle the extremes would lighten the aircraft appreciably.

Active controls offer the potential for shaping the spanwise wing load distributions to reduce the maximum bending moment and midspan shear loads during extreme maneuvers.

The flaperon-canard-elevator ride control system can substantially reduce vertical accelerations in the aircraft fuselage, maybe by as much as a third.

Commercial airplanes are generally designed to have a flutter margin of 20% in airspeed over the design flight boundary with a resulting weight penalty. A flutter control system could probably give the same safety factor with a smaller weight penalty.

Suppressing aeroelastic response by active control promises significant weight savings for advanced high-speed aircraft and a smoother ride for STOL aircraft. Langley has in progress two programs for developing such controls and ways of modeling them in transonic wind-tunnel studies.

The flutter suppression studies employ a control law based on an aerodynamic energy approach to mode stabilization using active control concepts developed by Nissim.[7] Using Nissim's control law, Boeing, under a NASA contract, studied flutter suppression systems for an early SST design.[8] They considered three pairs of leading- and trailing-edge controls located as shown on the right in Fig. 11. Two linear accelerometers sensed motion at a station located on the wing surface near the hinge line of each control surface's midspan. The flutter suppression systems raised flutter speed 11% for configuration C and 28% for configuration B. Another calculation combined configurations A and B in a single flutter suppression system as shown in the root locus diagram on the left in Fig. 11. Although the calculations included 12 modes, the diagram shows only the two modes involved in flutter. Working together the combination raised flutter speed more than 40%.

The wind-tunnel model for the flutter suppression program is a simplified semispan representation of an early SST airplane design. It has a single pair of leading- and trailing-edge controls similar to configuration B.

The previously described configurations rely primarily on the forces and moments exerted by control surfaces to oppose forces or motions measured by sensors near those surfaces. Accurate predictions of the forces which these control surfaces will exert, therefore, play an important part in control system design.

Some advances have been made in the automated calculation of the aerodynamic forces during subsonic flight on a planar wing with trailing-edge controls due to the oscillatory motions of both wing and control surfaces, including their elastic deformations.[9] The control surface on each half-span wing consists of from one to four segments, each with sealed gaps. The calculation relies on the subsonic lifting-surface integral equation relating downwash and lifting pressure on the wing. Experimental data[10] agree quite well with theoretical values.

NASA plans continuing support of both analytical and experimental programs for developing active controls technology. We believe the payoff for future STOL and high-speed aircraft merits such consideration.

Current NASA thrusts include acquiring experimental data and improved methods for predicting control-surface loads. Of particular significance in this area are the loads for future STOL and supersonic transport vehicles.

Short-range STOL vehicles show high promise for meeting the needs of the intercity traveler and alleviating airport congestion. As pointed out in the CARD study,[11] approximately 80% of all domestic air carrier movements cover less than 500 miles—thus, the need is pressing. From the structures point of view these vehicles have as their most important characteristics heavy engines and heavy and complex flaps combined with reduced cross-sectional areas in the wing boxes to carry these heavy components (Fig. 12). Hence designing the wings with adequate strength and stiffness to avoid flutter and other aeroelastic problems

	STOL	Conventional transport
Engine weight / Bare wing weight	1.5	0.4 to 0.8
Flap system weight / Bare wing weight	0.5	0.1 to 0.2

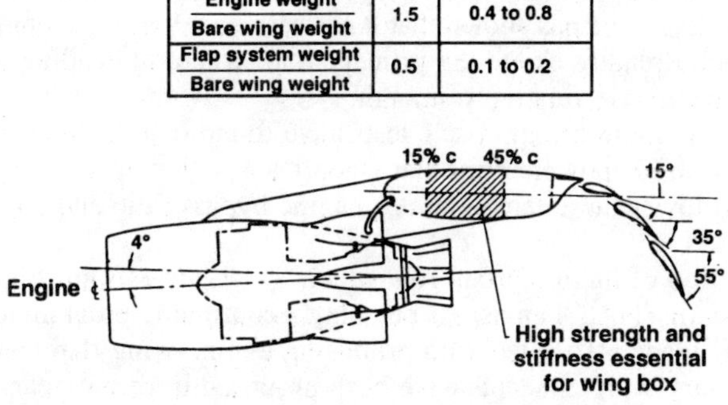

Fig. 12 Blown flap weight breakdown.

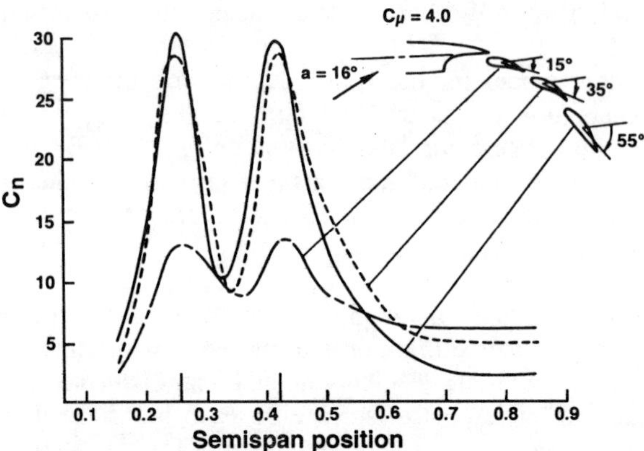

Fig. 13 Spanwise variation of flap normal force.

presents a challenge. To compound the problem no adequate database exists for either quasi-static or unsteady aerodynamic loads.

A partial solution consists of tailoring high-strength and high-stiffness composite materials to the structure and obtaining unsteady aerodynamic coefficients through combined analysis and experiment.

Jet STOL aircraft will also generally direct high-velocity, and usually high-temperature, jet exhausts against the flaps to gain powered lift. Despite a concerted effort in this area, near-term analytical predictions of these impinging loads, both static and dynamic, do not appear very accurate because of the complex flow fields. NASA has run experimental programs at Ames, Lewis, and Langley to derive realistic loads by testing small- and large-scale models. On the graph of the spanwise variation of flap normal force coefficients (Fig. 13), note that the peak loads occur at the spanwise locations of the engines and that the highest flap loads change from the rear flap to the forward flap as the wing tip is approached.

The dynamic loads for the first flap show that the wind stream does not determine the input pressures or the response. Engine jet impingement plays the major role. Experience has shown that loads associated with the engine-imposed pressure spectra readily excite the primary flap structural bending modes, even though the spectra are relatively smooth.

During a maximum lift approach, tests have found that the temperature where the jet cores strike powered lift flaps approaches that in the engine tailpipe (1000°F). To lower these temperatures, engine bypass ratio and jet mixing need to be increased.

In recognition of the important role which STOL aircraft must have in future transport systems, NASA envisions pursuing a continuing program to investigate the structural loads associated with promising engine-wing-flap configurations.

Several rotary-wing concepts have been proposed in recent years to combine the hover efficiency of a helicopter with the high-speed cruise efficiency of a fixed-wing aircraft. The tilt-proprotor craft has wing-tip-mounted rotors positioned vertically for helicopter flight and horizontally for high-speed airplane flight. These vehicles exhibit a wide variety of dynamic and aeroelastic phenomena unique to their configuration and significantly influencing their overall design. Successful development of a tilt-proprotor VTOL aircraft depends, in great measure, on establishing a firm dynamics technology base from which to assess design implications.

For several years, Langley has had under way a generalized dynamic and aeroelastic research program for proprotor craft. This program includes wind-tunnel investigations employing several aeroelastically scaled models and the development of analytical techniques and associated computer programs. Among other things, the experiments seek to learn: the parameters of whirl flutter; transient response during feathering and unfeathering of the rotor, simulating the stopping and starting of a folding proprotor; how to maintain stability at low (including zero) rpm and during blade folding; proprotor/wing response to gust excitation; the effects of Mach and Reynolds number on static stability and control characteristics; effects of rigid-body degrees of freedom on proprotor stability and, conversely, the effects of the proprotors on flight-mode stability; and compressibility effects on stability and response. Results will provide data for validating developed analyses and revealing configuration-related phenomena.

The proprotor, pylon, and wing assembly of a proprotor craft exhibit a wide variety of flutter modes as shown by the example plotted in Fig. 14. Rotor rpm here extend from zero to overspeed. The low-speed regime gives information important to the stability of a folding proprotor during feathering. Excessive vibration or oscillatory loads when operating near pylon-wing or blade resonances limited maximum airspeed at points marked by solid symbols.

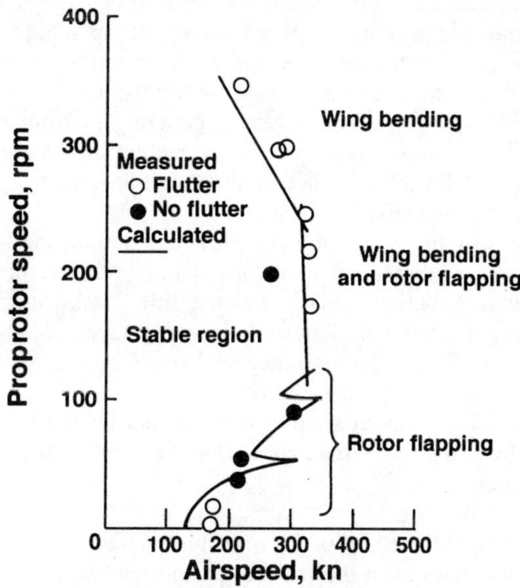

Fig. 14 Tilt-rotor flutter boundaries.

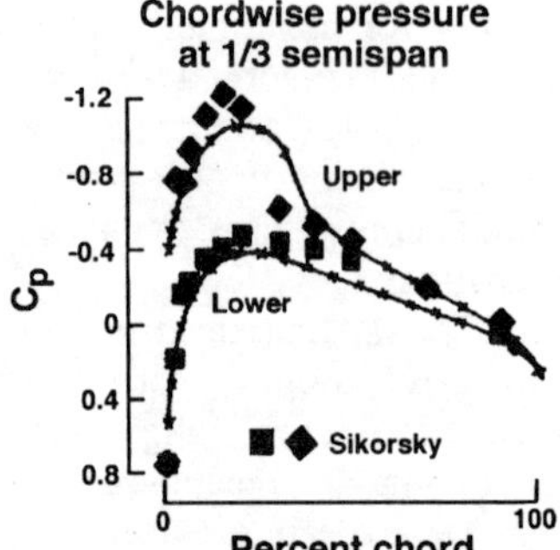

Fig. 15 Calculated transonic loads vs wind-tunnel values. For Mach 0.7, = 2.2 deg, *AR* = 8.

The calculated flutter boundaries were obtained by an analysis developed at Langley. Analysis supporting this research continues and will widen in scope to include additional degrees of freedom, airframe aerodynamics, and blade flexibility.

Loads on aircraft at subsonic and moderate supersonic speeds can be predicted with good accuracy on realistic three-dimensional configurations using linear theory. At hypersonic speeds, Newtonian impact theory or other techniques which depend simply on the local surface inclination to the stream give reasonable results. These theories have been applied to a variety of configurations and incorporated into computer programs in wide use throughout the aerospace industry. However, the design loads for many components of aerospace vehicles, including supersonic transports, the Space Shuttle, and high-subsonic and low-supersonic-cruise aircraft occur in the transonic speed range. Unfortunately, the previously mentioned techniques prove inadequate or inappropriate in the transonic range where they must account for nonlinearities and mixed flow with embedded shocks. These effects necessitate the use of nonlinear flow equations which present computers cannot solve for general three-dimensional configurations. Nevertheless, the past few years have seen important advances, particularly in applying various finite difference methods to two-dimensional, axisymmetric, and simple three-dimensional problems.

Recent efforts by many researchers allow calculating the following:

- Flow about two-dimensional airfoils using both the nonlinear, transonic, small disturbance equations (SDE) and the full, nonlinear potential equations (FPE), including a crude simulation of boundary-layer effects.
- Axisymmetric flows about pointed and blunt bodies, including tunnel wall effects in the case of the pointed body.
- Supercritical flows about simple wing geometries with either blunt or sharp leading edges by using finite difference relaxation techniques applied to the small disturbance equations.

Figure 15 compares a sample calculation made in 10 min on a CDC 6600 computer with unpublished Sikorsky wind-tunnel data. The calculations correlate well with the test data even though they do not account for tunnel wall effects.

In the near future, using existing computers, we can expect to extend these techniques to more general wing planforms and possibly some simple wing bodies. The more powerful computers coming along in a few years make prospects bright for predicting the transonic loads on more general configurations, comparable to those which can now be treated by linear theory. And they should permit efforts to proceed on coupling inviscid and boundary-layer flows for three-dimensional geometries. As a result, we anticipate solving these problems in the near future:

- Three-dimensional swept, twisted, and tapered wings, SDE.
- Axisymmetric bodies with tunnel wall effects, FPE.
- Two-dimensional shock/boundary-layer interaction.
- Simple wing-body combination, SDE.

The aircraft designer faces among the more difficult tradeoffs that between structural integrity for long service life and efficiency for high performance. Shortcomings which make decisions less certain include inadequate information to provide a basic understanding of fatigue; inefficient processes for coordination, storage, and retrieval of existing data; an inadequate database; and the inability to express functional relationships between the phenomena (or related data) and the design variables. Yet designers seem to generally agree they must treat fatigue and fracture in every phase of design—particularly in the early phases—to assure orderly progression in selecting materials and structural concepts needed for efficient aircraft.

NASA believes that important improvements can be made in treating fatigue and fracture and is working to improve understanding of them as well as to provide tools which will enable the designer to rapidly apply existing knowledge. Let me cite two examples.

The choice of material for an aircraft structure depends on the operating conditions of stress, desired life, and possible initial flaw size. Figure 16 pictures one way to express their relationship. The figure contains some familiar curves. The curve in the $(S/\rho,\ N)$ plane is the familiar S–N curve used for fatigue

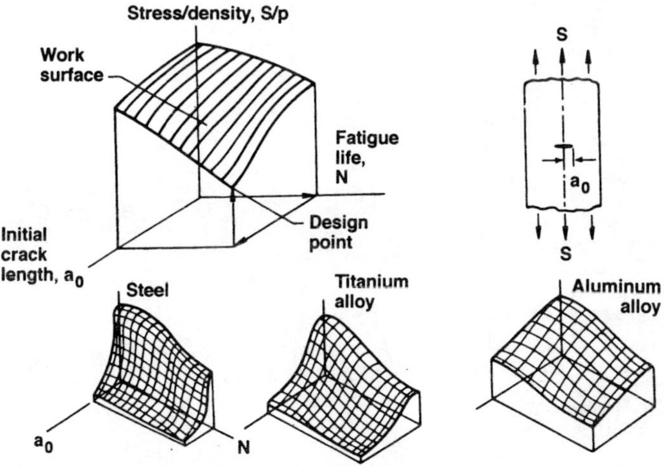

Fig. 16 Operating stress for specified life and initial crack size.

calculations. In the (S/ρ, a_o) plane, we have the familiar K_c curve used in fracture mechanics as a measure of material toughness.

More and more often for highly efficient structures, the size of a potential initial flaw and the rate at which the flaw grows in the material govern design stress. The stress must be low enough so the crack never becomes critically long during the design life. Consequently, to lengthen life the designer lowers the operating stress so that the crack grows very slowly. The loci of design points form an "operating surface." The equation for this surface depends on material properties such as crack resistance and fracture toughness. The surface defines the permissible stress/density ratio for all combinations of initial crack length and design life.

The three lower sketches in Fig. 16 illustrate the operating surfaces for three arbitrarily selected materials. High-strength steels have high specific strength and good fatigue performance provided no flaws are present. Currently used aluminum alloys generally have superior fracture strength and improved resistance to fatigue crack propagation. The titanium alloy used in this example has intermediate properties and could make a good choice for cases where small flaws may be present and short lives suffice.

Thus, there appears a reasonable chance of formulating equations which numerically characterize materials in terms of their combined fracture toughness and fatigue resistance. Such functional relationships would ease automation of the database and the inclusion of fatigue and fracture in an integrated design process.

NASA has worked out a program (Fig. 17) aimed at producing a fracture design module for IPAD in the coming decade. Such a module would help the

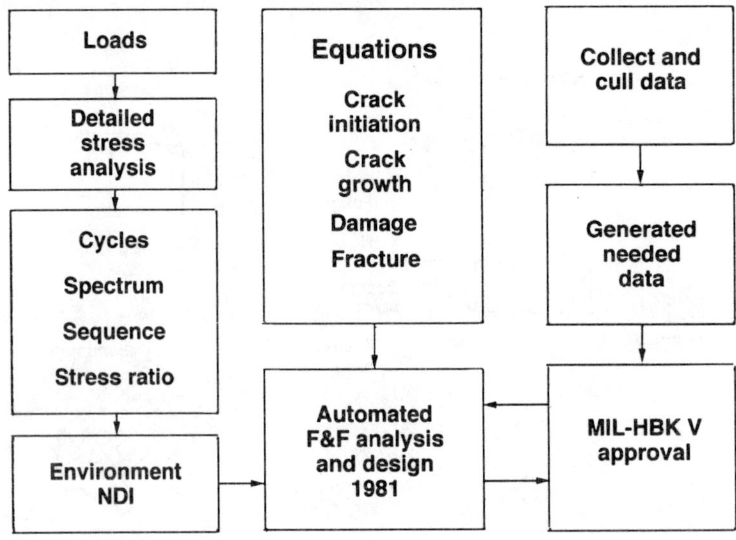

Fig. 17 NASA program fatigue-and-fracture design module for IPAD in 1981.

user draw on existing knowledge, identify specific gaps in the technology needed for his specific problem, and readily apply existing data in a systematic and timely fashion.

As spelled out by the center block, the first step is to formulate reliable equations for crack initiation, crack growth, damage, and fracture. A parallel need expressed by the blocks at left is to better predict and monitor the detailed stress field in all elements of the structure, that is, to understand and quantitatively state the total environment of the material. The blocks at right give the sequence of steps toward better characterizing the materials in the same sense that allowables are now given for static properties in MIL-HBK-5.

This program places a tall order, but NASA believes that the joint efforts of the industry and government, and the use of high-speed, high-capacity digital computers, will produce a high level of automation and a more rational basis for design for fatigue and fracture by 1981.

References

[1] Yaffee, M. L., "Use of Composites to Increase," *Aviation Week and Space Technology,* Sept. 11, 1972, pp. 87–94.

[2] Goble, R. L., "Composites, Lighter and Cheaper," *Astronautics & Aeronautics,* Vol. 10, No. 8, Aug. 1972, pp. 44–49.

[3] Becker, J. V., "Prospects for Actively Cooled Hypersonic Transport," *Astronautics & Aeronautics,* Vol. 9, No. 8, Aug. 1971, pp. 32–39.

[4] Anderson, M. S., and Kelly, H. N., "Materials Application to Civil Aircraft Structures in the Seventies and Beyond," *Vehicle Technology for Civil Aviation—The Seventies and Beyond,* NASA SP-292, 1971, pp. 179–192.

[5] Rainey, A. G., "Advanced Active Controls Technology," *Vehicle Technology for Civil Aviation—The Seventies and Beyond,* NASA SP-292, 1971, pp. 245–257.

[6] Yaffe, M. L., "New Controls to Shape Future Aircraft," *Aviation Week and Space Technology,* Oct. 16, 1972, pp. 46–50.

[7] Nissim, E., "Flutter Suppression Using Active Controls Based on the Concept of Aerodynamic Energy," NASA TN D-6199, March 1971.

[8] Rainey, A. G., Ruhlin, C. L., and Sandford, M. D., "Active Control of Aeroelastic Response," AGARD Flight Mechanics Panel Symposium on Stability and Control, Braunschweig, Germany, April 10–13, 1972.

[9] Rowe, W. S., Winther, B. A., and Redman, M. C., "Prediction of Unsteady Aerodynamic Loadings Caused by Trailing Edge Control Surface Motions in Subsonic Compressible Flow—Analysis and Results," NASA CR-2003, June 1972.

[10] Forsching, H., Triebstein, H., and Wagener, J., "Pressure Measurements on an Harmonically Oscillating Swept Wing with Two Control Surfaces in Incompressible Flow," AGARD Proceedings, No. 80—"Symposium on Unsteady Aerodynamics for Aeroelastic Analyses of Interfering Surfaces Part II," AGARD CP-80-71, Nov. 1970, pp. 15-1–15-12.

[11] Anon., "Joint DOT-NASA Civil Aviation Research and Development Policy Study—Supporting Papers," DOT TST-10-5, NASA SP-266, March 1971.

Structures—A Technology Overview

Roger A. Anderson, William A. Brooks Jr.,
Robert W. Leonard, and Joseph Maltz
NASA Langley Research Center

Lightweight reliable structure is a key requirement for a reusable Space Shuttle. To understand this requirement, an overview of the structures and materials aspects of this vehicle will be developed here, and some of the features that may present problems will be discussed. The major elements of a technology development program directed at solutions to these problems will then be described. The treatment of this subject in a single article must necessarily be selective and brief. Therefore, we draw the reader's attention to several reports and papers that discuss structures, thermal-protection systems, and materials for the Space Shuttle in greater depth.[1–3]

What characteristics of the Space Shuttle give impetus to a structures and materials technology development program? In a gross sense, the Shuttle consists of two large aircraft-type vehicles which undergo a spectrum of aerodynamic, acoustic, and inertia loadings plus a severe aerodynamic re-entry heating environment. It must meet a requirement for 100 uses with minimum-cost maintenance between flights. Existing design experience covers principal mission characteristics for expendable launch vehicles, orbiting manned spacecraft, and small test aircraft such as lifting bodies and the X-15 research vehicle. To incorporate the mission characteristics of all these vehicles into a single reusable vehicle system, however, produces a design challenge of unprecedented complexity.

Drawing upon available experience, preliminary design solutions have been proposed. For example, a possible structural arrangement of the booster consists basically of a conventional tank-airframe fitted with lifting surfaces for fly-back capability, as indicated in Fig. 1. Wing and tail may be unprotected, aerodynamically heated structures of nickel alloy or other heat-resistant metals, with provision for thermal expansion at the points of attachment to the cooler tankage. A segmented aerodynamic fairing is suspended from the tank-airframe. This thin-gage fairing, which serves as a heat shield, transmits the local pressure loadings to the tanks without carrying primary axial thrust and bending loads. Available nickel alloys appear adequate for its construction, except in localized stagnation-heating areas. The use of temperature-resistant external surfaces permits flexibility in selection of the flight profile and is tolerant of uncertainties in aerodynamic-heating predictions. When restrictions are placed upon staging velocity and the

Astronautics & Aeronautics, Vol. 9, No. 2, Feb. 1971, pp. 38–47.

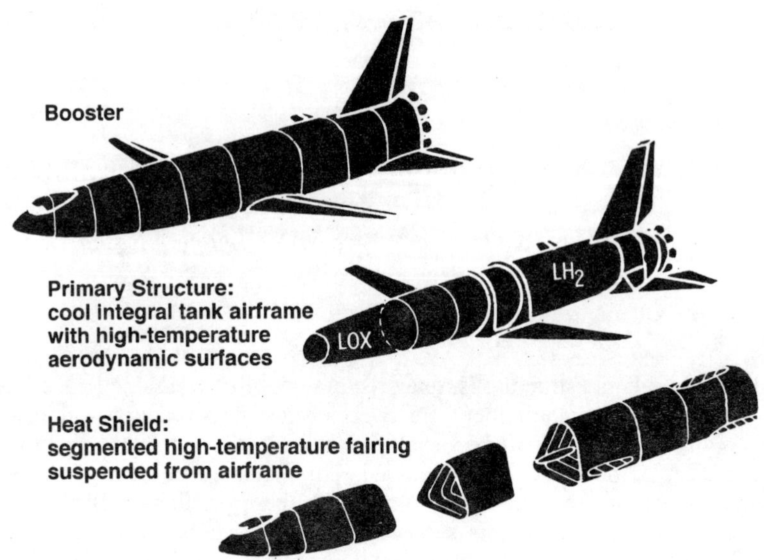

Fig. 1 Booster structural arrangement.

return maneuver, various heat-sink approaches to booster design can also be considered.

One approach to the delta-wing version of the orbiter stage uses a classic skin-stringer airframe sustaining aerodynamic, engine, and cargo loads, as indicated in Fig. 2. It is dominated by the large-volume cutout for the cargo bay; this complicates the placement and installation of the cryogenic propellant tanks. The drawing shows a straightforward nonintegral tank and airframe. Better internal volume utilization and, hence, a somewhat smaller vehicle would result if the circular were replaced by noncircular tankage. Tankage integrated with the airframe to the extent that it provides the primary load paths may permit further reduction of weight but leads overall to a more complex thermo-structural design problem. Because the orbiter is a very weight-sensitive vehicle, these options all require in-depth study before an optimal system can be selected.

Aerodynamic heat shields are required on the lower surfaces and stagnation-heating areas. With high-temperature alloys for the airframe, only minimal thermal-protection treatment may be needed on the leeward surfaces if the vehicle is restricted to high angles of attack during the entry-heating pulse. More-flexible mission operations are likely to dictate thermal-protection treatment on all exposed surface areas. In any event, heat shields for the orbiter will differ substantially in construction details and materials from those used for protection of the booster because of the more severe heating environment.

The complexity of the structural system of the Shuttle and the demands of its multiple-environment mission lead to major design and, hence, weight and cost uncertainties. Principal uncertainties include:

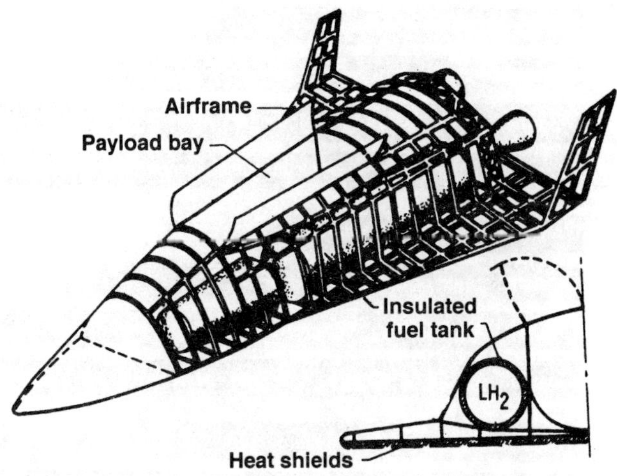

Fig. 2 Orbiter structural arrangement.

• *Accuracy of weight determination in the preliminary design phase.* This is particularly serious because of the sensitivity of gross liftoff weight and payload to changes in the dry weight of the orbiter; this sensitivity has been clearly indicated in both sponsored system studies and the independent studies made so far by industry. Such studies suggest that this sensitivity may be several times as great for the Shuttle as for high-performance aircraft.

• *The ability of thermal-protection systems to last for many missions with minimal costs for refurbishment.* There is general agreement that thermal-protection systems present perhaps the most critical technology deficiencies and contribute in a major way to vehicle operational costs.

• *The sequence of tests needed to establish initial flightworthiness and the inspection and recertification procedures during mission operations.* There is no precedent for establishing these procedures for a launch and entry vehicle.

To address these and other uncertainties and produce timely results that will have a beneficial effect upon Shuttle weight, reliability, and cost, NASA has instituted a major technology development program. Figure 3 gives an overview of the segment of this program covering structures and materials. The program is being carried out in both government and industrial laboratories. Major tasks, with the estimated time required for their accomplishment, are shown grouped into three categories: Structural Design, Thermal-Protection Systems, and Materials.

Before dealing with the content of these program tasks, certain characteristics which they exhibit in common should be mentioned.

1) They can be completed without restrictive assumptions as to vehicle aerodynamic configuration. Thus, they apply equally well to all configurations and mission modes.

2) They generally involve technology advances taking two to four years. Short-term, detailed design problems characteristic of ongoing hardware projects are not dealt with, nor are technology goals requiring 5–10 years.

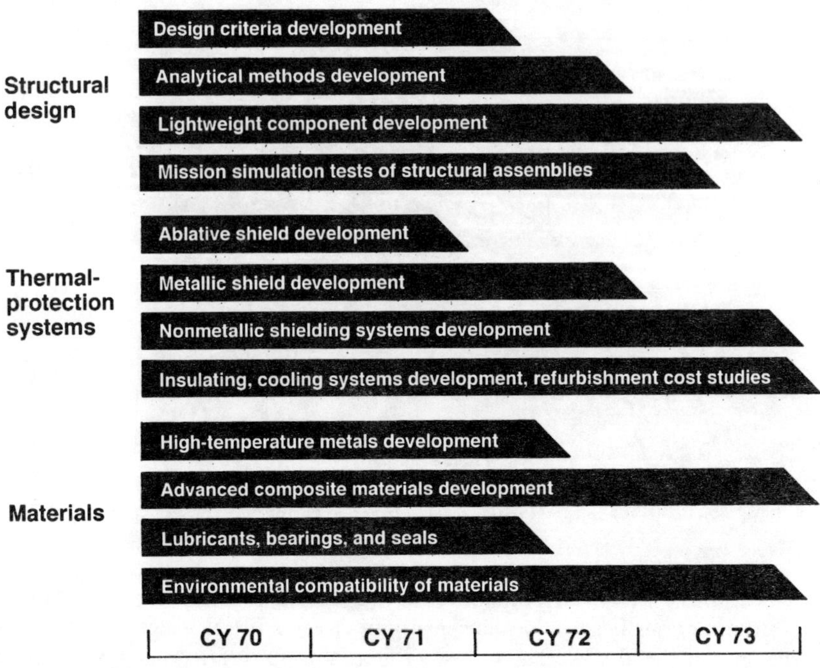

Fig. 3 Major elements of technology program.

3) The tasks in each area can be carried out without major dependence upon completion of those in the other areas. Thus, tasks are being pursued concurrently with cross-feed of information as it develops.

Structural-Design Technology

High priority is placed upon development of precise structural-design criteria because of the severe operational environments and the weight sensitivity of the Shuttle. With the aid of an elite panel of industry experts, NASA compiled a baseline structural-design criteria document.[4] However, numerous criteria questions remain. For example, the combination of many sources of repeated loads in a mission (typical mission outline in Fig. 4), the many-mission life requirement, and weight-critical design with high stresses may lead to fatigue-critical structures.

Precise definition of the Shuttle's life-cycle load spectrum is therefore needed. Other missing criteria information include proper definition of thermo-structural design factors, rational load-combination criteria, and the impact of fail-safe/safe-life design philosophies on Shuttle weight, cost, and integrity. There is little precedent or data from either aircraft or spacecraft-design experience to arrive at rational criteria in some of these areas. An additional year of work is forecast to generate the needed information.

A major aim of the analytical-methods task will be to extend fracture-mechanics design methods to the complex geometries and stress states expected in the

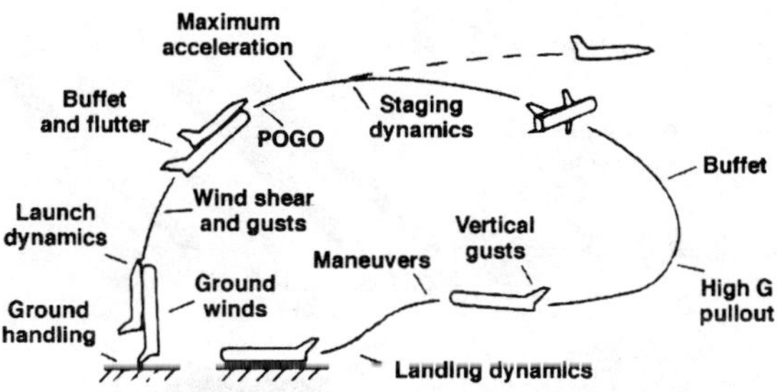

Fig. 4 Sources of repeated Shuttle loads.

pressurized tanks, which make up a large fraction of system weight. This task also covers development of adequate structural-analysis procedures for the whole airframe to support the unusual precision required for Shuttle design.

A particular focus will be further development of the general-purpose, finite element structural analysis computer program known as NASTRAN.[5] NASTRAN is user-oriented, nonproprietary, operational on IBM, CDC, and Univac computer systems and capable of analyzing entire vehicle systems for static stress and deflection, buckling instabilities, vibration modes and frequencies, and transient and random responses. It has been undergoing operational shakedown for over a year in selected laboratories and has now been released to interested users. Several improvements to NASTRAN are receiving priority because of Space Shuttle needs: namely, heat-transfer and thermal-stress analysis capability and improved, higher-order finite elements.

Still referring to Fig. 3, the lightweight-component-development task primarily concerns establishing design confidence in structural components embodying high-performance filamentary composite materials. This task is shown extending through CY 73 because of the exhaustive testing necessary to build confidence in new forms of structure.

Figure 5 depicts some of the structural components which might show weight reductions of the order of 20% through application of such materials. Experimental work is underway on stiffener and ring elements, skin-stringer panels, and filament-overwrapped tanks. Some of the parts of typical engine thrust structures are currently being designed for tests under simulated mission conditions. In the applications depicted, limited quantities of filaments are adhesively bonded to basically metallic structures in places where they will most enhance a desired structural property for a modest quantity of high-cost filaments. This selective-reinforcement approach stands in contrast to the somewhat more familiar all-filamentary composite structure, which uses metal inserts only at critical places, such as joints between major components.

Applying filamentary composites requires careful attention to design details which may affect fatigue life. Figure 6 presents fatigue results, obtained under NASA contract NAS1-8858, for the edge joint of a boron-reinforced titanium panel. Translating load from boron layers to the all-metal edge member is done by a series of steps. These specimens were subjected to completely reversed cyclic in-plane loads. The data indicate that the composite construction yields a

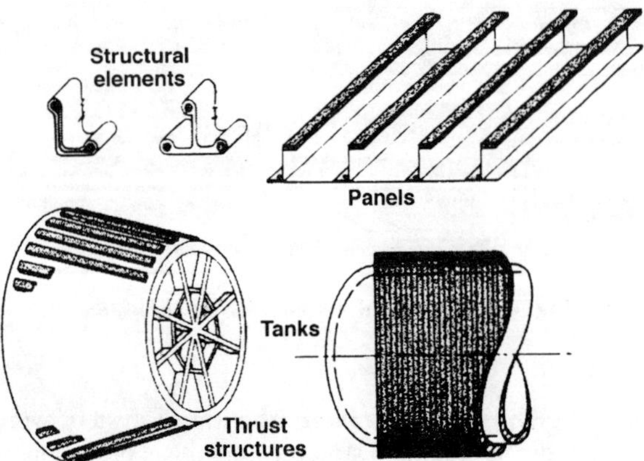

Fig. 5 Selective reinforcement of Shuttle components with filamentary composites.

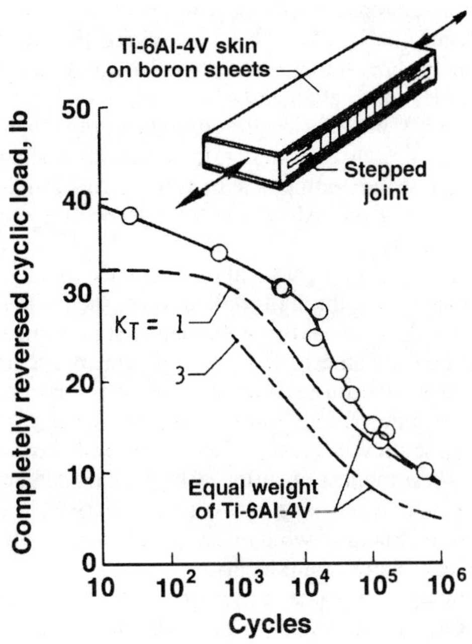

Fig. 6 Boron stepped-joint fatigue strength.

fatigue life that compares favorably to both notched ($k_T = 3$) and unnotched ($k_T = 1$) titanium plates of equal weight.

Returning to Fig. 3, mission simulation tests of structural assemblies refers to both static and dynamic tests of fairly large-scale structural models. Prototype sections of Shuttle tanks, structure, and thermal-protection systems are currently being designed, and the Saturn S-II stage static-test stand at the Marshall Space Flight Center is being modified to subject these models to simulated flight environments with loading and radiant-heating equipment. The purpose of these tests is to gain experience in the conduct of ground-qualification tests for complex Shuttle hardware and to identify problems in the performance of integrated hardware systems. Typical structural assemblies to be tested are approximately one-third-scale sections of orbiter and booster fuselage including cryogenic tankage and thermal-protection systems.[3] Four major tests will be completed early in 1973.

Facilities for dynamic simulation of flight environmental effects on prototype Shuttle hardware are in short supply. Figure 7 presents the expected flight trajectory for a re-entering orbiter superimposed upon the performance of the only existing aerodynamic facilities which provide true simulation of surface pressure and heat loads on large-scale structural test specimens: the 50-MW arc-heated Gas Dynamics Facility at the Air Force Flight Dynamics Laboratory and the 8-ft High Temperature Structures Tunnel (HTST) at NASA Langley. Simulation is available only at velocities well below those for which maximum heating occurs on the orbiter and then only for tunnel runs substantially shorter than the duration of aerodynamic heating. However, important checks upon the aeroelastic response of heat shields as well as upon heat-transfer effects due to full-scale protuberances, gaps, and shock interactions can be made.

Various types and sizes of test specimen can be accommodated by the 8-ft HTST. In the holder for aeroelastic testing of full-scale surface panels in a two-dimensional flow field, the pressure difference across the panel can be regulated through control of the pressure in a cavity behind the panel. In the Mach-7 high-

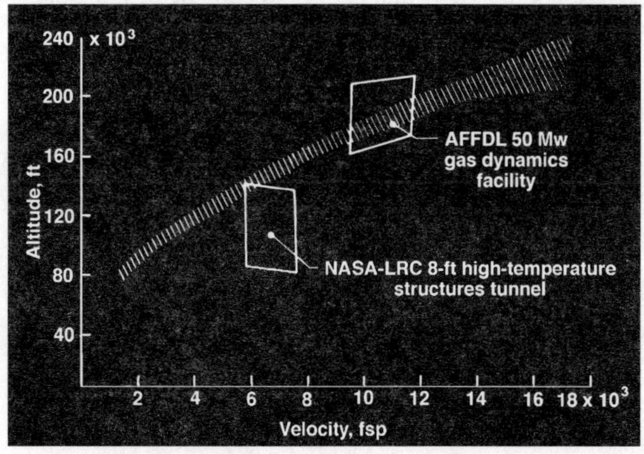

Fig. 7 Large aerothermodynamic facilities for testing orbiter structures.

temperature environment provided by the 8-ft HTST, any degradation in panel stiffness caused by in-plane thermal stresses will show up as warpage or possibly panel flutter. The more elaborate structural-test model with the rounded nose has segmented metallic heat shields attached to an insulated substructure. This model has survived static tests under radiant lamps and is being prepared for aerodynamic tests which will check its aeroelastic, structural, and thermal performance with flow through cracks and joints. A tantalum nose cap helps the specimen survive in the tunnel environment.

A test bed designed to be used in the 50-MW Gas Dynamics Facility can accommodate various thermal-protection system (TPS) stagnation-area concepts as well as surface heat shields. And it permits detailed measurements of thermal and mechanical response in repeated tests under the heating and loading conditions imposed by a Mach 12–14 flow field.

Thermal-Protection Systems Technology

As shown by the second group of bars in Fig. 3, activities related to three principal thermal-protection systems design concepts are being pursued concurrently. Each represents a different stage of technology readiness, as shown in Table 1.

The most highly developed and most thoroughly investigated reusable TPS materials are the metals. The technology is moderately advanced, but improvements are required for the dispersion-strengthened nickel-base alloys and coated refractory metals. Metallic heat shields are quite complex due to features used to minimize thermal distortions, join panels, seal joints, and insulate the primary structure. The metal systems show moderately good reliability and reuse potential. Some uncertainty exists regarding the dispersion-strengthened nickel alloys and the coated refractory metals.

A needed extension of prior art is to increase the allowable dimensions between expansion joints to minimize the number of parts on a vehicle as large as the Shuttle orbiter. Figure 8 illustrates one design suitable for thin-gage nickel alloys. It accommodates thermal expansion in one direction by bending of the corrugated elements and by slip joints and rotation of the support elements in the other direction. Design problems common to metallic shields are given in Fig. 8. One

Table 1 Technology readiness

Item	Metallic radiative	External insulation	Ablative
Technology status	Moderately advanced	Emerging	Well advanced
System complexity	Complex	Simple	Simple
Reliability	Moderately good	Unknown	High
Reuse potential	Moderately good	Promising	None

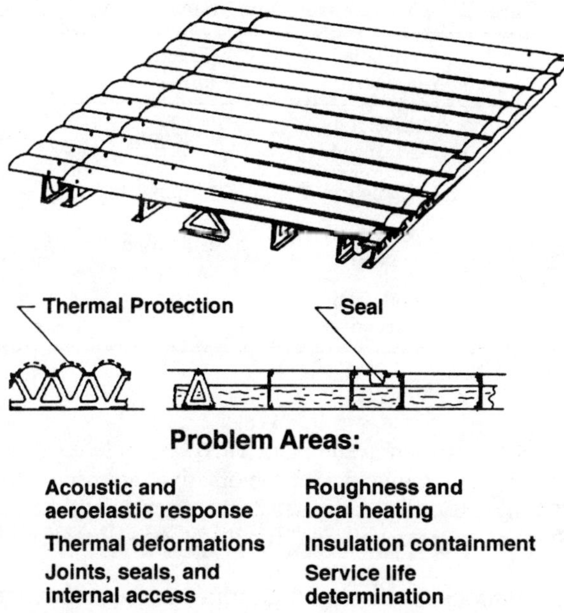

Fig. 8 Metallic heat shield design problems.

involves provision of a seal to control inflow of hot boundary-layer air. Tests are being performed to understand the significance of airflow through gaps and beneath the shields due to pressure differences on various areas of the flight vehicle. Control of this flow may require positive means for sealing gaps, as shown in the drawing, or alternately the use of inert-gas purge systems to provide a positive outflow through the gaps.

A major effort on metallic heat shields in the technology program involves a detailed assessment of coated columbium for use on the Space Shuttle. The evaluation will be performed by the design, analysis, fabrication, and static and hot-gas-flow testing of heat shield components and full-size panels. Some effort is also underway on coated tantalum systems but is confined to materials testing and does not include a technology scale-up to full-size panels.

Activities related to the dispersion-strengthened nickel alloys and the conventional superalloys consist primarily of the detailed determination of performance in the Shuttle entry environment. Included are creep, residual strength, and oxidation tests of the candidate materials. Contingent upon the results of such testing, a program similar to that for coated columbium will be established for the best available dispersion-strengthened nickel alloy.

The TPS technology program (Table 2) also covers the development of nonmetallic heat shield concepts. The concept of most interest is ''external insulation''—low-density refractory oxides, such as quartz, mullite, aluminosilicates, and zirconia, attached to the exterior vehicle surface and exposed to the mission environment. The materials are employed in the form of rigidized-fiber batts,

Table 2 Thermal-protection systems summary

Item	Metallic radiative	External insulation	Ablative
Technology status	Moderately advanced	Emerging	Well advanced
System complexity	Complex	Simple	Simple
Reliability	Moderately good	Unknown	High
Reuse potential	Moderately good	Promising	None

woven batts, and syntactic reinforced foams. This concept looks attractive because it promises many of the simplicities of the ablative approach while offering a multiflight capability. But the associated technology is still in an early state, and reliability and reuse potential are difficult to assess because of insufficient experience.

Several potential problems have been identified in the surface insulation concept. One is inherent brittleness and a resulting low-strain limit, although the material forms proposed are not as brittle as the usual ceramics. These materials characteristically have little strength and must be attached to structural substrates or directly to the primary load-carrying structure. Consequently, the strains introduced by deformation of the structural substrates must be accommodated by these materials without deleterious effects. Another substantial problem with this class of materials: Surface coatings must prevent moisture absorption, increase emissivity, and minimize erosion and handling damage. While neither of these problems has been solved satisfactorily, progress in the last few months has been encouraging and new concepts are being developed.

In addition to in-house research on material characterization, the TPS technology program on external insulation has three major contractual development studies which consider different material approaches and temperature ranges. The main thrust of these studies concerns assessment of materials suitability for the Shuttle, coating and materials development, and concept verification by the production and testing of full-scale panels.

Another nonmetal concept, the carbon-reinforced carbon composite, is being examined for application to high-temperature areas such as noses and leading edges. These systems are also brittle and in addition have high-temperature attachment problems. Clever designs and fabrication methods must be developed if these materials are to be used on the Shuttle. The carbon systems also experience oxidation on exposure to entry conditions. Carbon oxidation can be inhibited by mixing or diffusing metals into the carbon to form refractory carbides when heated. Another approach uses overlays of refractory oxides.

Figure 9 describes oxidation characteristics of conventional carbon and oxidation-inhibited carbon. For the case shown, silicon carbide was used as an inhibitor to raise the threshold temperature for oxidation to about 2700°F. At temperatures above the oxidation threshold, the carbon systems will ablate and, thus, tolerate only a limited number of entries.

Activities in the TPS technology program on carbon-carbon systems concern raising the oxidation threshold to 3500–4000°F. Other aspects include improvement of interlaminar shear strength and development of appropriate leading-edge designs.

The ablative systems are generally agreed to be the backup TPS for large surface areas. They could also be used as the primary system for local areas of high heating, with replacement after each flight. Recently, there has been increased interest in using the reliable ablative approach to achieve an early operational status for the Shuttle. Later versions of the Shuttle would then incorporate more reusable TPS, such as the external insulation or the metallic heat shields. Due to similarities in construction, dimensions, and thermal performance, the ablative and external insulation concepts are more compatible from the point of view of design.

Although performance improvements can be made to ablators for Shuttle application, a greater need is to reduce substantially the costs of ablative heat-shielding. Cost reduction is the principal goal of the current ablative heat shield technology activity. It is being vigorously pursued through development of appropriate fabrication processes and better understanding of the kind of manufacturing defects allowable without compromising performance. This work is far enough along to forecast that the ablative heat shield could be used as a reasonably cost-effective interim thermal-protection solution for an early version of the Shuttle orbiter. Estimated costs, based on actual fabrication experience, are shown in Table 3. These costs are probably an order of magnitude less than those associated with other NASA manned entry vehicles.

The fourth major TPS task (Fig. 3) includes several activities of smaller scope than the preceding three. Solutions are being sought to the packaging, retaining,

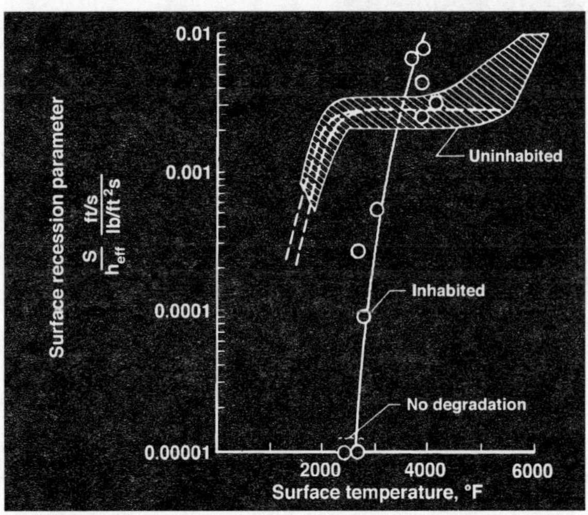

Fig. 9 Thermal degradation of carbon-carbon materials.

Table 3 Estimated costs

Number of flights	Purchase cost, $/ft^2	Cost/flight, $
1	220	1,800,000
10	75	600,000
100	50	400,000

and installation problems associated with high-temperature insulation for use primarily with metallic heat shield systems. At present, the cooling-systems work primarily involves assessment of potential weight-saving or performance improvements which may be realized from surface-cooling schemes for stagnation areas and from active and passive cooling approaches for use beneath heat shields. The last item covers refurbishment-cost studies.

Practically no sound data exist on the operational aspects of refurbishment, inspection, and recertification of TPS for subsequent flights. A contract study has been initiated to identify refurbishment costs. The first phase of the study covered selection of heat shield concepts and the development of a plan for the second phase. That will generate data from actual refurbishment exercises conducted on a large-scale mockup.

Some refurbishment-cost estimates were generated, however, as part of the first phase. A by-product of the cost-estimation study was the contractor's estimate of service life for several different types of heat shields, as given in Fig. 10. The estimated lives are keyed to the proposed applicable temperature ranges given in the third column. The nominal lives are indicated by the open symbols.

Perhaps even more significant is the life uncertainty, the extent of which is indicated by the end points of the lines in Fig. 10. If these estimated results are

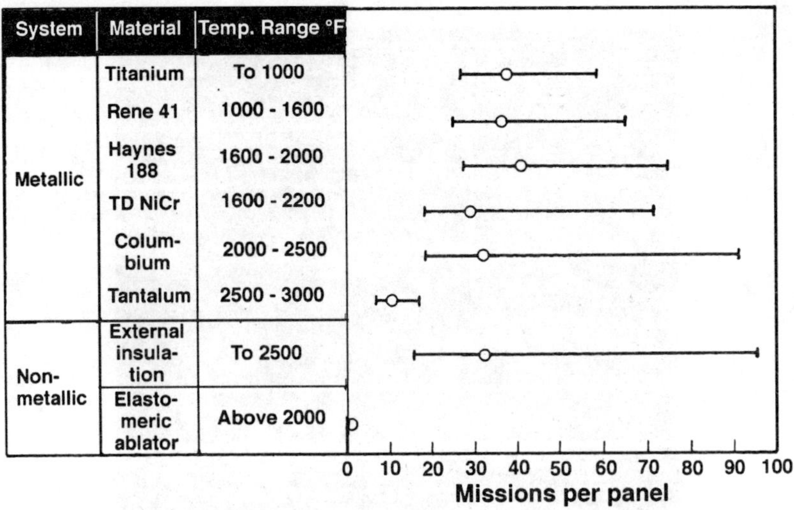

Fig. 10 Service life of heat shield panels: an estimate.

experimentally validated, periodic replacement and repair of the TPS is likely to be required over the life of the Shuttle, and the system must be designed with this operational feature in mind.

Materials Technology

As indicated in Fig. 3, materials technology concentrates on four key tasks. The largest of these tasks covers high-temperature metals development and characterization.

The demands on TPS candidate materials increase in severity with increasing temperature. Up to approximately 1800°F, we have enough experience to justify confidence in the ability of titanium-, iron-, nickel-, and cobalt-base alloys to do the job. There is some uncertainty about the upper-temperature limit and the design properties of each class of metal. This information is needed for reliable design.

It is doubtful that any material will serve for 100 missions above 2200°F. However, since only small areas are expected to be involved, occasional replacement is tolerable. Five columbium-base alloys, each provided with an optimized oxidation-resistant coating system, are being screened in a program which complements the heat shield technology efforts already discussed. Above 2400°F, coated tantalum alloys are being examined as an alternative to nonmetallic materials.

The intermediate range of 1800–2200°F presents the most serious problem, since the areas of the Shuttle experiencing these temperatures may be too large for economical refurbishment. Conceivably, coated columbium alloys could be used in this temperature range or the conventional superalloys could be extended to cover it. Another possibility is to use dispersion-strengthened nickel-base alloys.

Evaluation of these various candidates is being pursued with furnace tests and with arcjet tests in high-velocity airstreams. Figure 11 gives results of arcjet tests on TD-NiCr (Ni-20% Cr-2.4% ThO_2) and on a modification of it involving the addition of small amounts of aluminum and yttria. It shows the mass loss from

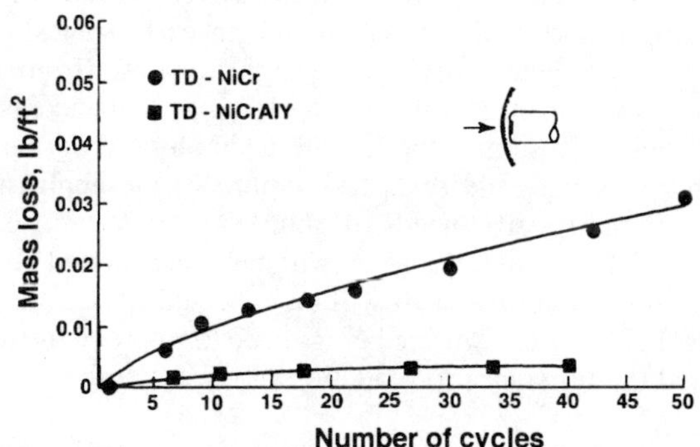

Fig. 11 Comparative oxidation rates for TD-NiCr and TD-NiCr AIY. Tw, 2200°F; pw, 14–15 torr; Ht, 2900 Btu/lb; cycle of 1800 s.

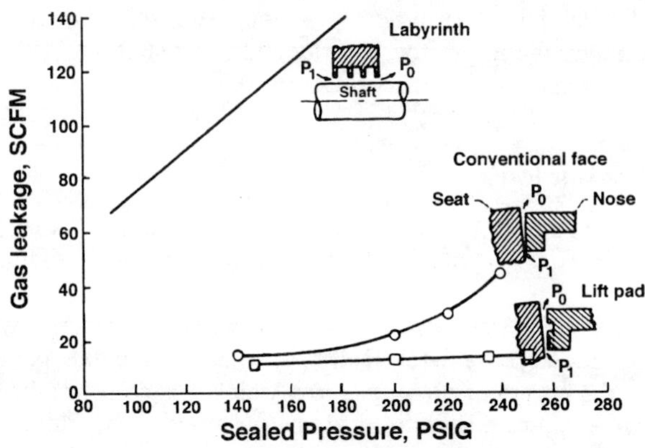

Fig. 12 Performance of seals.

small forward-facing disks of material subjected to simulated mission cycles of 1800 s.

Such data indicate that, at 2200°F, heat shields of the unmodified alloy may have to be replaced occasionally, whereas the modified alloy shows promise of lasting for the required 100 mission cycles. To assure the availability of dispersion-strengthened metal alloys, the technology program includes a process development contract aimed at producing good-quality sheet 60 by 24 by 0.010–0.040 in. in quantity by 1973. Modified compositions are also being developed, and modest quantities of sheet material are being produced for experimental purposes.

The second task in the materials-technology program concerns advanced composite materials. A small effort, it supplements the one on lightweight primary structural components already described. The rapidly changing field of composite materials presents a host of new and improved filaments, matrices, and lay-up procedures approaching technology readiness. Materials such as advanced graphites, polyimides, and polyquinoxalines as well as metal-matrix materials are being developed and evaluated. Although in early stages of development, composites offer such potentially great weight savings (from extended high- and low-temperature capability, increased strength, and increased modulus of elasticity) that their inclusion in the Shuttle technology program is justified.

The lubricant, bearings, and seals task embraces the application of existing research results to the specific needs of Shuttle systems.

For example, CaF_2 solid lubricants[6] will be incorporated into bearings for aerodynamic control surfaces which will be expected to operate at temperatures as high as 1600°F. Modifications are being made to improve the low-temperature performance and the resistance to oxidation of these lubricants at high temperatures.

Another development, the self-acting "lift pad seal," is being adapted to minimize leakage in high-pressure cryogenic hydrogen and oxygen systems while providing multimission capability. Figure 12 shows the favorable performance of this seal relative to other types.

The need for hydraulic fluids operative at 800°F is being met by modification and blending of polyphenyl ethers and polyphenyl thioethers.

The last task, environmental compatibility, extends technology in such diverse areas as hydrogen effects, corrosion, stress corrosion, and flammability to match the Shuttle requirements. Two contracts concern the effect of gaseous hydrogen at high pressures and at high temperatures upon container materials and aim at defining the limitations of materials in this aggressive environment.

Concluding Remarks

The tasks selected for funding in this technology program represent only a minor fraction of the total submitted for consideration by qualified government and industry personnel. Resources have been allocated to the mainstream problem areas. It is anticipated that these critical problems will be resolved in a timely manner and that R&D results will be applicable to the structural subsystem of all the Shuttle concepts currently under preliminary design study.

References

[1] Brooks, W. A., and Vosteen, L. F., ''An Assessment of Structures and Thermal Protection System Technology for a Space Shuttle System,'' AIAA/ASME 11th Structures, Structural Dynamics, and Materials Conference, Denver, CO, April 22–24, 1970.

[2] Riehl, W. A., and Cataldo, C. E., ''Materials Technology for the Space Shuttle,'' AIAA/ASME 11th Structures, Structural Dynamics, and Materials Conference, Denver, CO, April 22–24, 1970.

[3] ''Space Transportation System Technology Symposium,'' *Structures and Materials,* Vol. III, NASA TMX-52876, July 1970.

[4] ''Structural Design Criteria Applicable to a Space Shuttle,'' NASA SP-8XXX, Nov. 1970 (Advance Copy).

[5] ''The NASTRAN User's Manual,'' NASA SP-222, Oct. 1969.

[6] Sliney, H. E., ''Bearings, Lubricants and Seals for the Space Shuttle, Space Transportation System Technology Symposium,'' *Structures and Materials,* Vol. III, NASA TMX-52876, July 1970.

Aircraft Structures Technology

The NASA Structures and Materials Research Program for Supersonic-Cruise Aircraft

Paul A. Cooper and Richard R. Heldenfels
NASA Langley Research Center

Cancellation of the national SST program in 1971 left not only environmental and economic but also several major technical questions unanswered. In the structures area, the major unresolved problems concerned the poor flutter characteristics of the aircraft and the high empty-weight fraction, which adversely affected its economics. The Department of Transportation funded a follow-on technology program to complete selected tasks in the areas of flutter, titanium honeycomb-panel development, and fuel tank sealants. But it did not include advanced structural concepts or high-temperature composite materials, and it confined the flutter investigations to the delta-wing configurations studied in the national SST program.[1]

This article reviews the objectives and current status of the research elements in the structures and materials part of the NASA Supersonic-Cruise Aircraft Research (SCAR) program begun in FY 73.[2]

The SCAR structures and materials program emphasizes technology that may greatly reduce structural weight of large, flexible, long-life supersonic aircraft. The R&D effort treats advanced structural concepts, structural-design procedures, aeroelastic loads and response, and materials applications. Because of limited resources, only research areas with long-term potential for high payoff have been pursued; no attempt has been made to investigate all the structural problems likely to be encountered in the design and manufacture of advanced supersonic-cruise aircraft. Furthermore, the funds available permitted testing only small structural components.

In the SCAR program, NASA Langley Research Center has done most of the fundamental structures and materials research applicable to supersonic-cruise aircraft, with contributions from the Dryden Flight Research Center on ground and flight tests of structures, from Ames Research Center on fuel tank sealants, and from Lewis Research Center on high-temperature matrix materials for advanced composites.

SCAR emphasizes, as just mentioned, advanced structural concepts that will be applicable to high-performance supersonic-cruise aircraft and to the development, manufacture, and proof-testing of advanced titanium and composite components for both primary and secondary structures. It also stresses steady and unsteady loads and flutter-calculation methodology for large, highly flexible aircraft in the transonic flight regime and experimental definition of the long-wavelength

Astronautics & Aeronautics, Vol. 14, No. 5, May 1976, pp. 26–37.

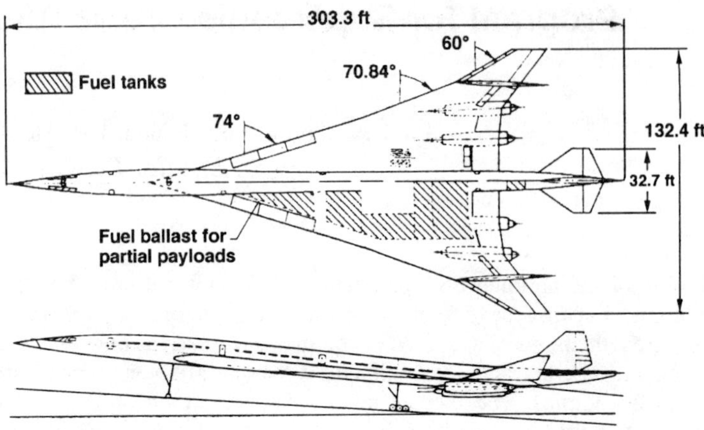

Fig. 1 Aircraft configuration and fuel tank location.

atmospheric turbulence that influences supersonic-cruise aircraft design. The program is divided into three research areas:

- Concept studies of the efficiency of candidate structural concepts under varying load indices expected in high-performance, large supersonic-cruise aircraft.
- Structures technology to provide analytical techniques, design methods, and empirical data to support efficient design of long-range supersonic-cruise aircraft.
- Applications of candidate structural materials in the operating environment of a supersonic-cruise aircraft to evaluate long-life performance and manufacturing techniques for small structural components.

Concept Studies

Technological areas where research progress should greatly aid supersonic-cruise aircraft design include active control systems, advanced composite materials, improved aerodynamic configurations, and advanced analysis and design methods.

To establish a realistic baseline, preliminary structural-design studies were conducted by Lockheed California Company and by The Boeing Commercial Airplane Company. Since supersonic-cruise aircraft tend to be large and flexible, the studies incorporated realistic aeroelastic considerations based on finite element structural analysis and sophisticated aerodynamic-loading analysis, both steady and unsteady. The strong interaction of the various disciplines prompted the use of computer-aided design methods to improve and expedite the aeroelastic and structural-resizing cycle. These studies appraised some of the computer-aided design methods now in use by the aerospace industry and defined areas requiring further development.

Langley chose for analysis an arrow-wing configuration (Fig. 1) derived from the SCAT 15F series.[3] It was to be sized for a takeoff mass of 750,000 lb, a payload of 49,000 lb (approximately 230 passengers), a cruise speed of Mach 2.7, and, if possible, a range of 4200 n mile. The engine characteristics were based on previous system studies. Boeing assumed the use of advanced-technology afterburning turbojet engines in the 50,000-lbf-thrust category, and Lockheed assumed duct-burning turbofan engines in the 90,000-lbf-thrust category.

Lockheed used a multidisciplinary analysis interfacing two large computer codes—NASTRAN (statics, dynamics, and structural stability) and FAMAS (aerodynamic loads, structural response, and flutter analysis).[4] The most promising skin-stiffening concepts and internal structural arrangements were selected and refined.

As shown in Fig. 2, three structural configurations consistent with an assumed 1980 technology readiness were investigated as primary load-carrying concepts for the wing. Two types of uniaxially stiffened panel concepts were considered:

- Panels supporting loads in the spanwise direction (spanwise stiffened) with a multirib substructure and widely spaced spars.
- Panels supporting loads in the chordwise direction (chordwise stiffened) with a multispar substructure and widely spaced ribs. The chordwise-stiffened concepts make use of structurally efficient beaded-skin designs which, when properly oriented in the airstream, provide acceptable aerodynamic characteristics and allow thermal expansion, thereby minimizing thermal stresses.

Of the three configurations, the chordwise-stiffened approach offered the maximum mass-saving potential when coupled with selective reinforcement of the basic metallic rib and spar caps with unidirectional boronpolyimide composites. This structural arrangement was used as the primary design concept for the wing, with the exception of the outboard wing panels which, because of stiffness requirements, used monocoque aluminum-brazed titanium honeycomb panels. The basic material was Ti-6A1-4V titanium alloy.

A detailed three-dimensional finite element model containing approximately 2200 degrees of freedom was developed for the final Lockheed structural analyses. Strength sizing and one resizing were conducted at six wing regions and four fuselage regions. Flutter characteristics were determined on the strength-designed structure for three Mach numbers (0.6, 0.9, and 1.85), and additional stiffness was added to correct flutter deficiencies.

Boeing used a multidisciplinary analysis depending mainly on the ATLAS program for static structural analysis, sizing, and dynamic behavior. (See the next section of this article for a brief discussion of the ATLAS code.) Static aeroelastic loads were determined with the FLEXSTAB program[5] using elastic properties and geometry supplied by the ATLAS code in a direct interface. Boeing used design concepts consistent with a 1975 technology base, and the concepts were initially sized based on design loads and environmental conditions from previous study results (SCAT 15F study[3]).

Boeing selected as the basic structural concept an aluminum-brazed titanium honeycomb. The analysts then generated an ambitious 8500-degrees-of-freedom three-dimensional finite element model, as shown in Fig. 3. The sketch in the upper part of the figure depicts the complete model, the wing and adjacent fuselage structure being modeled with two-dimensional elements and the remainder of the aircraft simulated with beam elements. Boeing decided that a model of this complexity was needed to assure a meaningful analysis of the engine beams,

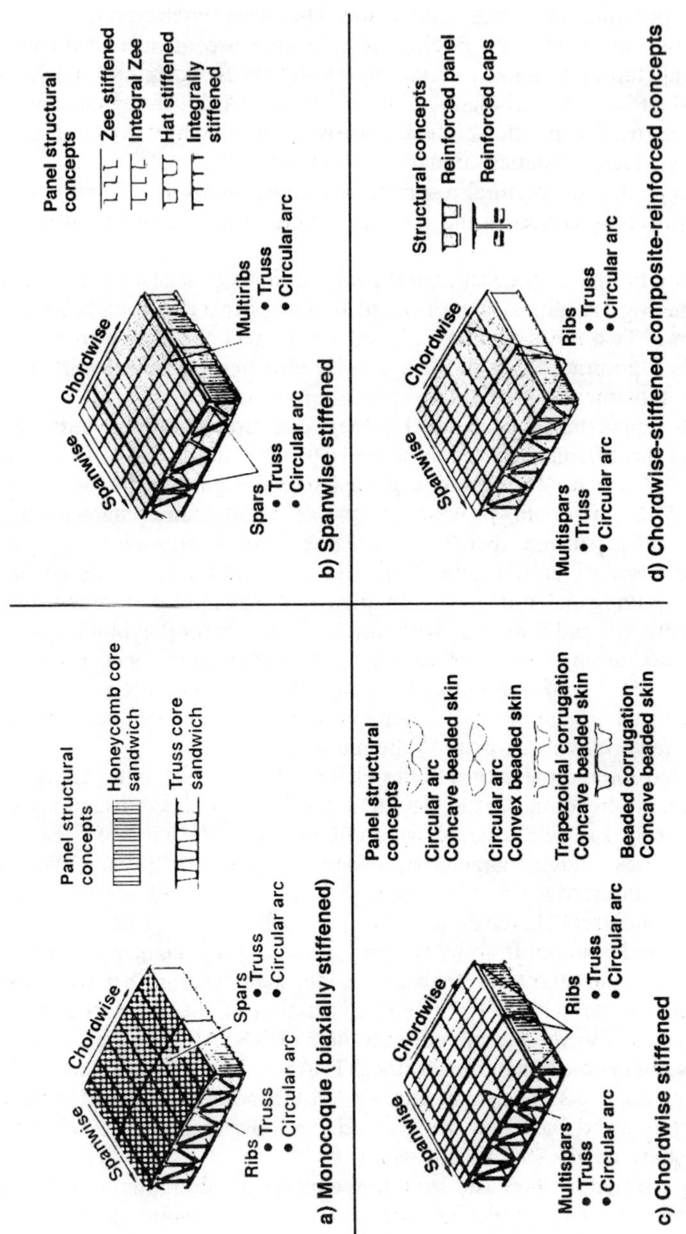

Fig. 2 Lockheed structural concepts for an arrow-wing supersonic aircraft.

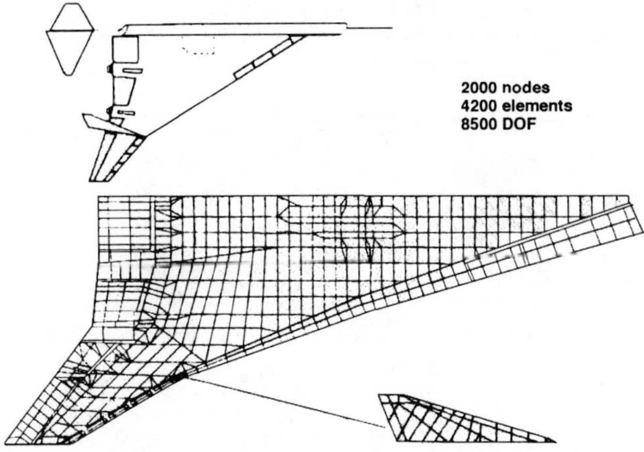

Fig. 3 Boeing finite element model of an arrow-wing supersonic aircraft.

leading- and trailing-edge controls, wing secondary structure, landing-gear and wheel-well cutout, and wing-mounted fins as well as wing primary structure. For dynamic analyses, it retained approximately 250 degrees of freedom. The wing was resized using an automated fully stressed resizing algorithm, and a check of the strength-sized structure showed that it had a flutter deficiency that was partially removed by additional resizing based on engineering judgment.[6]

Figure 4 shows the general critical design conditions for the final wing cover for the Boeing design. The wing schematic shows three distinct zones dividing both the upper and lower surfaces of the wing according to the three design considerations that dictated structural sizes. The tip structure was stiffness-critical

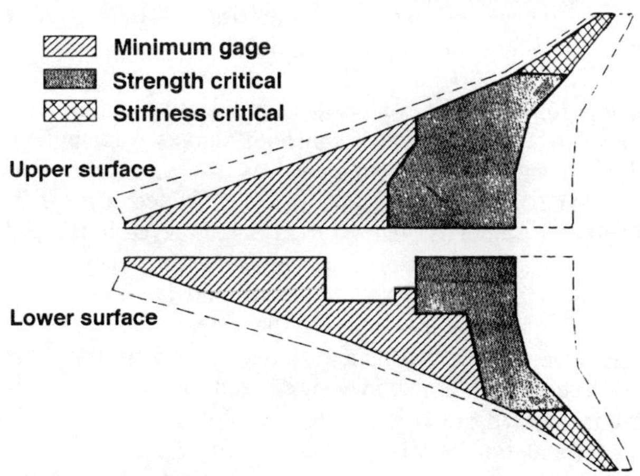

Fig. 4 Wing-cover critical design conditions.

Table 1 Arrow-wing group mass statement in lb m

Element		Boeing[a]			Lockheed[b]	
			%GTOW			%GTOW
Structure		224,400	29.9		201,300	26.8
Wing	95,800			90,600		
Horizontal tail	6,500			7,900		
Vertical tail	5,800			5,400		
Fuselage	56,100			42,100		
Main gear	37,300			27,400		
Nose gear	3,800			3,000		
Nacelle	19,100			24,900		
Propulsion		56,800	7.6		58,100	7.8
Systems		77,100	10.3		54,400	7.2
OEW		358,300	47.8		313,800	41.8
Payload[c]		49,000	6.5		49,000	6.5
Fuel		342,700	45.7		387,200	51.7
GTOW[c]		750,000	100.0		750,000	100.0

[a]Range of 4000 n mile. [b]Range of 4200 n mile. [c]Numbers held constant throughout study.

and sized to meet the flutter requirements. The aft box was strength designed to transmit the wing spanwise and chordwise bending moments and shears. The forward box structure consisted of minimum-gage surface panels and substructure components. Foreign object damage (FOD) governed selection of minimum gage. The Lockheed design evidenced the same structural characteristics with a slightly larger tip area defined by stiffness conditions and a slightly smaller area defined by strength considerations. Both aircraft designs have a large wing area at minimum gage, making the arrow-wing configuration a relatively inefficient structure on a mass basis, a situation typical for lightly loaded low-aspect-ratio wings.

Table 1 gives the group mass statements for the final aircraft. In making comparisons, the reader should be cautioned that the configurations were slightly different, and whereas Boeing used 1975 technology, Lockheed projected a 1980 technology that utilized composites for selective reinforcement in lighter structural concepts. The major effort in both studies was applied to the wing structural design; thus, the wing mass values are more refined than the other values shown. The structural concepts were evaluated using mass and range increments as merit functions, and the total ranges given in the table depend on projected engine efficiencies, which are somewhat speculative.

These design studies, using large, complex models, permitted evaluating advanced computerized design methods. For the arrow-wing supersonic-cruise configuration, static aeroelastic effects and flutter prove very important and should be considered as early as possible in design so that stiffness-constrained members will not be unnecessarily resized for strength. Such structural analysis requires generation and validation of a large, complex finite element model. Automated modeling methods and sophisticated graphics can greatly decrease both manpower expenditure and flow time for this task. Automated strength resizing takes much less time in the design cycle than manual methods. (In the Boeing study, automated resizing of the wing elements took only an overnight stress run, whereas manual resizing of the fuselage elements required an additional three man-weeks of effort.) Realistic automated strength resizing is an important factor in reducing design-cycle time because the finite element model can be generated more quickly by using unrefined initial estimates of member sizes. Several publications summarize the major activities under these studies.[4,6–8]

The Boeing study, continuing, will evaluate mass reductions that might be achieved by advanced composites and structural concepts expected in the 1980s: about 9% in operating empty mass and 18% in structural mass. Substantiation of these figures by integrated analysis and design tools will help guide research planning on advanced materials and design concepts. Detailed design and concept studies will consider representative sections of major components of the baseline structure so that results will be directly comparable with those obtained for the 1975 technology-baseline titanium structure.

Parallel structural tradeoffs are being made at NASA-Langley with aid from Vought engineering personnel using large-scale computer-aided tools.[9] These studies currently emphasize methods of redesigning strength-designed structures to remove flutter deficiencies and application of filamentary composite materials to supersonic structures.

Structures Technology

Analytical techniques, design methods, and empirical data required for the structural design of a long-range "aeroelastic" supersonic-cruise airplane; techniques for predicting steady and unsteady transonic loads; landing-load alleviation for highly flexible aircraft; in-flight measurements of atmospheric turbulence; and flutter analysis and experiment—all are being improved in this program area. Analytical design procedures for fatigue and fracture prevention, thermal stress, and flutter prevention are being advanced and consolidated into computer codes.

Advanced Design Codes

Two major integrated structural analysis and design codes and a study code are being developed for preliminary design of aircraft.

The code called ATLAS being constructed by The Boeing Commercial Airplane Company integrates structural analysis and design through a modular system of computer codes which employs a common executive and database framework.[10] Its system capabilities may be grouped into three categories (Fig. 5)—executive, technical, and data handling. Its control module and several precompilers permit the user to specify the sequence and mode of execution of the computational parts. A control deck takes input data directly or from a previously stored file, and it can interface with an external program. ATLAS can analyze stiffness, stress, mass, strength for resize, vibration, aerodynamics, and flutter, and it can interpolate calculations. Its data-handling capabilities include data storage in a database accessible to all system modules, data transfer to and from external programs, data transfer to and from restart tapes, and plotting and other forms of output. The common database gives automatic compatibility of data among modules and rapid access to data on disc storage.

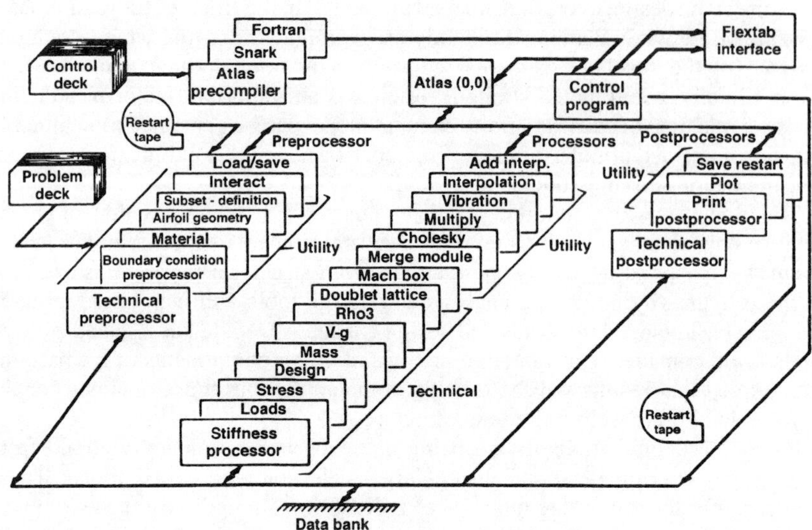

Fig. 5 ATLAS system architecture.

The second design code called SAVES (Sizing of Aerospace Vehicle Structures) was developed at NASA Langley to calculate aerodynamic loads in a flexible body and to size the structure for strength.[11] Initially, the program was organized in the conventional design sequence shown on the left in Fig. 6; three iterative loops yielded a converged fully stressed design. In Loop A, the aerodynamic loads were iterated until convergence; in Loop B, the loads were then applied to the structure until a converged set of resized elements was obtained. These loops were located within a large one, Loop C, and were passed through until the loads and resized members were both converged.

Because that method proved time consuming and costly, the SAVES program was restructured into the parallel iteration architecture shown on the right in Fig. 6, the aerodynamic loads and structural resizing being calculated at the same time.[12,13] The same structural analysis was used for both operations. This arrangement has only one loop, and the airloads and element sizes converge simultaneously. Its fewer iterations and analyses save about 80% in computer CPU time. Convergence is achieved more quickly and results are more accurate with the continuously updated airloads.

A computer code called WIDOWAC (Wing Design Optimization with Aeroelastic Constraints) has been developed at Langley and Technion. The code, a preliminary-design research tool, yields accurate minimum-mass sizing for finite element wing models that must satisfy multiple design constraints including flutter. The rigorous algorithm used in WIDOWAC has the generality to allow any number of different kinds of design constraints to be imposed during the design process. Minimum-mass designs can be obtained for multiple mechanical, thermal, and inertial loading conditions.[14–18]

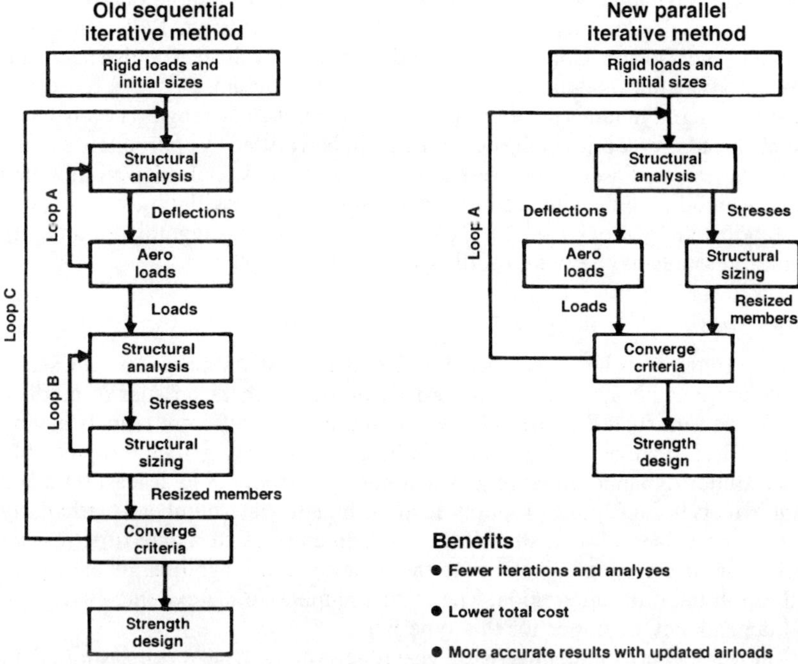

Fig. 6 New SAVES system for structural design.

Design Synthesis Procedures

Several studies are under way to improve structural-sizing procedures.

One completed by Lockheed examined some of the component processes involved in sizing structure to meet flutter requirements; these included efficient forms in which to cast unsteady aerodynamic parameters for use in repetitive flutter calculations, efficient methods for performing the flutter calculations, and the characteristics and efficiency of several procedures for minimizing structural mass subject to flutter constraints.[19–21]

An existing weight-estimating computer program developed by General Dynamics has been modified to include a fracture-and-fatigue module that calculates allowable stresses used in sizing a structure. A finite element program has also been developed by Lockheed to analyze cracked anisotropic sheets. For thermal-stress design problems, an improved, fully stressed design algorithm has been developed for sizing simple structural members subjected to combined thermal and mechanical loadings. The algorithm has been shown to converge in fewer iterations than ordinary fully stressed design for situations where thermal stress has comparable magnitude to or dominates mechanical stresses.[22]

Flutter Technology

Experience in the design of supersonic aircraft increasingly emphasizes the need for adequate consideration of aeroelastic effects during early stages of the design cycle. Dynamic aeroelastic instability, *flutter*, rates paramount attention *throughout* the design cycle. Contractual efforts with Boeing, Bell Aerospace, Lockheed, and Boston University will develop and evaluate sophisticated unsteady-aerodynamic modules for flutter-prediction programs.[23-28]

Unsteady-aerodynamics theories least satisfy the transonic range. The advanced transonic unsteady theories being developed will be verified by appropriate experimentally determined oscillating pressures. A three-dimensional sidewall-mounted oscillating model wing with leading- and trailing-edge oscillating controls has been fabricated by Lockheed for NASA wind-tunnel testing this summer to measure pressures as the wing oscillates in pitch and roll.

Flutter Tests

Several flutter tests have been performed under the SCAR program. One series evaluated effects of tunnel size and porosity on flutter characteristics of models; an SST-type semispan flutter model was tested in three different tunnels having ventilated test sections.[29] The studies indicated that wall porosity appreciably affected flutter dynamic pressure at Mach numbers from 0.8 to at least 0.92 and that the effects became more pronounced at the higher Mach numbers, particularly at the lower values of porosity. Other experiments tested a one-fiftieth-scale model of an arrow wing to determine the location and magnitude of the flutter speed dip in the transonic region. The results emphasize the extreme configurational dependency of flutter for this type wing.

All-movable horizontal tails with geared elevators have been proposed to reduce tail size and required deflections. Joint experiments by Boeing and NASA have provided transonic flutter characteristics of a typical advanced geared-elevator configuration for later correlation with analysis. At Mach numbers from 0.9 to 1.1 an increase of the gear ratio to 2.8 raised the dynamic pressure required for flutter about 17%.

Transonic Loads

The problem of accurate load prediction for large flexible aircraft becomes acute when critical design conditions occur in the transonic and supersonic speed regimes. In these regions, at typical angles of attack and control deflections, the predictions become more difficult due to nonlinear effects caused by flow separation, leading-edge vortex separation (Fig. 7), shocks, and mixed flow.

How well the best state-of-the-art theoretical techniques or combination of theory and experiment can account for these flow conditions is known in only a few circumstances. For typical low-aspect-ratio configurations and transonic flight conditions where various nonlinear phenomena become important, no satisfactory methods of incorporating experimental data from rigid models into aeroelastic solutions have been developed. Until a validated analytical or empirical approach has been developed, the field will need expensive and time-consuming wind-tunnel test programs simulating *each* flight design condition on a flexible aircraft.

An arrow-wing configuration employing both a twisted and a flat wing, as well as a variety of leading- and trailing-edge flap deflections, has been tested by Boeing in its transonic wind tunnel at Mach numbers from 0.40 to 2.5, with

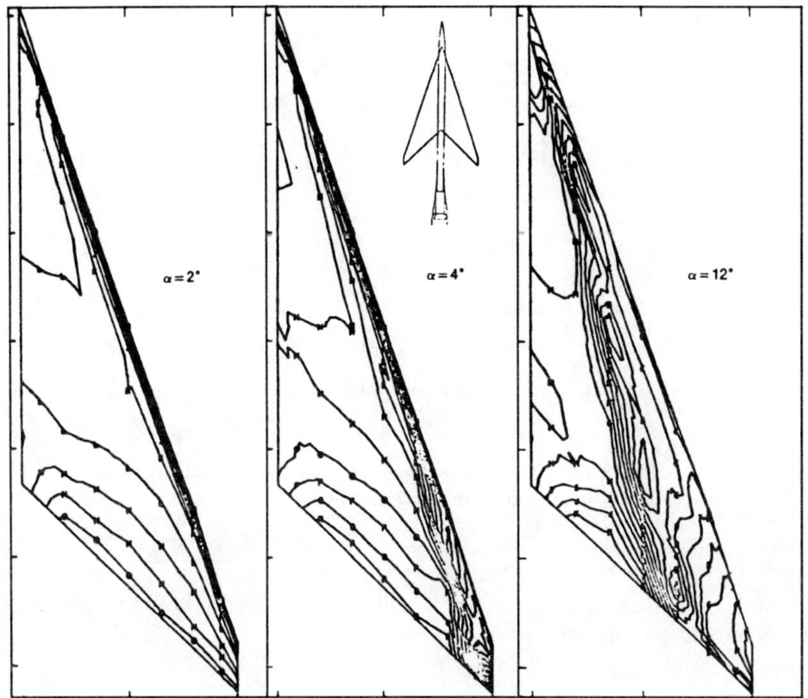

Fig. 7 Effect of separated vortex on upper-surface isobar configuration. At Mach 0.95.

angle of attack varying from −8 to 16 deg. Figure 8 shows a comparison of experimental and theoretical surface pressures at an angle of attack of 8 deg for a low-aspect-ratio wing with twist at $M = 0.85$. The potential flow theories used are based on inviscid theory and attached flow. The pressure distributions described by the solid lines were calculated using a unified subsonic/supersonic panel technique contained in the Boeing computer code FLEXSTAB. The distributions described by the dashed line were calculated using the Boeing computer code TEA-230 (this code is based on an analysis employing a panel solution to the exact incompressible potential flow equation satisfying boundary conditions of the exact configuration surface).

Comparisons are shown in the figure for an inboard wing station, where the twist is 1 deg and near the tip, which has a 4-deg twist. The calculated pressure distributions do not compare well with the experimental data at the outboard station. The presence of the leading-edge vortex has completely changed the nature of the flow over most of the wing, and the attached potential flow theoretical method cannot predict the experimental results.[30,31] Attempts to introduce empirical corrections to account for elastic deformations and, thus, improve predictions of the surface pressure distribution have been unsuccessful, since theoretical

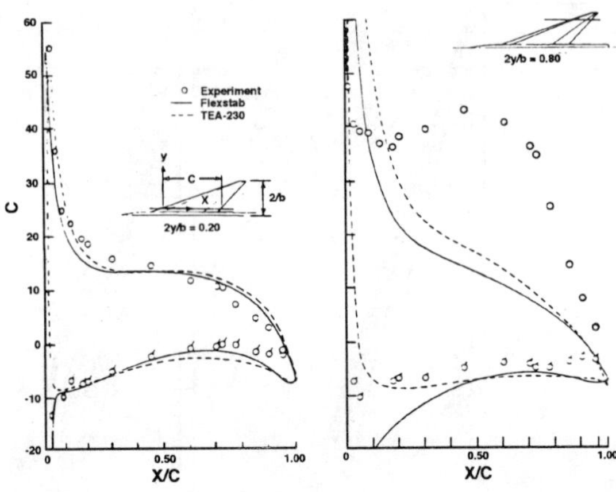

Fig. 8 Twisted-wing pressure distribution. At Mach 0.85, 8-deg angle of attack.

corrections are linear and work only in situations where the actual flow changes due to aeroelastic distortions are also linear.

Landing Loads

The impact and taxi phases of the ground-air-ground cycle of operations impose large structural loads on the airframe of an aircraft. Active-control landing gear might give supersonic-cruise aircraft longer operational life, safer ground-handling characteristics, and more-acceptable ride quality. A preliminary analytical study of such gear has been conducted for a hydraulic control actuator in series with a modified version (increased stroke and pneumatic volume) of a passive-gear design. As shown in Fig. 9, the active-control gear reduced the wing force by 26% with an increase in strut stroke of 15%. This reduction in force would increase the fatigue life for aluminum wing structure by a factor of approximately 4.5. Tests are being made to validate the active-landing-gear study results. A computer code for takeoff and landing analysis with aircraft flexibility effects has been formulated by McDonnell Aircraft,[32,33] and data on takeoff and landing loads has been obtained from a NASA YF-12 aircraft flown at NASA Dryden Flight Research Center.

Acoustic Loads

A joint NASA–Lockheed test was conducted in the Langley anechoic facility to study noise shielding with overwing engines and to determine fluctuating pressures in the engine jet-impingement area. Results indicate that both the maximum flyover noise and the noise exposure time can be reduced using overwing engines.[34] Overwing engines, however, can cause other problems, such as increased acoustic loading on the airframe.

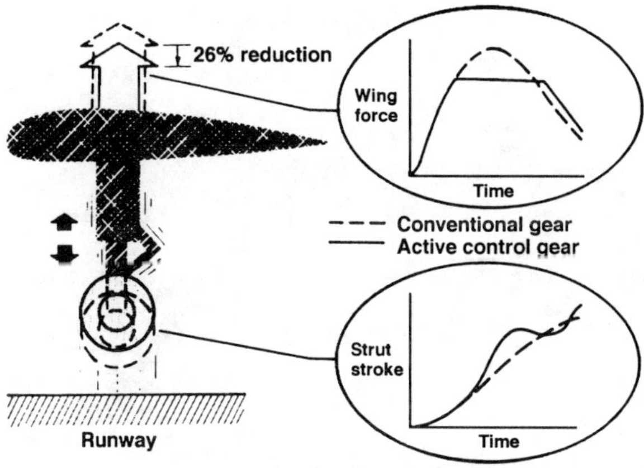

Fig. 9 Active-controls gear reduces wing force.

Atmospheric Turbulence

A program to measure gust velocity over a wide altitude range and in different meteorological conditions is nearing completion.[35] Principal meteorological conditions of interest include the jet-stream mountain waves and flight at low altitude (or in the Earth's boundary layer). The operational plan for this program was to locate and sample with instrumented aircraft atmospheric turbulence associated with specific meteorological conditions and to study the resulting time histories and spectra. For samplings up to 50,000 ft, the instrumentation system is installed on a B-57B aircraft. Plans call for installing the instrumentation in a B-57F for measurements above 50,000 ft. The aircraft seats a NASA test pilot and a meteorologist as an observer. After a flight the meteorologist analyzes the meteorological conditions encountered, makes turbulence forecasts, and assembles meteorological information from reports from the National Weather Service and Turbulence Plot messages from Northwest Airlines as well as pilot reports through FAA Air Traffic Control Centers.

In the von Kármán description of atmospheric turbulence spectra, a value L essentially defines the location of the "knee" in a curve of power spectral density vs frequency. For large supersonic-cruise aircraft, variations in L below values of about 2500 ft have a pronounced effect on aircraft response, whereas variations at higher L values have a very small effect. Figure 10 shows a power spectrum of vertical gust velocity from a flight at 43,000 ft in windshear associated turbulence. The von Kármán equation with L taken as 1000 shows excellent agreement with the measured results. Seventy-six data runs from a total of 46 flights are now being processed. Results from this program should help clarify the nature of the power content of atmospheric turbulence at low frequencies and so will be important for design of supersonic aircraft.

Materials Application

New materials offer the greatest potential for reducing the structural weight of supersonic aircraft. Successful application of new materials, however, requires

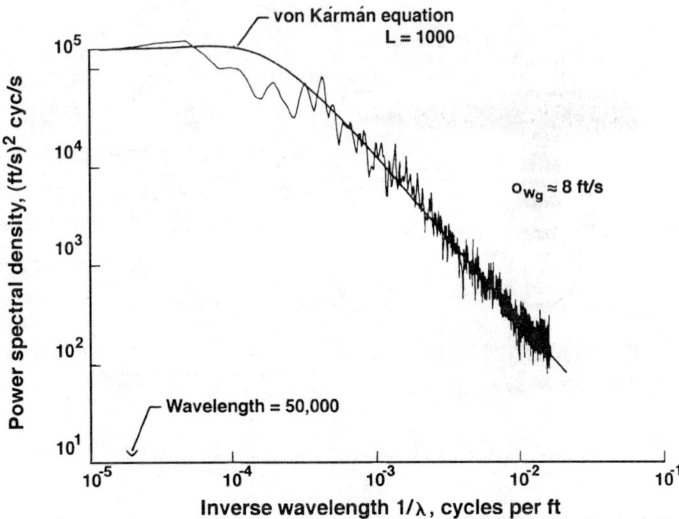

Fig. 10 Power spectrum of vertical gust velocity. Approximately 43,000-ft altitude under windshear turbulence.

extensive and detailed data on their performance under the long-time, high-temperature environment of supersonic-cruise flight and the development of economical and reliable manufacturing methods. In the SCAR program, new materials are being developed, and environmental testing, fatigue and fracture testing, and advanced fabrication and joining process developments are under way.

New Materials

Long-life high-temperature sealants for fuel tanks and long-life processible polyimide resins for the matrix of elevated-temperature filamentary composites lead the list of new material developments.

The objective of the fuel tank sealant program—to provide fully characterized, predictable fuel tank sealants—includes the synthesis, characterization, formulation, and curing of new elastomer candidates.[36] Identified degradation mechanisms of candidate sealants include reversion, characterized by shrinkage in joints above 400°F; low tear strength, caused by the formation of cracks and a resultant loss in tensile strength at high temperatures; and poor adhesion to titanium.[37] Actual-time simulated tank studies have been conducted at Boeing. An experimental fluorosilicone (DC 77-028) sealant was installed in a test chamber and subjected to pressures and flight-cycle temperatures. Test specimens were also installed in an inboard wing tank of the NASA YF-12A. In addition, new perfluoroether sealants are being developed with higher temperature stability than state-of-the-art sealants.[38,39]

Polyimides, one of the most promising classes of resin for high-temperature composite structures, have the potential of performing satisfactorily for long periods at temperatures in the 450–600°F range. Considerable difficulty has usually been encountered in applying these materials because of variable properties and complex manufacturing-process problems. Langley and the GE Research and Development Center are exploiting the newly discovered polymer called polyetherimide, a soluble aromatic polyimide that can be processed simply in an autoclave at nominal temperatures and pressures.

Studies on high-temperature resin development undertaken by Lewis Research Center stress improving processing characteristics while retaining useful mechanical properties to 600°F, particularly in autoclaving polyimides and polyphenylquinoxalines. Lewis has developed a class of readily processable high-temperature-resistant polyimides known as PMR (Polymerization of Monomer Reactants).[40] HTS graphite-fiber composites fabricated with a PMR polyimide, after 600 h of exposure in air at 600°F, show flexural strength 50% higher than that of a composite made with a commercial polymide. Of even greater significance, this work demonstrates the broad applicability to other polymer systems of the in situ polymerization characterizing PMR polyimides.[41]

Environmental Effects

Besides continuing some of the extensive work on titanium alloys conducted during the national SST program, SCAR has pressed research and testing on advanced composite materials. The principal environmental study, including fatigue resistance, of available composites materials is being done by General Dynamics/Convair.

In a two-phase effort, phase I evaluated existing data for a baseline material in each of five classes of composites and conducted environmental simulation for cumulative exposure to 10,000 h followed by mechanical property tests and material evaluations to determine exposure effects. In phase II, the experimental and analytical characterizations will be extended to cumulative exposure up to 50,000 h.

The filaments and matrices selected as baseline are as follows: 4-mil boron/5505 epoxy; AS graphite/3501 epoxy; 5.6-mil boron/P 105A polyimide; HTS graphite/710 polyimide; and 5.6-mil boron/6061 aluminum alloy, diffusion bonded. The B/PI material has been removed from the program because of excessive variability of matrix-controlled properties and rapid degradation of B/PI specimens during short time exposures.

Aging of B/E at 350°F in air at atmospheric pressure for 10,000 h produces a sizable decrease in 350°F tensile strength. Similar exposures at 250°F, atmospheric pressure, and at 350°F, 2-psi air, have no effect on 350°F tensile strength. The tensile-strength degradation at atmospheric pressure was caused by absorption of moisture by the epoxy systems. This significantly decreased short-time elevated-temperature strength. The results point out the need for a moisture-proof coating when these materials work long periods in ambient environments. Similar behavior was experienced by the G/E material. As shown in Fig. 11, static thermal aging of G/PI at 550°F in air for 5000 h produces a decrease in 550°F tensile strength of unidirectional material but has no net effect on cross-ply material. Similar exposures of G/PI at 450°F produced no significant changes in tensile strength.

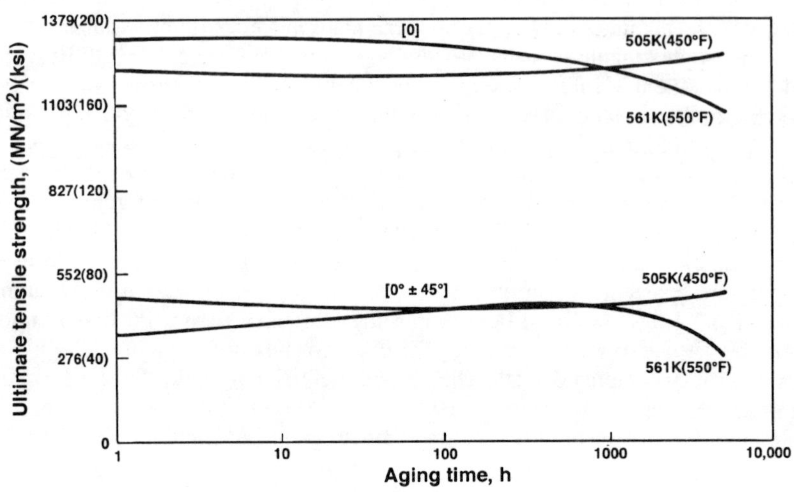

Fig. 11 Tensile strength of HT-S/710 graphite/polyimide. At 450° and 550°F after thermal aging at the same temperatures.

The SCAR program continues two items from the DOT/SST follow-on program on titanium:

1) Long-time and short-time exposure and subsequent testing of small titanium honeycomb-core specimens. Tests include laboratory exposures, subsonic-flight environment on Boeing 727 commercial transports, supersonic-flight environment on a YF-12 aircraft and engine exhaust. Short-time tests alternately expose specimens to aircraft fluids and elevated temperatures. Cumulative exposures up to 50,000 h are planned for some specimens.

2) A three-year flight service program involving titanium spoilers on the B-737 airplane. Boeing fabricated three wedge-shaped spoilers to demonstrate the feasibility of brazing tapered titanium honeycomb-core structure utilizing the 3003 aluminum braze material. Following flight service, the spoilers will be removed from the aircraft and subjected to nondestructive evaluation and limit-load testing by Boeing.

Fatigue and Fracture

The unknown effects of aerodynamic heating and long cruise times complicate predicting structural fatigue resistance of supersonic-transport materials and structures. For subsonic airplanes, a full-scale test usually verifies structural-fatigue strength. However, a full-scale fatigue test of a supersonic transport would be very expensive and time consuming since the cyclic thermal environment can be duplicated only in real-time. Two objectives of fatigue studies in the SCAR program have been determining real-time and thermal exposure effects on fatigue strength of candidate materials and structures and developing accelerated fatigue tests.

At the start of the DOT/SST program, in 1965, Lockheed–California Company undertook a study to determine the life of notched titanium-alloy coupons in both real-time and accelerated temperature/stress fatigue tests.[42] The study has continued under the SCAR program. All the tests entail flight-by-flight fatigue loading. Real-time tests simulated the real-time cyclic heating of wing-skin material. Three types of accelerated test were conducted with rapid cyclic temperature, constant elevated temperature, and constant room temperature. Fatigue life in real-time tests generally came within the range for the three kinds of accelerated test, but all the materials showed longer fatigue life in real-time tests than in accelerated tests with constant elevated temperature.

Boeing investigated the influence of thermal stress on fatigue life of structural elements and methods to shorten and simplify fatigue testing. As a structural element it used a flat sheet with a hat-section stringer attached with rivets. Nonuniform heating simulated aerodynamic-heating environments. Rapid thermal cycling and load cycles established the fatigue life and failure locations. The tests produced empirical stress/number-of-cycles (*S–N*) curves for the titanium structural elements and analysis that accounts for thermal-soak effects. Related work performed at the Langley Research Center on real-time and accelerated fatigue tests is described in references 43 and 44.

Design methods and structural concepts are also being developed for composite structures that can tolerate significant amounts of damage without failing catastrophically. The approach involves developing a fracture theory for cross-plied laminates that can be used to predict the strength of damaged laminates and also developing methods of analysis that can predict the influence of softening strips, stringers, and other damage-tolerant features on the residual strength of damaged structure. Several graphite/epoxy laminates of the (0/±45/90) family and several boron/aluminum laminates of the (0/±45) family are being fabricated for testing at Langley to determine the reduction in strength of cross-plied laminates due to crack-like flaws.

Manufacturing Technology

SCAR work includes developing economical and reliable manufacturing methods for metal and composite aircraft structures.

Advanced fabrication and joining processes for titanium[45] and high-temperature composite materials are being investigated with Lockheed–ADP as the prime contractor.[46] Full-scale structural panels are being designed and fabricated to replace an existing integrally stiffened shear panel in the upper-wing surface of the NASA YF-12 aircraft. The program includes ground testing and Mach 3 flight testing of five types of full-scale structural panels and laboratory testing of representative structural-element specimens. For each structural concept, element-type specimens are tested at room temperature after constant-temperature exposure or cyclic exposure. Specimens are exposed for up to 10,000 h at temperatures as high as 1000°F. For cyclic exposure, exposure times are limited to 1000 cycles, where each cycle contains a temperature hold at 600°F to simulate a 3-h flight with 2 h at supersonic speeds.

Two titanium panel concepts are being investigated:

1) A Z-stiffened skin-stringer panel (Fig. 12) made by weldbrazing, a joining process developed at NASA Langley.[47,48] Weldbrazing combines resistance spot-

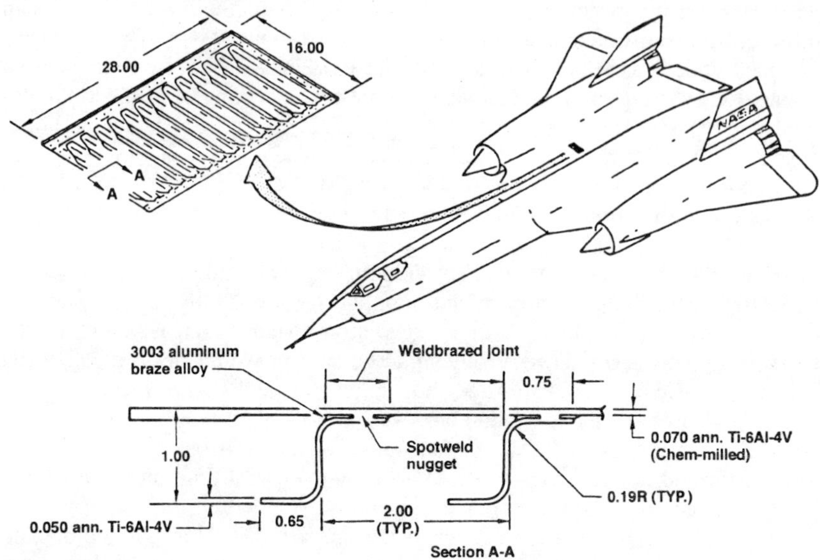

Fig. 12 Weldbrazed titanium skin-stringer panel. Measurements in inches.

welding to create a controlled gap at the faying surfaces and brazing to produce a continuous high-strength joint. The weldbrazing panel weighs 8.5 lb, the same as the integrally stiffened panel it is designed to replace. One panel has experienced approximately 200 flight hours with 60 h above Mach 2.6 and, of that, almost 47 h at Mach 3.0. No-exposure and 100-h-exposure tests at room temperature and at 600°F were conducted as "proof of design" before beginning flight service of a panel. The tests after cyclic exposure (1000 cycles at atmospheric pressure) and 10,000-h constant-temperature exposure revealed no degradation in shear strength.

2) A honeycomb-core panel by Rohrbond, a Rohr Industries, Inc., titanium-joining method that uses a proprietary liquid-interface diffusion process. The components to be joined are selectively electroplated with several layers of material which act as a eutectic to aid diffusion bonding. The panel consists of a titanium frame fusion welded at the corners, titanium honeycomb core, and titanium face sheets. The Rohrbond panel weighs 12% less than the weldbrazed and integrally stiffened panel. The flight-service Rohrbond panel has experienced 47 flight hours with 14 h above Mach 2.6 and, of that, almost 8 h logged at Mach 3.0.

Three composite-panel concepts are being investigated. McDonnell Douglas Astronautics Company–East is studying brazing and manufacturing methods for panels with boron/aluminum face sheets and a titanium honeycomb core. NASA Langley is studying fabrication methods for panels with graphite/polyimide face sheets and glass/polyimide honeycomb core as well as borsic/aluminum panels with titanium honeycomb core.[49] The borsic fiber is boron coated with silicon carbide to alleviate chemical reaction of the boron with the aluminum matrix at elevated temperatures. Weight-saving estimates for the composite-panel designs compared to the original YF-12 titanium panel vary from 30% for the metal-matrix designs to 55% for the graphite/polyimide design. Fabrication processes for these panels are being developed.[50]

Langley has been investigating optimum stiffened compression panel designs based on elevated-temperature operations (450°F) with borsic/aluminum metal-matrix composite material. Five different configurations have been evaluated analytically over a range of loadings: hat-stiffened, corrugated-stiffened, open corrugated, honeycomb sandwich, and hat-stiffened honeycomb sandwich panels. Results show more than 30% structural-weight decrease for an optimally designed stiffened borsic/aluminum honeycomb concept over corresponding titanium panels. Two experimental programs, one on materials characterization and the other on buckling allowables, are under way to verify results.

Concluding Remarks

In more than three years of work, the SCAR structures and materials program has generated significant new technology, identified additional needs, and provided several lessons on structural design of supersonic-cruise aircraft. The airplane configurations that are efficient aerodynamically have aeroelastic problems that must be alleviated by configuration changes, active controls, or additional structural mass. This problem must be defined early in the design process. Fortunately, multidisciplinary computer-aided design techniques are being developed that permit rapid evaluation of many structural and aerodynamic options in sufficient depth to identify solutions in a timely and economical manner.

Although the current supersonic-cruise aircraft research program includes work on loads, additional methodology must be developed for calculating steady and unsteady aerodynamic loads, particularly in the transonic-speed range. Of particular concern are nonlinear viscous effects on load distribution on highly flexible aircraft structures.

The highest potential for structural-weight reduction lies in application of advanced composite materials. Progress is being made in measuring the effects of the supersonic-cruise environment on both metal- and resin-matrix composites. The graphite/polyimide systems appear to be the most attractive for the 300–600°F temperature range.

Continuing research in SCAR structures and materials should bring high payoffs in the late 1980s. Characterization and development of high-temperature composites will be pursued along with additional effort to increase the availability of modules and integrated systems for computer-aided design. Other technology developments will be supported within resource limitations.

References

[1] Anon., ''SST Follow-On Program Phase II—Quarterly Progress Report Oct.–Dec., 1973,'' Contract DOT-FA-72WA-2893, Boeing Commercial Airplane Company.

[2] Cooper, P. A., and Heldenfels, R. R., ''NASA Research on Structures and Materials for Supersonic Cruise Aircraft,'' NASA TM X-72790, 1976.

[3] Anon., "Mach 2.7 Fixed Wing SST Model 969-336C (SCAT-15F)," Document Nos. D6A11666-1, 2, Contract No. FA-SS-67-3, Boeing Commercial Airplane Group, Seattle, WA, 1969.

[4] Sakata, I. F., Davis, G. W., Robinson, J. C., and Yates, E. C., Jr., "Design Study of Structural Concepts for an Arrow-Wing Configuration Supersonic Cruise Aircraft," AIAA Preprint No. 75-1037, Aug. 1975.

[5] Tinoco, E. N., and Mercer, J. E., "FLEXSTAB—A Summary of the Functions and Capabilities of the NASA Flexible Airplane Analysis Computer System," NASA CR-2564, 1976.

[6] Robinson, J. C., Yates, E. C., Jr., Turner, J. M., and Grande, D. L., "Application of an Advanced Computerized Structural Design System to an Arrow-Wing Supersonic Cruise Aircraft," AIAA Preprint No. 75-1038, Aug. 1975.

[7] Sakata, I. F., and Davis, G. W., "Arrow-Wing Supersonic Cruise Configuration Structural Design Concepts Evaluation," NASA CR-132575, Oct. 1975.

[8] Turner, M. J., and Grande, D. L., "Study of Structural Design Concepts for an Arrow-Wing Supersonic Cruise Configuration," NASA CR-132576, Dec. 1975.

[9] Anon., "Computer Aided Structural Methods with Application to a Supersonic Arrow-Wing Configuration," NASA CR-132551, July 1974.

[10] Miller, R. E., Jr., "Structures Technology and Impact of Computers," ASME Winter Annual Meeting, Houston, TX, Dec. 1975, published in *Integrated Design and Analysis of Aerospace Structures.*

[11] Giles, G. L., Blackburn, C. L., and Dixon, S. C., "Automated Procedures for Sizing Aerospace Vehicle Structures (SAVES)," *Journal of Aircraft,* Vol. 9, No. 12, Dec. 1972, pp. 812–819.

[12] Giles, G. L., and McCullers, L. A., "Simultaneous Calculation of Aircraft Design Loads and Structural Member Sizes," AIAA Preprint No. 75-965, Aug. 1975.

[13] Sobieszczanski, J., "Building a Computer-Aided Design Capability Using a Standard Time Share Operating System," ASME Winter Annual Meeting, Houston, TX, Dec. 1975, published in *Integrated Design and Analysis of Aerospace Structures.*

[14] Haftka, R. T., Starnes, J. H., Jr., and Barton, F. W., "A Comparison of Two Types of Structural Optimization Procedures for Satisfying Flutter Requirements," AIAA Preprint No. 74-405, April 1974.

[15] Haftka, R. T., and Starnes, J. H., Jr., *WIDOWAC (Wing Design Optimization With Aeroelastic Constraints) Program Manual,* NASA TM X-3071, Oct. 1974.

[16] Haftka, R. T., "Parametric Constraints with Application to Optimization for Flutter Using a Continuous Flutter Constraint," *AIAA Journal,* Vol. 13, No. 4, April 1975, pp. 471–475.

[17] Haftka, R. T., and Starnes, J. H., Jr., "Applications of a Quadratic Extended Interior Penalty Function for Structural Optimization," AIAA Preprint No. 75-764, May 1975.

[18] Haftka, R. T., and Yates, E. C., Jr., "On Repetitive Flutter Calculations in Structural Design," AIAA Preprint No. 74-141, Jan. 1974.

[19] O'Connell, R. F., Hassig, H. J., and Radovcich, N. A., "Study of Flutter Related Computational Procedures for Minimum Weight Structural Sizing of Advanced Aircraft," NASA CR-2607, Sept. 1975.

[20] O'Connell, R. F., Hassig, H. J., and Radovcich, N. A., "Study of Flutter Related Computational Procedures for Minimum Weight Structural Sizing of Advanced Aircraft—Supplemental Data," NASA CR-132722, Sept. 1975.

[21] O'Connell, R. F., Radovcich, N. A., and Hassig, H. J., "Structural Optimization with Flutter Speed Constraints Using Maximized Step Size," AIAA Preprint No. 75-778, May 1975.

[22] Adelman, H. M., Walsh, J. L., and Narayanaswami, R., "An Improved Method for Optimum Design of Mechanically and Thermally Loaded Structures," NASA TN D-7965, Aug. 1975.

[23] Weatherill, W. H., Ehlers, F. E., and Sebastian, J. D., "Computation of the Transonic Perturbation Flow Fields Around Two- and Three-Dimensional Oscillating Wings," NASA CR-2599, Aug. 1975.

[24] Morino, L., "A General Theory of Unsteady Compressible Potential Aerodynamics," NASA CR-2464, Dec. 1974.

[25] Morino, L., and Kuo, C. C., "Subsonic Potential Aerodynamics for Complex Configurations: A General Theory," *AIAA Journal,* Vol. 12, No. 2, Feb. 1974, pp. 191–197.

[26] Morino, L., Chen, L. T., and Suciu, E. O., "Steady and Oscillatory Subsonic and Supersonic Aerodynamics Around Complex Configurations," *AIAA Journal,* Vol. 12, No. 3, March 1975, pp. 368–374.

[27] Chen, L. T., Suciu, E. O., and Morino, L., "A Finite Element Method for Potential Aerodynamics Around Complex Configurations," AIAA Preprint No. 74-107, Jan. 1974.

[28] Morino, L., and Chen, L. T., *Indicial Compressible Potential Aerodynamics Around Complex Aircraft Configurations,* NASA SP-374, Sept. 1975.

[29] Ruhlin, C. L., Destuynder, R. M., and Gregory, R. A., "Some Tunnel-Wall Effects on Transonic Flutter," AIAA Preprint No. 74-406, April 1974.

[30] Manro, M. E., Tinoco, E. N., and Bobbitt, P. J., *Comparisons of Theoretical and Experimental Pressure Distributions on an Arrow-Wing Configuration at Transonic Speeds,* NASA SP-347, Sept. 1975.

[31] Manro, M. E., Manning, K. J. R., Hallstaff, T. H., and Rogers, J. T., "Transonic Pressure Measurements and Comparison of Theory to Experiment for an Arrow-Wing Configuration—Summary Report," NASA CR-2610, Oct. 1975.

[32] Dick, J. W., and Benda, B. J., "Addition of Flexible Body Options to TOLA Computer Program," NASA CR-132732, Sept. 1975.

[33] McGehee, J. R., and Carden, H. D., "A Mathematical Mode of an Active Control Landing Gear for Load Control During Impact and Rollout," NASA TN D-8080, Sept. 1975.

[34] Guinn, W. A., Bulena, F. J., Clark, L. R., and Willis, C. M., "Reduction of Supersonic Jet Noise by Over-the-Wing Engine Installation," NASA CR-132619, April 1975.

[35] Murrow, H. N., and Rhyne, R. H., *The MAT Project—Atmospheric Turbulence Measurements with Emphasis on Long Wavelengths, Proceedings of the 6th Conference on Aerospace and Aeronautical Meteorology,* American Meteorological Society, El Paso, TX, Nov. 12–15, 1974, pp. 313–316.

[36] Rosser, R. W., and Parker, J. A., "Fuel Tank Sealants Review," TM X-62401, NASA Ames Research Center, Dec. 1974.

[37] Paciorek, K. L., Kratzer, R. R., Kaufman, J., and Rosser, R. W., ''Syntheses and Degradations of Fluorinated Heterocyclics,'' *Journal of Fluorine Chemistry,* Vol. 6, 1975, pp. 241–258.

[38] Rosser, R. W., Parker, J. A., de Pasquale, R. J., and Stump, E. C., Jr., ''Polyperfluoroalkylene Ethers as High Temperature Sealants,'' in American Chemical Society Symposium Series No. 6, *Polyethers,* 1975.

[39] de Pasquale, R. J., Padgett, C. D., and Rosser, R. W., ''Highly Fluorinated Acetylenes,'' *Journal of Organic Chemistry,* 40, 810, 1975.

[40] Serafini, T. T., ''Processible High Temperature Resistant Polymer Matrix Materials,'' International Conference on Composite Materials, Geneva, Switzerland, April 7–11, 1975.

[41] Serafini, T. T., Delvigs, P., and Vannucci, R. D., ''In Situ Polymerization of Monomers for Polyphenylquinoxaline-Graphite Fiber Composites,'' NASA TN D-7793, 1974.

[42] Anon., ''Fatigue Strength Evaluation of Titanium Materials for the Supersonic Transport Under Flight-by-Flight Loading Spectra,'' FAA Technical Rept. SST-66-2, Jan. 1966.

[43] Imig, L. A., and Garrett, L. E., ''An Investigation of Fatigue Test Acceleration for Conditions of Mach-3 Flight Using Flight-by-Flight Loading,'' NASA TN D-7380, Dec. 1973.

[44] Imig, L. A., ''Crack Growth in Ti-8Al-Mo-V with Real Time and Accelerated Flight-by-Flight Loading,'' TM X-72754, July 1975.

[45] Ochicano, M. L., and Kaneko, R. S., ''Survey of Titanium Structural Shape Fabrication Concepts,'' NASA CR-132384.

[46] Webb, B. A., and Dolowy, J. F., Jr., ''Retort Brazed Bonding of Borsic/Aluminum Composite Sheet to Titanium,'' NASA CR-132730, Sept. 1975.

[47] Bales, T. T., Royster, D. M., and Arnold, W. E., Jr., ''Development of the Weldbraze Joining Process,'' NASA TN D-7281, June 1973.

[48] Bales, T. T., Royster, D. M., and Arnold, W. E., Jr., ''Weldbrazing of Titanium,'' *SAMPE Quarterly,* April 1974.

[49] Royster, D. M., Bales, T. T., and Wiant, H. R., ''Joining and Fabrication of Metal-Matrix Composite Materials,'' NASA TM X-3282, Sept. 1975.

[50] Hoffman, E. L., Payne, L., and Carter, A. L., ''Fabrication Methods for YF-12 Wing Panels for the Supersonic Cruise Aircraft Research Program,'' 7th National SAMPE Conference, Albuquerque, NM, Oct. 14–16, 1975.

The Next Careful Steps in Commercial Aircraft Structures

Paul L. Sandoz
The Boeing Company

Many technologies have contributed to a succession of commercial transports which have expanded world air travel from $11.8 billion revenue passenger miles in 1947 to $304 billion revenue passenger miles in 1971. Flight safety for U.S. scheduled airlines has improved from 1.25 fatalities per million passenger miles to a rate of 0.12 during this same period. In forecasting the structural design of future commercial transports, the message is clear: obtain the maximum benefit from new materials and fabrication processes, yet preserve the durability and fail-safeness of contemporary transports.

The value of durability shows up in low maintenance costs and airplane availability for high daily utilization rates. Both have a tangible influence on an airline's return on investment. Because of the high current level of structural reliability, commercial transport operators of domestic routes pay only 6% of maintenance costs for structures (Fig. 1).

High daily utilization rates (Fig. 2) characterize current scheduled airline operations. Daily utilization is defined as accumulated flight time divided by calendar days of airline possession of the airplane for the reporting period.

Flight lengths vary widely from model to model depending on range and customer usage. Boeing has as its fatigue design goal 20 years of service without significant cracking of structure. This degree of structural reliability must prevail for the service-proven utilization schedules of Fig. 2 regardless of flight length. For fatigue design purposes we defined typical discrete flight profiles (Fig. 3).

The various components of the airframe are designed to the most critical combination of flight length and number of flights, thus providing a reasonable degree of operational flexibility. Designing simply for average utilization would reduce flexibility.

Boeing has developed and put into use a way of matching operating requirements with fatigue capability. This design tool for validating durability performance has these advantages:

- Easily and directly applicable to detail design.
- Extensive visibility of probable fatiguc pcrformance.
- Capable of disciplined refinement.
- Demonstrated relationship to experience.
- High data compression for experience retention and application.

The fatigue design requirements can be condensed to a simple flight profile and used to select the critical profile for each structural component or major

Astronautics & Aeronautics, Vol. 11, No. 7, July 1973, pp. 38–47.

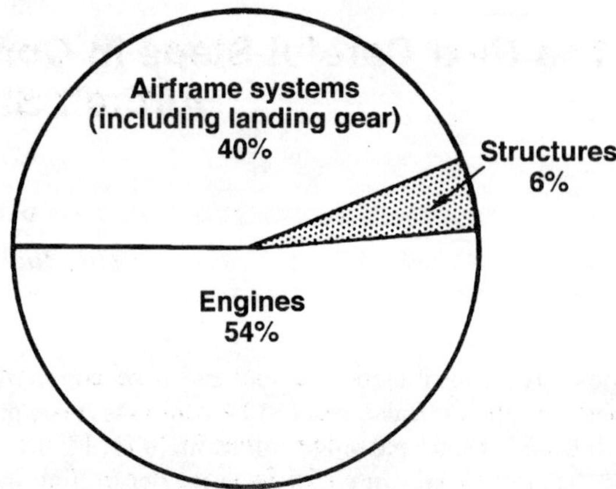

Fig. 1 Split in airline maintenance costs. For 707, 720, 727, and 737 from 1964–70.

subcomponent as shown in Fig. 3. A typical operating gross area stress profile for a wing lower-surface detail appears in Fig. 4.

The load conditions for each segment identify stress levels which contribute significant damage. The largest stress excursion per flight is called the master cycle, and the stress ratio ($R = f\ \mathrm{min}/f\ \mathrm{max}$) of the master cycle is determined. A clear understanding of the master cycle operating stresses and the load conditions that produce these stresses support these calculations.

Study of fatigue crack initiation in all jet-transport structure has shown that all details, regardless of quality and/or alloy, follow a general *S–N* curve shape. Although the life cycle scale depends on the detail under study, the curve has a nearly linear shape on a log-log alternating stress cycle scale and has a common slope between 10^4 and 10^6 cycles. Variations in mean stress or quality merely slide the sloped line to the right or left on the life scale. Use of these standard *S–N* curves can convert the master once-per-flight stress cycle to a damage ratio.

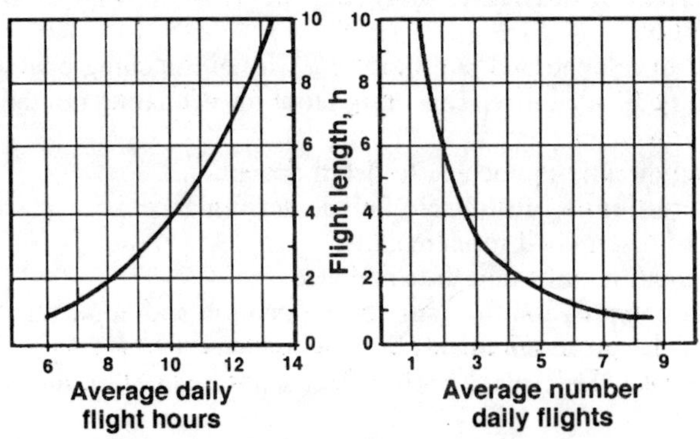

Fig. 2 Commercial transport utilization.

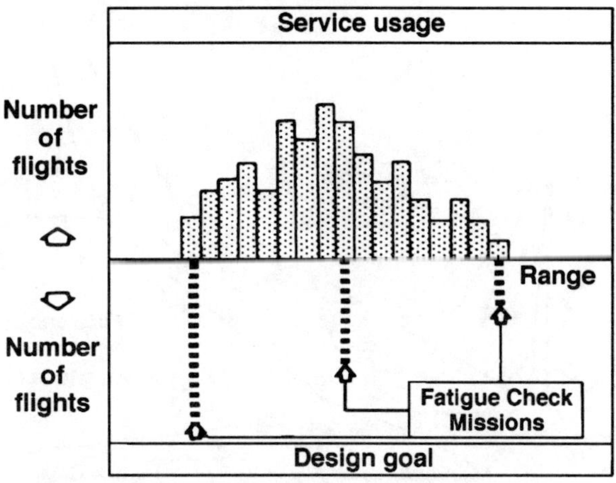

Fig. 3 Flight profile for design.

This damage ratio represents a percent total damage due to the largest stress excursion and collapses the spectrum problem into an equivalent number of well-defined master cycles in the design goal. For example, a damage ratio of 0.5 means the detail must be able to withstand twice the number of master cycle stress excursions during its life. Figure 5 gives damage ratio contours for a typical commercial jet transport.

Numerous variables affect airplane structural reliability. Establishing conservative criteria for each one can appreciably increase weight. But ignoring significant factors can raise maintenance costs and lower utilization rates. Adopting new structural concepts in commercial transports will require developing a suitable

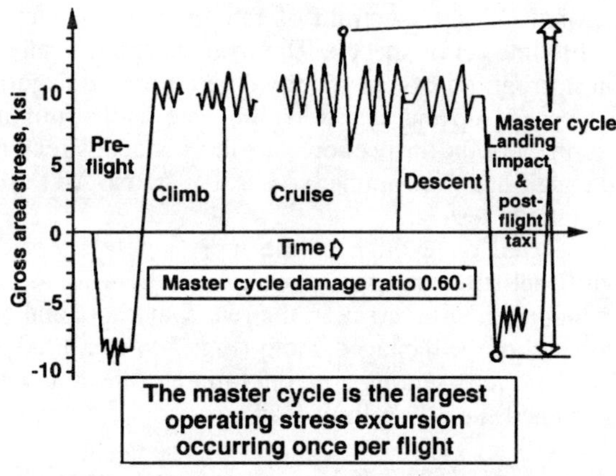

Fig. 4 Wing flight load spectrum for lower-surface detail.

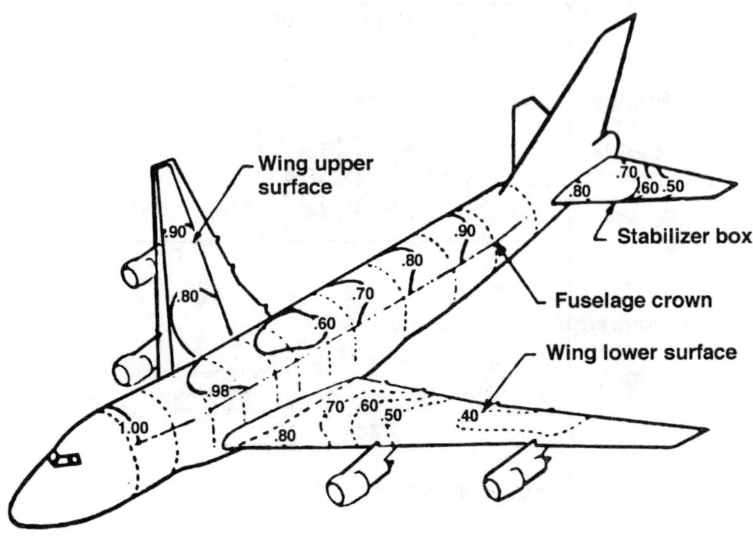

Fig. 5 Master cycle damage ratio contours.

substitute for large samples of fleet experience and well-understood test data to maintain present reliability.

Airworthy service experience has demonstrated the soundness of current industry fail-safe practice. The general criteria call for designing a structure to sustain 93 or 100% limit load after the loss of a single structural element or obvious partial failure of a member. The 747 fail-safe test program illustrates an interpretation of this general requirement.

The program proceeded in two phases: fail-safe cyclic loading and residual-strength demonstration. During the fail-safe cyclic phase, the airframe underwent 4000 repetitions of the flight spectrum of fatigue cyclic loading after having experienced one lifetime set of spectra. The load spectrum duplicated that used for the fatigue-test program, except for the deletion of fuselage pressurization cycles in order to speed up the cyclic loading rate and to minimize the time required to perform periodic inspections. In the residual-strength tests, seven 100% limit load conditions were applied to a test airframe. All testing took place in a full airplane fatigue testrig.

Damage was inflicted on enough of the most highly stressed locations to represent all significant structure. As an example of the wing locations (Fig. 6), cuts simulated a hidden 2.75-in. crack in the rear spar chord and a skin crack at a location just inboard of the damaged chord (Fig. 7). After 4000 test flights the cuts had grown to 3 in. representing a partial failure of the rear spar chord. This structure then sustained the 100% limit load stress of 35.5 ksi for the fail-safe demonstration.

At a second location (Fig. 8), a 3.8-in. cut in a rear spar web had grown an additional 3.7 in. after 4000 additional fatigue spectra. The original saw cut was

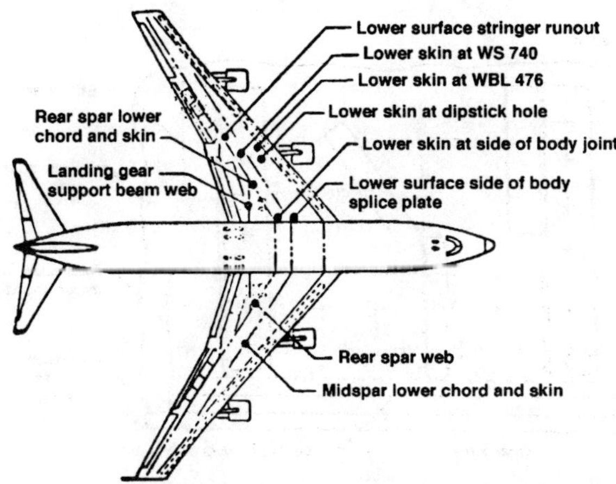

Fig. 6 Simulated damage to the most highly stressed locations in the wings before and during cyclic loading.

considered an obvious partial failure. Cyclic loading produced the desired crack propagation. The spar structure then sustained the 100% limit load condition of positive high angle of attack. Wing station 570 is a critically loaded section on the rear spar due to the concentrated load produced by the restraining action of the landing gear beam.

Fatigue and fail-safe tests early in the 747 program on a simulated fuselage structure validated the structure, which was critically loaded in hoop tension under pressurization. Additional fuselage structure as well as fin structure (Fig. 9) survived fail-safe testing.

A monocoque location (Fig. 10) became a chosen site because it is one of the areas most highly stressed by flight loading and pressurization. Also, it lies within the region of potential damage due to fragments from a partial engine disintegration. The saw cut severed stringer no. 6 and 12 in. of skin. Raising the

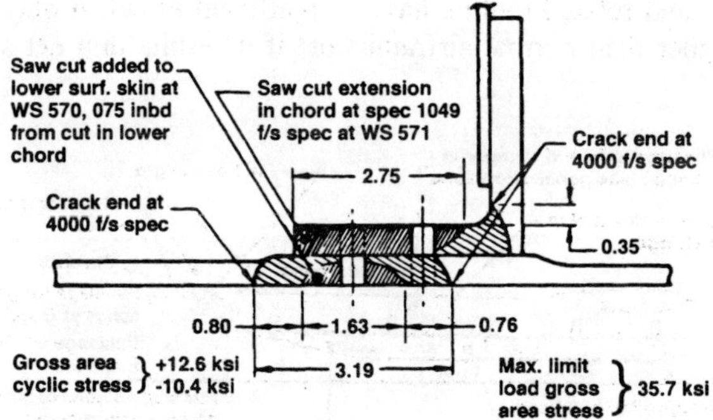

Fig. 7 Simulated cracks made by sawing cuts in a rear spar chord and the lower skin inboard of it.

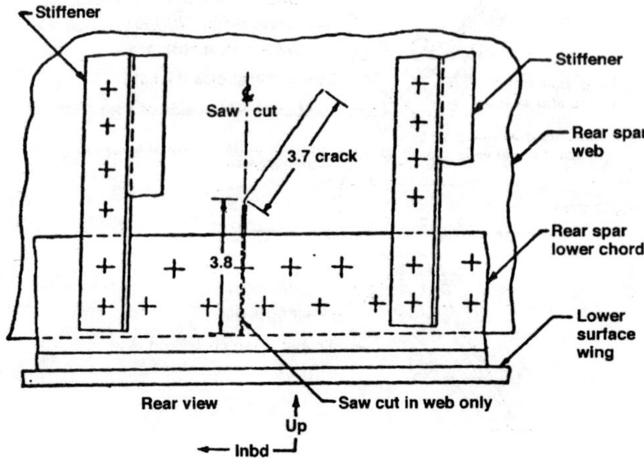

Fig. 8 Saw cut in web of rear spar.

body test moment at this station above the 100% design limit produced stress consistent with the combined loading from flight and pressurization. The crack grew until arrested at stiffeners 5 and 7 (Fig. 10). No further growth occurred during the application of two other limit load design conditions which significantly stressed this location.

New structures must benefit the airline or manufacturer financially. Weight reduction exerts the greatest leverage in improving airline economics. For the same payload and range, structural weight saving reduces necessary wing area and thrust to offer further potential weight cuts. A lighter plane uses less fuel which reduces gross weight even more. The airline will save on maintenance and fuel consumption on a "smaller" airplane with smaller engines.

The cost of the airframe depends on many things, most of which have to do with the design. Besides pounds of airframe, materials, fabrication processes, the structural arrangement, and details of design influence cost. The size of the production run and related tooling have a significant effect. It may be desirable to accept a higher than normal airframe cost if it results in a net saving due to

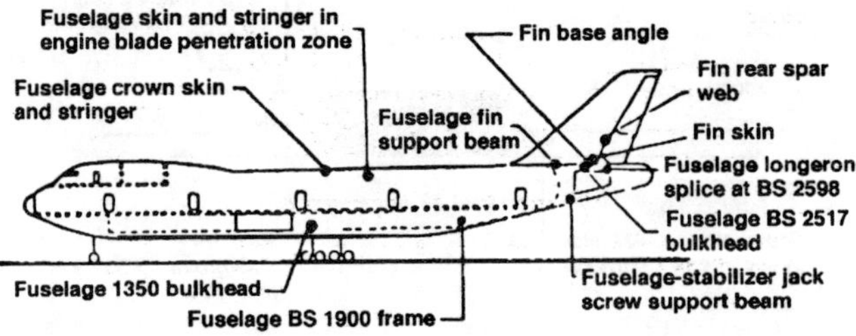

Fig. 9 Simulated damage locations for fuselage and fin.

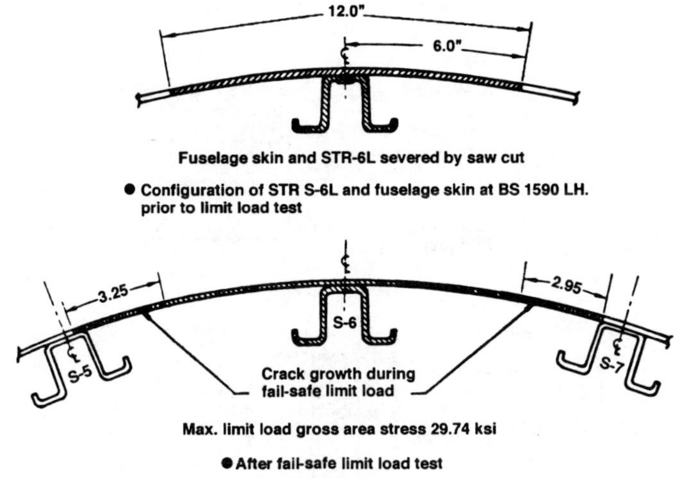

Fig. 10 Simulated damage to body crown structure.

a reduced operating cost. A supersonic transport can accept a higher airframe cost primarily because of the greater productivity speed gives it.

When a manufacturer schedules delivery of a transport relying on a maximized benefit from weight savings in preliminary design, he takes a maximized contractual risk. What if structural weight savings do not materialize as planned? It is axiomatic that a structural system must be well proven in terms of design and fabrication before being broadly applied on a new transport. In addition, its durability and fail-safeness must be predictable and equivalent to systems preceding it. The development program for a new structure must progressively increase confidence that weight savings will materialize, that manufacturing costs are understood, and that behavior in service will be acceptable.

Airframe manufacturers have used adhesive bonded structure for over 20 years. They make extensive use of adhesive bonding in secondary structure, but usually confine it to special applications in primary structure.

To take full advantage of the potential performance gains offered by adhesive bonded structure, the design concept must be integrated with manufacturing considerations. The advantages of honeycomb sandwich fuselage shell structure (Fig. 11) reflect such integration.

The weight savings comes from improved allowables in strength, fatigue, and fracture toughness. Cost saving comes primarily from reducing part-count and may come from increased automation of part-handling during preparation for bonding. The designer can trade the improved fatigue life for longer life and more-durable structure at the same weight or keep the same life and reduce weight by using the improved fatigue quality to raise stresses. For a constant-amplitude stress-level fatigue cycle, sandwich joints have a much longer bonded honeycomb life than conventional joints (Fig. 12). The additional life results

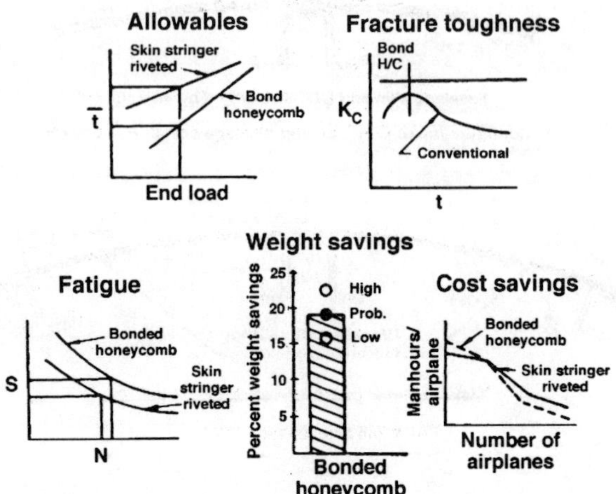

Fig. 11 Bonding honeycomb sandwich fuselage shell structure.

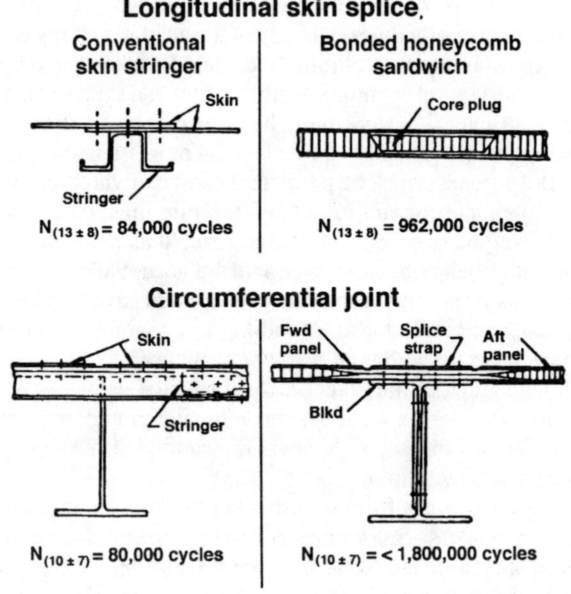

Fig. 12 Bonding gives longer fatigue life to fuselage joints.

Table 1 Test comparison for 2024-T3

	$\frac{pR}{t}$[a]	% Stiffening	Stress at crack tip	Skin gage
Skin stringer	18,000	20	12,950	0.064
Honeycomb	21,900	20	15,600	0.052

[a] $\frac{pR}{t}$ stress for 9.0 psi pressure and 128 in. radius.

from uniform load transfer, elimination of holes, and the reduction of the local eccentricity developed at the joint.

Tests simulating penetration of the fuselage pressure shell by a 15-in. blade illustrate the ability of adhesive bonding to maintain fail-safeness (Table 1). Adhesive-bonded honeycomb sandwich structure in the fuselage shell permits raising critical prt hoop stress by 4000 psi by stabilizing the crack tip.

Longitudinal straps can contain circumferential cracks in one of the fuselage skins of a honeycomb sandwich shell, with no weight penalty, since they act as part of the bending material of the shell. A design curve (Fig. 13) uses fracture-mechanics analysis to relate strap spacing requirements to ultimate design-stress level.

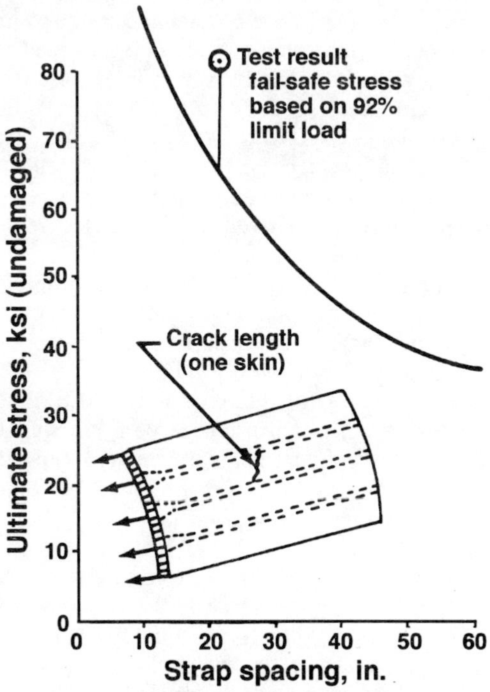

Fig. 13 Spacing related to strength.

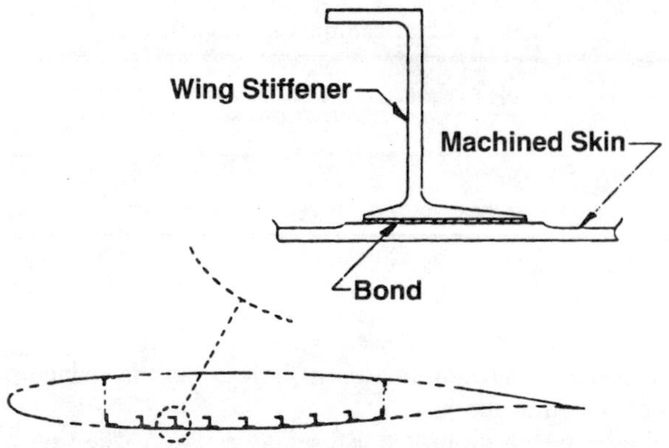

Fig. 14 Bonded wing raises allowable tension 7–12% by eliminating rivet and bolt holes.

Tests took the panel to the required fail-safe stress level at both room temperature and −67°F before continuing up to failure at room temperature.

Manufacturing feasibility panels established the realism of the projected cost saving. The panel assemblies show the complexity of fuselage shell design that makes estimating the cost of new structural concepts difficult. One panel represents a door and window cutout in the constant section and highly loaded part of a fuselage. The other has the same cutouts in the tapered part of a fuselage carrying significantly lower loads. The construction of these panels differ. In the first, machining the skins reinforced the structure around the cutouts. In the second, laminating the skins provided reinforcement. Each construction type is optimum for its area in the fuselage.

The bonded skin-stiffener wing structure realizes its main structural advantage by eliminating stress concentrating rivet and bolt holes. It allows raising the stress in tension 7–12%. In compression it furnished some increase in effective skin width adjacent to stiffener flanges. The increase diminishes as skin and stiffener gages become greater and the failure mode changes from local instability to general instability. The complete lower surface of the wing including the front and rear spar chords is a bonded assembly (Fig. 14).

A bonding fixture substitutes for the extra long (70 ft) autoclave otherwise required. The bonding fixture, in an ordinary room environment, applies both heat and pressure. Since the stiffener and skin elements are relatively thick (up to 0.30 in.), conformance (fit up) during bonding must be controlled principally by machining tolerances. A variance limit of 0.007 in 3.00 was established for the study.

Bonded wing structure would pare 800 lb from the weight of an airplane at a savings of $72 per pound as a consequence of the large number of fasteners no longer required. The application of bonding to wing structure looks very attractive and should prompt an early start on the required development.

We see these structural advantages from bonded fuselage and wing:

- Increases 2 W effective width on skin-stringer wing structure.

- Makes full use of material in shear (honeycomb body).
- Permits higher stress allowables (static strength and fatigue).
- Slows the growth of fuselage cracks.
- Has high buckling stability for low end loads.
- Allows use of higher strength alloy due to reduced stress concentration.
- Increases critical crack length in fuselage.

Adhesive-bonded primary structure could appear in commercial aircraft in the late 1970s if made a goal. U.S. aircraft manufacturers, particularly The Boeing Company, have made significant progress in this direction. Yet several areas require additional investigation to insure low risk and effectiveness before making a firm commitment. Three main areas need development programs: the characterization of the long-time service performance of surfaces, adhesive systems, and materials used in bonded primary structure; electrically continuous bonded structure; and the quantification of service damage resistance and the cost of repairing bonded structure.

If the projected weight and cost savings hold up in production, then the airlines can look forward to ROI gains shown in Table 2.

Advanced composites offer another opportunity for significantly improving the structural efficiency of commercial aircraft. Their higher specific strength and stiffness should save weight, thus money. But they will do so only if maintenance costs stay at an acceptable level, they do not compromise the daily utilization rate, and preliminary design sizes the airplane to take full advantage of weight saving. Numerous industry and governmental programs have demonstrated that a high percentage of the potential benefits of these materials can be realized in flight-quality hardware. Look at the rapid transition advanced composites have made in the last decade from lab samples in the early 1960s to military airplane production components in the late 1960s.

Much work remains to enable a commercial airframe manufacturer to make a large-scale commitment to extensive use of advanced composites. Boeing favors introducing one discrete structural component at a time to production. This approach enables airline personnel to adjust to any impact on maintenance, including having to buy new tools or equipment and training maintenance

Table 2 Airline savings from bonding

	Bonded aluminum honeycomb fuselage				
	Weight increment	% Body struct. wt	Fabric cost $/lb wt saved	% Change doc	Incremental ROI, %
Fwd fusclage	−970#	——	——	−0.40	+0.20
Aft fuselage	−2450#	——	——	−1.20	+0.45
Total	−3420#	9	0	−1.60	+0.65
	Bonded aluminum wing lower surface				
	Weight Increment	% Participating weight	Fabric saving $/lb wt saved	% Change doc	Incremental ROI, %
Wing	−800 lb	8.4	$72/lb	−0.48	+0.18

personnel. Should significant problems arise, they are confined to the single component, and both the manufacturer and the airline can concentrate on correcting them quickly. The confidence of manufacturer and customer will grow over a long period, and when the so-called all-composite commercial transport appears as a paper design, it will have a much greater chance for acceptance.

Designers can follow different avenues to incorporating advanced composites. They can selectively reinforce metal structure so that each pound of composite material applied saves maximum weight. Or they can minimize the total structural weight by integrating design arrangement and composite materials, which results in a nearly all-composite structure. Each approach has merit for specific applications, and a complete structure might very well use both.

In selecting candidates for the first application of advanced composites in commercial airframes, Boeing has concentrated on structural components that are easily inspectable, easily replaceable, functionally simple, subject to a minimum number of design requirements, simply loaded, and noncritical to flight safety. Redundant control surfaces, for example, satisfy these criteria, and we developed a 707-320 boron epoxy foreflap for flight-service evaluation. After obtaining FAA certification, two such foreflaps have gone into commercial service and have each accumulated over 8000 flight hours without incident or degradation.

Boeing is performing a program with similar objectives under NASA contract NAS1-11668. In the program, 117 graphite epoxy 737 spoilers will go into commercial service for five years with periodic inspection and removal of selected spoilers for testing. The 737 provides excellent test exposure. Each aircraft accumulates approximately 3000 flight hours per year. It makes relatively short flights to provide maximum environmental cycling. Twenty-seven aircraft from five airlines, including domestic and foreign carriers, will take part.

Advanced composites are expensive. Several considerations, however, help offset the high raw materials costs. Because they save weight, fewer pounds of structure must be fabricated to do a given job. Improvement in airplane performance may allow paying a premium for structure. Innovations in joining techniques, such as widespread use of adhesive bonding, can help reduce manufacturing costs. Innovative design can reduce part-count and waste of expensive materials.

Any projection of composite prepreg depends on many factors, such as materials type, volume of production, and tightness of the material specification. The material cost projection formed the basis for calculating fabricated costs of structure having tensile load-carrying capability equivalent to 7075-T6 aluminum and using all composite and composite reinforced aluminum. Since reinforced structures use about one-sixth the composite of ones made with all composite, the reinforced band is narrower. Weight savings for the metal reinforced concept is 20–25%, consistent with the metal structure sized for limit load capability. All-composite design saves 40–45%. However, these savings reflect no penalties for electrical compatibility, fail-safeness, or joints. By comparison, the projection assumed fabricated aluminum structure costs an average $45 per lb.

The composite effectiveness factor (CEF)—pounds of weight saved per pound of composite used—typically equals 2 for reinforced structure and 0.8 for all-composite structure. Reinforced structures appear more desirable until all-composite thinking can enter right at the preliminary design stage. Cascading weight savings not represented by the cost of composite structure equivalent to one pound of 7075-T6 aluminum structure would then come into play.

Despite the remarkable progress in advanced composites in the last decade, they will not reach their full potential unless several advances are made in the state of the art. In manufacturing, reliable and economical automated processes must replace many of today's manually oriented ones. Further reductions in materials cost are needed to stimulate expanded use. Simplified design tools, methods of analysis, and requirements comparable in accuracy and ease of application to those employed for metal structures are needed.

Also, polymer matrix composites and adhesive bonding must successfully resist contaminants, extremes of temperature, and humidity to see wide use in primary structure. We must better understand the role of the matrix and the interface between matrix and fiber. Composite structures must adequately resist lightning strike and provide electrical continuity. Flight-service programs must validate overall durability. Quality assurance must at least compare to that attainable in metal structures. Gaining it may require using combinations of process controls, nondestructive testing, and sample destructive testing.

Except in a few cases, most aircraft applications of advanced composites to date have been in components substituted for ones of conventional materials. Maximum payoff with advanced composites will come when manufacturers commit them for broad application in preliminary design. Comprehensive studies should develop and verify techniques for evaluating the cost-benefit of new materials and structural concepts to all classes of aircraft. Government and private organizations need this information to establish rational levels and priorities for funding materials and structures R&D.

Extensive use of advanced composites in commercial transports depends upon development programs responding to the aforementioned needs. Fortunately, most of the work can be clearly defined and programmed. The bill will be high, but must be paid for commercial transports to gain the full benefit of advanced composites.

Early engineering trade studies for a commercial SST which set speed (structural temperatures) as a variable and traded among several material systems showed titanium provided the lowest weight structure to carry a given payload (Fig. 15). In this comparison the aluminum alloy 2618 would limit the cruise Mach number to approximately 2.1. Both the stainless steel and titanium alloys could withstand temperatures to Mach 3.0. The lower *L/D* of supersonic configurations magnifies the importance of low structural weight. As further design studies progressed on the U.S. supersonic transport, a fixed-wing Mach 2.7 airplane evolved.

This configuration had as one of its major impacts on structures and materials technology forcing the development of a titanium honeycomb sandwich system for wing cover panels. Three technical reasons exerted the force. First, because of generally low spanwise and chordwise end loads over a major portion of the wing, sandwich material provided the most efficient structure. Second, flutter testing showed a need for high wing stiffness, and again sandwich structure could most efficiently provide it. Third, liquid fuel in integral fuel tanks would come in direct contact with the wing structure; sandwich structure would provide thermal insulation for the fuel.

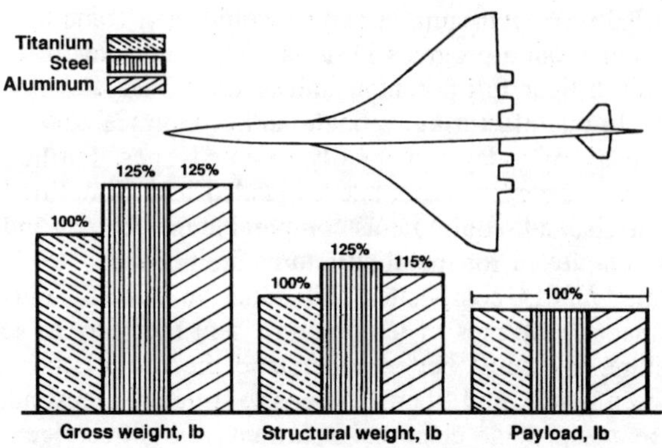

Fig. 15 How metals choice affects weight of SST.

By one measure of productivity, seat miles per unit of flight time, a projected Mach 2.7 supersonic transport would surpass any previous aircraft (Fig. 16). This speed would heat the aircraft to a higher operating temperature; so a portion of the increased productivity must pay for a higher cost structure.

Although titanium sandwich structure will cost quite a bit more than conventional aluminum structure, the need for minimum structural weight for a supersonic transport justifies its use. The final prototype design called for approximately 16,000 sq ft of titanium sandwich (Fig. 17). The design did not use titanium honeycomb sandwich for the most highly loaded center section of the wing nor the relatively deep (up to 11 in.) wedge structure on both leading and trailing edges of wing and tail surfaces because process development could not have been carried out in time to meet the planned prototype manufacturing schedule. We have this development work under way under a Department of Transportation

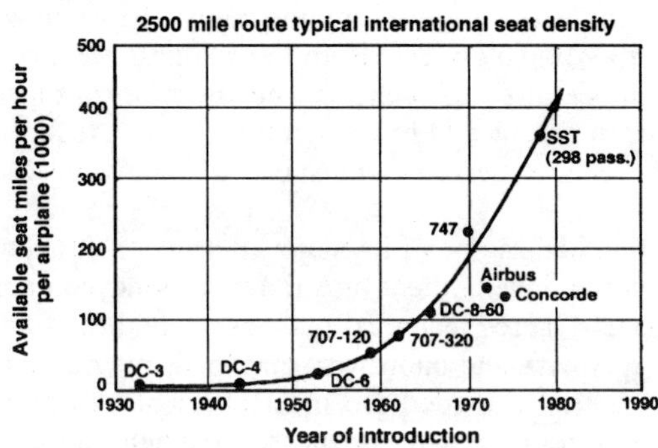

Fig. 16 Trend in airplane productivity. For 2500-mile route, typical international seat density.

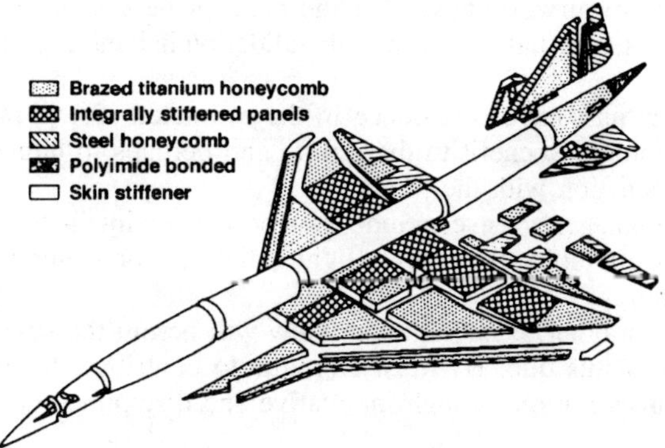

Fig. 17 Types of structure on SST.

contract, DOT-FA-55-71-12. With process development, heavier gage face sheets and newly designed edge members will permit using titanium sandwich structure in areas of load intensities up to 30,000 lb/in.

Titanium alloy developments can bring further structural development to supersonic aircraft. The present workhorse alloy of the industry is titanium 6A1–4V. Preliminary lab testing of proposed new alloys has shown appreciable possible improvements in both strength and fracture toughness (Fig. 18).

Further titanium alloy research in the next two to five years should ready commercially available alloys with improved properties for aerospace structural applications. The benefits from these alloys will depend somewhat on the specific configuration of a future SST. However, noting the ultimate strength increase for a constant fracture toughness gives a qualitative assessment.

Commitment of a production program to a new structural concept depends on acceptance by three key managements, and any development program must aim at getting that acceptance. The technical manager must believe that he can

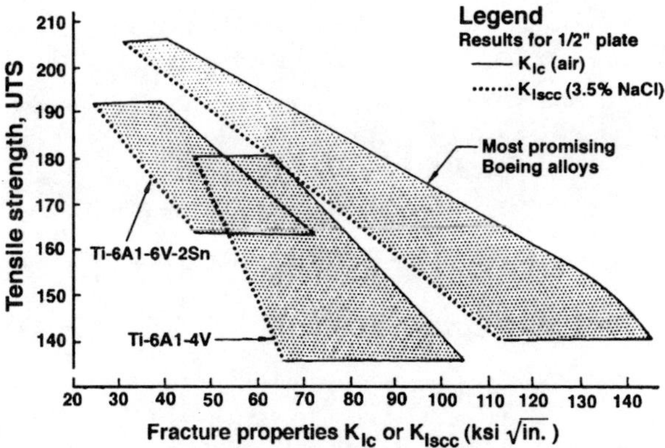

Fig. 18 Properties of new titanium alloys.

maintain long-term airworthiness, that the concept best satisfies the airplane's total requirements, and that he has enough validation in hand to insure the success of the concept.

The producer must have confidence in the estimated cost to manufacture, in the eventual economic benefit to the airline and perhaps to himself, and in the customer's satisfaction with the product.

The airline manager must conclude that he can maintain the new structure with normal effort, hold its present flight reliability or better it, and gain an operating saving.

No advances in other technologies in view will permit the structural designer to maintain the status quo. He must continue to contribute to the potential of commercial transportation through innovative and disciplined design.

Tomorrow's Structural Engineering

D. S. Warren
Douglas Aircraft Company

What sort of future can aircraft structural engineers now practicing and prospective ones in college anticipate? A bright and challenging one. One that will see a rapidly growing world market for civil air transports; significant technology advances, particularly in fatigue criteria, damage tolerant design, composites, and computer-aided design; and an emphasis on lowering the cost of manufacture.

An estimate of ICAO future world seat-mile need (excluding the USSR) (Fig. 1) unmet by the existing fleet and planes on order gives a general picture of the coming market.[1] This projection takes into account overall and airline-by-airline requirements and includes considerations of load factor histories, route structures, scheduled future deliveries, probable retirement of current aircraft, use by second owners, and possible government actions.

An estimate of need by aircraft type (Fig. 2) reflects in more detail the route structures and requirements of individual airlines. A further breakdown would be highly speculative, but these general observations seem probable:

• Wide-body designs will constitute the mainstay of production. Manufacturers will compete keenly for this market of 4000 aircraft and invent numerous derivatives of the current designs (DC-10, B-747, L-1011, and A-300) to avoid the high-investment risk of a completely new design. At least one significant new technology increment will emerge.

• Producing STOL and SST aircraft will entail major engineering and development efforts which will benefit significantly from government programs.

• Structural engineers will play a key role in responding to market demands both because strategies for designing derivatives react sensitively to structural considerations and because new technology will affect structures directly, as through the use of composites, and indirectly, as through the more efficient structure allowed by the supercritical wing.

Every area of structural technology can expect significant developments in the next decade. A supersonic transport and aircraft incorporating innovations such as high-lift devices and a supercritical wing will make new demands on structural design. In addition, designers will want to develop greater capabilities to take advantage of new resources, particularly continued rapid evolution in computers and operating systems. The following paragraphs summarize the anticipated technology escalation in five major fields of structural engineering.

Automation by computer has had a major impact on procedures for predicting static and dynamic loads, stresses, deflections, and flutter. These methods have

Astronautics & Aeronautics, Vol. 11, No. 7, July 1973, pp. 48–55.

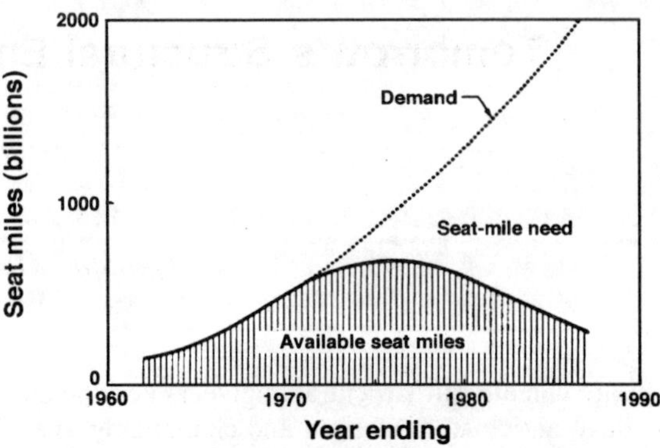

Fig. 1 ICAO world seat-mile demand. For all passenger services excluding USSR.

improved accuracy and reduced design development time. Figures 3 and 4 show the state of the art in matrix structural analysis of commercial airframes. The mathematical model pictured in Fig. 3 formed the basis for final analysis of internal loads (stresses) for the DC-10 using the Fortran Matrix Abstraction Technique (FORMAT).[2] In its complete form it included approximately 100,000 internal forces and was analyzed by multilevel joining of 77 substructures. The analysis considered a large number of different loading conditions and positions of the control surfaces. Fail-safe design examined numerous single simulated failures.

Preliminary analyses based on simple representations of the various components completed soon after the start provided detailed design information in time to influence the early states of design. Detailed data for each of the substructures remain on file for use in developing derivatives. Analyses for two key cross sections (Fig. 4) agree quite closely with test data, coming within 5% of the

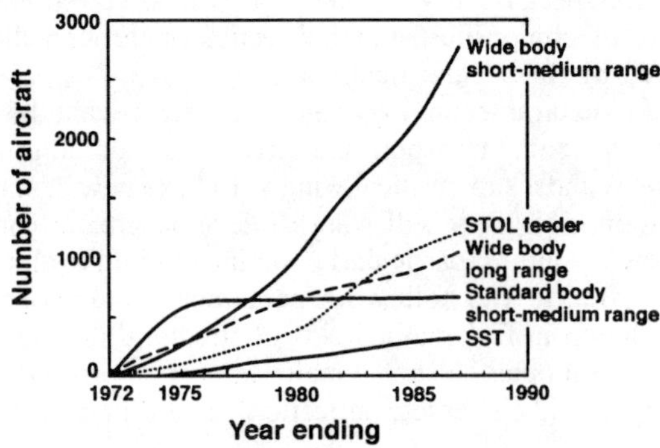

Fig. 2 Aircraft demand by type. For ICAO, all passenger services excluding USSR.

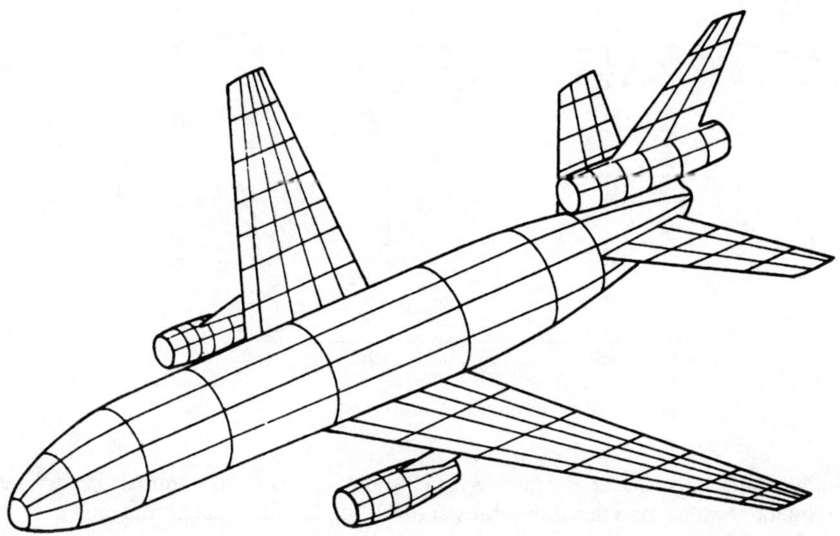

Fig. 3 Finite element aeroelastic loads analysis.

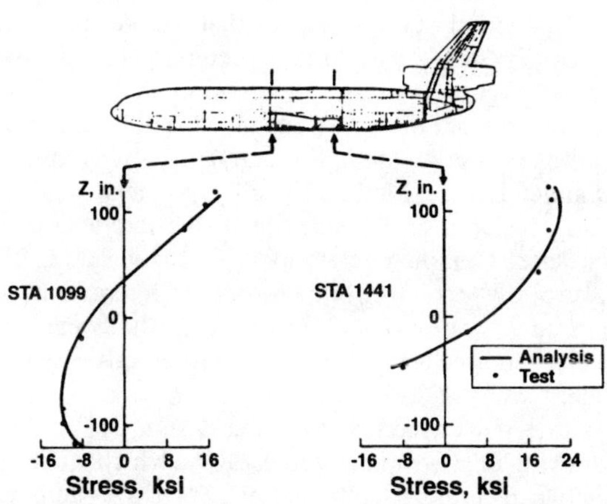

Fig. 4 Computer matrix analysis of stress in the DC-10 airframe.

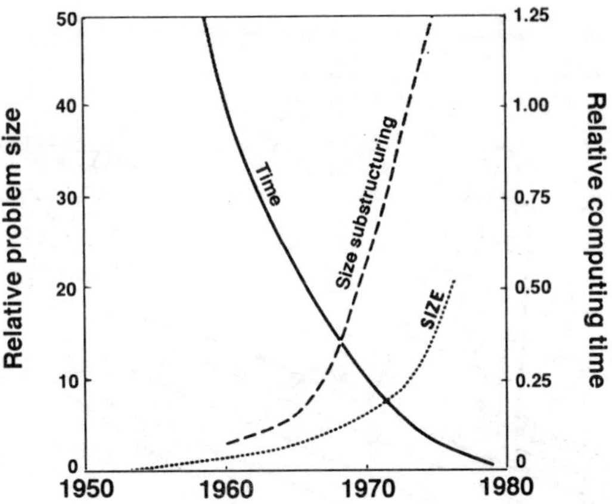

Fig. 5 Development of matrix structural analysis.

highest measured stress, despite significant deviations from nominal theory due to major contours and discontinuities at the wing, landing gear cutout, and access doors.

Matrix structural technology will probably find even wider application for reasons made clear by Fig. 5. The efficiency of single applications will increase as computing speed grows, computing algorithms improve, and structural theories gain precision. More elaborate substructuring will increase the detail in representations of complete airframes as strategies for managing analyses gain sophistication in three ways: by taking full advantage of expanded computer data storage, by growing more responsive to airframe development schedules, and by scheduling concurrent and coordinated analyses of wings, fuselage, and other components.

Three other areas—aeroelasticity methods that include quasi-static loads, integrated design/analysis systems, and direct structural synthesis—will come under intense development. Figure 3 outlines the trend in analysis of aeroelastic loads. The new methods result from marrying advanced static structural analysis to the new family of finite element methods for analyzing aerodynamic pressures.[3] For example, the Doublet Lattice Method reliably predicts steady and oscillatory pressures on lifting surfaces to account for all the interactions between planar and nonplanar surfaces.[4] Complementary methods for analyzing pressure distributions on bodies also allow predicting body forces and interactions between aerodynamic forces on bodies and surface.[5] The new methods can predict complete aeroelastic load distributions accounting for total aircraft equilibrium, including thrust and nonlinear drag forces.

The next big stride in commercial airframe design and analysis will be integrated systems drawing on a computerized database which will minimize communications lags between aircraft technologies. Such a system will fit into the present airframe organization and assume a role defined as follows:

- The technical know-how of existing engineering teams represents the most valuable resource. The integrated design/analysis system must support, not replace or upset, this skill.

• The integrated system has as its basic purpose to improve communication between technologies during individual and group action.

• The technology groups should own the system jointly.

• Management must see progress in applying the system for itself as well as through reports from technology groups.

The Integrated Program for Analysis and Design (IPAD),[6] under development by NASA, will constitute a valuable and fundamental step forward in automation of structural design/analysis.

A mathematical model of the primary structure of a possible supersonic transport gives an idea of the state of the art in structural synthesis for automated design (Fig. 6). Variations in this structure were studied using a structural optimization module in FORMAT called Automatic Redesign and Re-analysis for Optimum Weight (ARROW).[7] ARROW optimizes a structure for strength after the user sets minimum allowable gages and deflections. The user further specifies a "game plan" by choosing among these options: fully-stressed redesign, feasible design step (upward scaling to satisfy deflection and stress constraints), modified optimality criteria search, and unconstrained minimization search.

ARROW plays a key part in a special procedure for preliminary design of advanced aircraft systems. The complete procedure includes automated modeling using interactive graphics, determining aeroelastic loads from finite element aerodynamic and structural modeling, ARROW, and flutter optimization. The flutter optimization minimizes weight and allows varying the structure and moving masses. Further developing these systems to fully account for all essential considerations in final design presents a major challenge to structural engineers and computer system specialists.

Fracture control is very expensive and will become more so. The present approach incorporates double redundancy. First the designer sizes the primary structure to make sure it gives one lifetime of service before fatigue-cracking starts. Then detail design insures that a visible crack will take at least two major inspection intervals to grow to critical length. Finally, crack stoppers are incorporated to arrest a crack after fast fracture. Later, relatively expensive and detailed development tests verify the adequacy of these provisions. The trend

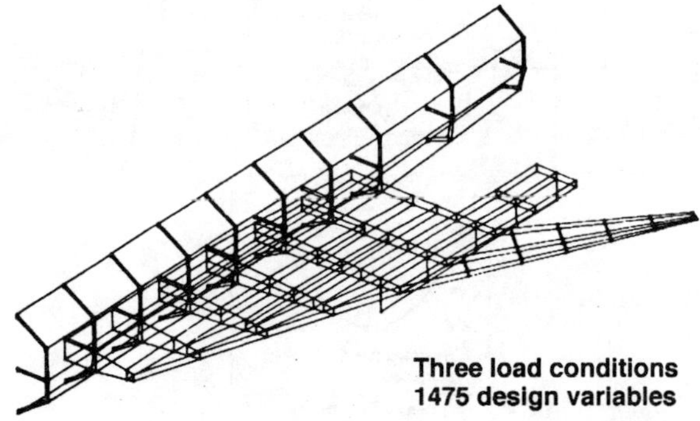

Fig. 6 Mathematical model of structure for an SST.

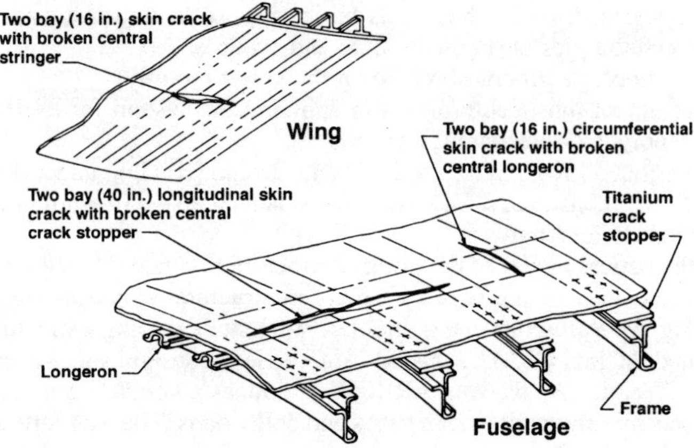

Fig. 7 Damage tolerance criteria.

for military aircraft is toward even more stringent criteria which will undoubtedly push commercial programs toward more-complicated and expensive development. Current criteria for damage tolerance in commercial airframes (Fig. 7) combine U.S. and European certification requirements.

Recent years have seen significant progress in improving methods for designing in adequate residual strength. A relatively advanced approach (Fig. 8) provides for the effects of stiffness of fasteners and eccentricity of the stiffness of stringers.[8] It incorporates into the residual-strength equation a factor (R_{ct}) determined by finite element analysis and reflecting the effects of local stiffness on crack-tip stresses in a semimonocoque structure. The analysis gives a highly reliable prediction of residual strength after final failure of either skin or stiffener.

Current procedures and published data present several distinct problems. Confused testing standards head the list. For instance, tests with panels less than

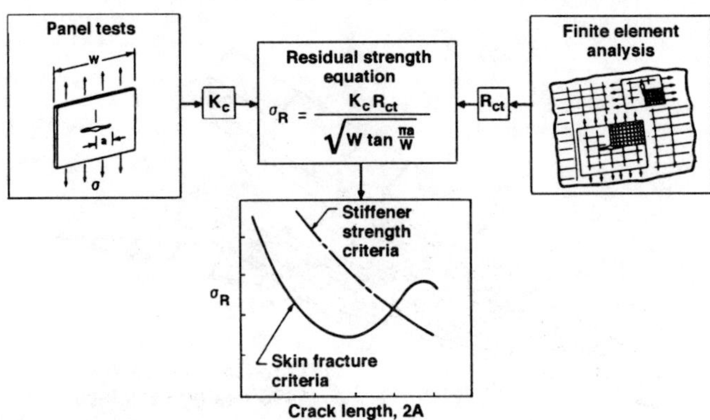

Fig. 8 Residual-strength analysis of cracked structure.

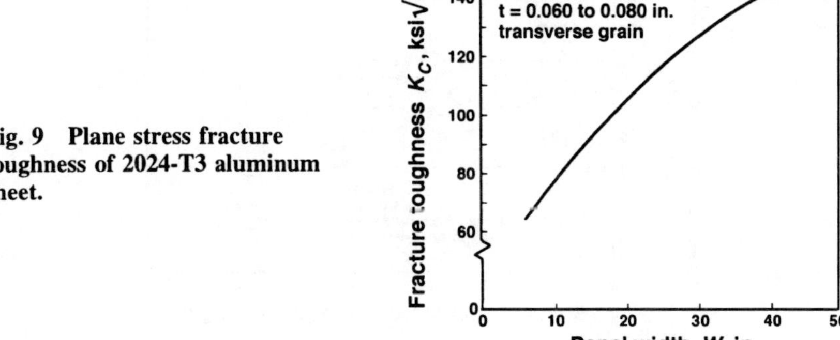

Fig. 9 Plane stress fracture toughness of 2024-T3 aluminum sheet.

48 in. wide made of 2024-T3 aluminum sheet picture this material as having a low fracture toughness (Fig. 9). On the contrary, this material performs extremely well in fail-safe design applications. Data based on narrow panels are misleading and should not guide decisions.

To date, tentative standards have been established only for specimens similar to the "compact tension specimen" designed to measure plane strain fracture toughness, K_{IC} (Fig. 10). Ninety percent of the structure of a current wide-body transport falls into the "plane stress" or "mixed mode" thickness classifications for fracture-toughness testing for which no testing standards exist. In addition to the misleading narrow panel data, the lack of testing standards has led to publishing other erroneous information. For example, testing thin sheets without guides (Fig. 10) allows buckling of the free edge along the crack which induces additional stress at the crack tip, which may not happen under actual conditions.

The influence of military design criteria will probably lead to greater dependence on analytically predicting crack growth. But present methods cannot predict

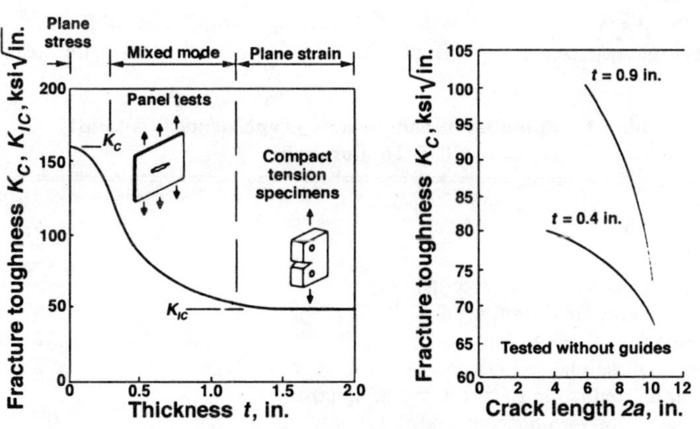

Fig. 10 Fracture toughness of 2024-T3 sheet and plate.

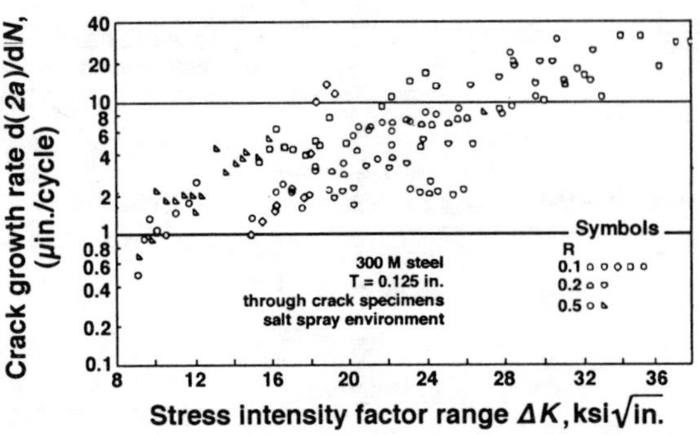

Fig. 11 Crack growth.

crack-growth rates reliably. The best current approach, Forman's Equation, assumes that the stress intensity factor range (ΔK) defines crack-growth rate (da/dN).[9] However, as the data in Fig. 11 show, such an assumption is extremely simplistic and inadequate for reliable long-life predictions. The data plotted here have a scatter factor of 10, suggesting other factors at work. Also, these data represent only constant-amplitude cyclic loading. To account for spectrum effects, including crack-growth retardation due to overloads, will take an additional level of understanding.

High potential weight saving will assure development of composites for military use. Their use in commercial aircraft depends more on operational economics.

A comparison of graphite/epoxy with 7075-T6 aluminum (Table 1) outlines the promise of composites. Relative specific strengths for primary failure modes all run significantly higher for tailored graphite/epoxy components. The potential weight saving for military designs based on these values ranges from 20–40%. The 25% estimate of the potential weight saving for commercial transport airframes is probably conservative.

Using the ATA definition of direct operating costs,[10] the impact of a 25% weight savings approximately offsets the added costs of composites (estimated

Table 1 Potential of composites: graphite/epoxy against 7075-T6 aluminum

Relative specific strength (S_{comp}/S_{alum})	
Tension	1.3–1.9
Shear	1.2–2.2
Compression (panel stability)	1.5
Fatigue (bolted joint at $N = 10^7$)	3.8
Bearing	2.2
Potential weight savings	25%
Potential reduction of direct operating costs	
Current design concepts ($150/lb)	0%
Future design concepts ($100/lb)	10%

average in-place structural cost of $150 per lb). New design concepts under development will take advantage of simplified fabrication (e.g., fewer parts) to substantially reduce in-place structural costs. When the overall average cost drops to approximately $100 per lb, advanced composites will offer savings of about 10% on direct operating costs, which will prompt their introduction into commercial airframes.

Taking full advantage of composites will call for considerably better methods of analysis. The material is extremely brittle and unforgiving of high local stresses overlooked by simplified analysis. Laminates are generally not isotropic, so require more sophisticated analysis. And the material's low interlaminar strength requires controlling out of plane loads.

A comparison of stress-strain data for two representative composite patterns with those for 7075-T6 aluminum (Fig. 12) shows the complete lack of ductility in high-strength graphite/epoxy. Consequently, analyses for composite designs will require far greater detail. Current design of multifastener joints in aluminum takes advantage of nearly complete plasticity at ultimate load. In contrast, multifastener joints using composites have no plasticity. A test of a nine-bolt joint (Fig. 13) shows what happens as a result.[11] Strain gauges measured the load distribution among the bolts at failure of the joint. The laminate put almost all the load on a few bolts, whereas a plastic material would have spread the load more evenly among them. Here the load in the end bolt agreed within 7% of the value predicted by a linear elastic analysis. Similarly precise analysis will have to establish bolt patterns, bolt sizes, and laminate tapering for all major joints in a composite design.

The selection of a laminate pattern offers an extreme range of options (Fig. 14). In addition, the structural engineer may select blends of numerous available

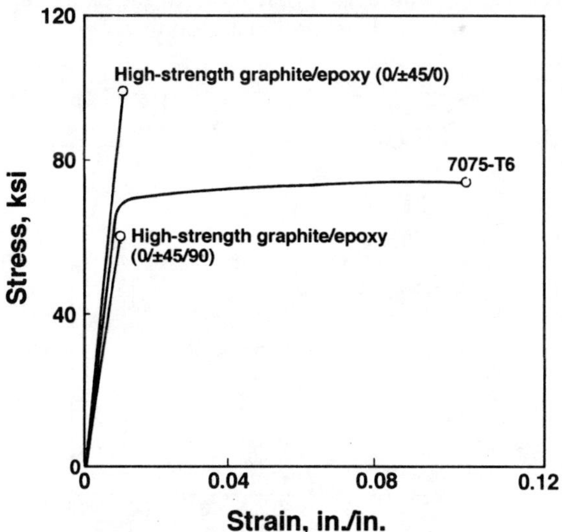

Fig. 12 Stress-strain data for two composite patterns and 7075-T6 aluminum.

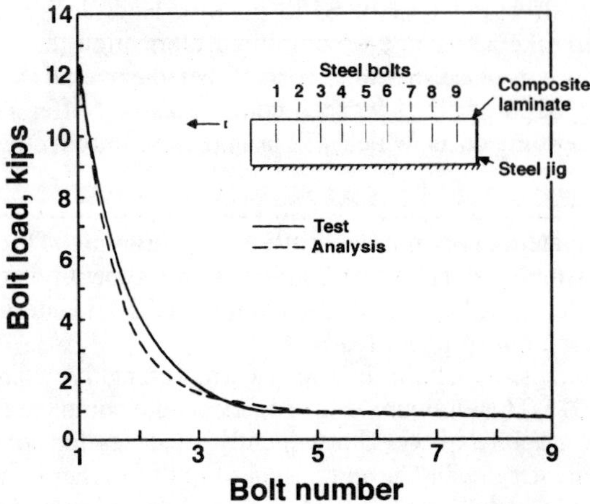

Fig. 13 Test results for a nine-bolt joint.

materials. He will need to have a high level of technical skill to make the right choices.

Cost-effective use of advanced composites will depend upon new design concepts such as the example in Fig. 15.[12] This design has extremely high potential for low cost because of its few parts. The trusscore arrangement allows introducing control-surface loads without using bulkheads; molding eliminates clips, stiffeners, and similar hardware. Softening strips in the skin panels along the bolt lines minimize stress concentration due to bolt holes in the skin. Large-scale stress concentrations such as access doors are treated similarly.

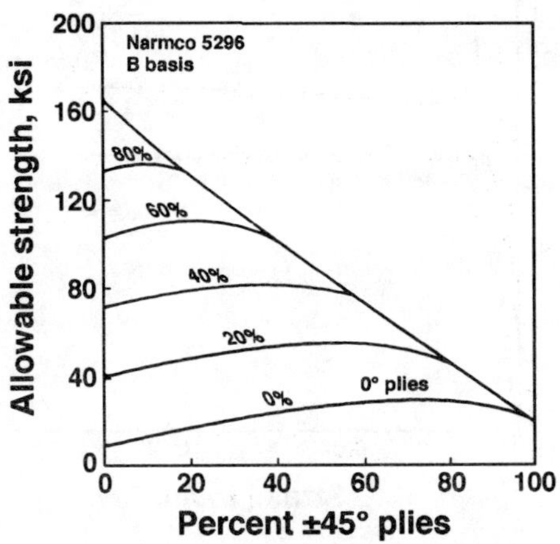

Fig. 14 Laminate ultimate tensile strength.

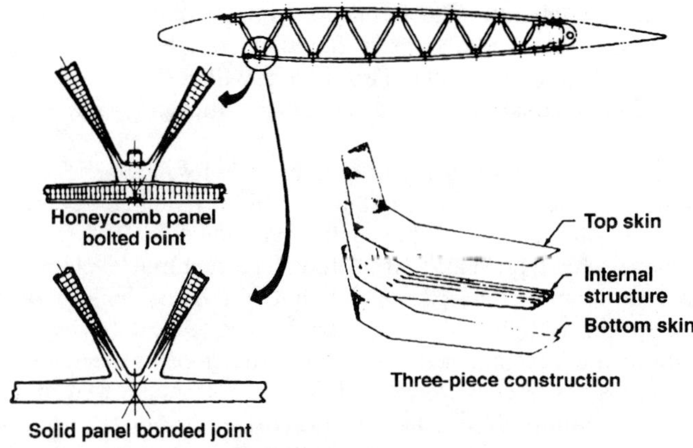

Fig. 15 Advanced concept for a composite wing.

Active controls may revolutionize criteria and load technologies (Table 2). In the extreme case airframe designs may have as their primary aim taking advantage of active controls. Basic aerodynamic stability will be provided by the combination of conventional aerodynamic stability and the active control system response, which will allow relaxing static stability requirements. Canards will provide additional longitudinal stability, and a small vertical stabilizer, in conjunction with a small forward-mounted vertical fin, directional stability. A radically different configuration offering dramatic potential weight savings will emerge. Basic load distributions, critical failure modes, and design criteria for such configurations may differ greatly from those for conventional airframe geometry. For example, the flight conditions that typically design specific components of conventional airframes may not be critical for controls-configured designs.

Maneuver load control makes use of an automatic system to control steady airloads. In its simplest form, segmented control surfaces will redistribute spanwise loading, thereby reducing the static strength needed in critical sections of the primary airframe, such as the wing root, and significantly cutting total airframe weight.

Gust-load alleviation and flutter suppression require control of unsteady airloads. Unsteady airload control will call upon the complete control system acting

Table 2 Impact of active controls

Features	Impact
Relaxed static stability	Radically different configuration, new failure modes and design criteria
Maneuver load control	Steady airload control, spanwise load redistribution, reduced static strength requirement
Gust-load alleviation and flutter suppression	Oscillatory airload control, reduced stiffness requirement, increased fatigue life, improved passenger ride comfort

in response to an elaborate set of sensors strategically located throughout the airframe. The control system transfer functions will be designed to minimize gust response and suppress flutter. This system will act as an artificial means of avoiding a "flutter penalty," increasing fatigue life and improving passenger ride comfort.

Current criteria and methods do not fully provide for the use of active controls. New ones must be developed to capitalize on their potential benefits.

A supersonic commercial transport will expose the manufacturer to a new risk. Figure 16 shows why. It presents a simulated SST structural development schedule derived by scaling up the DC-10 schedule to eight years, the approximate upper limit for economic viability. The full-scale fatigue test establishes the costs the manufacturer can expect under the warranty provisions of the sales contract for the 50–100 airframes already assembled. The two years allowed in the schedule is a practical upper limit for determining this cost. Unfortunately, current technology for high-temperature fatigue testing allows compressing the time scale by only a factor of 2, meaning that the SST fatigue test would take 10 years, by which time the manufacturer would have made commitments on several hundred airplanes. To avoid this potentially disastrous risk, airframe manufacturers must learn to compress the time scale of high-temperature fatigue testing by a factor of 10.

The success of all aircraft rests largely on costs. James Watt, the famous Scottish practical engineer, supposedly said, "The purpose of engineering is the making—not of mechanisms—but of money." This quotation should remind the structural engineer that he plays a basic role in assuring prosperity for the industry and his company.

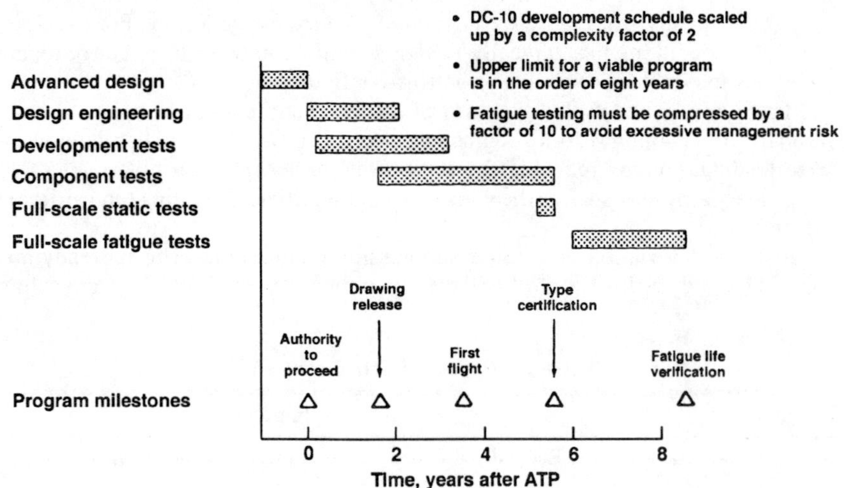

Fig. 16 Time-compressed fatigue testing on an SST. Assuming a DC-10 development schedule scaled up by a complexity factor of 2, the upper limit for a viable program is in the order of eight years, and fatigue testing must be compressed by a factor of 10 to avoid excessive management risk.

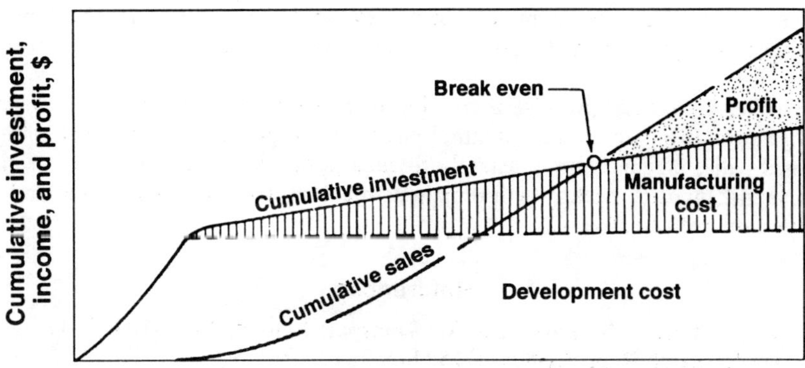

Fig. 17 Overall program economics.

In the typical program (Fig. 17) cumulative investment comprises the total of development and manufacturing costs. Development costs are nonrecurring and include engineering, laboratory testing, flight testing, and tooling. Manufacturing costs are recurring and represent the direct costs of labor and materials for producing airplanes. The manufacturer turns a profit only after cumulative sales exceed cumulative investments beyond the breakeven point.

Thus, a company starts a commercial program only after it has a reasonable guaranty of a profit. Any unnecessary costs it incurs defer the breakeven point and reduce the profit and, thereby, reduce its ability to undertake subsequent programs. Income derives from aircraft sold at a unit price negotiated well in advance of delivery. No alternate funding exists to offset unnecessary costs.

Structural engineers have a major influence on the cost of developing and manufacturing commercial aircraft. The primary and secondary structures account for approximately one-third of the sale price, and structural engineers have a primary responsibility for this cost. In addition, structural engineers share in the responsibility for costs of structural aspects of onboard systems, ground and flight tests, and design engineering. Structural engineers could substantially influence the number of new designs for satisfying future needs (Fig. 2). Individual structural engineers could contribute by doing the following:

- Maintaining an awareness of market requirements, structural efficiency, and costs.
- Recognizing that many decisions involve structural technology and cost and that a structural engineer with a proper cost perspective makes the best authority but that without a proper cost perspective his opinion has very limited value.
- Realizing that drawings and specifications establish an impregnable cost floor or minimum for manufacturing costs. For example, stipulating a high-cost surface finish on a drawing precludes following many manufacturing cost-saving options.
- Appreciating the powerful engineering lever on both nonrecurring and recurring costs. Nonrecurring costs such as testing and tooling are generally proportional to engineering costs, and an engineering simplification may drastically

reduce the total development costs for a new or derivative design. The general engineering approach and design concepts establish the basic level of recurring costs.

Business forecasts promise a growing market for commercial passenger and cargo aircraft on which the structural engineer can practice his art. Technology trends will challenge him constantly to enlarge his skills. And an old skill that keeps this industry healthy will remain as much in demand as ever—the ability to design cost effectively.

References

[1] DAC Staff, "The Passenger Air Transport Market, 1972–1987," Douglas Market Research Rept. CI-804-2775, July 1972.

[2] Pickard, J., et al., "FORMAT—Fortran Matrix Abstraction Technique," AFFDL-TR-66-207, Vol. V–VII, Dec. 1968.

[3] Rowan, J. C., and Burns, T. A., "Quasi-Static Aeroelasticity," McDonnell Douglas Rept. MDC-J5550, Jan. 1972.

[4] Kalman, T. P., Rodden, W. P., and Giesing, J. P., "Application of the Doublet-Lattice Method to Nonplanar Configurations in Subsonic Flow," *Journal of Aircraft,* June 1971.

[5] Giesing, J. P., Kalman, T. P., and Rodden, W. P., "Subsonic Steady and Oscillatory Aerodynamics for Multiple Interfering Wings and Bodies," *Journal of Aircraft,* Oct. 1972.

[6] McComb, H. G., Jr., "Automated Design Methods in Structural Technology," *Proceedings of Conference on Vehicle Technology for Civil Aviation,* NASA Langley Research Center, Nov. 1971.

[7] Dodd, A. J., "Specification for a Static Structural Optimization Capability," McDonnell Douglas Rept. MDC-J5442, March 1972.

[8] Swift, T., "The Application of Fracture Mechanics to the Design of Damage Tolerant Stiffened Aircraft Structure," American Society for Metals, 1972 Westec Conference, Los Angeles, CA, March 13–17, 1972.

[9] Forman, R. G., Kearney, V. E., and Engle, R. M., "Numerical Analysis of Crack Propagation in Cyclic Loaded Structures," *Journal of Basic Engineering,* Transactions of ASME, Sept. 1967.

[10] Air Transport Association of America, "Standard Method of Estimating Comparative Direct Operating Costs of Turbine Powered Transport Airplanes," Dec. 1967.

[11] Lehman, G. M., "Structural Joint Design and Development for a Graphite-Epoxy Horizontal Stabilizer," Conference on Fibrous Composites in Flight Vehicle Design, Dayton, OH, Sept. 1972.

[12] Schjelderup, H. C., and Purdy, D. M., "Advanced Composites—The Aircraft Material of the Future," Douglas Paper 5883, AIAA Third Aircraft Design and Operations Meeting, Seattle, WA, July 1971.

Trends in Aerospace Structures

Michael F. Card
NASA Langley Research Center

Recent developments indicate that we may be on the threshold of a revolution in aerospace structures, embracing new concepts and significantly better structural-weight fractions. The next two decades will surely see significant advances as compared to those experienced over the past two. Increases in allowable operational stress levels; utilization of high-strength, high-toughness materials; and new structural concepts will highlight this advancement. Improved titanium and aluminum alloys and high-modulus, high-strength advanced composites, with higher specific properties than aluminum and high-strength nickel-base alloys, are expected to be the principal materials.

Significant advances in computer technology will cause major changes in the preliminary design cycle and permit solutions of otherwise too-complex interactive structural problems and, thus, development of vehicles and components of higher performance. Vehicle economics will force more emphasis on cost-effective design. The energy crisis will have an impact on material costs and choices and spur the development of more-weight-efficient structures. Finally, there will be significant spinoffs of aerospace structures technology, particularly in composites and in design/analysis software.

Substantial Weight Savings Predicted

For aircraft structure alone, improvements in materials and design may bring structural weight savings of 20–30%, as indicated in Fig. 1. Although only limited improvements appear possible for aluminum alloys, they will probably continue to be widely used because of the extensive experience with them and their attractive cost. However, adhesive bonding will enable advanced structural concepts to be developed with aluminum alloys. Titanium alloys now offer greater opportunities for weight reduction. Moreover, advanced composites have even higher potential than the titanium alloys (Fig. 2) and are most promising for the rest of the century.

On the horizon today are also several new structural concepts. Bonded honeycomb sandwich combined with bonded-in extrusions and a bonded frame assembly (Fig. 3) represents an example. This concept permits fewer frames and stringers and yields a 20% weight reduction. Production of this type of structure will require improvements in fabrication techniques and a higher degree of automation. Other examples include use of welded and cast titanium structures and adhesively bonded advanced fibrous-composite designs.

Astronautics & Aeronautics, Vol. 16, No. 7/8, July/Aug. 1978, pp. 82–89.

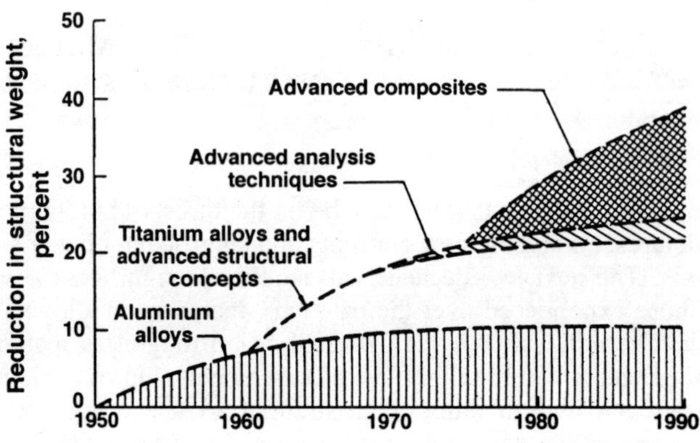

Fig. 1 Prospects for structural weight savings.

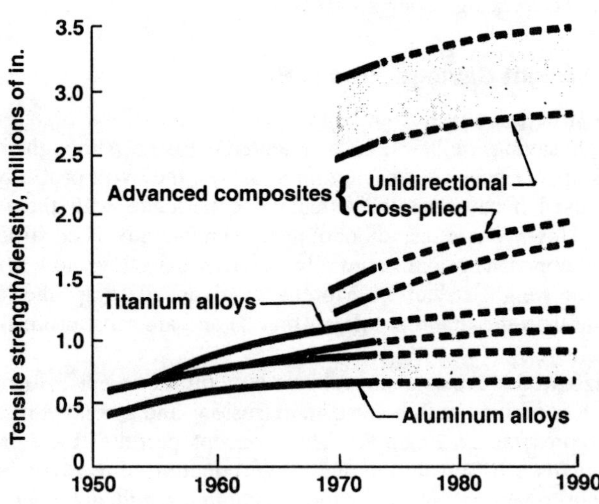

Fig. 2 Prospects for improvement in strength-to-density ratio.

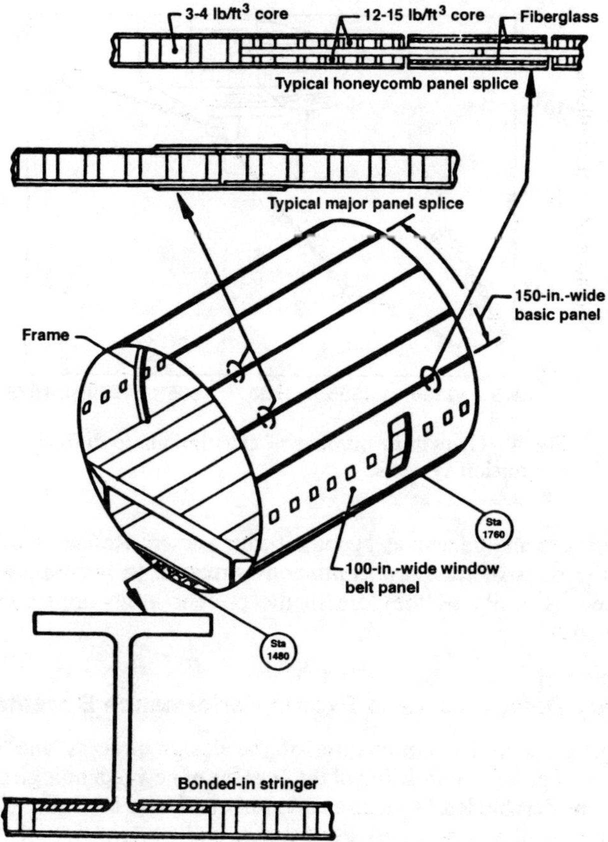

Fig. 3 Bonded honeycomb sandwich combined with bonded-in extrusions and a bonded frame assembly.

The U.S. Air Force's primary adhesively bonded structure technology (PABST) program represents a significant effort on structural design and fabrication of advanced fuselage structure. It seeks to eliminate mechanical fasteners in future cargo aircraft. The focus for studies is a large section of the YC-15 Advanced Medium STOL Transport. A large aluminum fuselage section has been fabricated and is being prepared for structural tests. Program aims include 15% weight reduction coupled with 20% cost reduction.

The next two decades may see the construction of components or large structural sections incorporating networks or lattices of high-strength filaments contained in a matrix, rather than the mechanical assembly of smaller parts. This could be accomplished by a comprehensive computer analysis that would derive the critical load paths and control the laminating or winding machine to build a structural fiber network optimized for the design loads. Such a concept will probably evolve over a period of years as the chemical and physical properties

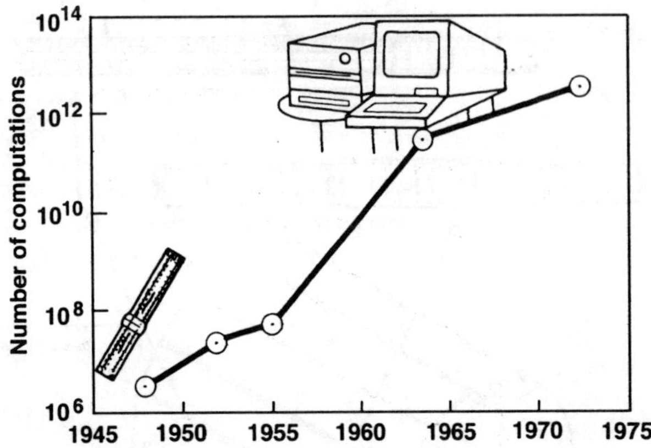

Fig. 4 Growth in number of calculations to design propulsion systems.

of new materials and adhesives become better known. Research will be needed to permit the parts of this new continuous structure to be inspected, repaired, and replaced as easily as they are in the present multipart structures and at acceptable costs.

Preliminary Design Tools to Exploit Performance Benefits

Major progress in the automation of the design process and in computer technology will permit a cascading of the benefits of new technologies in advanced vehicles. Considerable R&D backs an attempt to achieve both automated preliminary design as well as automated detail design and fabrication tools.[1–4] The value of such activities will be to accelerate the design process and permit more accurate assessments of technical interactions in real-time.

The expansion of automated analysis and design has been required by a complex of design conditions and objectives. As an example, Fig. 4 compares the number of mathematical calculations required to develop a modern propulsion system vs systems of 30 years ago. In the 1950s it took 2–4 million calculations; now it takes some 14 trillion. The slide rule of the 1950s has already been replaced by complex computer systems. The number of mathematical calculations required in future systems will be staggering, but fortunately computational capacity has been developed at a comparable rate.

In many vehicle and component studies (e.g., Space Shuttle and supersonic transport) stringent performance requirements require the vehicle to be highly "tuned." This forces a more intimate knowledge of interdisciplinary design interaction, especially among structures, aerodynamics, and controls. The incorporation of advanced technologies—such as materials, active load controls, or low-drag aerodynamics—in automated preliminary design will be vital to identify preferable configurations.

Studies of advanced structures technology, such as advanced composites, have shown considerable size reduction in vehicles, major configuration changes, and dramatic performance benefits to be possible. The studies include advanced technology (Mach 0.95) transports (ATTs), V/STOLs, large cargo aircraft, future SSTs, fighters, and RPVs.

A study of a proposed Mach 0.95 commercial transport shows that advanced composites can reduce weight some 22,000 lb on a 381,000-lb-gross-weight airplane. This saving equals about one-quarter the design payload. Performance benefits include a 5.5% reduction in the direct operating cost (DOC). Reconfiguring and resizing the airplane to exploit the full potential of composites would reduce its weight further.

Studies of other types of vehicles show similar or even more substantial benefits from the use of advanced composite structure. Total life cycle cost of such advanced structure will define its real value. An ATT study shows an economic payoff even for a 1980s aircraft.

Design-to-Cost Technologies

Emphasis on design-to-cost will surely influence the trend in structures; for in at least some cases of advanced design, engineers have met challenges of newness and performance with too much sophistication.

In concert with the long-term factors of monetary inflation, market-wage rates, and materials costs, Fig. 5 displays the increases in cost we are paying for higher performance and advanced technologies.

Although the military aircraft in Fig. 5 do not reflect radical changes in structural concepts or materials, they do represent growing structural sophistica-

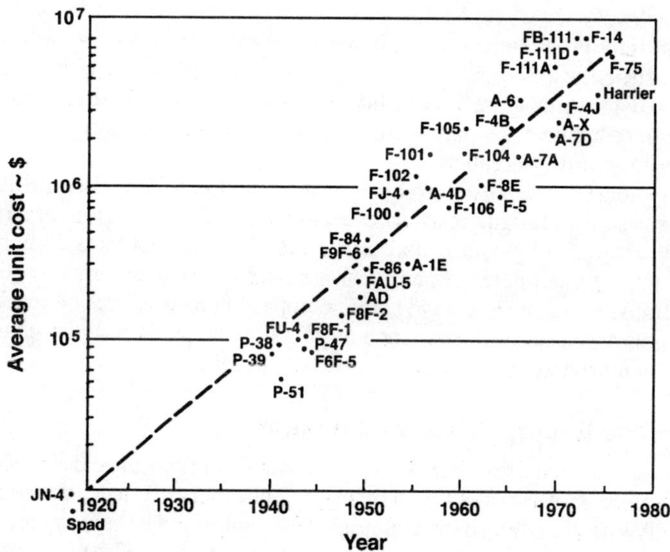

Fig. 5 Climb in cost of military aircraft.

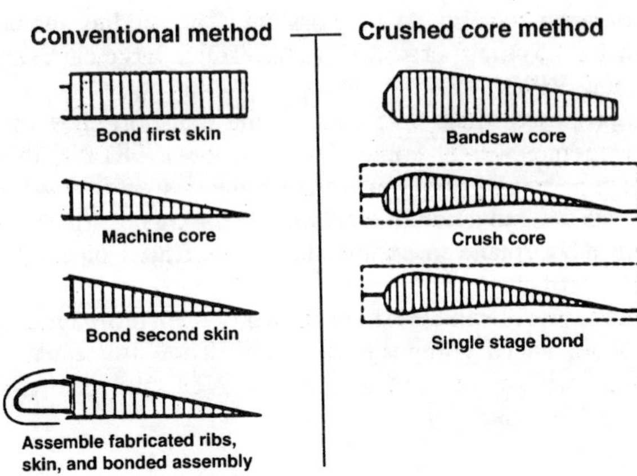

Fig. 6 One-stage bonding of flap.

tion and associated increasingly high manufacturing costs. A significant cost-factor in structures is parts-count. In itself this may not seem surprising; but structural designers have little, if anything, in the way of tools to guide them in this aspect of design. Engineers must be motivated to develop these tools and apply them to future aerospace structures.

Figure 6 describes an example of cost reduction. The conventional design has relatively high cost but low weight. The alternative reduces cost by 35% without increasing weight. To realize measurable cost reduction in structural designs, engineers must learn to generate ideas to reduce cost similar to the ideas to reduce weight.

In the design of commercial aircraft, and large aircraft in general, primary and secondary structures account for about a third of the sales price. Structural engineers have a primary responsibility for this cost. Clearly, they must play a major role in reducing costs in the future. If they fail, advanced structural systems will take a long time to realize.

Not only must the cost of production be considered, but also maintenance costs. Corrosion and fatigue resistance are important design factors of structures using new alloys, composites, and adhesives. Future structures will have to withstand the high temperatures and acoustic intensities of supersonic flight for long durations. Additional demands will be imposed on structures by innovations in propulsion and aerodynamics (such as VTOL) that will radically change the operational environment of civil and military aircraft.

Impact of the Energy Crisis on Structures

Every structure requires energy for its construction and may influence either the production or consumption of energy. The energy crisis, therefore, will undoubtedly influence future structures and in complex and not always predictable ways. However, some trends that are likely to continue are already apparent. In an environment of increasing energy costs and decreasing supply of fossil fuels, structures will be affected by changes in material cost and availability, changes in materials costs relative to other costs, energy-efficiency considerations in the end use, and new requirements for structures required for new energy sources.

Higher costs for energy will certainly tend to make materials that require large amounts of energy to produce (e.g., aluminum, titanium, and magnesium) less attractive in relation to those with lower energy requirements (steel, for example). Conversely, less availability of some raw materials (aluminum and iron ore, for example) will favor materials produced from precursors plentiful in supply (glass, graphite, and magnesium from sea water). This tradeoff will be complex, as many raw materials are found in a range of purities with differing energy requirements for refinement. A particularly complex situation applies to materials such as plastics, which are normally produced from petroleum and natural gas, also key factors in the energy supply.

A likely outcome will be the continued use of a wide range of materials, with several being competitive for many applications. It seems probable that in relative terms the use of aluminum will decrease, displacement of other materials by plastics will be reduced, and the relative importance of ceramics will increase.

It is possible that some "low technology" and "traditional" materials will become more important (e.g., wood, leather, brick, concrete). Renewability and recyclability will also be considerations. New materials and methods will be stimulated by changing economic pressures. It seems very probable that the cost of materials will increase relative to other costs such as labor (including design, analysis, and testing) and computation. This tends to favor more-complex and highly engineered structures with higher efficiencies and less surplus weight but with equal or greater reliability and performance.

The development of any new energy source will require the construction of appropriate structures. These may have novel requirements, some far beyond the state of the art. The feasibility of space solar power depends on the construction in orbit of collectors of enormous extent. Many proposed schemes for energy production envision the use of low-grade energy sources: sunlight, wind, waves, tides, ocean thermal gradients, etc. These have the common feature of requiring huge structures to serve as collectors. To be competitive these structures must also be cheap but resistant to the operating environment (solar UV, sea water, windstorms, rain, etc.). In such structures the field faces unprecedented challenges.

Spinoff of Aerospace Technologies

The aerospace industry routinely contributes advanced technology to the nonaerospace community. In structures, two major contributions can be expected in the next decade: lightweight structural concepts and computerized analysis and design tools.

Today we see aluminum structures, a spinoff from the aerospace industry, in buildings, trains, and other commercial products. Composite materials, such as glass-fiber-reinforced plastics, are also quite widely used today in various commercial products. Graphite-fiber-reinforced thermosetting and thermoplastic resins are appearing in nonaerospace products, such as golf club shafts, tennis rackets, prosthetic devices, gear wheels, heddle frames for textile machinery, racing bicycles, push rods of engines, and transmission shafts. The volume potential for graphite-fiber applications in these commercial markets will considerably influence the price of this high-performance material and, in turn, will increase the aerospace applications of such composites. In fact, we already can see a significant impact of increased production in driving the cost of these materials down. The nonaerospace applications of graphite-fiber-reinforced plastics will also accelerate the development of mechanized methods of manufacture and nondestructive testing.

A second area of technical spinoff involves computerized analysis and design. Several large-scale analysis tools such as NASTRAN developed by the aerospace-structures community have found a large community of nonaerospace users. Because of the value of computer software and services, a thriving commercial business has emerged, with many former aerospace engineers playing key roles. Many future codes developed to assist computer-aided structural design and computer-aided manufacturing should find wide application in other industries.

Technology Status

Over the past six years, significant changes have been evolving in design criteria for both military and commercial aircraft. For example, the U.S. Air Force has decoupled structural-safety and durability requirements in new structural specifications. To gain structural safety, all U.S. Air Force aircraft must now be damage tolerant either through the use of fail-safe or slow-crack-growth design concepts. The U.S. Air Force no longer accepts the past "safe-life" approach of applying a scatter factor to full-scale fatigue test results to establish aircraft-life limits.

The field still requires full-scale fatigue testing, but more in the nature of a developmental as opposed to a proof-of-compliance test. It identifies critical areas and demonstrates the basic durability of the airframe. Also, full-scale fatigue testing serves as a primary tool for establishing the future structural-maintenance needs for the aircraft.

Commercial aircraft design has also seen changing requirements, such as an increase in the fail-safe residual strength from 80 to 100% limit load and the need to identify safe periods of unrepaired usage subsequent to encountering significant structural damage.

Both military and commercial requirements can be expected to continue to place increasing emphasis on the importance of production quality control, in-service inspections, crack-propagation analysis, individual aircraft tracking, and crack-propagation and residual-strength testing.

Further, it can be expected that a greater emphasis will be placed on improving the structural durability and ensuring structural safety of gas turbine engines, landing gears, and other structural subsystems.

Loads Analysis Methods

Advances in external-load prediction techniques will likely occur in three primary technical areas. First, improved aerodynamic methods will make possible a more detailed representation of the distributed fluid-dynamic loads acting on the vehicle. Second, more-accurate consideration of the vehicle's internal structural response to applied external loads and the rates of loading will be incorporated into analyses. Third, better understanding of atmospheric turbulence will make possible a more realistic analytical representation of its three-dimensional characteristics and the influence of these on the response of the airframe and pilot.

The aerodynamic work of several investigators on new "panel techniques" to determine pressures has already paid dividends in the definition of aerodynamic loads. The finite element distribution of aerodynamics forces available from these methods have been used as an internal part of the latest loads developments, such as NASA/Boeing FLEXSTAB and McDonnell Douglas MALS.

The major property that makes these methods so valuable is the ability to determine three-dimensional steady and unsteady aerodynamic loadings on nonplanar multibodied vehicles with previously unattainable accuracy. Future improvements in these methods, along with new theories which accurately account for viscous effects, transonic shock onsets, and other nonlinear flow properties, will certainly find their way into load methods, providing the calculation economics are kept at a reasonable level.

Structural-analysis systems that directly interface internal and external loads are well developed and are being exploited to understand better the detailed repeated as well as the maximum loading environment for the structure. Investigations into both fracture occurrence and crack growth show that crack initiation and growth depend on sequence of loading as well as the detailed, rather than the general, stress field, thus requiring more attention to combined loadings and their phasings.

More comprehensive modeling has been replacing idealization of the airplane and propulsion system as a stick model of one-dimensional beams and rings for use in steady-state transient and dynamic aeroelastic analyses. Structural influence coefficients and/or stiffness matrices available from finite element structural methods provide a detailed representation of the structural elements to external and internal loads. These structural matrices are compatible with the new aerodynamic and heat-transfer techniques for use in all forms of structural and aeroelastic problems, such as those entailing significant internal thermal gradients, centrifugal loads, flow disturbances, vibratory responses, and aeroelastic deformations.

Active controls for improved maneuvering in fighters and for load alleviation, relaxed stability, and structural-response suppression in larger aircraft already see important applications (B-1, C-5A, L-1011), and these will expand.

Just as the work on flow fields is leading to supercritical wing sections and other external-configuration improvements, work in the materials field is introducing the possibility of tailoring stiffness and aerodynamic response by use of either metal or resin/composite materials. The future should see an effort to analyze more closely some innovative structural concepts seeking to achieve optimal loads design by tailoring the detail distribution of material. A large part of today's active-control technology was aided by the structural dynamicist's knowledge of the deformation and transient response of the airframe. Future work of a similar type will be required to tailor material to optimum structural configurations since aeroelastic stability as well as strength is involved.

Structural Concepts

Advanced aircraft systems present numerous challenges to the structural designer. Composite structures—candidates for almost all advanced systems—are

Table 1 Current composite structures design efforts

Component	Type of structure
DC-10 rudder	Graphite rib-stiffened skins, graphite spars
L-1011 ailerons	Kevlar sandwich skins; graphite/kevlar spars
B-727 elevators	Graphite sandwich skins; graphite spars and ribs
L-1011 vertical fin	Graphite hat-stiffened skins; graphite spars
B-737 horizontal tail	Graphite I-stiffened skin panels; graphite spars
DC-10 vertical fin	Graphite honeycomb skin; graphite spars, ribs, and intercostals
A-7D wing panel	Graphite/boron sandwich skins; graphite and aluminum spars and ribs
B-1 horizontal stabilizer	Graphite monolithic skins; multiple graphite spars
AV-8B wing box	Graphite monolithic skins; graphite spars and ribs

expected to evolve from secondary to empennage to fuselage to wing structures to propulsion-system blading (see Table 1). Several complex components are under development in NASA and DOD programs.

More-advanced aircraft will offer some unique problems. The low-drag, energy-efficient laminar flow control (LFC) aircraft must be designed with maintainable porous wing surfaces supported by lightweight load-bearing substructure. An advanced, more economical SST will require higher payload and range and hot-structure capable of operating up to 600°F. Giant cargo aircraft could require huge wings with wing-stored cargo bays.

The new generation of missiles, such as MX, will also make extensive use of composite materials to minimize weight and improve performance. Kevlar/epoxy will probably be used for the design of motor cases. Graphite/epoxy will give metals strong competition for the design of interstages, guidance bays, and many other support structures. Carbon-carbon materials should find a place in motor nozzle and re-entry vehicle nosetip applications. The use of these composite materials as primary structures will require continued improvement of analytical methods and material characterizations. These improvements are needed to provide design confidence without extensive subscale and full-scale testing. Demanding service-life requirements are also expected to necessitate the development of accelerated-aging methods to identify and quantify the life-limiting failure mechanisms associated with the use of composite materials.

For small near-term spacecraft, structural concepts will likely become more modular and standardized to reduce costs. The ability to retrieve and repair a spacecraft will probably affect some configurations. Graphite-composite structure should find many applications in high-stiffness, low-expansion structures. Metal-matrix structures may be needed for applications calling for high thermal conductivity.

By far the greatest challenge to the space-structures community will come in the development of large structures for erection in space. Potential structures range in size from 100 to several thousand meters across for antennas, work platforms, and solar power stations.

The most significant research problem in this area will be the development of a practical method of on-orbit construction. The design of antenna reflector

structure will prove particularly challenging because of accuracy requirements; it should evoke innovative solutions. We can expect new concepts to appear in precision erectable or deployable structures and in structures space-fabricated from bulk materials. The interaction of attitude and surface-control forces with flexible structure will pose new problems.

Structural-Analysis Methods

As suggested previously, computer advances are significantly affecting devel opment of both analysis and design tools. Finite element analysis will be upgraded to handle composite structures. More economic modeling and faster solution times will be possible through the use of microprocessors and order-of-magnitude advances in a new generation of large digital computers. It is possible that unique electronic hardware for stiffness-matrix formulation and manipulation may greatly expand the ability to make detailed structural analysis.

Nonlinear mechanics has been emerging as a research area. There appears to be some interest on the part of certifying agencies to do more intensive analysis of crashworthiness. We see need for more rapid analysis techniques to investigate transient dynamic-loading effects. The field needs better conceptual models, capable of capturing gross distortion effects with relatively crude modeling. Finally, significant progress is being made on active load alleviation. A research need here may be an analysis formulation which will permit better communication between specialists in controls and structures.

Testing

Full-scale structural ground tests will continue to be necessary elements in developing satisfactory aircraft structural systems. Past design programs have demonstrated fairly conclusively the necessity for testing to obtain the required structural strength, damage tolerance, and durability. Although analysis has significantly improved, particularly with the advent of automatic computers, the complexity of structural systems of the future may require full-scale test programs until adequate databases can be established.

Test programs on past structural systems have not always been timely in surfacing costly redesign problems as related to production and in-service scheduling. It follows that considerable emphasis is being given to evaluating structural integrity before production decisions. This means that structural ground test programs will be conducted under tighter, more-rigorous schedules. For example, Air Force programs now require that full-scale static, damage-tolerance, and durability testing (for one lifetime) be complete before a commitment to manufacture production hardware.

The highest research priority for development and change in structural testing involves damage-tolerance and durability testing, which in the past has been lumped under the category of fatigue testing. Unfortunately, a significant number of structural systems had fatigue-test programs that did not properly protect the operating fleets.

The purpose of damage-tolerance and durability testing is to demonstrate the ability of the system to withstand damage without catastrophic failure in flight, to surface the critical areas vulnerable to day-to-day usage, and to establish the economic life of the system. Past programs have employed simplified block-loading schemes primarily at room temperature with no simulation of environment. The emphasis, now changing, will go to more-realistic loads and environmental simulation. Loads will be applied to the structural systems on a random flight-by-flight basis, and both thermal and chemical environments will have to be accounted for where significant. To illustrate, advanced composite materials raise the question of proper preconditioning and environmental exposure. Considerable development will be required to simulate properly moisture and temperature effects which accumulate in real service over a period of several years.

The greater complexities for damage-tolerance and durability testing are at odds with the emphasis to reduce costs in developing aircraft systems. Both schedule and funding problems bring into sharp focus the need to reduce test spans and test costs. Unquestionably, test procedures and facilities will continue to be oriented toward automatic control. A number of technical questions must also be addressed, including load truncation, accelerated environmental conditioning, equivalent cyclic rates, and effective instrumentation and nondestructive inspection procedures. Finding answers to the problems arising out of time-dependent phenomena will require major technical efforts.

Component testing will continue as an important element of the design development process and will certainly have an important role in developing full-scale test procedures which address the environmental and time-related problems mentioned above. In addition, element and small-component testing may very well be the only expedient alternative in the near term to establishing wearout rates, moisture effects, change in failure modes, and other problems related to the time domain. Relating the data from such tests to the full-scale airplane evaluation process poses major questions that must be addressed. Large component and subcomponent tests should also be considered in lieu of full airframe test articles, particularly for large vehicles costing a great deal.

Proof testing of complete airframe structures has been used in major programs (F-111 and B-52) to assure flight safety. The principle behind this type of testing is the screening of flaws in the structural system by stress fields. Where it has been used, the proof test was shown to be a viable alternative to the necessary nondestructive inspection. This type of testing is not likely to become a widely used approach, but it has been shown to be feasible in certain situations.

Although considerable work in the past has addressed the statistical aspects of fatigue data, very few design programs now use the statistical approach in lieu of a deterministic scheme. A practical methodology of design and evaluation testing which includes test plans and data interpretation statistically based would make a major contribution to the structures-integrity process. The problem most frequently encountered is cost; namely, the number of specimens required for both elements and major components is prohibitive. Nevertheless, the need is there, and indeed certain systems have been designed, at least on a semistatistical basis, as far as structural reliability is concerned. The challenge appears to be finding the proper relation between small-component tests and the behavior of the operating fleet.

For unmanned spacecraft, current testing philosophy is just opposite to aircraft trends. To reduce program costs, there are significant pressures to reduce all large-scale testing. A current practice eliminates the structural-qualification test model and instead uses selected tests (usually vibration tests) on flight hardware. Naturally, this type of qualification forces greater reliance on analytical models. Very large space structures, moreover, introduce a unique research problem: how to obtain the maximum benefit from ground and flight tests of scaled or unscaled sections of structure and how to extrapolate this knowledge to the full-scale system.

Concluding Remarks

The trends in aerospace structures outlined here portend significant challenges for research. Achievements in advanced design and flight safety over the past decade have demonstrated the pre-eminence of the aerospace-structures field. It will not, we predict, lose its sway. By striving for basic understanding of advanced structures and by maintaining a pragmatic outlook in applying advanced methods, the aerospace-structures engineer will be able to make both novel and practical contributions to future flight systems and, thus, freshly support and expand the work of the engineering field at large.

References

[1] Heldenfels, R. R., ''Automating the Design Process: Process, Problems, Prospects, Potential,'' AIAA/ASME/SAE 14th Structures, Structural Dynamics, and Materials Conference, AIAA Preprint No. 73-410, Williamsburg, VA, March 20–22, 1973.

[2] Goran, R. C., ''1976 Technology Options for Cost-effective Design of Aircraft,'' *Proceedings of the AIAA/ASME/SAE 17th Structures, Structural Dynamics, and Materials Conference,* King of Prussia, PA, May 5–7, 1976.

[3] Price, J. L., Gershon, I. J., McKenny, L. O., and Meed, C. E., Jr., ''Time-Phased Development Methodology—The Key for Reliable Engines in Future Military Aircraft Weapons Systems,'' ASME 78-GT-167.

[4] Price, J. L., ''Trends of Future Turbine Engine Life Prediction: Time Phased Automated Analysis and Test Verification,'' NATO AGARD 50th Panel Meeting/Symposium, Ankara, Turkey, Sept. 19–23, 1977.

Space Structures Technology

Technology Assessment and Outlook

Michael F. Card, Edwin T. Kruszewski, and Angelo Guastaferro
NASA Langley Research Center

The past four years have seen a significant number of preliminary studies of large space systems. Published studies suggest many exciting potential applications, limited only by the artist's imagination. Indeed, the competition among large space systems has often been described as the "Battle of the Airbrushes."

Unfortunately, many of these studies lack depth in the development of true system requirements. A credible system application requires much interaction between technologists and the user community to define a "real" system. An unfortunate characteristic of large space systems is that approval for their operation in space will involve regulating agencies in communications, environmental protection, politics, and foreign affairs. Thus, barring an unforseen military crisis, we must expect these systems to come about at an evolutionary rather than explosive pace.

The technology for large space systems, however, must be guided by system expectations. A credible set of mission requirements must first be defined and some preliminary systems design work performed. Next will be needed a rationale for selecting one configuration over competing methods of construction. Approached properly, these studies will point the directions for technology programs. The technology studies will then produce design information for future systems.

In a search for credible large space structures, studies have been conducted of the requirements for antennas.[1] Of particular interest to the NASA Langley Research Center is the possibility of extending radiometry technology so that to the current coastal monitoring could be added the ability to make measurements sufficiently accurate to detect soil moisture. Such measurements could be the foundation of a technology for improving agricultural productivity around the world. As a technical venture it, thus, appears to have high potential value. The required antenna reflector measures about 800 m in diameter and requires stringent surface accuracy. Langley has begun a program to identify technology requirements and define an interdisciplinary focus for research on this reflector.

The magnitude of the technology challenge inherent in large space structures leads us to predict a gradual evolution in space-construction capability in the 1985–95 time frame. When the Space Shuttle becomes operational, it is expected that a series of construction experiments will be needed to provide data that cannot be obtained in ground test programs. These experiments will likely be conducted with the Shuttle as a base and probably will include both erectable (members joined in space) and deployable (prepackaged, self-erecting) platform structures.

Astronautics & Aeronautics, Vol. 16, No. 10, Oct. 1978, pp. 48–54.

For antenna structures, it has been proposed to use the Shuttle as a space testbed to monitor large precision reflectors under orbital conditions. It is anticipated that Shuttle pointing may not be adequate for some systems and that large appendages may cause safety concerns. These pressures will probably result in free-flyers evolving in the 1990 time frame. Once the technology gained from these flight experiments can be assimilated, and proven cost-effective, major large space-structure-construction projects should be feasible about the middle 1990s.

Developing the technology to make large space structures feasible presents significant challenges to the structures, dynamics, and materials community. In the structure area, sheer size will require development of analysis techniques to prevent computer saturation. Scaling techniques will be vital to allow extrapolating data on subsections or models to full-scale articles. Practical design concepts will be needed that produce satellites structurally effective yet transportable in high-density packages.[2] Since past structures have been largely designed to meet transportation loads, a re-evaluation of structural design criteria must be undertaken. The abundance of applications requiring low distortion suggests a need for some innovative thinking on surface-distortion correction. Finally, the problem of large-scale assembly concepts using members with practical joints needs to be addressed.

Options for assembling large space structures range from manned and remote-manipulator operations to fully automated robotics. It is expected that initial capital cost and reliability will be major drivers in the methods selected. Preliminary experiments have been conducted in NASA Marshall Space Flight Center's neutral buoyancy chamber to determine the astronaut's ability as an assembler. For early space-construction work, manned assembly will likely become a baseline largely because of cost.

Construction techniques involving automation will be required on large systems, but cost will be a serious consideration. Remote-manipulator arms will require considerable development and modification before they are suitable for joining operations. However, some interesting advances are being made in programmable robot assembly stations. The potential of such systems cannot be ignored.

Reliable modular deployment techniques will be required to make such deployable structures competitive with erectable ones. Analyses of inertial forces and displacements, especially during "imperfect" or nonsymmetric deployment, could be useful in design and ground testing.

Two areas important in control technology will be distributed-control and fuel-efficient methods. There appears to be a foundation for modern distributed-control theory, but the field needs an assessment of the practicality of systems with a small number of sensors and controllers. Fuel-efficiency tradeoffs are needed for control systems with and without consumables.

Finally, two aspects of dynamics are likely to become important. As advanced assembly techniques are developed, the dynamics of assembly will become a design driver. Another important consideration will be the possible use of active or passive damping to make flexible (low-frequency) structures practical.

In the materials area, the most difficult problem currently identified concerns the long-life requirements of large space structures. The technologies required are accelerated ground test techniques to account for degradation under ultraviolet and proton and electron bombardments and some theoretical or semiempirical techniques of life prediction. As suggested, ground predictions need to be confirmed by having multiyear flight exposure tests in environmentally severe orbits. An environment experiment carrier is needed which can be visited periodically by the Shuttle for removing or adding test specimens.

A self-generated environmental hazard is associated with electrical conductivity and charging. Reliable discharging mechanisms are needed to prevent charge buildup. The emergence of large structures with large planar sheets as reflectors, shields, or sails suggests that thin-film technology—production and handling of thin coated films—will be needed. Lastly, the desire for nondistorting structure will likely bring a resurgence of interest in zero-expansion materials such as graphite or Kevlar and in how closely minimum expansion properties can be controlled.

To meet the technology challenge inherent in large space systems, the NASA Office of Aeronautics and Space Technology has begun a Large Space Systems Technology (LSST) program. Although managed by Langley, through the LSST Office, this program embraces a multicenter effort with participation by Goddard Space Flight Center, the Jet Propulsion Laboratory, Johnson Space Center, Marshall, and Langley. To assure that the developed technology meets all NASA's mission needs, an Inter-Center Working Group, composed of representatives from the participating centers, will review LSST program plans and progress.

Success of the LSST program, therefore, depends on the broadening of narrow, individual mission requirements shown by a variety of missions. The technology development needed for these missions has now been identified from NASA, DOD, industry, and university inputs and integrated into the following appropriate technology areas: *systems*, *structural concepts* (erectable, deployable platform; deployable reflector; and space fabrication), and *associated technologies* (analyses and interactive design, assembly, control systems, electronics, and advanced materials).

System studies will be used to identify the performance and corresponding technology requirements of candidate missions needed to achieve national and NASA goals. These identified systems will then be used as focal points to integrate, evaluate, and continually update requirements and give priority to the needs for the technology being developed in all areas of the LSST program.

The structural-technology areas are grouped by the method used to construct the system in space. The same approach is used in each of these areas to develop the technology. Investigations will be conducted to define candidate concepts appropriate for each requirement. Initially, concepts and techniques will be investigated by means of analyses, ground testing, engineering studies, feasibility tests, and pilot models. These will be used to select and develop the most promising concepts. Finally, the developed technology will be integrated to define baseline structures, construct large test models, demonstrate proof-of-concept, and scope the applicability of the structural system. Every effort will be made to conduct technology experiments in ground facilities. Where ground simulation of the space environment proves inadequate, flight experiments may be performed.

The category of associated technologies closely couples with the structural-concept areas. The analyses and interactive-design area was established to develop the predictive ability required by the hostile space environment and the large size of the structures. This ability will permit an accurate and rapid assessment of discipline interaction in design, both conceptual and final.

The assembly area will define, develop, and evaluate methods and timelines for joining structures in space.

Three kinds of control systems (shape, attitude, and orbit transfer and stationkeeping) will be developed.

Electronics needed for operation and control will be developed also. Three problem areas will be examined: signal conditioning and data acquisition, power distribution, and data-channel interference and multipaction effects.

The final technology area in this category, advanced materials, will see the development of 1) accelerated laboratory testing and analyses techniques to predict environmental effects of both metals and composites, 2) long-life dimensionally stable matrix composites, and 3) advanced thermal-control techniques.

The challenge of the LSST program will be to develop the technology in these closely coupled areas, or disciplines, so that systems for a multitude of future missions can be designed, delivered, and constructed in space in a cost-effective and timely manner. The multicenter character of the program allows the matching of expertise with technology tasks. The multidisciplinary character coupled with the systems approach assures that synergistic effects among the technology areas can be evaluated as early as possible in a conceptual development. The use of the systems to evaluate and give priority to technology tasks should optimize program funding and assure responsiveness to national and NASA needs.

In short by using the collective technology resources of NASA and its industrial and academic partners, before long we hope and expect to make the construction of large space structures in space a reality rather than a dream.

References

[1] Hedgepeth, J. J., "Survey of Future Requirements for Large Space Structures," NASA CR-2621, Jan. 1976.

[2] Bush, H. G., Mikulas, M. M., Jr., and Heard, W., Jr., "Some Design Considerations for Large Space Structures," AIAA/ASME 18th Structures, Structural Dynamics, and Materials Conference, AIAA Preprint No. 77-395, March 20–23, 1977.

Orbit on Demand: Structural Analysis Finds Vertical Launchers Weigh Less

Allan H. Taylor, Jeffrey A. Cerro, Christopher I. Cruz, L. Robert Jackson, J. Chris Naftel, and Kathryn E. Wurster
NASA Langley Research Center

Orbit-on-demand vehicles have no history. Thus, we had no opportunity to extrapolate historical weight data. We had to go back to fundamentals and base our structural analysis on loads, failure modes, and design features called for by the mission. Many of the governing loads were encountered right on the ground or in the first few minutes of flight—sitting on the runway on alert for long periods fully loaded with fuel, accelerating along the runway for takeoff, hitting gusts, and pulling up into a steep ascent. By avoiding some of these conditions, vertical-takeoff vehicles came out lighter.

Aerodynamics

Detailed aerodynamic analyses for V-2 and H-2-Sub Phase 2 orbiters used the Aerodynamic Preliminary Analysis System II. This preliminary system incorporates the unified distributed panel program for subsonic and supersonic calculations and the hypersonic arbitrary body program, including skin-friction calculations, for hypersonic analyses. The resulting aerodynamics for both orbiters proved to be satisfactory for meeting a 1100-n-mile entry cross-range requirement, with a maximum hypersonic lift/drag ratio (*L/D*) of 1.8 for the H-2-Sub orbiter and 1.4 for the V-2 orbiter.

The effect of center of gravity (CG) on wing size and shape plays an important role in the analyses. The V-2 orbiter has a mildly aft entry CG (69% of body length). A relatively small wing with an unswept trailing edge allows hypersonic trim. The H-2-Sub orbiter, on the other hand, has a 71% entry CG that forces a considerably larger wing for hypersonic trim than needed for landing field length. Also, the wing must be swept, which increases structural weight. Both V-2 and H-2-Sub designs have adequate hypersonic trim.

Propulsion

Propulsion systems include liquid oxygen/liquid hydrogen (LOX/LH) engines based on the Space Shuttle main engine at full power (109% of normal power), either with a single nozzle with an expansion ratio of 150 or two-position nozzle with expansion ratios of 40 and 150. For dual-fuel vehicles, a LOX/RP-1 engine used hydrogen as an auxiliary fluid for cooling and as fuel for a hydrogen-rich

Aerospace America, Vol. 23, No. 2, Feb. 1985, pp. 58–61.

gas generator. Engine thrust was scaled as needed for each concept. These engines call for extension of the technology for the two-position nozzle and hydrocarbon combustion. Technology advances assumed in engine materials would reduce engine weight 20% from current engines of similar design. Important advances in engine life, condition monitoring, and reliability would be needed for the vehicle to operate as planned, such as little need for maintenance.

Air-breathing propulsion included subsonic turbofans with technology typical of commercial transport engines expected in the early 1990s. Turbojet engines for use to Mach 3 were based on supersonic-transport engine designs. For Mach 5 staging, the same turbojet engines supply thrust until subsonic-combustion ramjets take over at Mach 3.5. All air-breathing engines except the ramjet use normal jet fuel. Ramjets use hydrogen. Such air-breathing engines require development to reach the projected 80,000-lb sea-level static thrust per engine.

A dual-expander, tripropellant engine illustrated the impact of a more advanced rocket technology. This engine has a LOX/RP-1 chamber surrounded by an annular chamber for LOX/LH. Using this engine in the V-1 concept reduced the gross weight about 30%.

Trajectories

The Program to Optimize Simulated Trajectories was used to develop the ascent and entry trajectories for this study. Mission and thermal-protection-system needs were prime considerations in trajectory development. We maximized the weight of each orbiter and payload for insertion into a 50-by-100-n-mile polar orbit. The total ascent acceleration was limited to 3 *g*, and dynamic pressure was limited to 1500 psf, with a further limitation of 1000 psf at staging for the multistage vehicles. The V-2 orbiter never reaches the 1500-psf dynamic pressure, but the H-2-Sub orbiter does at 60 s from staging. Both orbiters reach 3 *g* and hold this limit by throttling until the thrust drop at transition from dual fuel to LOX/LH.

Horizontal- and vertical-takeoff concepts differ much in the ascent loads applied to the structure. The horizontal-takeoff concept has a 1.5-*g* normal acceleration during flight, compared to 0.4 *g* for vertical takeoff. Horizontal takeoff incurs greater gust loads and runway dynamic loads.

Each orbiter (Phases 1 and 2) was called on to perform a gliding, unpowered entry from 160-n-mile polar orbit. Planform loading and vehicle aerodynamics, both specific to the vehicle configurations, were used in defining boundaries of the entry-flight envelope. We assumed a single-pass return and a 1100-n-mile cross range. This scenario necessitated entry at lower angle of attack for orbiters with less hypersonic *L/D*. Leeside heating may become an issue for such orbiters. In each case, the peak heating rate, duration of heat pulse, angle-of-attack history, and bank-angle profile were chosen to reduce the total heat load—subject to the temperature constraints of the thermal-protection system (TPS) and the 1100-n-mile cross-range requirement. In general, the temperature capability of the TPS materials was not a constraint, thus allowing leeway for more severe entry maneuvers.

Entry trajectories principally affected vehicle designs through TPS weight. The size of some orbiters and the desire to reduce ascent drag, however, cause leading-edge radius to be much smaller than for the Shuttle. Small leading edges, high heating rates, and shock-wave interactions produce temperatures exceeding the capacity of passive thermal protection. Heat-pipe leading edges can withstand these conditions by transferring heat from stagnation regions to cooler upper surfaces, and such edges were used for our designs. Heat pipes also avoid rain erosion.

Structures

On-demand launch imposes many constraints on the vehicle's structure, especially in rain, snow, and gusty wind. Horizontal takeoff, with the orbiter fully loaded with propellant, imposes additional conditions, such as runway loads and vertical-gust loads during ascent. Vertical-takeoff vehicles are affected by horizontal gusts, but less than horizontal-takeoff systems. Staged concepts also experience aerodynamic-interference loads.

An aeroshell structure selected for the orbiters embodies a purge space and allows periodic tank removal for inspection. The purge space permits the vehicle to be held on alert in adverse weather without moisture or ice buildup. The orbiters have wall construction using warm (600°F) organic-matrix composition airframe insulated with the durable TPS and a metal-matrix-composite cryogenic tank with a high-temperature (500°F) foam insulation. Structural design also includes heat-pipe leading edges and an advanced carbon-carbon hot structure for the lightly loaded control surfaces. A concerted effort will be needed to develop these structural technologies. Proof-of-concept tests should be possible before starting a full-blown vehicle program and will help get the vehicle into early service.

Major airframe structural components were sized to withstand the principal design loads with a safety factor of 1.5, and an allowance was added for such items as joints, fittings, and access panel. Because reusable flight-weight cryogenic tanks have not yet been demonstrated, the tank designs received more extensive analysis. The propellant tank was initially sized for the burst-pressure load which includes a safety factor of 2.0. The burst pressure includes the ullage pressure and inertially induced hydrostatic-head pressure. The resulting wall thickness was evaluated for cycle life based on fracture-mechanics properties of the selected material, initial flaw size, stress level, temperature, and design life of 500 flights. Next, the tank bending loads were calculated and stiffeners added in areas where the tank skin would buckle when unpressurized. The fuselage ring-frames were sized in a similar fashion. The wing weight was determined by analysis of over 50 rectangular segments. The wing structural thickness was calculated as the local mold line less the required TPS thickness. Insulation and TPS were sized to maintain materials within reuse temperatures. Finite element analysis verified that the designs had acceptable natural frequencies and deflections.

Because of a combination of more-severe ascent loads from runway dynamics, wind gusts, and aerodynamic pull-up, horizontal-takeoff orbiters must have heavier structures. In addition, the slender bodies and thin wings to improve ascent aerodynamics of horizontal-takeoff vehicles prove less efficient structurally. For example, the V-1 concept and the Phase 2 H-2-Sub concept have nearly equivalent wing areas, but the horizontal-takeoff concept has double the wing weight. The weight difference of over 5000 lb equals the Phase 1 mission payload. The H-2DT and V-2DT orbiters have the same length, 47 ft, but the horizontal-takeoff design's bending moment caused by aerodynamic pull-up exceeds the vertical-takeoff bending moment for wing gusts by a factor of 4. The horizontal-takeoff body structure, thus, weighs 50% more.

We looked at whisker composites as a measure of more-advanced structural technology. Composites with long-whisker reinforcement, analysis revealed, might reduce the gross weight of the V-1 concept about 30%. So far, long whiskers can only be produced in experimental quantities, and thus this technology was not entered into the baseline designs.

Conclusion

From these analyses, we conclude that several vehicle concepts meet the needs of an on-demand launch system. The vertical-launch vehicles, for example, with their lower loading conditions, prove inherently lighter than the more heavily loaded horizontal-launch vehicles. But further R&D will be essential to refine the technologies that will bring about workable orbit-on-demand vehicles.

Structures and Subsystems

A. H. Taylor, I. O. MacConochie, L. R. Jackson, and J. A. Martin
NASA Langley Research Center

Over the past four decades, space vehicles have matured from relatively small and simple expendable ballistic systems to large and complex reusable hybrid systems. During this period, technology improvements have greatly reduced the cost of delivering a payload. Among the pacing technologies have been new and improved materials combined with efficient structural design concepts. These and other developments have shaped the present Space Transportation System (STS), and now, for the replacement of the STS in the year 2005, new materials and novel designs are being studied to identify the technologies that *should be developed* for a low-cost future space transportation system (FSTS).

Structural Goals and Concepts

Structural design goals included weight, economy, and serviceability. Minimum weight had to be traded against technical risk to arrive at the lowest development and production costs. Long life, serviceability, and all-weather launch ability were key goals required to reduce fleet size, maintenance cost, and schedule delays and thus minimize life cycle cost.

Three basic structural arrangements (Fig. 1) were considered for the FSTS orbiter.

The first, a nonintegral-tank arrangement, has the cryogenic propellant tanks suspended within the airframe by ball-jointed links to handle differential thermal growth while transmitting inertial loads. A structural bulkhead of a grid of beams integral with the fuselage carries thrust into the airframe.

The second, an integral-tank arrangement, makes the cryogenic tanks the primary structure of the fuselage, and this design requires external panels to serve as a thermal-protection system (TPS) and aerodynamic surface. The wing is attached to the tank by struts that provide simple supports and permit unrestrained differential thermal growth between the cold fuselage/tank structure and the warmer wing. The rocket thrust is transmitted directly into the tank structure through an integral conical thrust structure.

The third arrangement, a hybrid of the first two, has an integral-tank and thrust structure encapsulated in a separate wing-fuselage structure—an ''aeroshell''—and supported within it by aft trunnion and forward hinged supports.

Each of the three arrangements can employ either a hot or an insulated structure.

The current Space Shuttle and several experimental rocket-propelled airplanes have nonintegral-tank arrangements. The integral tank has been used for the X-15

Astronautics & Aeronautics, Vol. 21, No. 6, June 1983, pp. 50–56, 62.

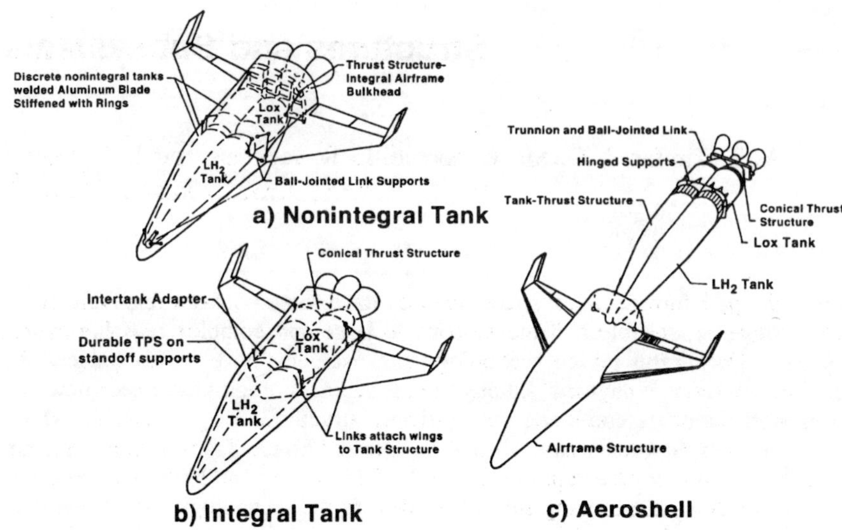

Fig. 1 Orbiter structural concepts.

and for most of the expendable rocket boosters such as Atlas, Centaur, and Saturn. The aeroshell is a novel dual-structure concept.

Wall Constructions

Various wall constructions can be used for each structural arrangement. Figure 2 shows three representative arrangements. Each employs a blade-stiffened aluminum tank with reinforced closed-cell foam cryogenic insulation. Similar to the one used on the Saturn S-IVB, the foam has a temperature limit of 175°F and is not compatible with liquid oxygen.

The insulated-wall construction (Fig. 2a) could be used for either the nonintegral tank or aeroshell. A durable TPS, now being studied for the Shuttle,[1] insulates its airframe from the aerothermal environment. It consists of either foot-square metallic or yard-square carbon-carbon packages (tiles) containing fibrous insulation, depending on the expected surface temperature. Slip-jointed or pivoted simple supports mechanically attach the tiles, thus eliminating load-induced strain in the TPS or thermal stress between the TPS and structure. The TPS must maintain the tank insulation, tank structure, and airframe below their temperature limits: 175°F for the cryogenic foam insulation, 300°F for aluminum-alloy structure, 350°F for graphite/epoxy structure, and 600°F for graphite/polyimide structure.

Figure 2b shows a wall construction using the same durable TPS for the integral tank. Here, the tiles are mechanically attached to stand-off supports on the tank surface instead of the airframe.

The other wall construction (Fig. 2c), a hot-structure airframe for either the nonintegral tank or aeroshell, uses advanced carbon-carbon (layers of carbon cloth in a carbon matrix called ACC-4)[2] or hybrid carbon-carbon (layers of carbon

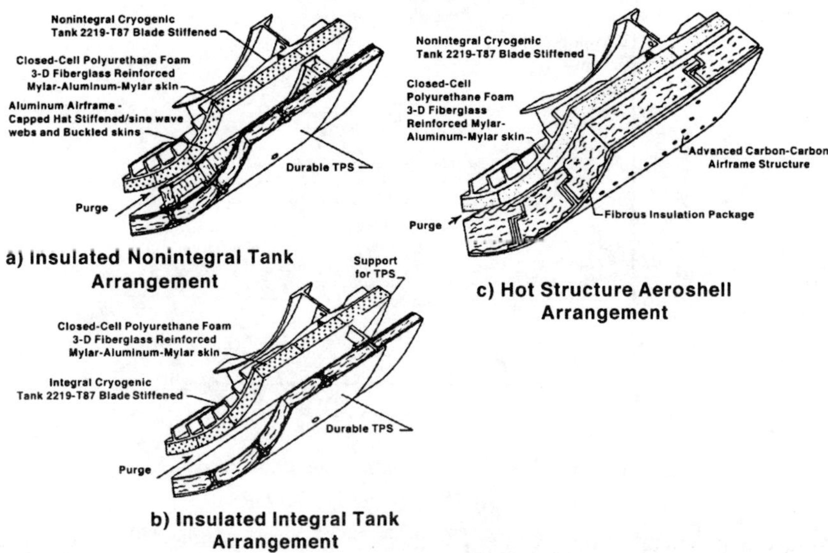

Fig. 2 Wall constructions.

cloth and carbon filaments in a carbon matrix), which has a greater specific strength.

Figure 3 compares carbon-carbon with the other candidate materials. The carbon-carbon can withstand the maximum lower-surface temperature of 2800°F on the FSTS orbiter. A rapidly developing material, carbon-carbon still entails a very high technical risk (and cost) because of limited materials tests and no experience in using it as a structure.

The carbon-carbon structures have no external TPS, but they require a packaged fibrous insulation (microquartz, contained in a foil-gage nickel-alloy package) between the hot structure and the tank to keep the aluminum tank and cryogenic foam to allowable temperatures. To prevent icing in inclement weather during periods of ground hold, the space between the tank and the airframe is purged with heated dry air for all three-wall constructions.

Primary Structures

A box-section stiffener (Fig. 2a) consisting of flanged inner and outer caps spaced by foil-gage, sine-wave webs was selected for metallic structures. Honeycomb-core sandwich was selected for highly loaded graphite-composite structures and zee-stiffened skin panels for the lightly loaded composite structures. Either truss webs or full-depth, sine-wave shear webs were selected for the wing structures.

The carbon-carbon surface panel structures (Fig. 2c) used in this study are relatively simple and generally flat skins with angle stiffeners. Truss webs were used for wing spar and rib webs in the hybrid carbon-carbon concept because a

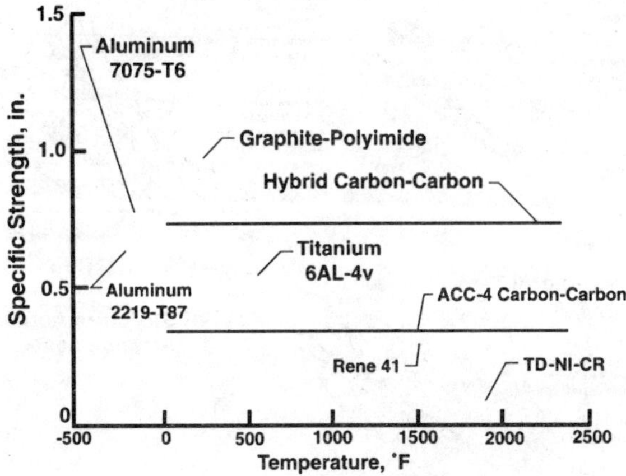

Fig. 3 Carbon-carbon vs other materials.

truss web can support the relatively high shear loads at considerably less weight than an angle-stiffened-skin web.

Further research is needed for the carbon-carbon materials. Currently, the inherent advantages of the basic materials are limited because of the relatively thick minimum gage, low strength and ductility, and poor oxidation resistance. Also, manufacturing and fabrication techniques are very labor intensive.

Concept Comparisons

The various structural arrangements and wall constructions were compared for structural weight, life/economy/technical risk, and serviceability.

Structural Weight

Tables 1 and 2 compare the structural weight and unit weight of each of the principal structural components: tank, fuselage, wing, and thrust structure for a matrix of structural arrangements and wall constructions. The techniques for the weight analyses are presented in Reference 3.

Internal pressure is the primary tank load. An intersecting cylinder tank with a blade-stiffened membrane wall between the lobes is used to minimize tank weight. Because the nonintegral tanks support the same internal pressure but have a lower bending moment than the integral or aeroshell tanks, the nonintegral tanks are about 10–12% lighter than the tanks of the other concepts. Aluminum (2219-T87) was selected for the tanks because it has moderate structural efficiency, is weldable for construction, is compatible with both propellants, and has excellent resistance to crack growth.

The fuselage weight of the integral-tank structure is 30% of the nonintegral-tank and aeroshell concepts, primarily because the integral tank serves as the fuselage. The other concepts use a separate (redundant) fuselage. The integral tank is more efficient for supporting the aerodynamic pressures acting on the flat lower surface than the long straight ring-frames on the lower surfaces of the

Table 1 Orbiter structural-weight comparisons

Structural concept and material, component	Structural arrangement							
	Nonintegral tank		Integral tank	Integral tank/thrust structure/aeroshell				
	Insulated structure, nonbuckled skin, aluminum	Insulated structure, buckled skin, aluminum	Insulated structure, nonbuckled skin, aluminum with graphite/polyimide wing	Hot structure, nonbuckled skin, ACC-4 carbon-carbon	Insulated structure, buckled skin, aluminum	Insulated structure, nonbuckled skin, graphite/epoxy	Insulated structure, nonbuckled skin, graphite/polyimide	Hot structure, nonbuckled skin, hybrid carbon-carbon
Tank	62,800	62,800	70,700	69,300	69,300	69,300	69,300	69,300
Structure	56,500	56,500	64,000	63,000	63,000	63,000	63,000	63,000
Insulation	6,300	6,300	6,300	6,300	6,300	6,300	6,300	6,300
Fuselage	109,300	103,900	52,000	90,100	97,700	88,300	82,600	72,400
Structure	77,000	72,300	18,500	67,000	66,100	47,800	48,100	49,300
TPS	31,600	31,600	33,500	23,100	31,600	40,500	34,500	23,100
Wing	40,000	38,800	29,700	59,000	38,800	29,100	28,700	19,800
Structure	30,000	28,800	20,100	59,000	28,800	16,800	19,100	19,800
TPS	10,000	10,000	9,600		10,000	12,300	9,600	
Thrust structure	29,700	29,700	9,000	10,000	10,000	10,000	10,000	10,000
Total structure	241,800	235,200	161,400	228,800	215,800	196,700	190,600	171,500
Weight comparison	1.50	1.46	1.00	1.42	1.34	1.22	1.18	1.06
Development and production cost comparisons	1.15	1.15	1.00	1.29	1.09	1.08	1.10	1.06

Table 2 Orbiter unit-weight summary for FSTS orbiter structural concepts

Structural concept and material, component	Structural arrangement							
	Nonintegral tank		Integral tank	Integral tank/thrust structure/aeroshell				
	Insulated structure, nonbuckled skin, aluminum	Insulated structure, nonbuckled skin, aluminum	Insulated structure, nonbuckled skin, aluminum with graphite/polyimide wing	Hot structure, nonbuckled skin, ACC-4 carbon-carbon	Insulated structure, buckled skin, aluminum	Insulated structure, nonbuckled skin, graphite/epoxy	Insulated structure, nonbuckled skin, graphite/polyimide	Hot structure, nonbuckled skin, hybrid carbon-carbon
Cryogenic tank	4.0	4.0	4.4	4.3	4.3	4.3	4.3	4.3
Structure	3.2	3.2	3.6	3.5	3.5	3.5	3.5	3.5
External foam	0.5	0.5	0.5	0.5	0.5	0.5	0.5	0.5
Internal foam	0.3	0.3	0.3	0.3	0.3	0.3	0.3	0.3
Fuselage	4.5	4.3	2.6	3.8	4.0	3.7	3.4	3.1
Structure	3.2	3.0	1.2	2.8	2.7	2.0	2.0	2.1
TPS	1.3	1.3	—	—	1.3	1.7	1.4	—
Fibrous insulation	—	—	—	1.0	—	—	—	1.0
Wing	5.1	5.0	3.6	7.4	5.0	3.6	3.6	4.2
Structure	3.7	3.6	2.4	7.4	3.6	2.1	2.4	4.2
TPS	1.4	1.4	1.8	—	1.4	1.5	1.2	—

other concepts. Trusses for support of the TPS and struts for attaching the wing to the integral tank form the entire fuselage structural weight listed in Table 1. These trusses and struts are considerably lighter than the separate fuselage structure of the nonintegral-tank and aeroshell concepts.

For the aeroshell concepts, the graphite-composite fuselage structure is considerably lighter than aluminum or ACC-4 carbon-carbon fuselage structure, and the hybrid carbon-carbon hot fuselage structure weighs the least.

Since the air loads during entry are the same as during ascent for the FSTS, graphite/polyimide offers a lighter wing structure than graphite/epoxy because the graphite/epoxy at its temperature limit of 350°F requires more insulation than the graphite/polyimide concept at its temperature limit of 600°F, making the latter the lighter-weight system. All the insulated-structure concepts could use the graphite/polyimide wing structure for a further weight reduction. The graphite/polyimide wing structure is about one-third the weight of an ACC-4 carbon-carbon wing and about two-thirds the weight of a hybrid carbon-carbon wing, but the composite wing requires considerable TPS weight, thus reducing the efficiency of the graphite-composite structures.

A direct load path for thrust through the conical structure to the LOX tank of the integral-tank concept and integral tank and thrust structure of the aeroshell concept proves more efficient than transferring the thrust from the engines to the airframe by a grid of beams, as with nonintegral tanks. The weight of the thrust structure of integral-tank and aeroshell concepts is about 30% of that of the nonintegral tank concepts (Table 1).

Hot-structure (carbon-carbon) aeroshells have insulation only inside the fuselage, and the wings require essentially no internal insulation. Thus, the total TPS weight of the hot structures represents only 60% of the weight of the insulated structural concepts. Each concept uses equally efficient insulation because each insulation has the same density-conductivity product.

Life, Economy, and Technical Risk

The primary structure of all the insulated concepts should exceed a goal of 1000 missions. Fracture-mechanics analysis of the tanks indicate a through-the-thickness crack life of 400 cycles (for an initial flaw size of 0.050-in. radius). The life requirement eliminates a honeycomb structure and mandates a single-wall membrane structure for the aluminum tanks of all concepts.[3] The oxidation life of the carbon-carbon materials may not exceed 300 missions.[2] Thus, oxidation life of carbon-carbon needs a factor of 3 improvement to meet the 1000-mission goal. Since cycle lives of cryogenic foam and durable TPS packages are not known, tests are needed to determine the life of these insulation systems.

Development and production costs (Table 1) and the technical risk of carbon-carbon hot structures exceed those for the insulated concepts that use carbon-carbon only as heat-shielding on lower surfaces. However, carbon-carbon technology is maturing rapidly and, in time, may achieve lower cost and risk.

Serviceability and All-Weather Operations

To remove the tanks for inspection, the insulated nonintegral-tank concept requires a splice in the fuselage and removal of the TPS around the splice. The integral-tank concept requires removal of many heat shields to inspect the tank. The tank for the aeroshell structure can be inspected by removal from the integral tank and thrust structure. Disconnecting these two structures from each other allows lifting the aeroshell off in a vertical position.

Hot aeroshell structures are watertight. Insulated concepts, on the other hand, might ingest water between and under the watertight TPS packages. However, such water ingestion may not pose any operational constraints because heated purge gas prevents icing for all concepts.

Structural-Concept Ratings

The integral tank and aeroshell concept (using hot hybrid carbon-carbon) have the best ratings. The aeroshell concepts offer better serviceability than the integral-tank concept. Of the aeroshell concepts, the insulated aluminum structure offers the best near-term performance because this technology has the least risk. The graphite-composite concepts are attractive because of their lower weight, but they entail higher technical risk. A hot hybrid carbon-carbon concept has the lowest weight of the aeroshell concepts and better all-weather characteristics than the insulated aeroshells, but it has higher costs and technical risk. Consequently, the hybrid carbon-carbon hot-structure aeroshell potentially offers the best solution for the FSTS orbiter.

Technology Requirements

This study identified several technologies which, if developed, would improve the FSTS. A better closed-cell cryogenic foam insulation that has a high cycle life and a high reuse temperature (at least 400°F) is required. The ability to detect flaws smaller than the current 0.050-in. criterion is needed to minimize weight and improve the cycle life of structures. The development and characterization of new materials such as carbon-carbon and composites (including metal-matrix composites, not considered in this study) are needed to exploit their fundamentally superior properties. Higher shear strength, smaller gage, better oxidation resistance, more-efficient structural concepts, and rapid-fabrication techniques are required for these advanced materials.

Subsystems

The structures and thermal-protection systems constitute only half of the dry weight of the FSTS orbiter. The other half includes the propulsion system and other subsystems. Subsystem technologies were selected not only for performance and weight, but also operational advantages, such as greater reliability and longer life.

Propulsion

Modified Space Shuttle main engines (SSMEs) power the orbiter. The modified engines have a thrust of 120% normal power (a projected improvement for the SSME). Also, the modified engine employs a two-position nozzle (expansion ratios of 50:1 and 150:1) to improve both sea-level and high-altitude performance.

Besides the liquid oxygen/liquid hydrogen (LOX/LH_2) main engines, the FSTS orbiter uses an integrated propellant system for the orbital-maneuvering system (OMS), the reaction-control system (RCS), the fuel cells, and the propellant payload delivered to orbit. (Such commonality of propellants for propulsion was proposed during early Shuttle studies of a fully reusable two-stage system.)

The Shuttle now uses storable propellant for its OMS and RCS (monomethylhydrazine for fuel and nitrogen tetroxide for oxidizer). The proposed LOX/LH_2 OMS for the FSTS operates at specific impulse of 480 s and is about 50% more efficient than the storable system.[4] A cryogenic RCS offers only a small advantage because of the high dry weight of the system compared to the energy demand. The RCS engines use gaseous, as opposed to liquid, oxygen and hydrogen.[5]

LOX/LH_2 propellants are less toxic and less corrosive than storables. This reduces hazards of propellant handling and lessens turnaround time.

Prime Power and Distribution

On the Shuttle orbiter, prime power includes LOX/LH_2 fuel cells and hydrazine gas-turbine-drive hydraulic pumps. The fuel cells provide all the power for the avionics, whereas the hydraulic system powers the rocket-engine gimbal and the control-surface actuators. On the FSTS orbiter, the fuel cell powers the avionics, but batteries and brushless samarium-cobalt electric motors drive all actuators. These motors are quadrupally redundant in the windings and operate through a single (but highly reliable) electromechanical system. (A study of electric vs hydraulic systems for the Shuttle estimated a savings of 3000 lb by using such motors.) The all-electric actuation system eliminates hydrazine and the hydraulic fluids and their distribution systems.

Life Support and Environmental Conditioning

The cabin size and oxygen and nitrogen supplies are smaller than the current Shuttle because a two-man crew is proposed for the FSTS. The Freon radiators used on the Shuttle are replaced with a water flash evaporator used during the entire orbital phase of the mission. (The Shuttle uses a flash evaporator from 100,000-ft altitude until opening of the cargo-bay doors.[6]) The extended use of the flash evaporator is only attractive because of the short duration of the mission and the relatively low heat load compared to the Shuttle's.

Centralized Fluid Management

Having baselined the propulsion systems as LOX/LH_2, eliminated the Freon in the environmental-control system and the hydraulic fluid in the actuation system, and substituted batteries for hydrazine gas turbines, we can consider centralized fluid management by applying the principles discussed in References 7 and 8. On the orbiter, for example, the fuel cells utilize LOX/LH_2 (propellant-grade fuel cells are required); the crew gets its oxygen supply from the fuel-cell LOX dewars; the fuel cells give off water; the flash evaporator uses water; and the heat-exchange loop uses water in the cabin and cold plates under the electronics. In addition, the system needs only nitrogen for the cabin atmosphere and ammonia for the cooling system below 100,000 ft on entry. The integrated propellant-management system includes a series of concentric baffles and screens within the main propellant tanks to control and maintain the quality of the ever-diminishing supply of LOX/LH_2 during launch, orbit, and entry.

A cross-feed system transfers reserve, residual, and intentionally transported propellant surpluses. Another feature of the system is a cross-feeding from the fuel-cell dewars for contingency or emergency return propellants. This emergency propellant would be used in the OMS and RCS in the event of a malfunction in the central propellant-management system.

Avionics

Significant improvements in avionics are projected for the FSTS for autonomy and reduced operational costs. The FSTS requires only half the average and peak power of the Shuttle. CRTs will be replaced with solid-state, flat-panel displays needing only a fraction of present power. Fault-tolerant systems will reduce redundancy and laser gyros should bring a weight reduction of 25% over electromechanical gyros.

The avionic subsystems will improve reliability and lifetime and monitor the flight system better. Both orbiter and booster make extensive use of built-in testing equipment with fault isolation to identify and isolate problems and make system checks. The test equipment requires reliable, intelligent sensors (compatible with the local environment) and advanced computers to store and diagnose the state and health of each subsystem.

Concluding Remarks

Many subsystems selected for the FSTS are already under development or being incorporated as mature hardware into the Space Shuttle or other flight vehicles.

As pointed out, a centralized fluid-management system would simplify operations by reducing the number of fluids that must be handled and eliminating hazardous storable propellants and hydraulic fluid.

Projected avionics would reduce redundancy and weight while increasing reliability and operational capacity. Avionics would exert a powerful influence by helping maintain subsystems readiness.

With advanced technology, it is projected that the overall subsystem weight can be reduced by 30% compared to the current Shuttle, while still improving performance and operational simplicity.

References

[1] Hays, D., "An Assessment of Alternate TPS for the Space Shuttle Orbiter," NASA CR-165790, Feb. 1982.

[2] While, D. M., "Development of Advanced Carbon-Carbon (ACC) Composites," NASA CR-165842-2, Vol. 2, July 1982.

[3] Taylor, A. H., et al., "Structural Concepts for Future Space Transportation Systems Orbiters," AIAA Paper No. 83-0210, Jan. 1983.

[4] "RL-10 A-3-3 Rocket Engine—Design Repeat," NASA CR-10920, Feb. 28, 1966.

[5] Bensen, R. A., Shaffer, A., and Burge, H. L., "Space Shuttle High-Pressure Auxiliary Propulsion Subsystems Definition," NASA CR-115162, March 31, 1971.

[6] "Space Shuttle Environmental Control/Life Support Systems," NASA CR-1981, May 1972.

[7] Aydelott, J. C., and Symons, E. P., "Reduced-Gravity Fluid Management Technology Program," 1980 JANNAF Propulsion Meeting, NASA TM-81540, Monterey, CA, 1980.

[8] Stark, J. A., "Low-*g* Fluid Transfer Technology Study," NASA CR-134911, May 1976.

Structures and Mechanisms—Streamlining for Fuel Economy

Michael F. Card
NASA Langley Research Center

Being evolutionary, a space station will pose new problems in structural design. To make transportation costs as low as possible, station structure must be packaged efficiently in high-density modules taking the minimum of Shuttle volume fractions. Once in orbit, these elements must be deployed and assembled with a high degree of reliability. The inherent mechanical complexity of deployable structures must be traded against risking man in less-complex erection activities. NASA and industrial R&D has shown the practicality of a crew erecting space structures.

In terms of architecture, a master plan must be created which exploits near-term hardware yet can grow into a large complex. Basic issue: how to build into components the means for structural growth without compromising the initial system. One possible solution is an initial configuration becoming part of a stiff closed cell in the final configuration. A stiff configuration is expected to reduce control-system problems.

How much should be invested initially to permit configuration growth? In terms of structural design, growth can be achieved by increasing initial design safety factors, introducing dynamic isolators to prevent loads from reaching the initial components, or preplanning refurbishment of initial structure with stronger elements. Intelligent design tradeoffs require definition of on-orbit loads (including docking and maneuvering) translated into load spectra for estimating fatigue life.

Structural components of the station will likely include tanks for long-duration fuel storage, manned capsules, large (possibly inflatable space-hardened) hangars, and large solar-array appendages. This ensemble will be drag-prone. Indeed, drag in low-Earth orbit makes configuration a key parameter in system tradeoffs. A 150-kW sun-pointing solar array in a 220-n-mile orbit requires over 120,000 lb of fuel for 20-year life *without drag reduction.* A sun-pointing array trimmed to a trailing attitude in shadow needs a lesser amount of fuel—something over 85,000 lb. A fixed trailing array (about twice the size of a sun-pointing array) can cut fuel requirements to about one-fifth (26,000 lb) of the requirement for the untrimmed sun-pointing array. The desire to reduce drag could make cable stiffening attractive.

Because of the evolutionary nature of the space station, an archive of engineering design data will be needed to track configurations. Databases would be alterable at the analyst's or designer's convenience after stringent verification of

Astronautics & Aeronautics, Vol. 21, No. 3, March 1983, pp. 66–68.

proposed new data. The database would be transferable to different computer operating systems, thus avoiding the need to convert computational codes.

Since full-scale "mated" static and dynamic tests will not be feasible as the station grows, designers will have to rely on analysis for qualification of structures. Space-station complexity will pose a need for improvements in thermal, structural, and controls analyses—model simplifications, reduced computation times for heat-transfer calculations, methods of handling structural nonlinearities (e.g., in solar arrays and articulated mechanisms), and improvements in modal synthesis and system identification.

A pathfinder model will be needed as soon as the basic configuration is known; it will be used to identify and resolve test and fundamental analysis issues. A replica-model program parallels full-scale hardware development. Unlike past programs, this model would continue as long as the space station grows. It will be the only source of mated vibration test data before orbital assembly. In parallel with this replica program, component and element full-scale testing would be done. Finally, on-orbit, vibration tests would check ground test data and analyses and measure characteristics to be entered into the archival database.

The long-life requirements of a space station will produce unique problems. Improvements in materials must be made to reduce contamination, slow materials degradation, and extend the life of coatings. Paints show continuous degradation, whereas secondary reflective surfaces appear to stabilize after initial degradation (probably due to early contamination). More data on degradation is needed to characterize a broader range of materials, particularly composites. If deterioration cannot be prevented, simple methods of repair must be devised to restore desirable qualities.

The long-term integrity of structures must be preserved. Structures will need localized protection of critical elements against space debris and micrometeoroids. Redundant, fail-safe structures are likely to be employed with some form of onboard fault detection, especially for tankage. We may need to apply small forces to observe changes in characteristic "signatures" of on-orbit structures to test their integrity.

To minimize on-orbit maintenance, long-life lubrication systems and advanced sealing concepts must be developed. Solid-film lubricants and self-acting seals developed for engine applications need further research to determine their long-time integrity in applications such as docking latches and cryogenic pumps. Advances must be made in the reliability of mechanisms to avoid costly errors in detail design. Computer-aided kinematic analysis of real mechanical systems (including effects of tolerances, interference, and flexibility) should improve design quality.

The spectacular growth of microprocessor technology suggests the development of "smart" mechanisms—programmable mechanisms having sensor feedback and capable of performing multiple functions. Through a combination of local actuators and sensors, they can be programmed to retract and deploy and take positions in any desired configuration. Designers will likewise need reliable moving joints and latches with minimum deadband, low coefficient of expansion, long life, and high functional reliability.

To complement this development, a family of fluid couplings and electrical connectors needs to be developed and integrated with mechanical subsystems for module hookups, satellite maintenance, and refueling operations.

Finally, we will need new mechanisms for docking. Concepts are being investigated to reduce docking loads (''soft'' docking with final velocities less than 0.1 ft/s) and to berth inert payloads.

Technical concerns and a desire to develop experience in space-station operations will likely lead to early experiments using the Shuttle. One logical candidate for a pathfinder experiment will be space construction. Problems in the dynamics and control of large flexible structures should also lead to prestation flight experiments; basic issues are ability to predict and control vibrations on orbit and to perform vibration testing, practical systems for orbit control, and usefulness of adaptive control.

NASA is studying practical approaches for performing low-cost experiments. A reusable pallet, called the Space Technology Experiments Pallet, would minimize integration cost and would allow reuse of expensive electronics. Plans so far would have the pallet used specifically for structures, structural-dynamic, and controls experiments that can be packaged compactly to allow convenient payload manifesting.

The development of a space station requires creative structural engineering to evolve a simple power module into a complex permanent base. The viability of this ''infinite life'' spacecraft will reflect the ingenuity of materials-and-structures engineers in solving durability and maintenance problems.

Structural Dynamics Technology

Needs and Trends in Structural Dynamics

George Morosow
Martin Marietta Aerospace

Michael Dublin
General Dynamics

Eldon E. Kordes
NASA Dryden Research Center

This article covers dynamic analyses of aerospace vehicles, dynamic testing of aerospace vehicles, application of dynamic analyses and testing to nonaerospace fields, education of dynamicists, and related surveys. Anyone using the information here should properly integrate it with the trends of planned hardware (e.g., crew and passengers, cargo, speeds, pressures, temperature, noise levels, schedules, and costs) and related disciplines—materials, control systems, structures, aerodynamics, thermodynamics.

The article does not attempt to identify priorities, since they are clearly identified otherwise with vehicles that will be built. Points made in different sections of the article may seem contradictory; they reflect an honest difference of opinion between committee members on how to handle some controversial problems. In such cases, opposing versions of the needs and trends are listed, and only time will reveal actual future directions.

Dynamic Analyses of Aerospace Vehicles

Items covered in this section include self-induced and forced oscillatory loads, approaches to dynamic modeling and analysis, nonlinear analyses, integrated dynamics design and optimization, and other pertinent topics (i.e., requiring improvements in accuracy of analysis to improve performance and reduce costs).

Self-Induced and Forced Oscillatory Loads

Unsteady aerodynamic forces play an important role in the solution of aeroelastic and aerothermoelastic problems such as flutter, buffet, and active-control-system design. Aerodynamically induced oscillations such as flutter and buffet, of course, pose serious performance restrictions in the design of aerospace vehicles.

Structural failures in vehicle structures, engines, and components due to forced oscillatory loads (e.g., those due to buffet, gust, acoustic fatigue, engine vibration, landing, and armament operations) impose maintenance and cost penalties.

Astronautics & Aeronautics, Vol. 16, No. 7/8, July/Aug. 1978, pp. 90–94.

Experiences reveal the following aerospace-vehicle needs:

- Flutter prevention.
- Prevention of engine failure due to aeroelasticity.
- Flight at high angles of attack in all speed regimes.
- Optimized structure for minimum weight and higher performance.
- Less dependence on testing and less conservatism and, thus, ability to make more-reliable analyses that will lower costs.
- Counteracting the degradation of vehicle performance and maneuverability in the transonic-speed range ($0.9 < M < 1.5$) due to buffet.
- Preventing failure of vehicles due to forced oscillatory loads.

Meeting these needs requires devising new and better theories for unsteady aerodynamic forces (i.e., self-induced oscillatory loads) at low and high angles of attack in all speed regimes (subsonic, transonic, supersonic, and hypersonic) for real-life configurations (external vehicle configurations and internal vehicle configurations), plus verifying the results experimentally. New numerical methods of computational fluid dynamics will help do so. To check new theories, measurements of unsteady pressure will be made on oscillating wings with oscillating control surfaces, wing-body combinations, and rotor-blade sections. The new and sophisticated unsteady aerodynamics prediction techniques will give greater assurance that aeroelastic problems will be avoided, with minimum weight penalty in design, and permit more-efficient flutter- and buffet-suppression devices to be integrated into the vehicle.

Two distinct approaches can be used to solve forced oscillatory load problems:

1) Development and validation of reliability and damage-tolerance characteristics of structures, conventional and composite, that encompass the integration of fatigue-crack propagation analysis based on fracture-mechanics theory, the character of the operational service loading, inspection frequency, and probability of crack-detection optimization, and service-use management.

2) Estimation of service life by conservative criteria for peak fatigue loads and the "early fatigue test and fix" philosophy.

Since these differing philosophies are being used on aerospace vehicles, the service experience at or near the end of their design life will much influence the subsequent use of either. Over the next two decades, the field, thus, needs to compile structural integrity data that:

- Identify the vehicle, its design service life, the method of analysis and test for forced oscillatory loads, and the approval agency (Air Force, Navy, Army, NASA, or FAA).
- Identify the service failure experienced, the vehicle downtime, and the cost of repairs. These must relate specific failure to the known local strain history, its mission, and configuration characteristics.

Whether or not fatigue prediction criteria can be generated that fit the entire vehicle and operational spectrum remains to be seen. (Launch vehicles may have an operational life measured in minutes; Space Shuttle around 200 h, fighter aircraft around 6000 h, and transport aircraft around 50,000 h.)

Along with methods of developing and validating service life, two distinct vibration-analysis trends treat the response of primary and secondary structure and components: 1) random vibration analysis and 2) sinusoidal analysis. Again, both are being used on current aerospace vehicles. These analytical philosophies are closely related to test philosophies (i.e., random testing or sinusoidal testing) and the resulting costs. Over the next two decades, data from service experience will compare these two philosophies of vibration analysis.

Dynamic Modeling and Analysis Approaches

Over the past decade, significant improvements have been made in the technology of dynamic analyses. Principally, these involve structural and liquid models using finite element approaches, description of flexible structures under applied forces, and use of the computer to solve complex technical problems. This has led to a concentration on large computer models.

A prime example is hydroelastic modeling. In the past, hydroelastic analyses primarily attempted definition of longitudinal instabilities. Asymmetric boost configurations such as the Space Shuttle, however, required models not restricted to the axisymmetric. They require consideration of liquid transverse sloshing, tank bulging and bending, and longitudinal oscillations and, thus, prompt the use of computer programs employing many-degree-of-freedom models, with little if any thought to the magnitude of the overall design cost. Large finite element models can easily promote excessive use of the computer. The cost-effectiveness of such practice may be questioned.

To be more rational and economical, the future trend will need to speed existing computer programs, perhaps through frequency- rather than time-domain analysis, to rescope design techniques to provide more realistic or simpler design criteria and specifications, and to build a better understanding of the dynamic load/response phenomenon, in the following ways:

- Dynamic loads are now treated as equivalent static loads, but for very short times many structures can tolerate transient loads exceeding static strength. If the load/failure phenomena were better understood, this effect might permit more efficient designs.
- The field urgently needs efficient means of mathematically characterizing complex, fully elastic structures to speed up assembly, set-up, and completion times. Additionally, ways of formulating large-scale models as subsets of equations will clarify the work of analysts.
- Significant cost savings can be achieved by determining rational equivalence of sine-wave excitation to simplify response and loads analysis; by rationalizing, simplifying, and automating mathematical modeling of structures; and by the avoidance of overgeneralized computer programs. Along this line, a new approach to handle aircraft gust penetration, using a discrete approach (vs the spectral approach) with statistical evaluation of the time-varying response coefficients, is being investigated here and in Europe.
- Random response of structures with large-amplitude deformation, nonlinear characteristics, and methods to handle structural response under nonstationary random excitation—all need additional exploration and experimental verification.
- Analytical methods need improvement for statistical modeling and for risk assessment of configurations with high structural dynamic response as well as sensitivity to production tolerances.
- Use of new-generation computers should be investigated; for instance, vector hardware routines applied to many-degree-of-freedom structural problems can significantly reduce cost.

Nonlinear Analyses

Linear analysis solves engineering problems with some degree of success, but accurate analysis of most engineering problems must treat nonlinear characteristics. The field needs a better attack on nonlinear problems, especially in dynamics.

The engineering systems in question are of two types. One, the inherently nonlinear, typically involves large deflection structures (e.g., solar arrays), on-off control systems, truss-membrane-joint free-play structure, and spring systems with variable-restraint or bottom-out conditions. Such structures frequently encounter nonlinear problems such as buckling, material inelasticity, crack propagation, creep, and panel flutter. The other deliberately uses nonlinear devices to obtain a beneficial effect such as improved performance or accuracy or smaller geometry. This fruitful area of work needs faster development. Examples of systems benefiting from such devices include nonlinear shock absorbers to reduce the size of the absorber and nonlinear components in automatic control systems to improve response time.

Thus, the future trend will be to increase emphasis on both analytical and experimental techniques to analyze nonlinear dynamics.

Integrated Dynamics Design and Optimization

For a high-performance vehicle, the dynamics analysis must be integrated early in design. This entails consideration of dynamic loads and the coupling of aeroelastic response with control, material-structural design, and performance requirements. In the application of high-stiffness, low-weight advanced filamentary composite materials, the stiffness distribution may be tailored to achieve a desirable aeroelastic deformation while at the same time providing adequate strength and stiffness for prevention of flutter and to carry design loads. Stiffness distributions for aircraft-structure composites are usually orthotropic with several stiffness characteristics. Favorable aeroelastic deformations could improve performance by increasing lift in maneuvers and reducing drag by allowing lower angle of attack in sustained maneuvers.

With the development of new analytical models, especially in unsteady aerodynamics in the transonic-supersonic regimes, new-generation computers can be exploited to define aeroelastic response of complete multiwing-body vehicles. The analysis of aerodynamic interactions of fuselage, wing, tail, engines, and tanks will be possible as analysts develop the appropriate theories.

Additionally, the emphasis being given active control systems for suppressing aeroelastic response generates a need for accurate control-surface aerodynamics. Comprehensive dynamics requirements combined with other requirements (such as performance and control design) must be integrated early in the design phase to assure meeting design goals at minimum cost.

The field, therefore, needs to develop optimization techniques that consider all dynamic factors, such as environment, mission constraints, control and aeroelastic interactions, and thermal loads.

Additional Dynamic-Analysis Topics

Additional dynamic-analysis topics require improvement in accuracy of analysis or test data to improve performance and/or reduce costs—in particular:

1) **Aeroelasticity**
   a) Variable-chamber devices.
   b) High-speed thermal environment.
   c) Rotorcraft.
   d) Empirical vibration modes and frequencies.
2) **Loads**
   a) Applied-load time histories (i.e., for prediction of shock, impact, vibration, and acoustic environments and resulting transient responses).
   b) Cavity-flow dynamic loads.
   c) Vehicle vibrating responses to pressure fields.
   d) Buffet and boost-flight dynamic loads under atmospheric turbulence.
   e) Vehicle crashworthiness.
   f) Short-duration transient loads.
   g) Acoustic dynamic loads.
   h) VTOL loads.
   i) Quasi-steady air load distribution on tactical missiles.
   j) Determination of bounds rather than costly computer simulation of loads.
   k) Methods to establish degree of correlation between analysis and test results.
3) **Isolation**
   a) Optimize active and passive isolation techniques (i.e., weight, space, cost, and power requirements on active systems).
   b) Interaction of vehicle and acoustic environments to integrate and optimize isolation and suppression devices with vehicle structure.
   c) Attitude-control/flexible-spacecraft interaction.
   d) Aircraft-control-system/airframe interaction.

Future trends will be aimed at improving the accuracy of such analyses and obtaining the required test data.

Dynamic Testing of Aerospace Vehicles

This section covers integrated test philosophy, test facilities, and other testing topics (i.e., requiring improvement in accuracy and techniques to improve performance and reduce costs).

Integrated Test Philosophy

Everyone wants to reduce the high cost of dynamic testing; for the past decade, we lacked the coordination (project-to-project, agency-to-agency, or company-to-company) necessary to simplify, minimize, and otherwise improve efficiency of dynamic testing.

The design, development, and testing of complex structural systems have typically been based on analysis and testing of full-scale hardware. Scale-model testing has been restricted usually to relatively simple structural concepts that lend themselves to accurate modeling—such as low-frequency modal surveys of Saturn tanks and Shuttle and Nimbus solar panels, specialized tests pertinent to sloshing, and fatigue tests of simple structures. It should be noted that the complex aeroelastic models used for flutter studies of aerospace vehicles will usually accurately represent many dynamic modes, not just low-frequency modes. The development of this technology into qualification and development testing of complete and complex structural systems should allow large savings in time and cost during development of advanced aerospace concepts, as well as readily allow testing of combined dynamic and static loads on large structural systems.

The trend should, therefore, be to develop an integrated analysis and test (all-phases) program that will lead to minimum risk, minimum weight, and minimum cost. Particular emphasis will be placed on the following:

- Design criteria to cover future missions and extended service life.
- Parametric analytical studies to identify marginal conditions and potential troubles.
- Early component testing to determine potential problem areas and fixes.
- Accelerated service testing and incorporation of results of such tests in new design criteria, if warranted.
- Inspection, maintenance, and component-replacement schedules.
- Integration of dynamics tests with tests made for other disciplines.
- Reduction of nonflight or prototype hardware and redundancy of tests and methods for rapidly and efficiently reducing results of test data into directly usable form.

Furthermore, because of the large expense invested in unused hardware, the future trend will be to evaluate the test philosophy and develop a rational engineering methodology minimizing test requirements and allowing use of the prototype test hardware in flight tests with high confidence of success.

Test Facilities

To minimize costs, a need exists to examine existing dynamic test facilities from the standpoint of automating its equipment, data reduction, and presentation of results. Test data in usable form are needed promptly. The greatest need for this occurs in testing at high temperatures and testing for dynamic loads. Large-scale, remotely piloted, free-flight modules may provide the needed capabilities at low overall cost.

Emphasis must also be placed on improving transonic wind tunnels used in studying flutter and other dynamic aeroelastic phenomena. Improvement, primarily in the form of increased density of the test medium, will become increasingly critical as active controls and advanced composite structure begin to enter designs.

In test facilities we should, therefore, expect to see full automation—from facility operations to usable end-product test data.

Additional Testing Topics

Other testing areas requiring improvement in accuracy and technique to improve performance and reduce costs include the following:

- Improvement in dynamic testing through minicomputer technology.
- Use of random-vibration testing.
- Development of realistic shock simulation and testing.
- Mechanical impedance test/criteria techniques.
- Rotorcraft dynamic tests, including flight-test procedures and techniques for measuring rotating frequencies and mode shapes.
- Subcritical flutter-testing techniques.

• Accumulation of data on life assurance and vibration testing to determine which analytical and test philosophies are most economical.

• Dynamic properties of in situ soils for use in constitutive equations of motion for studying ground-penetration events.

Application of Dynamic Analyses and Testing to Nonaerospace Fields

Dynamic analysis methods, testing philosophy, and techniques can be readily applied to nonaerospace fields. Typical applications include the following:

1) **Ground transportation**
   - **a)** Investigating stability at high speeds ($v > 200$ kn) for tracked vehicles (includes interaction with track design).
   - **b)** Structural optimization for minimum weight.
   - **c)** Improving ride qualities.
   - **d)** Crashworthiness analysis.

2) **Water transportation**
   - **a)** Investigating stability at high speeds ($v > 50$ kn).
   - **b)** Investigating transient loads.
   - **c)** Reduction of noise levels.

3) **Medicine**
   - **a)** Fourier analysis in cardiac and brain-wave diagnosis and its effect on body/organ model response.
   - **b)** Development of devices to aid handicapped people.
   - **c)** Analytical models of human structure and closed-loop response.

4) **Civil engineering**
   - **a)** Optimization of civil engineering designs considering Earth models, composite material utilization, and environmental hazards.
   - **b)** Detailed atmospheric models for application to pollution investigations.
   - **c)** Aeroelasticity in turbomachinery to account for spin body dynamics affecting elastic behavior.
   - **d)** Effects of wind loads on buildings and structures.

5) **Nuclear powerplants**
   - **a)** Flow in pipes.
   - **b)** Impacting objects.
   - **c)** Seismic accelerations.
   - **d)** Thermal-stress problems.

The trend should see application of aerospace technology, including dynamics, to these and other nonaerospace fields.

Education of Dynamicists

At one end of the spectrum, in the future, dynamicists will need to make extensive use of literature and technology produced outside of the United States to reduce costs of R&D here.

At the other end, grade and high schools should place more emphasis on improving the comprehension of science and engineering by teaching more and better problems in dynamics. The country also needs more counselors with scientific and engineering training, or at least awareness, to direct more students into our field. At the college level, emphasis on dynamics should be increased and more real-life problems treated, with specific examples in both aerospace and other fields. The future, in short, should emphasize counseling in engineering at all levels.

Related Surveys

In using this article the reader will also want to draw upon the following related surveys on dynamic needs and trends:

1) Barton, M. V., et al., ''Important Research Problems in Missile and Spacecraft Structural Dynamics,'' NASA TN D-1296, May 1962.

2) *Proceedings of the DOD Structures Technology Conference,* Battelle Memorial Laboratories, Columbus, OH, April 1974.

3) Hedrick, I. G., ''Future Trends of Structures and Materials for Spacecraft, Aircraft, and High Speed Surface Vehicles,'' AIAA 7th Annual Meeting in Houston, TX, Oct. 19, 1970. Available from Grumman Aerospace.

4) ''Nondestructive Evaluation,'' DOD Materials Advisory Board, Division of Engineering, Rept. MAS-252, Dec. 1968.

Background

At a meeting January 9, 1973, members of the AIAA Structural Dynamics Committee considered the desirability of preparing a concise report on ''Needs and Trends in Structural Dynamics.'' In later meetings, they decided to prepare such a report, as useful to anyone in industry, government, and the academic world who plans or executes structural-dynamics work, whatever its character—research, design, development, or service operation of hardware. This article represents the result of iterative reviews of memoranda and comments subsequently prepared by the committee. It summarizes the major needs and trends in structural dynamics as seen by the committee but does not attempt to catalog all structural-dynamics work involved in research and production of specific vehicles.

Structural Dynamics for New Launch Vehicles

Joyce Neighbors
Martin Marietta Manned Space Systems

Robert S. Ryan
NASA Marshall Space Flight Center

Assured access to space depends on launch vehicles that can reliably place payloads into Earth orbit within the launch window constraints of widely varying mission requirements. Space Station Freedom, Mission to Planet Earth, science observatories, communications platforms, Mission from Planet Earth, and defense applications are some of the activities that will dictate these requirements. Such projects also involve a broad range of orbits, including low Earth, geosynchronous, sun synchronous, lunar, and planetary.

Launch vehicles in the current inventory include the Delta, Atlas, Titan, and Space Shuttle. Both the aerospace industry and government labs, under NASA and Air Force leadership, are studying the next generation of launch vehicles to determine which design parameters will best assure reliable, operable launch capabilities with the lowest feasible development and operational costs.

Key considerations in this assessment are the dynamics problems that can arise during the few minutes of transition from the static configuration on the launch pad to the attainment of orbital velocity. The history of spaceflight is replete with interesting design solutions that offset the potential safety hazards of this period. These problems can appear during launch because of new and sometimes unanticipated issues in the design, analysis, and/or testing of the dynamic loads. Difficulties have generally arisen because of unusual or unforeseen coupling between the structure and other system features such as fluids, thermal, aerodynamic, and rotary dynamics and thermal and control phenomena.

An early example of such coupling involved a liquid-propelled Jupiter missile that broke up when the sloshing of the propellant coupled through the control system, driving the vehicle unstably during maximum dynamic pressure. Engineers then devised an effective, quick remedy known as the "beer can fix" (Fig. 1). It consisted of long, perforated cylinders with flotation spheres covering the propellant surface. Analytical characterization of the sloshing problem led to the incorporation of ring baffles, which also added to the tank's structural stiffness (Fig. 2). These are now a standard feature of liquid propellant tanks.

Pogo was another early example of coupling between the structure and the propulsion system (and, to a lesser extent, the payload dynamics). In pogo, the longitudinal structural vibration is excited by combustion in the propulsion system, setting up a pressure wave in the propellant. The propellant lines amplify

Aerospace America, Vol. 30, No. 9, Sept. 1992, pp. 26–29.

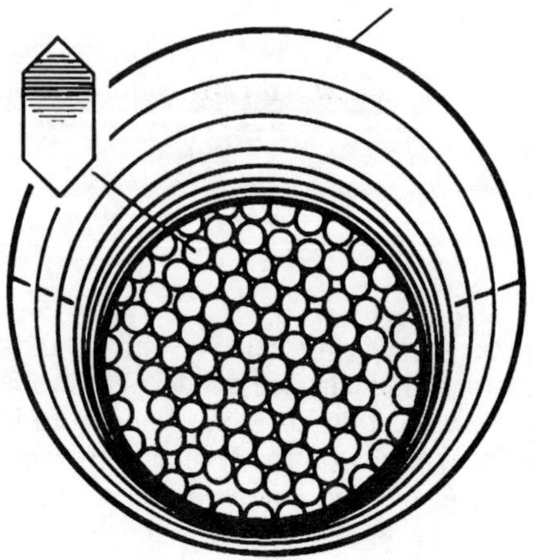

Fig. 1 The "beer can fix" involved the use of flotation spheres to attenuate propellant motion.

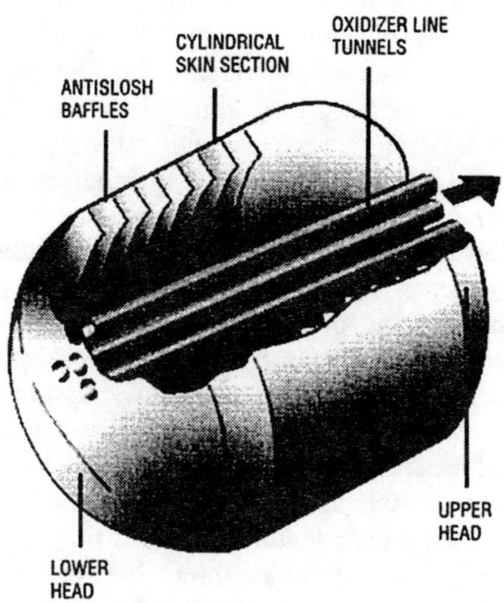

Fig. 2 Ring baffles, now a standard feature on liquid propellant tanks, control sloshing and add structural stiffness.

this pressure wave and transfer it to the propulsion system; this results in a thrust oscillation, which further drives the longitudinal vibration of the structure. The pogo suppression system for early Saturn vehicles involved adding an accumulator in the fuel system to detune and desensitize it.

In both the Shuttle and Titan IV programs, the use of water injection is effectively mitigating the problems of acoustics environment and ignition overpressure. The latter problem occurs at ignition of the solids where the air is displaced in the flame bucket by the high momentum exhaust of the solid motors. The overpressure is a function of thrust-rise rate, thrust magnitude, thermal conditions, and density. The phenomenon can lead to difficulties such as large elevon deflection and high accelerometer readings near the overpressure. Both Shuttle and Titan IV employ a water curtain suppression system for solving this problem.

Another measure for reducing structural dynamic effects at maximum dynamic pressure during launches is pitch and yaw biasing of the trajectory. This technique has a long history dating back to launchers such as the Saturn V. For that vehicle, the predicted probability of launching Skylab on a given March day was increased from 65 to 98% through the use of wind biasing.

Yet another means of reducing dynamics problems is the use of autopilot design for achieving load relief. This is accomplished through reducing the angle of attack by using accelerometers in the control loop. In addition, a near-real-time flight computer provides updated meteorological data for pitch and yaw biasing. At maximum dynamic pressure, in its most demanding configuration, Titan IV is designed for an angle of attack of 1 deg or less; Titan 34D Transtage is designed for 2.5 deg; and Titan III E for 1.8 deg.

In-flight measured loads on the Space Shuttle's orbiter wing were much higher than predicted and led to trade studies between performance, loads, and launch probability. The studies found that an optimum balance among these effects occurred at an angle of attack of –6 deg on the orbiter wing. Although the external tank was not designed for this angle of attack, analysis showed that it could be operated at this condition, which pushed its performance to the edge of the envelope.

Interesting liftoff transient loads occur in the Shuttle because of large asymmetrical coupling between the static and dynamic configurations. As the main engine thrust on the orbiter ramps up to full power, the vehicle is stretched and bent over because its attachments do not permit it to deform freely. At approximately 6.5 s, the solid rocket boosters ignite, timed to minimize the additional elastic energy stored in the Shuttle. These twang loads are complicated by variations in thrust alignment and balance between the solid motors, ground winds, and structural uncertainties. Sensitivity variations to these twang loads have demonstrated potentially serious effects on payload loads and require complex analyses.

A current example of design solutions comes from the Titan IV, a heavy-lift expendable vehicle that can carry 10,000 lb to geostationary orbit. Its structural design must enable 85% launch availability during the worst quarter of the year, when the aerodynamic environment is most demanding. The Titan IV has a 10-ft-diam core vehicle; the payload fairing is up to 86 ft long and measures 16.7 ft in diameter. Thus, it can accommodate the large tank of the Centaur G, which is 40 ft long and 15 ft in diameter.

To control Centaur/payload deflection, the structural design has a forward thrust bearing and features load-sharing between the Centaur and the payload fairing. The inner Centaur load branch carries about 40% of the load, and the payload outer branch carries about 60% of the air load. The deflection of the Centaur and payload within the fairing is, thus, controlled to a minimum clearance of 1 in. That figure is based on a design with a combined payload fairing, forward bearing reactor, and Centaur stiffness of 63,580 lb/in. Control is achieved by using the stiffer of the two boattail designs and a stiffness of 2610 lb/in. in the forward bearing reactor design.

Previous and existing vehicles provide a rich experience base for the design of a new launcher. However, unanticipated problems will likely arise and challenge the structural dynamics experts in new and unusual ways. Both government labs and industry are now studying the new National Launch System (NLS), whose central requirement for achieving lower costs and improving operability is robustness. Whereas initial studies of future launch vehicles focused on maximum use of Space Shuttle hardware, the current configuration involves replacing the orbiter with a cargo carrier and adding Shuttle main engines to the external tank. This design evolves to an in-line configuration during the first decade of the next century.

Simultaneous studies have focused on an Advanced Launch System (ALS) family, whose clean-sheet design does not assume use of any currently available hardware. Instead it would use the Space Transportation Main Engine (STME), which is now in preliminary definition. The configuration currently favored is a blend of the ALS and Shuttle-C called the Derivative family. Its in-line configuration uses the external tank as a core and includes the new STME, some features of the ALS, and Shuttle solid boosters. The architecture of the Derivative is the most cost-effective with respect to up-front funding but can evolve to higher capabilities as requirements dictate.

To achieve a robust structural design, the NLS studies have established very conservative criteria for loads calculation. For prelaunch, the unfueled vehicle is assumed to be on the pad with a 99% omnidirectional wind whose peak speed is 74.5 kn. For the prelaunch flight-readiness firing, assumptions include a vehicle fully fueled for 24 h and a 99% omnidirectional wind whose peak speed is 47 kn; the aerodynamic loads are to be multiplied by 1.5 to account for von Kármán vortex shedding effects.

Transient analyses conducted for the flight-readiness firing assume an engine out with shutdown-rebound; the dynamic loads so obtained are multiplied by an uncertainty factor of 1.5. For liftoff, assumptions include a 2-sigma occurrence of parameters such as thrust level, thrust mismatch, ignition timing, and ignition overpressure forces in a worst case on worst case combination. An engine-out condition is assumed, as is a 95% omnidirectional ground wind that lasts an hour and has a peak speed of 34 kn.

The dynamic transient analyses are to be conducted with multipoint hold-down release, and the dynamic loads are multiplied by an uncertainty factor of 1.5. At maximum dynamic pressure, nominal thrust and mass are assumed, with no load relief, and assumptions are very conservative. In the current cycle of analysis, assumptions include a 95% yearly scalar wind envelope that is omnidirectional. This includes 99% wind shears and 7.65-m/s^2 gusts, added air loads at maximum dynamic pressure to account for aerodynamic uncertainty; engine-out is also assumed. Finally, at maximum acceleration, the heavy-lift launch vehicle assumes a 4.0-*g* acceleration vehicle, and the stage-and-a-half assumes a 4.5-*g* vehicle.

If, indeed, a cost-effective vehicle can be designed to these conservative assumptions, it should be highly operable and reliable, as should any new launch vehicle. The structural dynamics lessons learned thus far are enabling the selection of design features that will maximize robustness in the NLS. Conservative loads, high design margins of safety, increased stability margins, and increased stiffness requirements are areas that hold potentially large payoffs.

A typical attempt to achieve robustness involves creating a design in which the first bending mode will have a high enough frequency to allow decoupling between the autopilot design and the flexible body dynamics. This would simplify the design and reduce the flight-by-flight support cost. These structural design studies will likely yield a broad understanding of the tradeoffs between weight-to-orbit, operability, reliability, and cost.

Noise, Vibration, and Aircraft Structures

Arthur A. Regier
Langley Aeronautical Laboratory, NACA

Introduction

Ever since multiengined aircraft have been in general use, there has been the problem of making the structural areas near the propellers and jets sufficiently strong to avoid fatigue due to noise. The much larger powerplants becoming available have greatly aggravated the problem of noise fatigue to the extent that noise damage is becoming an important consideration in the determination of aircraft configurations.

A rough idea of magnitudes of the sound pressures that are of current interest is illustrated in Fig. 1. The range of pressures is expressed as decibels and as pounds per square foot. Gun blast pressures are included in this figure because they represent experience with pressures of a different order of magnitude, but one which is being approached by some of our modern propulsive systems. Pressure fluctuations in the order of 180 dB or 0.25 atmosphere are produced near present propulsive systems and on aerodynamic surfaces subjected to shock buffeting.

Most aircraft skins and secondary structures will show fatigue damage after many cycles, when subjected to pressures ranging from 140 to 150 dB and permanent set or rupture after short exposure to 180 dB.

Input-Output Relations

In this paper the inputs considered are the noise pressure loadings; the outputs are the stresses in the structure. Noise inputs may be divided into three general classes: 1) discrete frequency—for example, propellers; 2) random frequency, as from jets; and 3) impulsive, as from gun blast or starting of rockets.

Discrete Frequency Input

An example of the significant input-output relations for a discrete frequency input is shown in Fig. 2. Most discrete frequency sources do not produce a pure tone of single frequency, but one rich in harmonics. As illustrated at the top of the figure, the pressure amplitude of each frequency component is indicated by the height of the bar at that frequency. The admittance of the structure is illustrated in the second graph. This is often referred to as the response curve; the peaks

Aeronautical Engineering Review, Vol. 15, No. 8, Aug. 1956, pp. 56–60.

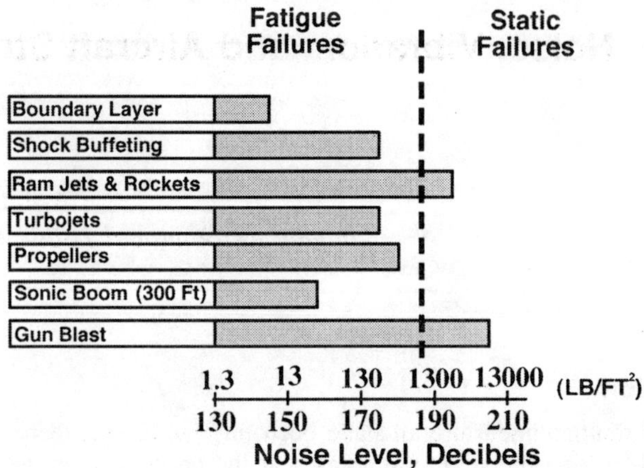

Fig. 1 Levels of noise loading on aircraft structures.

in the curve correspond to the various resonances in the system. The height of the admittance curve at any frequency relative to its value at zero frequency is the dynamic amplification. This dynamic amplification may go as high as 50 or more. The admittance may be considered as the stress at a particular point in the structure per unit pressure applied at the input point or area. The output is the product of input and the admittance. It, too, is a line spectrum for, if the system is linear, it can vibrate only at the frequencies at which it is forced by the input. The magnitude of the stress at a given frequency depends on both the input and the admittance, and from a practical standpoint much can be done to minimize stresses by changing both the input and the admittance.

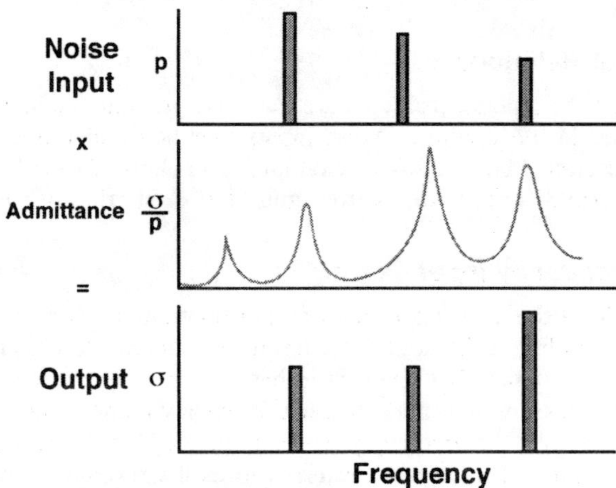

Fig. 2 Input-output relations for systems subjected to discrete frequency loading.

Random Inputs

In considering the response to random frequency inputs, one deals entirely in probabilities. (A classic problem along these lines was worked out by Lord Rayleigh.[1])

Recently there has been considerable activity in the application of statistical properties of random functions to such problems as gust response, buffeting, and response to random noise.[2,3] This type of analysis is usually referred to as generalized harmonic analysis, and some of its concepts are illustrated in Fig. 3. There is considerable similarity between this figure and the preceding one. One of the differences is that now one deals in squares of the quantities—that is, one might consider that the problem is cast in terms of energy rather than in terms of force or pressure. The input expressed as pressure squared per unit band width varies only gradually with frequency, having all frequencies present. The output also has all frequencies present. The value of the mean-square stress is given by the area under the output curve.

Contrary to the case of discrete inputs, for random inputs, small changes in the shape of the input spectrum or in the admittance curves have only secondary effect on the area under the output curve—the mean-square stress. Damping reduces the stresses—i.e., directly as the square root of the damping. However, the problem of obtaining damping materials and devices to withstand the high noise levels encountered has not been satisfactorily solved.

As was discussed with respect to Fig. 2, Fig. 3 might be considered the stress at a point due to a random load at another point. In order to find the total stress at this point, one should know the loads at other points of the structure, the pertinent admittance curves, the phases of the admittance, and the space-time correlations of the input.

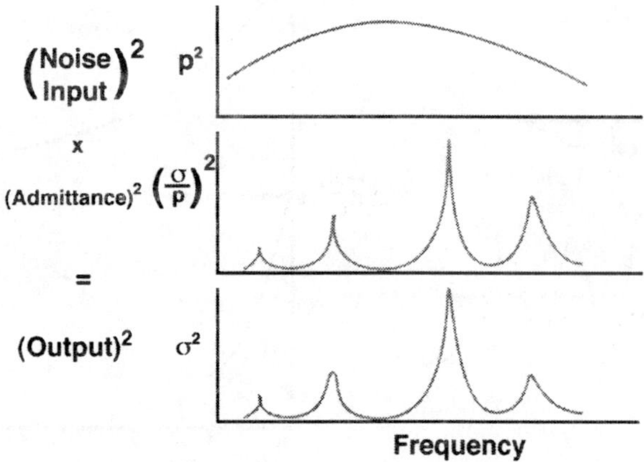

Fig. 3 Input-output relations for systems subjected to random frequency loading.

Impulsive or Blast Loading

Impulsive loadings are applied to aircraft structures by gun blasts, rocket ignition, shock waves from other aircraft such as sonic booms, and blasts from bombs or warheads.[4,5] Some of the characteristics of impulsive loading and the response of the structure are shown in Fig. 4. This information is taken from Reference 4, which gives a comprehensive treatment on structural damage due to gun blasts. The upper left-hand part of the figure illustrates a typical time history of a blast: a positive phase having a peak pressure by P_0 and then a negative phase of longer duration but lower peak value. The lower left illustrates a smoothed curve of the strain at a point in the panel. This shows that the response of the panel is a damped oscillatory transient of a period usually corresponding to a fundamental frequency of the panel.

The dynamic amplification of the panel is illustrated on the right side of the figure as function of the ratio T/τ. The dynamic amplification is defined as the ratio of the maximum vibratory stress at the first positive peak (A) to the stress the plate would have if the peak positive blast pressure P_0 were applied as a static pressure. The lower curve refers to the negative peak stress relative to the same pressure P_0.

Near-Field Noise

The noise of most aircraft sources is usually different in frequency content, and varies differently with speed, for a point near the source as compared with a point a few wavelengths from the source. This is principally due to two things: 1) aircraft sources are usually distributed, and 2) the various pressure components decrease with different power of the distance.

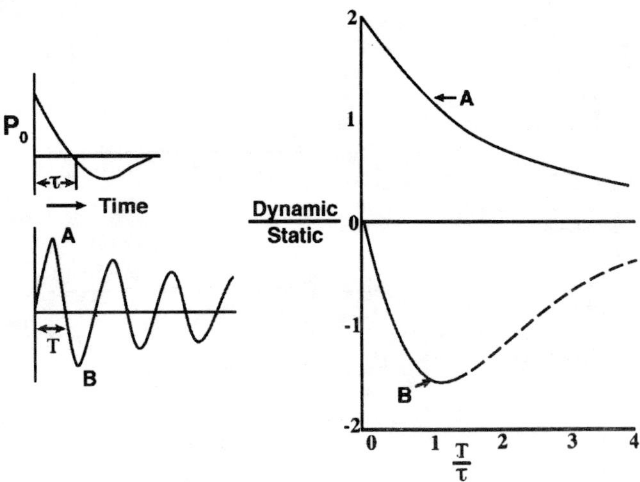

Fig. 4 Input-output relations for systems subjected to impulse loading.

To illustrate the first point, the sounds one hears in the percussion section of an orchestra are different from those one hears in the violin section, and they are still different from the sounds one hears at a choice seat in the auditorium. In a similar manner, the noise very near a jet nozzle is different from that very near the jet farther downstream where large-scale turbulent mixing occurs—and it is still different some distance away where one obtains the integrated effect of all regions of the jet.[6]

The second reason for differences between near and far fields lies in the fact that a large portion of the pressures at the source are balanced by the inertia of the fluid and are not radiated as compression waves in the fluid. Thus, for a dipole source (source-sink), much of the pressure field varies as the inverse square of the distance, whereas the radiation pressures vary as the inverse first power of the distance from the source. For a double dipole or quadrupole, the pressure varies as inverse distance cubed and squared, as well as the inverse distance for the radiation pressure.[7]

Propellers

Some practical implications of these factors are illustrated in Fig. 5 which compares the pressure variation with propeller-tip speed at distances in the order of 1/10 diameter from the propeller tip with those one might expect a few diameters or more from the propeller.[8] A word of caution needs to be said with respect to the ordinates in this two-part figure. The near-field pressures are plotted to a different scale and are much higher than the far-field pressures. The symbol m is the order of the harmonic—1 for the fundamental, 2 for the second harmonic, etc. B is number of blades. Thus, mB 2, 4, 6 may be construed as the fundamental frequency of a two-, four-, or six-blade propeller or the first three harmonics of a two-blade propeller. The main point of interest is that, for constant power, at several diameters from the propellers (far field), the pressures rise rapidly with increasing tip speed, whereas very near the propeller, the pressures may increase or decrease with tip speed depending on the value of mB.

Jets

Some near- and far-field comparisons for the noise from model jets[9] are shown in Fig. 6. Here is plotted the exponent of the jet velocity (which relates the noise energy to the velocity) as function of the distance to the nozzle. The direction of the field point is 90 deg with respect to the jet axis. Far from the nozzle, the noise energy varies as the eighth power of the velocity, as predicted by theory.[10] However, near the nozzle, the noise energy (specifically $p^2/\rho c$) varies only as the third power of velocity. (See also Reference 11.) Thus, the noise pressures near a jet will vary less drastically with velocity than they do at some distance away from the jet.

Further discussion of the near-field pressures from jets and the turbulence in jets may be found in References 6 and 9–18. Currently there is considerable interest in the space-time correlation for near-field jet noise, and work along these lines is proceeding at NACA.

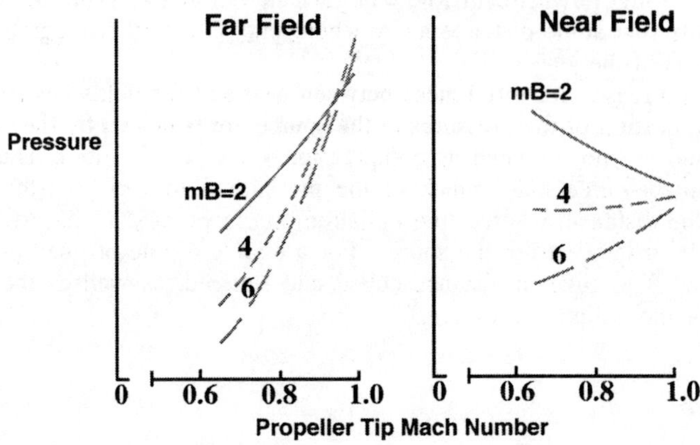

Fig. 5 Comparison of near- and far-field propeller noise.

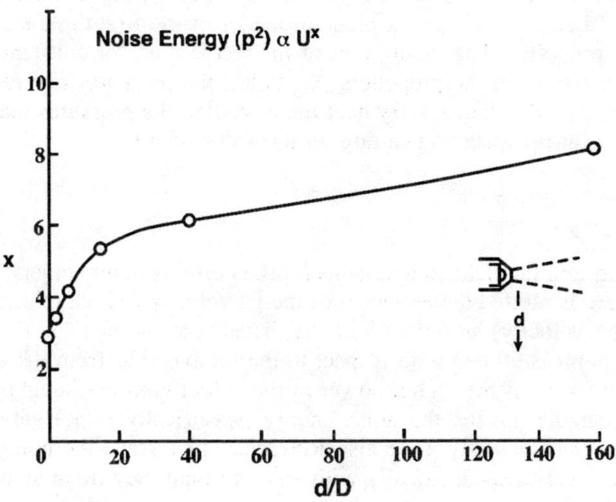

Fig. 6 Variation of the exponent relating sound energy to jet velocity as function of distance from the jet.

Aerodynamic Noise and Buffeting

There are reasons to believe that, for jet aircraft at high-flight speeds, the principal source of structural excitation will be that due to aerodynamic noise and buffeting. How serious this problem may become at transonic and supersonic speeds has not been determined. Cases of structural failures are known to have occurred in regions of flow separation, such as behind spoilers and wing fuselage junctures.

Some values of the pressure fluctuations in the boundary layer of subsonic aircraft are presented in Reference 19. These values were obtained by correcting sound reading in the aircraft for wall transmissibility.

Vibration Responses

The effect of vibration-mode shape on the response of a structure to a plane wave of sound is illustrated in Fig. 7. It represents the case of a plane wave traveling perpendicular to a beam on simple supports. For the noise to do work on the structure, the pressure must have a component in phase with the velocity of the structure. In the fundamental mode, all parts of the structure are moving in the same phase, and a maximum energy can be absorbed. In the second mode, half of the structure is moving out of phase with respect to the other half, and no energy is absorbed. The plane wave cannot excite this vibration mode. In the third mode, two loops cancel each other, and the structure can absorb only one-third the energy it would if all were moving in phase. In general, one might expect less and less response with increasing order of vibration mode, when the excitation is a plane wave in phase over the area. However, when the noise is not in phase over the structure, one may well have conditions in which maximum stresses occurred in some higher vibration mode.

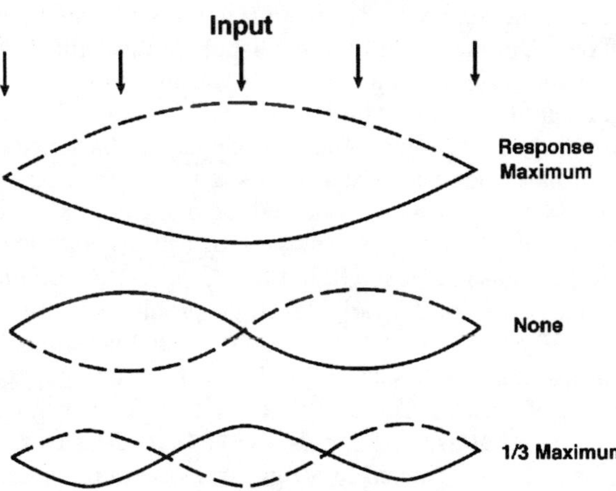

Fig. 7 Schematic diagram illustrating effect of vibration mode shape on response to plane wave noise excitation.

One of the questions that is often raised with regard to the response of aircraft structures is: How important is air damping as compared with structural damping? Some theoretical studies made for various modes of vibration of cylinders are presented in References 20 and 21. The case of panels, pistons, and numerous other shapes is treated in standard acoustic books such as Reference 22. An application of this analysis to aircraft structures shows that, because of three-dimensional effects, the reaction of the air on the structure is mainly as an added mass and that the radiation damping is usually negligible. This is contrary to what one would expect from plane wave acoustics, in which all the air reaction results in radiation damping.

When subjecting a complex aircraft structure—such as a cylindrical fuselage—to random noise, many vibration modes are excited to some extent, and, by making a narrow band frequency analysis of a strain-gage recording, many modes can be identified. So far, we have not been able to predict which modes will produce the highest stresses. It had generally been thought that only the lowest modes were important from a stress standpoint, yet when we extend the frequency range of our recording equipment to higher frequencies, we find more of the larger stresses at the higher frequencies. What part these higher frequencies may play in fatigue has not been determined.

Fatigue

The field of fatigue is an extremely old field of research, yet, in spite of its age, one can talk of result only in terms of probabilities and wide scatter bands. This is true because microscopic variations in a specimen may have a profound effect on its fatigue life. Results of noise fatigue generally follow in the same pattern, and it is not yet clear what the new elements in the problem are.

Although a considerable amount of information undoubtedly exists on noise fatigue and fixes, there has been little published on this subject. Perhaps one of the most systematic studies along these lines was made by The Glenn L. Martin Company with regard to seaplane hull plating fatigue. Noise fatigue also has been studied by Northrop Aircraft, Inc.; gun blast damage is discussed in Reference 4. The Boeing Airplane Company has done extensive work in noise fatigue, and some of the techniques they have pioneered are being used at other laboratories.

From a perusal of these studies, one is struck by the fact that stress raisers have little effect on ultimate strength but a large effect on fatigue.

In order to study some of the variables in the problem of fatigue due to noise, the Langley Laboratory of NACA has recently tested a number of panels with an intense noise of discrete frequency produced by an air chopper. The test set-up is similar to that used by The Boeing Company. An illustration of a fatigue failure obtained on a flat panel (11 by 13 by 0.032 in. 2024T3 Alclad) is shown in Fig. 8. The panel was fastened to a rigid frame with round-head screws. By looking closely at the picture, one can see the fatigue damage in various stages of progress, from the incipient cracks around the rivets to large pieces blown out of the panel. This latter is a characteristic of noise damage if the tests are not terminated soon enough.

Fig. 8 Sample flat-panel failure from siren tests (t = 0.032 in.).

Conclusions

We have presented a brief review of the problem of noise, vibration, and aircraft structures. One finds many complications of such dimension that it is doubtful whether predicting noise fatigue will ever become possible to better than an order of magnitude. From a study of the input-output relations, one can often obtain valuable guidance for obtaining fixes to specific problems. For simple panels, some success at stress prediction in random noise fields has been obtained.

Although some progress has been made, the field has hardly been explored. Much more information is needed on the magnitude, phases, and various correlations of the inputs of near-field noise, particularly for the random sources. Information is needed on the response of structures and the critical stress areas for each mode of response, and, lastly, a great deal of information is needed on the fatigue of skin structure and its supports when subjected to the type of loading imposed by noise.

References

[1] Lord Rayleigh, *Theory of Sound,* Vol. 1, 2nd ed., Article 42a, The Macmillan Company, London, 1929.

[2] Miles, J. W., "On Structural Fatigue Under Random Loading," *Journal of the Aeronautical Sciences,* Vol. 21, No. 11, Nov. 1954, pp. 753–762.

[3] Hess, R. W., Lassiter, L. W., and Hubbard, H. H., "A Study of the Response of Panels to Random Acoustic Excitation," NACA RM L55E15c, July 1955.

[4] Fiszdon, W., Lotn, I., Kay, W., William, D., and Kirby, W. T., "20 mm. Cannon Blast Effects on Stressed-Skin Structures," ARC Technical Repts. R and M 2085 (7195), July 1943.

[5] *The Effects of Atomic Weapons,* Rev., Superintendent of Documents, U.S. Government Printing Office, Washington, DC, Sept. 1950.

[6] Hubbard, H. H., and Lassiter, L. W., "Experimental Study of Jet Noise," *Journal of the Acoustical Society of America,* Vol. 25, No. 3, May 1953, pp. 381–384.

[7] Hubbard, H. H., and Regier, A. A., "The Nature of Aircraft Noise Source," NACA-University Conference on Aerodynamics, Construction, and Propulsion, *Aerodynamics,* Vol. 2, Oct. 1954.

[8] Hubbard, H. H., and Regier, A. A., "Free-Space Oscillating Pressures Near the Tips of Rotating Propellers," NACA Rept. 996, 1950.

[9] Lassiter, L. W., and Hubbard, H. H., "The Near Noise Field of Static Jets and Some Model Studies of Devices for Noise Reduction," NACA TN 3187, July 1954.

[10] Lighthill, M. J., "On Sound Generated Aerodynamically," *Proceedings Royal Society, A,* Vol. 211, 1952, pp. 564–587.

[11] Mawardi, O. K., and Dyer, I., "On Noise of Aerodynamic Origin," *Journal of the Acoustical Society of America,* Vol. 25, No. 3, May 1953, pp. 389–395.

[12] von Gierke, H. E., "Physical Characteristics of Aircraft Noise Sources," *Journal of the Acoustical Society of America*, Vol. 25, No. 3, May 1953, pp. 367–378.

[13] Greatrex, F. B., "Jet Noise," IAS-RAeS Fifth International Aeronautical Conference, June 1955.

[14] Tyler, J. M., and Perry, E. C., "Jet Noise," Trans. SAE, Vol. 63, 1955, pp. 308–320.

[15] Lassiter, L. W., and Hubbard, H. H., "Some Results of Experiments Relating to the Generation of Noise in Jets," *Journal of the Acoustical Society of America,* Vol. 27, No. 3, May 1955, pp. 431–437.

[16] Laurence, J. C., "Intensity, Scale, and Spectra of Turbulence in Mixing Region of Free Subsonic Jet," NACA TN 3561, Sept. 1955.

[17] Veneklasen, P. S., "Noise Characteristics of Pulse-Jet Engines," *Journal of the Acoustical Society of America,* Vol. 25, No. 3, May 1953, pp. 378–380.

[18] Lassiter, L. W., "Noise from Intermittent Jet Engines and Steady-Flow Jet Engines with Rough Burning," NACA TN 2756, 1952.

[19] Rogers, O. R., and Cook, R. F., "Aerodynamic Noise and the Estimation of Noise in Aircraft," WADC TR 52–341, Dec. 1952.

[20] Bleich, H. H., and Baron, M. L., "Vibration of Shells in an Acoustic Medium," *Journal of Applied Mechanics,* Vol. 21, No. 2, June 1954, pp. 167–178.

[21] Junger, M. C., "The Physical Interpretation of the Expression for an Outgoing Wave in Cylindrical Coordinates," *Journal of the Acoustical Society of America,* Vol. 25, No. 1, Jan. 1953, pp. 40–47.

[22] Morse, P. M., *Vibration and Sound,* McGraw–Hill Book Company, Inc., New York, 1948.

DAMVIBS Looks at Rotorcraft Vibration

Raymond G. Kvaternik
NASA Langley Research Center

Excessive vibrations have plagued virtually all new rotorcraft since the first such vehicle went into production in the United States over 40 years ago. Although vibration levels have fallen considerably in production aircraft since that time, the problem continues to occur even in modern rotorcraft designs. With few exceptions, it is not identified and addressed until flight test, and solutions at that stage are usually add-on fixes that adversely impact cost, schedule, and performance.

The helicopter industry generally uses the finite element method of structural analysis to calculate airframe static internal loads and for the usual checks on frequencies. These calculated static loads are routinely used in design for sizing structural members. However, vibration predictions based on this analysis method have not been widely used until recently because they were deemed unreliable as a basis for making design decisions.

For analysts charged with predicting helicopter vibrations, the problems are formidable. The rotor system generates complex periodic aerodynamic and dynamic loads that are transmitted to the airframe through both mechanical and aerodynamic load paths. The largest vibratory forces transmitted to the airframe are usually those that come mechanically through the mounting system. These forces occur at the so-called blade passage frequency, which equals the product of the number of blades and the rotor rotational speed. This frequency is typically in the 10–20-Hz range.

Complicating the structural dynamics problem is the light weight of helicopter airframes. These shell-type structures have multiple large cutouts and support several relatively heavy components. Even with the advanced analysis capabilities of finite element methods, airframe structural designers have until recently achieved only limited success in designing airframes that exhibit adequate vibratory response. A major deficiency has been incomplete understanding of the modeling requirements for vibration analysis of complex helicopter structures.

There has emerged within the helicopter industry a consensus on the need for more effective use of airframe finite element vibration models during the design process for purposes of remedying the vibration problem. Thus, during the late 1970s, industry advisory groups began urging NASA to work with them on improving the predictive capability of such models. As lead center for structures research, NASA-Langley was asked to define a research activity aimed at addressing the industry's needs in this area.

Aerospace America, Vol. 30, No. 9, Sept. 1992, pp. 26–29.

The culmination of this joint effort was the definition of a multiyear, industry-wide program aimed at meeting long-term needs. The program's objective was to establish the technology base required for developing an advanced finite element-based dynamics design analysis capability for vibrations; thus, the program came to be called DAMVIBS, or design analysis methods for vibrations.

The DAMVIBS program began in 1984 with the award of task contracts to the four major helicopter airframe manufacturers—Bell Helicopter Textron, Boeing Helicopters, McDonnell Douglas Helicopter, and the Sikorsky Aircraft Division of United Technologies. Under this program, industry teams have formed finite element models, conducted ground vibration tests, and made extensive test/analysis comparisons of metal and composite airframes.

Unique experimental/analytical studies have been conducted to identify an airframe's "difficult components," or those that require better representation in the finite element model for improved correlation with test results. Industry codes for comprehensive analysis of the vibrations of rotor and airframe as a coupled system have been evaluated. Studies aimed at establishing the potential role of structural optimization in airframe vibrations design work have also begun.

Five government/industry workshops have been held to review and discuss the results of these efforts. Because DAMVIBS is being phased out, the last meeting included a special session devoted to an assessment of the program and its benefits to the industry.

A finite element model was determined to be essential to any design effort aimed at developing a helicopter with low inherent vibrations. It was clearly established that such a model, one suitable for the computation of both static internal loads and vibrations, can be formed early enough in a new helicopter development program to influence the airframe design.

The cost of such a model was shown to be about 5% of the total airframe design effort, of which 4% is already usual for the static model; the vibration model represents the other 1%. Comparing the results of finite element analyses with results from ground vibration tests of both metal and composite airframes has demonstrated an improved capability for predicting vibratory responses up through about 20 Hz. However, agreement was found to be generally unsatisfactory at higher frequencies.

Subsequent "difficult component" studies conducted on airframes have shed new light on the importance of such components in vibratory response at the higher frequencies. These studies show that considerably improved correlation is possible if secondary effects that were historically considered unimportant are taken into account when the finite element model is being formed.

This finding means that for vibration analysis, such models must be substantially more detailed than the usual static model, contrary to previously held beliefs. Studies have revealed that support system dynamics, including the prestiffening effects due to gravity, may have to be included in the finite element models employed in correlations with ground vibration test results.

A comparative evaluation of the industry codes used for comprehensive analysis of coupled rotor-airframe vibrations has shown that these codes are not yet accurate enough to be relied on fully during design. This evaluation also examined formal, nonlinear programming-based optimization techniques. It showed that these could play a major role in airframe vibrations design work if the design

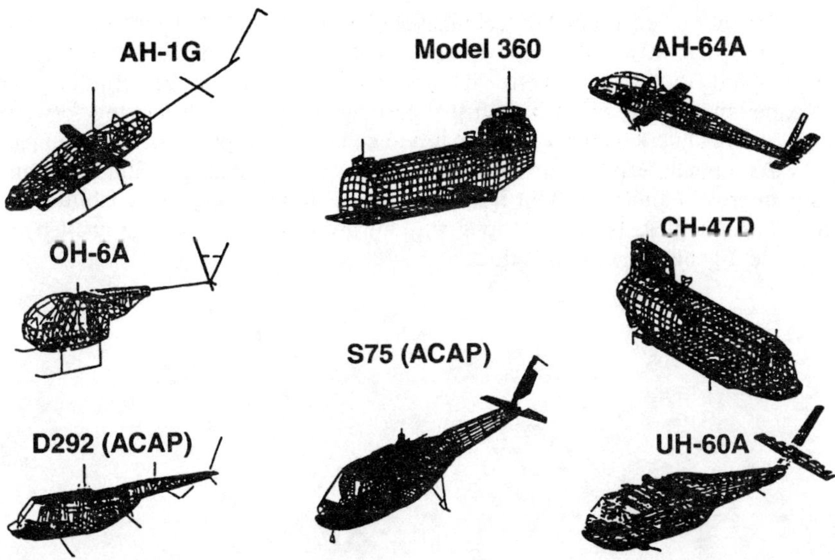

State-of-the-art finite element vibration models of both metal and composite airframes were formed by industry teams in an industry-wide assessment of basic modeling techniques.

models required by optimization algorithms can be made to reflect the nuances of airframe design.

The assessment also identified several key structural dynamics challenges that must be met if industry is to achieve a "jet smooth" ride for helicopters. There is a need to extend the predictive capability of finite element models up through the 25–30-Hz frequency range to encompass the predominant excitation frequencies of new and planned helicopters that have rotors with blade passage frequencies above 20 Hz.

Better and more-practical methods for improving models at the finite element level using ground vibration test data are also needed. Analytical techniques that more realistically account for structural damping are required, as are mathematical models of damping that can readily be incorporated in airframe finite element models.

The predictive capability of current comprehensive codes for analysis of coupled rotor-airframe vibrations needs improving. Also required are simplified rotor mathematical models suited for supporting airframe vibrations design work. There is a need for defining further the role of structural optimization in the airframe-design process and for developing computational procedures tailored for vibrations design work. To augment the use of basic design techniques in reducing vibrations, new, improved methods for actively and passively controlling airframe structural response must be developed.

DAMVIBS has provided both a unique leadership role and a focal point for rotorcraft structural-dynamics research in government, industry, and academia.

The effort has enabled notable technical achievements and changes in industrial design practice, all of which have significantly advanced industry's ability to use and rely on finite element vibration models during design. Building on the experience of DAMVIBS, industry participants have been moving forward aggressively under company sponsorship to enhance their prowess in the subject areas. As a result, expectations are that the industry will emerge with a substantially improved finite element-based dynamics design analysis capability for vibrations, one that should go a long way toward meeting the design challenges of the next generation of rotorcraft.

The Cautious Course to Introducing New SDM Technology into Production Systems

Richard N. Hadcock
Grumman Aerospace Corporation

Introduction of new structures, structural-dynamics, and materials (SDM) technology is becoming long and difficult. It must comply at the minimum with the three preconditions outlined in Fig. 1.

First and foremost, there must be a well-defined need for the new technology. This need may reflect the inability of existing technology to provide the desired level of performance, unacceptably high procurement or maintenance costs, or even basic deficiency in design or material.

Second, benefits must be well defined in terms of performance improvement, cost reduction, and risk. This implies that the new technology has achieved a reasonably high level of maturity and availability and that sufficient hard data are available to perform cost-vs-performance-vs-risk tradeoffs. Introduction of a new technology almost always increases risk. It takes between 10 and 30 years for new SDM technology to reach maturity.

Opportunity represents the third precondition for introducing new technology. The introduction of at least some new technology into existing systems has, in the past, been a very effective way of improving performance while simultaneously advancing the state of the art. Taking advantage of the *interaction* of new technologies during preliminary design, however, produces the greatest benefits. Through interaction in design, performance can be increased or vehicle size and associated procurement and operating costs can be reduced by a significantly greater degree than is possible by introducing new technologies into an existing system.

Through SDM advances the next generation of military aircraft, commercial aircraft, helicopters, and spacecraft should all enjoy the benefits of lower procurement or operating costs, less weight, and, in some cases, aeroelastic tailoring.

Technology needs have changed during the past 50 years, especially for fighter aircraft. Speeds have risen from 234 mph for the Boeing P26A to 1650 mph for the McDonnell Douglas F-15A. This sevenfold increase has resulted in wings with significantly thinner sections to reduce drag at supersonic speeds. The wings on piston-engined fighters of the 1930s and 1940s had thickness-to-chord ratios four times higher than those of current fighters. Such thin wings forced extensive sculpturing of wing covers and substructure to minimize weight. In addition, new materials had to be developed to withstand aerodynamic heating.

Over the past 20 years, service life has increased from a few hundred hours to many thousands of hours and, thus, has helped offset procurement costs. Also, long-life requirements pressed development of better corrosion- and fatigue-

Astronautics & Aeronautics, Vol. 18, No. 3, March 1980, pp. 31–33.

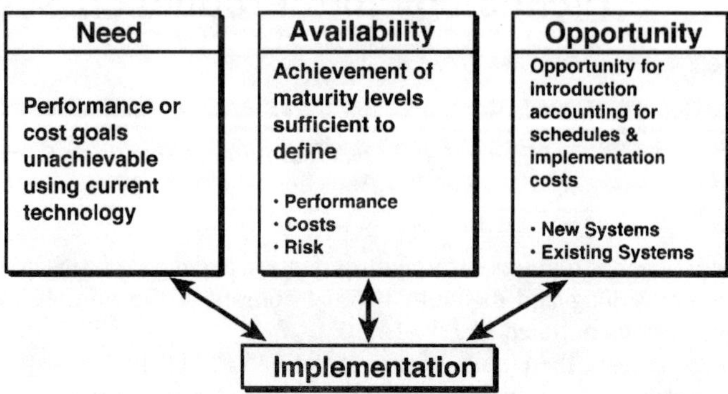

Fig. 1 Preconditions for introduction of new SDM technology.

resistant materials and better methods for fatigue and fracture design and analysis. Figure 2 notes the rise of some SDM technology requirements since the early 1950s.

Since the early 1960s, much effort has been expended on the development of advanced-composite structures and materials. Fairly recently these have finally become a production reality. However, the maturity cycle for full application (projected to be about 50% by airframe weight) will be 25–30 years (Fig. 3).

Meanwhile, we have had many fewer opportunities for introducing new SDM technologies into new systems. Figure 4 shows historical data on new military and commercial aircraft. In terms of the number of prototype first flights per year, a peak of 12 aircraft per year was reached between 1930 and 1940. Since then, the number of first flights has steadily declined. For the remainder of this

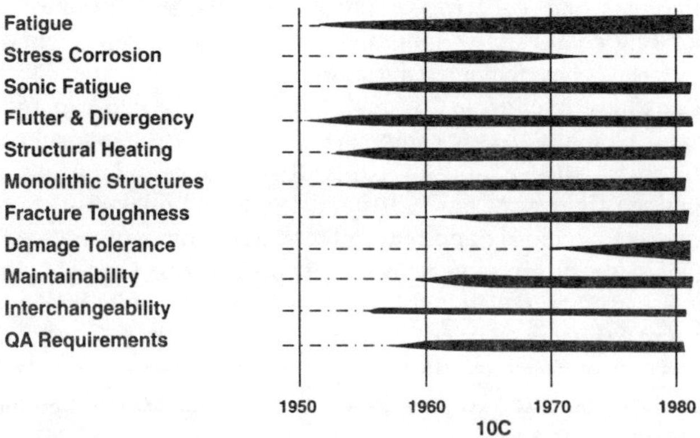

Fig. 2 SDM technology requirements.

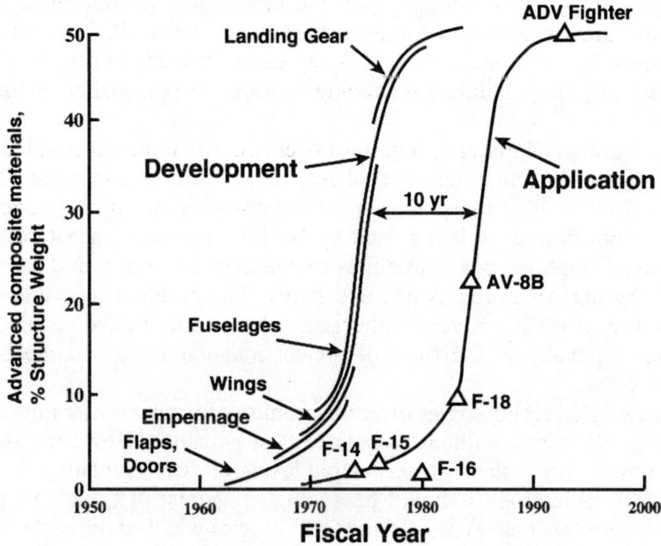

Fig. 3 Time between development and application of advanced composites to aircraft.

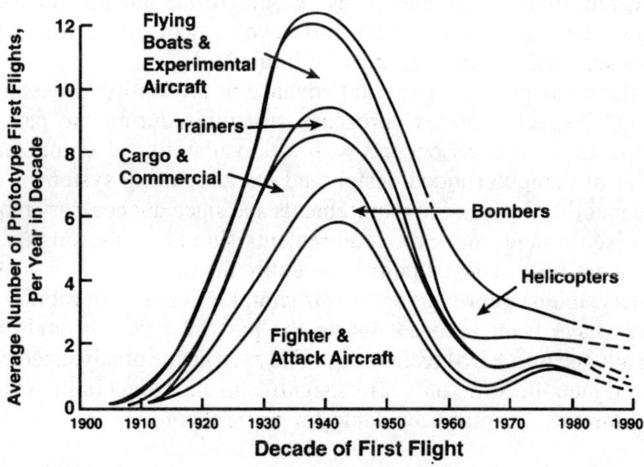

Fig. 4 First flights of prototypes of U.S. military and commercial aircraft.

century, the number will probably not exceed two per year. There will, therefore, be few opportunities for applying new SDM technology to new production aircraft during the next decade. There may be some opportunity for application to prototype or demonstrator aircraft and spacecraft.

Another problem associated with opportunity arises because appropriate new technologies are selected very early in the preliminary design phase of a new military aircraft. Go-ahead to initial operating capability (IOC) takes from six to eight years, but most major technical decisions are made during the *first year.* This means that the candidate technologies must be reasonably mature at go-ahead.

Since technological maturity requires in-service experience in conjunction with a sound design and manufacturing database, the only way to achieve an acceptable level of maturity will be to use these new technologies on existing systems. This approach is not new. It is being used by NASA to generate a database for the application of graphite/epoxy structures to commercial aircraft and has similarly been used by the Air Force, Navy, and Army. The problem with this approach is that the user is often not very enthusiastic about applying the new technology if it means a penalty in the form of greater manufacturing, maintenance, and inspection costs.

Some new SDM technologies offer the promise of further major improvements in structural efficiency without associated cost penalties. However, steps must be taken now to bring them to acceptable levels of development.

New SDM technologies should be evaluated against the needs of projected systems. A very few new materials and technologies which show real promise for improving performance and reducing costs should be selected for a concentrated development effort to generate qualification, design, and processing data for production. Specifications should be prepared to define the constituents and properties of specific composites or new alloys. Full-scale parts must be ground tested for certification and put into service as replacement parts.

Such an approach requires a well-defined *gate* between research and development. Research should continue in its present form, but the number of new technologies which graduate into the full development cycle, where both costs and requirements are escalating, must be limited.

Finally, there has been a significant advance in the ability to design, analyze, fabricate, and inspect complex aerospace structures during the past 20 years. Much of this advance is associated with the availability of computers and the development of computer-aided design and manufacturing systems.

Since "capability" can breed requirements and since unnecessary requirements can stifle advancement, any new requirements should be thoroughly evaluated and their impact fully defined before we apply them.

In summary, major improvements in structural efficiency, durability, and cost-effectiveness have been realized during the past 50 years through a series of continuing advances in SDM technology. The emergence of advanced composites and advanced metallic structures offers significant additional improvements, provided opportunities are made available for using them.

Special Structures for Aircraft and Spacecraft

Structural Configurations

Structural Comparison of Perforated Skin Surfaces with Other Means of Effecting Boundary-Layer Control by Suction

Charles B. Cliett
Mississippi State College

Introduction

The problem of stabilizing a laminar flow over a surface has been the subject of a large amount of theoretical research. Long before the first flight by man, this subject was receiving the attention of some of the best research scientists. The first recorded recognition of the inherent instability of flow over a surface where discontinuities exist in the velocity was published in 1845.[1] As early as 1904, Prandtl[2] recognized the possibility of influencing the boundary layer by suction. Since that date, considerable aerodynamic research has been conducted along these lines, but little progress toward actual application has been made. The structural side of boundary-layer suction has caused considerable concern but has been the subject of little research. This paper is an effort to fill a small portion of the gap between the progress made on the aerodynamic considerations and the lack of progress made, because of little effort, on the structural considerations. A second purpose, and the more important one, is to stimulate the interest of structures-minded people and, therefore, lead to serious study of the subject.

Aerodynamic Considerations

The first method advanced for controlling the boundary layer by suction was by the use of single or multiple slots. Many papers have been published dealing with this process of control. Among the more important is one by Pfenninger,[3] who reported the use of a series of spanwise slots spaced along the chord from the leading to the trailing edge. Each slot was connected to an individual chamber through which the air was sucked. The individual arrangement made possible some saving in suction power since the suction flow required of such an arrangement varies with the position along the chord. Including the power lost by suction, he found drag reductions of from 33 to 50%. The range of low drag coefficient more than doubled, and the glide ratio was 1/200. These are the results of tests made at a relatively low Reynolds number. Tests at a higher Reynolds number failed to show such favorable results. This, Pfenninger attributed to an increase of tunnel turbulence at higher velocities.

Aeronautical Engineering Review, Vol. 12, No. 9, Sept. 1953, pp. 46–54.

Airfoils especially designed for boundary-layer control by suction through slots have been advocated by Griffith. These were discussed extensively by Goldstein[4] in his 1948 Wright Brothers Lecture. The basic idea involved is that of producing a favorable pressure gradient over the entire region from the leading to the trailing edge. Many difficulties have been encountered during experimental procedures, but this method may well prove significant in the future.

In considering the design of a boundary-layer control system, it is important to maintain a laminar flow up to each slot, for, if the flow becomes turbulent, a great amount of suction is required to remove the turbulent boundary. For this reason, there exists a critical spacing of slots which is less for the higher Reynolds numbers than for the low Reynolds numbers.

A boundary-layer stabilization method that offers near infinitely close spacing of suction sources is the porous area suction method. In addition to fulfilling the demands for stabilization at high Reynolds number, this method also reduces the destabilizing influences of surface defects, of foreign particles adhering to the wing, and of surface waviness. A porous material has the advantage that it provides stabilization of the laminar boundary with the least sensitivity to disturbances. However, unless the porosity is made variable according to the local requirements along the chord, a large power expenditure for suction will be required. A possible solution is to divide the suction area into smaller areas, each of which has a separate suction pressure. This method, however, complicates the control of suction pressure, as well as the internal wing structure.

Perforated surfaces, consisting of closely spaced holes running spanwise in rows spaced the critical distance apart, offer a possibility of controlling the porosity by the size of the holes and by their spacing chordwise and spanwise. In the research reported in Reference 5, the laminar boundary layer was maintained to the trailing edge on the top surface of a classical airfoil with exceedingly small suction power.

Structural Considerations of Various Methods of Suction Boundary-Layer Control

The use of multiple slots as the mode of suction for boundary-layer stabilization would present several structural difficulties. The problem of torsional stability would of necessity have to be solved by an entirely new process. Since the suction power required varies along the chord, individual compartments with separate internal pressures would be needed to keep the power requirement low so that an overall efficiency gain could come from the boundary-layer stabilization. The interior of the wing would then become a series of individual compartments critical in design because of the duct loss factor. Because of the many cutouts and sharp bends that would be required, the fatigue problem would also become critical. With these thoughts in mind, it would seem that any design capable of meeting the aerodynamic requirements should result in a strength-weight ratio so low that the use of this method of stabilization is hardly practical. It is the author's opinion that use of multiple slots is not feasible without a serious weight penalty. Single slots, such as those used near the trailing edge, could be handled successfully.

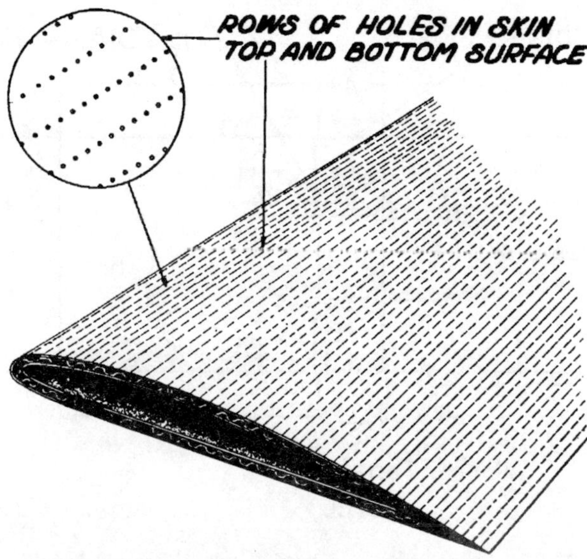

Fig. 1 Typical wing section showing hat sections used as ducts for suction air.

Porous metals (like sintered bronze)[6] which have been developed offer little hope of becoming structural materials for aircraft. They are not only heavy, but have very low strength. Sintered bronze cannot be machined if it is to be used for suction because the pores fill readily. Although the porosity can be changed by manufacturing procedures, it presently cannot be controlled within practical limits. For model purposes, its strength is probably sufficient, but in actual application it would probably be merely excess weight added to the structural components of a wing.

Perforated sheet materials have this structural advantage over slots: Torsion can be carried by the sheet. Since the use of many separate chamber pressures can be eliminated by varying the porosity, an additional advantage in structural simplification results. In fact, the technique of laminar boundary-layer stabilization using a perforated surface may involve little additional structure over the conventional airplane wing design.

The use of double skin surfaces with a relatively small air space between these surfaces would make possible easy adaptation of existing aircraft for boundary-layer control, if the control is applied through perforated sheet materials (Fig. 1). Since this structural condition is used on some of our current aircraft, they can be adapted for early study and application of boundary-layer control with a minimum of structural revision. It is for these reasons that this research on the structural properties of perforated sheet materials was undertaken.

Structural Properties—Perforated Sheet Materials

The design of all test specimens and the testing procedure conformed to the standards of the American Society for Testing Material (ASTM)[7,8] for all cases

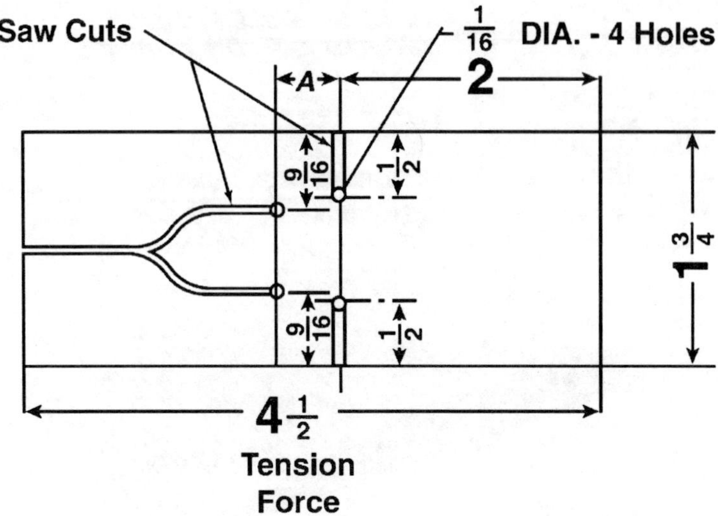

Fig. 2 Sheet aluminum tension-type shear specimen.

for which standards have been established. The tension-type shear specimen of the ASTM was used to determine the shear properties of both the perforated and unperforated plywood specimens. Since no standard exists for similar tests on aluminum sheet materials, it was necessary to design a specimen for these tests. The specimen as designed is shown in Fig. 2. The aluminum-fatigue specimen was designed such that, when perforated, the test section remained three dimensional rather than just a single failure cross section.

In all comparisons of predicted ultimate failure loads and experimental ultimate failure loads, the predicted ultimate failure loads were calculated from the probable ultimate failure stress. Practical application of the theories that follow would dictate the use of the minimum guaranteed properties for the materials concerned rather than probable failure stresses.

Ultimate Tensile Strength—Perforated 24ST-3 Aluminum-Alloy Sheet

For the case of uniform tension, the mathematical theory of elasticity indicates that a single hole placed in the center of a plate of isotropic material of infinite width produces a stress concentration at the edge of the hole of three times the value that would be produced if no hole were present. Various experimental methods, such as photoelasticity and the brittle material method, have shown that such a concentration does exist and that the magnitude is of the order previously stated.[9]

The chief interest herein is, of course, beyond the point where actual yielding of the material occurs. For years it has been assumed, and rightly so, that the stress concentration does not control the ultimate failure of such ductile materials as structural steel. Holleman[10] has verified this by tests on annealed X4130 steel. His experiments show that no reduction of strength occurs when the tensile stress produced is calculated on the basis of net area. In other words, regardless of the size of the hole or of the number of holes placed in a single transverse row, there is no reduction in the ultimate net area stress. Consideration of the conventional stress-strain diagram for this material leads one to realize that these results were to be expected. Beyond the yield point, this material yields, for all practical purposes, without any increase in stress. Therefore, even though yielding at the hole edge occurs early in test, no stress concentration remains when ultimate failure occurs.

Study of the stress-strain diagram for the less ductile 24ST-3 aluminum alloy reveals that different conclusions must be drawn for materials of its type. Certainly the point at which it yields is not well defined, and, of more importance, the stress required beyond yield, for any definite strain, increases until ultimate failure occurs. This means that some stress concentration exists even at the instant of ultimate failure and that this material's efficiency (based on net area) will be less than 100%. Results by Holleman,[10] Stevenson,[11] and by this author (Fig. 3) verify the conclusions drawn for the case of a single hole placed in the center of the test specimen.

As can be seen in Fig. 3, the percent of original strength based on net area decreases rapidly with an increase in hole size until approximately 18% of the cross-section area has been removed. Here a steady increase in the percent of original strength based on net area becomes apparent as the area removed increases. If one keeps the actual stress-strain diagram in mind, a probable explana-

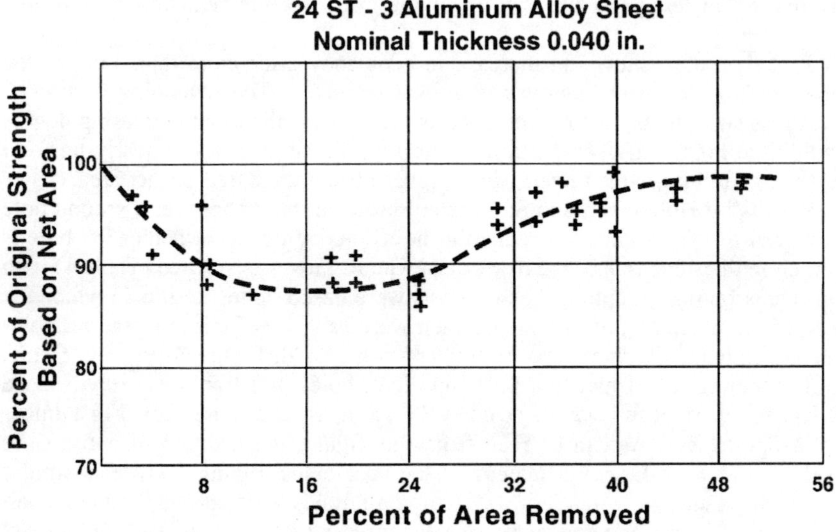

Fig. 3 The effect of a single hole placed in the center of the test specimen on its ultimate tensile strength.

tion of this phenomenon may be had from the theory of elasticity. The equation for the normal unit stress on a cross section, which includes a single hole in the center, in terms of the unit stress on a cross section sufficiently removed so that no effect of the hole is present, is

$$S_t = \frac{S_0}{2}\left(2 + \frac{r^2}{x^2} + 3\frac{r^4}{x^4}\right)^9$$

where

r = the radius of the hole

x = the distance to any point on the cross section measured from the center of the hole

S_0 = the unit stress on a cross section unaffected by the hole

It can, thus, be seen that the section near the hole yields long before other portions reach the yielding stage. Of course, once yielding has occurred, the equation is no longer valid, since the assumption of elastic behavior no longer holds. It is probable, however, that the portion below yield stress is stressed by a similar equation. We know that, for the aluminum alloy, continued yielding does not occur without additional stress. Thus, the procedure continues to failure with a stress concentration that becomes more intense as the point in question is moved from the edge of the sheet to the edge of the hole.

The portion of the curve (Fig. 3) which shows a definite decrease in efficiency has in all likelihood a slow process of failure. The stress concentration is on a limited portion of the cross section, and actual failure probably occurs here when the rest of the cross section is still capable of carrying the entire load. Therefore, the rupture slowly proceeds out from the hole until the remaining portion can no longer carry the entire load. This theory was apparently verified during actual testing, since an actual slow process of yielding did occur over a period of time, decreasing in length as the portion of the cross section removed by the hole increased.

That the curve shows an increase in efficiency after more than 18% of the cross section has been removed is as was expected. The remaining portion of the cross section, as the area removed is increased, falls to an increasing degree within the highly stressed portion. Therefore, the strain is more nearly uniform at time of failure, which results in a higher efficiency based on net area.

Results by Holleman[10] and Stevenson[11] indicate that, whenever a second hole is placed in the specimen, a reduction in efficiency occurs, whether the hole is placed in the same transverse row or elsewhere. This was verified (Table 1) with one encouraging exception. Their reports were based on test results in which the size of the holes placed in the specimen were relatively large. In general, holes for attachments by rivets and bolts were considered. Figure 4 presents results from specimens that had four 0.018-in.-diam holes in a transverse row, spaced 0.100 in. apart. Rows were spaced 0.100 in. apart and were varied in number from one to five. As can be seen from this figure, the addition of entire rows had no effect on the net efficiency of the specimens. Figure 5 presents similar results for a case in which four 0.031-in.-diam holes were spaced 0.100 in. apart in transverse rows. We can therefore say conservatively that the addition of extra rows of holes can be made with no decrease in efficiency provided the hole diameter is no greater than 0.031 in. and if the holes are spaced at least 0.100 in.

Table 1 Effect of a second hole not in the same transverse axis of the first[a]

No.	Hole diameter	Specimen description	Net area stress in lb per sq in.	Reduction in efficiency
1	0.0595	Single hole in center	60,700	—
2	0.0595	Second hole added, 0.12 in. above first in center of specimen	58,600	3.4
3	0.0595	Same as No. 2	58,600	3.4
4	0.0595	Two holes centered on 45 deg diagonal, 0.20 in. apart	56,400	7.2
5	0.0595	Same as No. 4	55,700	8.2
6	0.1200	Single hole in center	54,200	—
7	0.1200	Single hole 0.24 in. above first in center of specimen	49,100	9.5
8	0.1200	Same as No. 7	49,700	9
9	0.1200	Two holes centered on 45 deg diagonal, 0.20 in. apart	44,000	19
10	0.1200	Same as No. 9	43,200	20.3

[a]24ST-3 aluminum-alloy sheet, 0.032 nominal thickness.

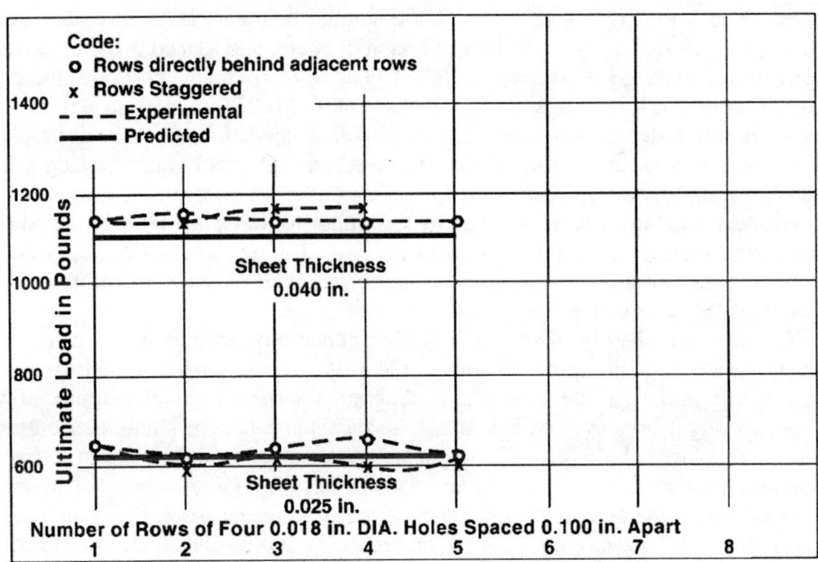

Fig. 4 The effect of adding additional rows of holes 0.100 in. away—0.018 in. diameter.

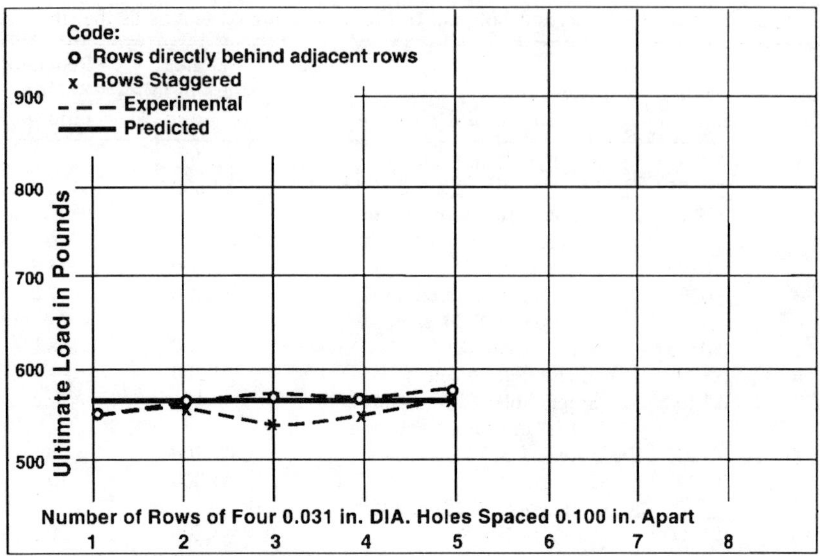

Fig. 5 The effect of adding additional rows of holes 0.100 in. away (diameter of holes 0.031 in.) sheet thickness 0.025 in.

apart. Successful flight tests, in which the laminar boundary layer was stabilized by suction through holes 0.018 in. and less in diameter and spaced 0.100 in. apart, have been conducted at Mississippi State College.[5] For practical purposes, then, it would seem that knowledge of the fact that holes of 0.031 in. in diameter can be used without a decrease in efficiency is all that is needed. Even more desirable, however, is an understanding of the phenomenon that controls the spacing and, therefore, the limitations upon it.

Unsuccessful attempts were made by the author to find a theoretical method of predicting ultimate failure in tension for perforated sheet, but a reliable empirical method based on the test results from specimens with only a single hole in the center of the test section was found.

Articles published by Childs and Kelly[12] concerning studies of the effect of abrupt changes of the cross section on the unit stress of rubber models proved interesting. Although the stress-strain diagram for rubber deviates appreciably from a straight line, even at low stress, at least some idea of strain distribution emerges. A single hole placed in the center of a rubber model results in a strain distribution that can be closely approximated by a straight line—that is, of course, a maximum at the hole and a minimum at the edge of the sheet. If, on the same transverse axis, an additional hole is placed in the model and if we assume that the strain distribution for a single hole can be expressed by a straight-line relation, the strain distribution—by the principle of superposition—becomes constant over the portion of the cross section between the two holes, and a failure carries a stress equal to the ultimate tensile stress of the material.

By means of photoelasticity, Coker[13] has studied the stress concentrations on tension specimens in the elastic range which result from symmetrical fillets formed by semicircles at the plate edge. Where these fillets remove 25% of the cross section, the stress concentration amounts to only 1.37 of the average stress, and the area over which the stress concentration exists is a relatively small portion of the total cross section. As the percentage of the area removed by the fillets decreases, both the area affected by the concentration and the degree of concentration become smaller. Hence, the conclusion may again be drawn that, for all practical purposes, when the holes are of the type desired for boundary-layer control by suction, the stress is constant over the cross section between the holes. Although it may seem that the use of such reasoning would lead to slightly unconservative results, such is not the case as will be shown later.

If the above theory can be applied, there remains to be considered only the portion of the cross section which lies between the last hole in the row and the edge of the sheet. This information can be secured directly from tests in which only one hole was placed in the center of the sheet.

A combination of the theory of constant strain between two holes in a single transverse row and the experimental results from the effect of a single hole has been used to predict the ultimate failure of 94 test specimens. A comparison between the predicted ultimate failure loads and the experimental failure loads is shown in Fig. 6. Specimens containing holes of diameters of 0.031 in. and of 0.018 in. were used. The number of rows was also varied from one to five. In some cases the holes were placed directly behind each other in the transverse rows, and in others they were staggered in such a way that the offset was exactly half of the transverse spacing. Sheet thicknesses of 0.025 in. and 0.040 in. were used for all cases.

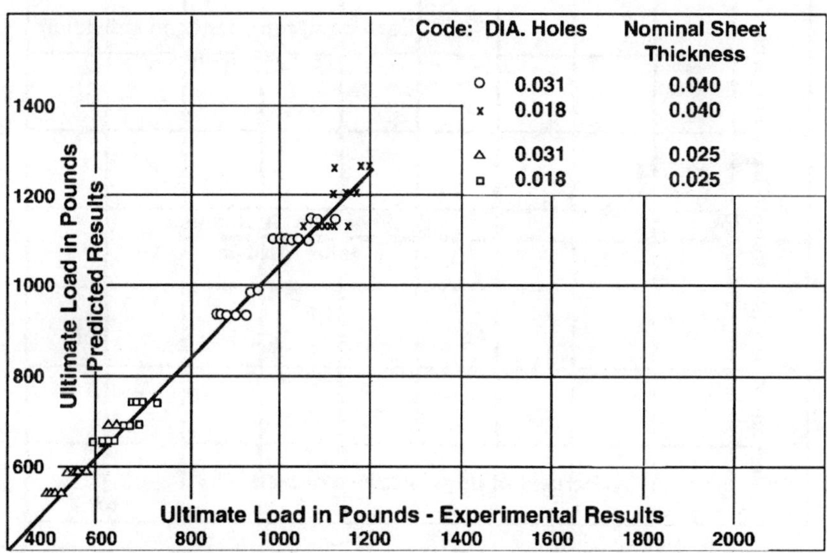

Fig. 6 Comparison of predicted and experimental ultimate tension loads—perforated 24ST-3 aluminum-alloy sheet.

As illustrated, variations between the predicted and experimental loads are relatively small. A variation of as much as 5% was the exception, and in only a few cases did one occur.

Ultimate Shear Strength—Perforated 24ST-3 Aluminum-Alloy Sheet

Figure 7 presents the results of single-hole shear studies on 0.032-in. 24ST-3 aluminum-alloy sheet. In these studies, the effect of a single hole drilled in the center of the test area was determined. The hole diameters were changed so that the portion of the failure cross-section area removed varied from zero to about 50% of the original cross section. Throughout this range, the remaining material proved capable of carrying the ultimate shear load of the unperforated material. Each of the 72 specimens developed an average ultimate shear stress above the minimum guaranteed ultimate shear stress of the unperforated material.[14]

Assuming all remaining material capable of developing the full ultimate shear strength of the unperforated material, the ultimate failure loads of the 36 perforated specimens were predicted. A comparison of these predicted failure loads and the corresponding experimental failure loads is shown in Fig. 8. The close agreement between these shows that this method of predicting the ultimate shear failure is reliable for this perforated material.

This method of predicting ultimate shear failure for perforated materials was checked by testing perforated 24ST-3 aluminum-alloy tubes in pure torsion. As was the case for the sheet materials, the predicted ultimate failure and the experimental failure were in close agreement.

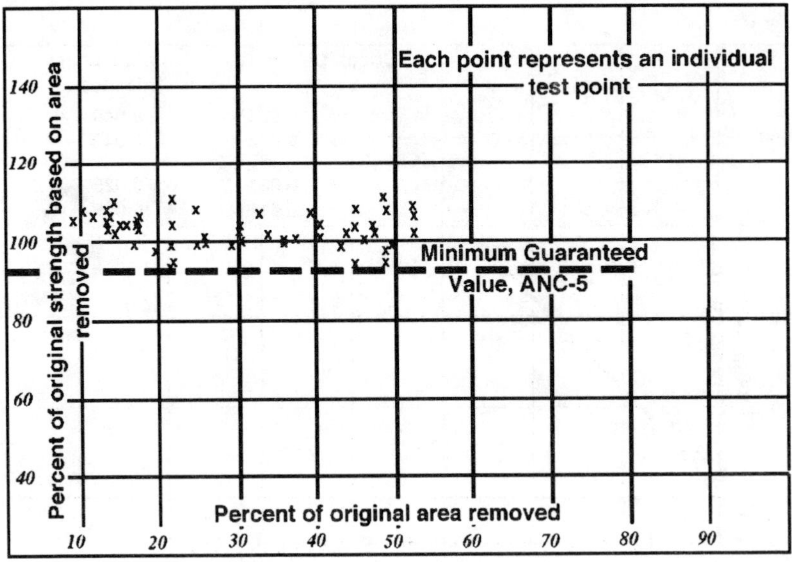

Fig. 7 The effect of a single hole placed in the center of the test area of the test specimen on its ultimate shear strength—0.032 in. thick 24ST-3 aluminum-alloy sheet.

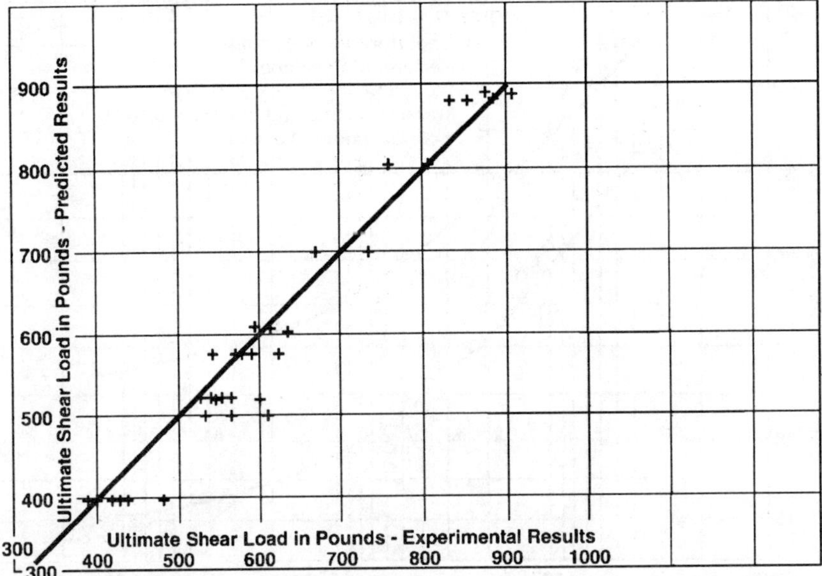

Fig. 8 Comparison of predicted and experimental ultimate shear failures—perforated 24ST-3 aluminum-alloy sheet.

Flexure Fatigue of Perforated 24ST-3 Aluminum-Alloy Sheet

The results of an investigation to determine the effect of perforations on the flexure strength of 24ST-3 aluminum-alloy sheet are shown in Fig. 9. The perforated specimens were prepared by drilling transverse rows of holes 0.018 in. in diameter with spacing equal to 0.10 in. The spacing between the rows was also 0.10 in. Material of nominal thickness, 0.040 in., was used throughout the investigation, and the loading was always applied parallel to the direction of rolling of the sheet.

The applied stress values shown for the extreme fibers of the perforated specimens are the average stress that they carried for any given loading. Below the elastic limit of the aluminum alloy, there certainly exists a large variation in the stress pattern between the holes, but no attempt was made to determine these localized values since they have no effect on the final conclusions to be drawn from the investigation.

The rate of flexure was 1800 cycles of completely reversed stress per minute. The loading continued uninterrupted until complete failure occurred.

As was to be expected, Fig. 9 shows considerable reduction in life expectancy for the perforated specimens. This reduction, however, is somewhat smaller than would be expected if only the stress concentration factors normally produced at circular holes were considered.

Often the figure of 2,000,000 completely reversed cycles of stress is used in comparing the flexure-fatigue-carrying abilities of various types of materials. If this life expectancy is considered, the investigation shows that the unperforated

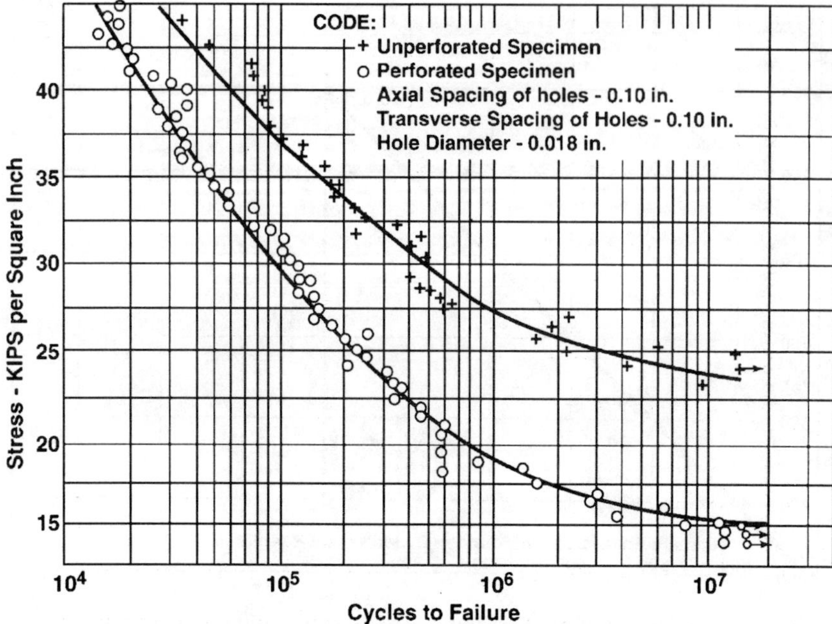

Fig. 9 Comparison of flexure fatigue stengths of perforated and unperforated 24ST-3 aluminum-alloy sheet—nominal sheet thickness 0.040 in.

materials can be stressed to approximately 26,000 lb per sq in. while the perforated materials can be stressed to only about 17,000 lb per sq in. These figures represent a loss in load-carrying ability of about one-third. If the life demanded of a specimen is greater than 2,000,000 completely reversed cycles, the reduction in load-carrying ability will be greater than one-third as compared to the load-carrying ability of the unperforated specimens, but, if the life demanded is less than 2,000,000 cycles, the reduction in load-carrying ability will be less than one-third.

Analysis Procedures—Plywoods

The method of least squares for observations of equal weight was applied to all experimental data received from tests involving plywood materials, but only the arithmetical mean and the probable error of a single observation were determined.[15] All data from plywood tests which are later presented in this report will, thus, be given in the form of $M \pm r$: M is the arithmetical mean, and r is the probable error of a single observation. Whenever the data were involved in computing such percentages as percent of original strength based on net area, only the arithmetical mean was considered.

Stress Concentration—Woods and Plywoods

The intensity of the stress concentrations produced by a hole placed in the center of a sheet of wood that is under tension is dependent upon many variables.

Large variations in Young's modulus, the shear modulus, and Poisson's ratio result in large differences in the value of the concentration factors produced in any given wood material as the direction of the applied tensile load is varied in relation to the grain direction. Since the relationships that exist between the moduli and Poisson's ratio vary from specie to specie (see Table 2-9, Reference 16), a pattern determined for any given wood would likely be quite different from what would exist in another specie under similar conditions. Just as the intensity of the concentration at the hole varies, so does its effect on the material at any given distance from the hole.

Since the grain of the individual plies of plywood varies as to direction (generally aircraft plywoods have the grain of adjacent plies at right angles to each other), a more uniform strength and elastic property condition exists. As the tensile load is applied at different angles to the face grain direction, however, a considerable variation in the stress concentration factor still occurs. (The application of Equations 2.14 and 2.15 in Reference 16 to any plywood material will readily verify this statement.) Also, differences in the stress concentration pattern around a hole will be found for similar plywoods constructed from different species of wood. For these reasons the results to be presented later must be considered to apply only to the plywoods in question and to these plywoods under the specified conditions.

The performance of other plywoods under similar conditions surely can be expected to be similar in nature, and general conclusions can be drawn concerning them, but direct use of the data contained herein is not possible.

Ultimate Tensile Strength—Perforated Plywoods—Face Grain Direction Parallel to the Loading Direction

The results of single-hole studies on ⅛-in. mahogany-yellow poplar plywood are shown in Fig. 10. A single hole was drilled in the center of the test specimen, and its effect upon the ultimate tensile strength of the specimen was determined. The loading in every case was parallel to the face grain direction.

It is obvious from this plot that the stress concentration produced by the hole controls to a large degree the ability of the material to resist the ultimate tensile load. The conventional stress-strain diagrams for the two woods involved are a key to this action. The steepness of their stress-strain curves, even after the proportional limit has been exceeded, means that a large stress concentration must still exist at the time of ultimate failure. That is, even though the large stress concentration factor produces local yielding at the hole early in the test, this yielding does not continue without a continued increase in the local stress. Therefore, at the time of first local failure and instantly thereafter when total failure occurs, an uneven stress distribution exists, and, thus, a reduction in efficiency based on net area occurs.

As the percentage of the area removed by a single hole increases from zero, a rapid reduction in efficiency based on net area occurs. This rapid reduction continues until about 25% of the cross-section area has been removed, and then the effectiveness of the remaining material is about 60%. When from 25 to above 50% of the cross-section area has been removed, the efficiency of the remaining

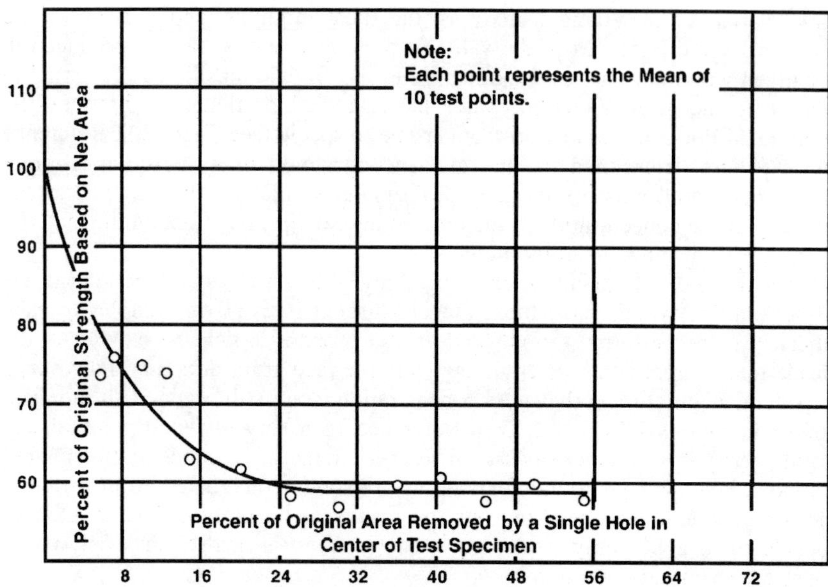

Fig. 10 The effect of a single hole placed in the center of a test specimen on its ultimate tensile strength (1/8 in. mahogany-yellow poplar plywood) grain direction of face ply parallel to loading.

material for all practical purposes is constant at 60%. No tests were conducted where more than 55% of the cross section was removed by a single hole.

A greater reduction in efficiency occurs in the plywood than in the aluminum alloy. Also, in the range from 0 to 25% cross-section area removed by a single hole, the rate of reduction is much higher for the plywood. This reduction in efficiency and its rate can at least be partially explained by two factors. First, the stress concentration factors produced in the core of the plywood material while it is still stressed in the elastic range is much larger than that produced in the aluminum alloy. If no relief is considered to be received from the face plies, this factor is 5.48 as compared with 3.0 for the aluminum alloy. The face material itself, considering no increase from the core, has a factor of 2.35. In any event, the overall concentration considerably exceeds that produced in the aluminum alloy. Second, the extreme steepness of the tensile stress-strain curve beyond the elastic limit continues much of this concentration until ultimate failure occurs.

That the curve fails to continue its downward trend beyond the point where 25% of the cross-section area was removed was to be expected. As the area removed increases, the remaining portion falls to an increasing degree within the high-stressed portion and, therefore, is more nearly equally stressed at the time of failure. If the tests had been continued beyond 55% cross-section area removed, very likely an increasing efficiency based on net area would have been found. Possibly no earlier increase was found because of the fact that the material some distance from the hole was little affected by the stress concentration produced by the hole.

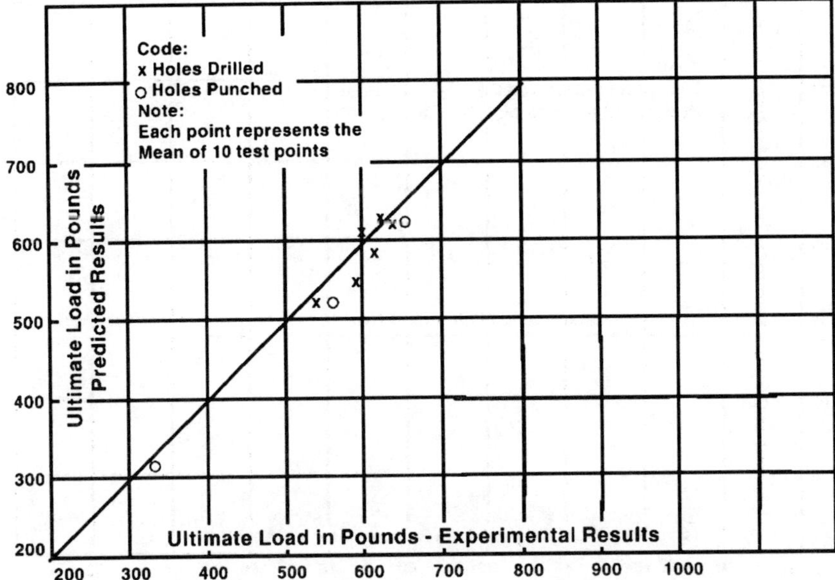

Fig. 11 Comparison of predicted and experimental ultimate tensile strengths (plywood tension tests) loading parallel to grain of outside ply.

If, at the time of ultimate failure, the theory of constant strain between two properly spaced holes holds for perforated plywoods, only the results of Fig. 10 are necessary for predicting the ultimate failure load in tension for this material. On the assumption that all material between two adjacent holes is capable of carrying the ultimate tensile stress of the unperforated material and that the portion between the last hole in the row and the edge of the sheet is capable of carrying only that percentage found in the single hole studies (Fig. 10), the ultimate failure loads have been predicted for 90 test specimens. Figure 11 gives a comparison of the predicted and experimental failure loads for these specimens, and, as shown, good agreement exists. Figure 11 also shows predicted and ultimate failure loads for perforated 0.070 mahogany-mahogany plywood. Good agreement also exists for this material.

Ultimate Shear Strength—Perforated ⅛-in. Mahogany-Yellow Poplar Plywood—Loading Direction Parallel or Perpendicular to the Face Grain

The results of single-hole shear studies on ⅛-in. mahogany-yellow poplar plywood are shown in Fig. 12. The effect of various size holes drilled in the center of the test area on the specimens' ultimate shear strengths was determined. For these studies the face grain was always perpendicular to the direction of loading, and, therefore, failure was always across the face grain.

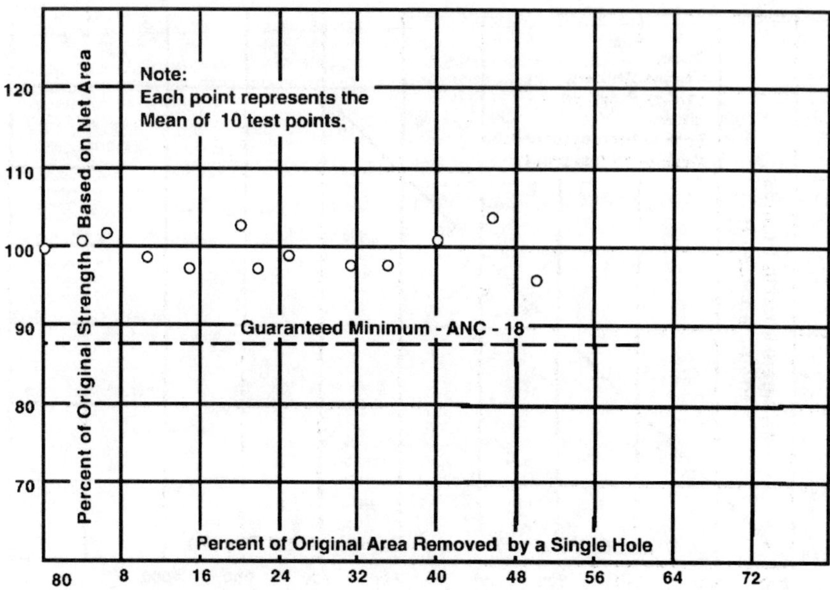

Fig. 12 The effect of a single hole placed in the center of the test area of a specimen on its ultimate shear strength (⅛ in. mahogany-yellow poplar plywood) shear failure perpendicular to grain of face ply.

As shown in Fig. 12, the percent area removed by a single hole was varied from 0 to 50%. For all practical purposes the load-carrying ability of the remaining material did not diminish or increase, regardless of the size of the hole drilled in the test area. In all instances, the experimental average net area stress exceeded the guaranteed minimum strength for this material as published in Reference 16.

With the preceding results on this material and the results from the aluminum shear tests as a guide, it was assumed that the ultimate shear strength of this material, when perforated, could be predicted by considering the remaining material fully capable of carrying the ultimate shear stress of the unperforated material. A comparison of predicted and experimental ultimate failure loads for six variations of perforated ⅛-in. mahogany-yellow poplar plywood with the face grain perpendicular to the loading direction is shown in the first half of Table 2. Close agreement between theory and experiment is shown, and, as a result of this, the ultimate failure loads for similar specimens whose face grain was parallel to the loading direction were predicted. The second half of Table 2 gives a comparison of these predicted values with the experimental failure loads and shows that the method of predicting failure is reliable in this case, as well as when the loading was perpendicular to the face grain.

Conclusions

1) From a consideration of both the aerodynamic and structural aspects, indications are that any early application of boundary-layer control by suction can

Table 2 Ultimate shear strength perforated ⅛-in. mahogany-yellow poplar plywood

Test no.[a]	Drill diameter, in.	Transverse spacing, in.	Longitudinal spacing, in.	No. of rows	No. of holes in row	Experimental ultimate failure	Percent original strength[b]	Predicted ultimate stress
			Loading Perpendicular to Grain[c]					
1	0.018	0.1	0.1	3	2	198	104	190
2	0.025	0.1	0.1	3	2	177	96.5	185
3	0.031	0.1	0.1	3	2	176	98.5	179
4	0.018	0.1	0.1	3	3	187	102	183
5	0.025	0.1	0.1	3	3	171	98	174
6	0.031	0.1	0.1	3	3	161	96.5	167
			Loading Parallel to Grain[d]					
7	0.018	0.1	0.1	3	2	203	104.5	194
8	0.025	0.1	0.1	3	2	183	97.5	188
9	0.031	0.1	0.1	3	2	192	105	183
10	0.025	0.1	0.1	3	3	190	101.5	187
11	0.025	0.1	0.1	3	3	189	107	177
12	0.031	0.1	0.1	3	3	173	101.5	170

[a]Each test represents mean of 10 test values. [b]Percent original strength based on net area. [c]Probable ultimate strength failure—1640 lb per sq in. (based on 10 tests). [d]Probable ultimate strength failure—1670 lb per sq in. (based on 10 tests).

most advantageously be applied through use of perforated lifting surfaces of conventional design. All other methods previously advocated result in less efficiency from either the aerodynamic or structural point of view or from both.

2) The theories presented in this report for predicting the ultimate tensile and shear strengths of perforated 24ST-3 aluminum-alloy sheet can be used for design purposes. Although the accuracy compares well with the accuracy of other stress analysis procedures, the use of a factor of 0.95 of the predicted value would guarantee conservative analysis. The decrease in the strength-weight ratio of the perforated aluminum alloy as compared to the unperforated alloy should by no means offset the aerodynamic improvements received from boundary-layer control by suction.

3) The theories presented for the prediction of the ultimate tensile and shear strengths of perforated plywood are suitable for design purposes. The percent reduction in strength-weight ratio is slightly higher for the plywoods than for the aluminum alloy because of the larger effect of the perforations on the ultimate tensile strength of the plywoods. This can be explained by the types of stress-strain curves possessed by the materials concerned. Beyond the yield point, the slopes of the stress-strain curves decrease very little, and, therefore, the degree of the stress concentration produced at the hole when the material is stressed below the yield point decreases very little as the material is stressed to ultimate tensile failure.

4) The fact that additional holes of the type needed for boundary-layer control by suction can be added without a reduction of efficiency, as far as the tensile strengths of the materials studied are concerned, is of primary importance. Without further investigation, the addition must be limited to holes of no greater diameter than 0.031 in. and minimum spacing equal to 0.10 in.

5) The effects of fatigue will in all probability be the most serious structural requirements imposed on the perforated materials. Although considerable reduction in life expectancy does occur under the action of flexure fatigue, this loss is much less than would be expected if only the stress concentration factors normally produced at a circular hole are considered. If the figure of 2,000,000 completely reversed cycles of stress is demanded, the loss in load-carrying ability for the perforated material is about one-third. If the life demanded is greater than 2,000,000 cycles, the reduction in load-carrying ability will be greater than one-third, but, if the life demanded is less than 2,000,000 cycles, the reduction in load-carrying ability will be smaller than one-third.

6) The theories previously discussed must be applied only to the aluminum alloy and the plywood materials studied. Generalities concerning the characteristics of other aluminum alloys and other plywoods can be drawn, but direct application of the data presented is not possible.

References

[1] Stokes, Sir George G., "On the Theories of the Internal Friction of a Fluid in Motion," *Cambridge Transactions,* Vol. 8, 1845.

[2] Prandtl, L., *Ueber Flussigkeitsbewegung kei sher kleiner Reibung,* Verhandlung III, Intern. Math. Kongress, Heidelberg, 1904, reprinted in Vier Abhdnlg. zur Hydrodynamik, Göttingen, 1927.

[3] Pfenninger, W., ''Experiments on a Laminar Suction Airfoil of 17 Per Cent Thickness,'' *Journal of the Aeronautical Sciences,* Vol. 16, No. 4, April 1949, p. 227.

[4] Goldstein, S. ''Low-Drag and Suction Airfoils,'' *Journal of the Aeronautical Sciences,* Vol. 15, No. 4, April 1948, p. 189.

[5] Raspet, A., ''Mechanism of Automatic Trailing Edge Suction,'' The Engineering Research Station, Mississippi State College, Research Rept. No. 1, 1951.

[6] Preston, H. H., and Rowcliffe, B. A., ''Note on Sintered Metal with a View to Its Use as a Porous Surface in Distributed Suction Experiments,'' Aeronautical Research Council, Technical Rept. 9383.

[7] *Paint, Naval Stores, Wood Adhesives, Paper, Shipping Containers,* American Society for Testing Materials, Standards, Part 4, 1949.

[8] *Non-Ferrous Metals,* American Society for Testing Materials, Standards, Part 2, 1949.

[9] Seely, F. B., *Advanced Mechanics of Materials,* Wiley, New York, 1948.

[10] Holleman, C. H., ''Tension Joints in Aircraft Structures,'' *Journal of the Aeronautical Sciences,* Vol. 10, No. 8, Oct. 1943, p. 295.

[11] Stevenson, C. H., ''The Effects of Open Holes on the Tensile Strength of Some Aluminum Alloys,'' *Journal of the Aeronautical Sciences,* Vol. 13, No. 8, Aug. 1946, p. 395.

[12] Childs and Kelly, ''The Resistance of Materials, the Effect of Sudden or Abrupt Changes in the Section on the Distribution of the Uni-Stress,'' *Railway Mechanical Engineer,* March, April, and May, 1919.

[13] Coker, E. G., ''Engineering Problems of Stress Distribution,'' *Proceedings, First International Congress for Applied Mechanics at Delft,* 1924.

[14] *Strength of Metal Aircraft Elements,* Munitions Board, Aircraft Committee, ANC-5, June 1951.

[15] Bartlett, D. P., *The Method of Least Squares,* Harvard Cooperative Society, Cambridge, MA, 1933.

[16] *Design of Wood Aircraft Structures,* Munitions Board, Aircraft Committee, ANC-18, June 1951.

Integrally Stiffened Structures

Paul Sandorff and G. W. Papen
Lockheed Aircraft Corporation

Introduction

Description of Integrally Stiffened Structure

The need for reduced production costs and increased structural efficiency in aircraft construction has led to the investigation and development of large sections of integrally stiffened surface structure in place of the usual multipiece skin-and-stiffener structure.

The name integrally stiffened structure is applied to construction in which the skin and skin-stiffening elements are made of one part. This is in contrast with conventional structure, which is built up of many individual fabricated sheet-metal parts held together by a multitude of attachments. There are several methods of making integrally stiffened structure: It may be machined from plate or billet, or it may be forged or rolled as a sheet product or extruded or cast—or a combination of these processes may be used.

The immediate advantages of such a structure are the reduction in number of parts and attachments and the reduction in handling expenses and assembly tooling. Other advantages also are generally obtained, such as reduced weight, improved surface smoothness, and simplified sealing.

Functional Aspects of Aircraft Structure

Aircraft surface structure must conform to contour, shape, and clearance requirements while supporting specified loads. Generally, these loads include normal pressure or air loads; overall bending (wing or fuselage bending), which causes axial stresses in the surface elements; and overall torsion, which places shear in the skin.

Conventional skin-stressed structure has been developed to meet these requirements with high efficiency in terms of weight. Essentially such structure consists of: 1) the external sheet or skin; 2) a system of stiffeners that rigidize the skin locally and, together with the skin, carry the major surface stresses; 3) a system of regularly spaced internal supports that establish the column length or the bending length of the skin-and-stiffener combination and also transfer shear load; and 4) still another system of elements perpendicular to the last (that is, parallel to the skin stiffeners) whose purpose is chiefly to preserve shape or carry beam shear. Generally there is a taper in size and strength of all elements along the length of the assembly.

Aeronautical Engineering Review, Vol. 9, No. 2, Feb. 1950, pp. 30–38.

Inherent Disadvantages of Conventional Construction

Conventional construction methods leave much to be desired toward satisfying the functional, structural, and manufacturing requirements. For instance, production consists of fabrication of many individual parts, extensive subassembly, and the attachment of many structural elements running at right angles to each other, which require clips, cutouts, joggles, and discontinuities. Skin and stiffener sections are tapered by reduction in gage, requiring splices and joints. Doublers and reinforcements are necessary for structural efficiency, and weight is determined by compromise with production. Aerodynamic smoothness requirements are difficult to meet; skin surfaces often lack local rigidity; and costly tooling and special techniques are necessary. The airplane is a product of severe compromises, which leave all parties unsatisfied, even when the best available solution is found to the problems of function vs cost.

Recent Developments in Solving the Problems

Engineers in practically every aircraft plant and the military air forces have been working on various means of solving these problems. Among the more outstanding developments are:

- Tapered sheet. The Air Force is sponsoring a program for the development of commercial methods of producing tapered sheet at the rolling mill. Experimental quantities may be obtained today from commercial sources. Several machine shops are now specializing in the production of tapered sheet by machining from plate.
- Tapered extrusion. While not as important from a weight saving or general design viewpoint, tapered extrusion offers tremendous possibilities in material conservation. It reduces machining and, in some instances, reduces the number of splices and parts required for section taper of stiffeners and beam caps.
- Tailored sheet. An effort to obtain the weight advantages of tapered sheet with the reduction in number of parts available by incorporating doublers and other minor section changes with the skin is made through the so-called tailored sheet. Some aircraft companies are setting up machines for producing this type of element, and there are several machine shops specializing in this type of machining.
- Integrally stiffened structure. Each of the previous developments attacks only a small phase of the problem. Integrally stiffened structure attacks all phases. It combines the advantages of tailored sheet with integral stiffeners, thus further reducing the number of parts, number of attachments, and number of splices and joints. Figure 1a illustrates the number of parts required to produce a tank panel by conventional construction. Figure 1b is the integrally stiffened structure answer to the same design requirement.

Machining

Machining from plate or billet is the only method immediately available for producing integrally stiffened sheet of satisfactory structural properties.

Machining has the advantage of great flexibility in design. High-speed equipment and methods now used extensively in milling aircraft parts have greatly

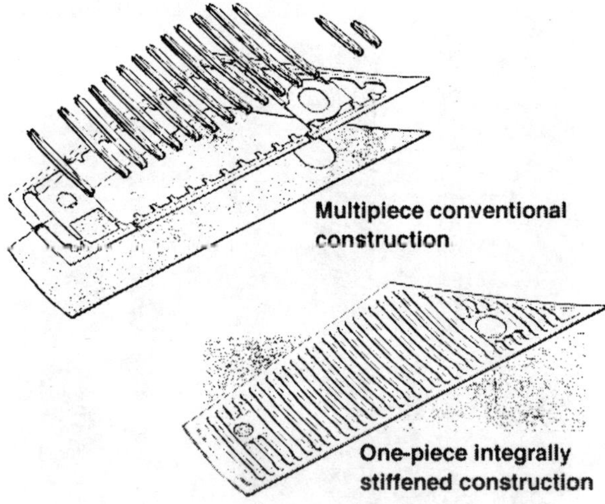

Fig. 1 Typical wing panel for high-speed aircraft: conventional vs integral stiffening.

increased the practicality and potentialities of machining as a primary method of fabrication. In addition, it may be used to duplicate any other type of integral stiffening for experimental study or for prototype construction, and when extruded and forged integrally stiffened designs become available, it may be desirable to apply the same machining methods to reduce skin thicknesses and to incorporate taper, local trim, and reinforcement pads. The great disadvantage of machining from plate is that too large a percentage of the raw material is wasted in chips.

End Milling

The most readily applied method of machining integrally stiffened sheet is that of end milling (axis of the cutting tool perpendicular to the machined surface).

No elaborate equipment is required. Figure 2 shows a machined bulkhead rib for an integral fuel tank. This was cut from ⅞-in. plate on a small Milwaukee

Fig. 2 Experimental bulkhead rib: web, stiffeners, flanges, and doubler pad machined integrally.

Fig. 3 Finished leading-edge panel machined from ¼-in. plate and formed to contour.

milling machine with standard cutters and accepted methods. The finished part is equivalent to a 13-piece assembly in conventional construction. By using equipment especially adapted for such work, such as high-speed cutting heads and special beds, the machining time can be considerably reduced.

Figure 3 shows an experimental leading-edge structure, equivalent to that used on the Constellation outer wing, which was machined from ¼-in. plate. These parts were made on a Cincinnati Hydrotel milling machine. The skin is of 0.040-in. nominal thickness, and all edges incorporate an integral doubler pad 0.072 in. thick to simplify attachment. Forming to the curvature shown was accomplished after machining by repeated bumping on a power brake; rolling has also been used to form such sections.

Slab Milling

For rapid removal of a large mass of metal, slab milling is much better suited than fly-cutting or end milling. The advantages are apparent chiefly for quantity

Fig. 4 Machined wing-surface panels ready for assembly.

production because heavier equipment and higher horsepower are required and the process is generally less flexible in application.

For the machining of integrally stiffened surfaces, Onsrud spar-milling equipment has been applied with excellent results. Typical surface panels with integral doublers and attachment pads which were machined on the Onsrud spar mill from $\frac{7}{8}$-in. plate are shown in Fig. 4. The skin thickness of the finished parts tapers from 0.091 to 0.045 in. The depth of cut is cam guided and is accurate to within a few thousandths of an inch. Taper in skin thickness as well as local reinforcements and integral pads for end attachments is readily obtained. These panels, when assembled, form the upper surface of an experimental outer wing for the Constellation airplane. Similar construction was used in the lower surface, which is shown in the assembly jig in Fig. 5; the completed wing panel is shown in Fig. 6.

Machining to Finished Shape

The discussion thus far has been limited to the production of flat integrally stiffened panels. Most applications, however, will undoubtedly require contoured surfaces. The forming of flat integrally stiffened panels to contour poses, in some cases, production difficulties. Machining to required contour has, therefore, been investigated. It has also been suggested that plate of the required cross-section thickness be formed to contour and then machined to final shape. This naturally would require less machining time and would conserve material over machining from a solid billet large enough to encompass the curvature. The type of equipment required to machine to the contoured shape must of necessity be a high-speed duplicating-type machine based on the Keller or Hydrotel principle. The best that is expected of equipment of this type is only a slight improvement on the end-mill-type operation; consequently, the machining operation would be slow and costly.

Whether or not machining from billet or rolled flat stock to the finished shape proves feasible and economical, it will probably be desirable to do some trimming and finish-machining on contoured parts. It appears feasible to forge integral panels to finished contour with subsequent machining for joints and other areas requiring closer tolerances than can be maintained by forging. It is believed that high-production equipment capable of machining a finished formed integrally stiffened part should be developed.

Fig. 5 Lower-surface structure of machined outer wing in assembly jig.

Typical Machined Surface Structure

The experimental wing structure described above was made interchangeable with the long-range Constellation outer wing, which serves as an integral tank carrying 550 gal of fuel. With data obtained in the construction of this wing, a complete study was made of the value of the machining process in its present stage of development, as compared with conventional methods for producing this type of structure. The following conclusions were reached:

• Structure. The external surfaces of the machined structure are exceptionally smooth and free of rivets. No skin buckling occurs under any design load condition. All splices and bulkhead flange attachments are made with a single row of rivets through thick skin pads that are integral with the panels. On removable doors, a ⅛-in. thick flange is provided. There are no chordwise splices over the entire 20-ft span. The machined taper in skin thickness and stiffener height provides optimum structural properties at all stations.

• Weight. With the machined integrally stiffened structure, a weight saving of 230 lb per airplane was realized. This amounted to 25% of the weight of the structure subject to redesign. Half of this was due to structural weight saved by elimination of splices and by more-efficient use of material; half was due to weight of fuel-tank sealing compound saved at stiffener-skin attachments.

Fig. 6 Experimental outer wing: section forward of main beam made with machined integrally stiffened surfaces.

Table 1 Cost comparisons for machined structure

	Percent of total cost of conventional design	
	Conventional structure	Machined integrally stiffened design
Fabrication	7	17
Assembly	38	20
Tooling	21	29
Raw material	16	25
Tank sealing	18	14
Total	100	105

• Cost. The approximate costs, estimated on the basis of production of 50 items, are compared in Table 1. Assembly cost—the major cost item in conventional airplane construction—is greatly reduced by integrally stiffened design. But current methods of machining, it is seen, involve increased fabrication, tooling, and material costs, which at present overbalance the saving in assembly.

Future Possibilities

The data of the previous section clearly show that further development should be directed at 1) reducing raw material cost and 2) speeding up the machining process.

Reducing cost of raw material. In machining surface panels from thick aluminum-alloy plate, 80–95% of the original stock is cut away in chips and trim. Chips have low salvage value. There is, therefore, an advantage in machining from a roughly shaped section, which might be prepared, for example, by forging or extrusion.

Reducing cost of machining. Machining time and cost of machining would be considerably less if equipment more suited to fabrication of surface panels were available. For example, in applying the spar-mill type of machine to quantity production, gang cutters should be used, and the full width of the panel should be cut at one pass. The bed width should be increased to accommodate the largest available plate sizes, reducing overall handling time and eliminating splices. A cutter driving motor of about 500 hp, together with considerably greater machine rigidity, would be desired.

The advantages of automatic control for quantity production have only begun to be exploited. Once the work is fixed to the bed plates, all operations should be sequenced, indexed, and controlled automatically. Automatic equipment is available to some degree on milling and profiling machines in general use today.

Machining other types of integrally stiffened sheet. Current developmental work in the art of extrusion and forging, if carried to successful conclusion, will make available integrally stiffened shapes of much greater structural advantage than those machined from thick plate. These other types are described in later sections.

To use such extruded and forged shapes to best advantage, some machining will be necessary. Skin thickness will be reduced; stiffener sections will be sized; and tapers in thickness and depth will be incorporated, leaving doubler pads for end attachments and local reinforcements. These machining operations are similar to those used in machining plate stock, and essentially the same type of equipment is required.

Rolled-Ribbed Sheet

Developments to Date

One of the most promising methods of making integrally stiffened sheet is by rolling stiffening ridges into the sheet stock as it is processed in the mill. The problem to date has been the determination of sizes, proportions, and materials that would be most satisfactory for structural applications and would still be reasonable to produce by rolling. Sections suitable for rolling are illustrated in Fig. 7.

Rolled-ribbed sheet is far superior in rigidity to unstiffened sheet of the same total weight. Because of limitations of the rolling operations, however, it cannot be made to compare in efficiency with a built-up skin-and-stiffener structure. Its advantages, therefore, lie only in applications where adequate skin stiffening by ordinary methods is impossible or impractical. A large portion of the exposed surface of every airplane falls in this category—trailing-edge surfaces, fairing surfaces, cover panels, control surfaces, etc.—where applied loads are so light that skin gages are the minimum as determined by handling requirements, stiffeners are as close as production expense will allow, and surface roughness and ''oilcanning'' still remain a problem.

Development work on the equipment and technique of rolling ribbed sheet stock is currently in progress at Reynolds Metals Company under a contract with the U.S. Air Force.

Applications

Application of rolled-ribbed sheet to aircraft structures has been under continued investigation at Lockheed, and several experimental designs have been constructed using ribbed sheet made by machining from plate.

As a result of these studies, the section shown uppermost in Fig. 7 has become standardized. This section has 0.016-in. thick skin, with integral stiffeners 0.125 in. deep at 1-in. spacing. The total weight is the same as that of plain 0.020-in. sheet, but its rigidity is more than 20 times as great. This one section, it has been found, can be applied much as ordinary sheet stock in place of the 0.016-, 0.020-, and 0.025-in. skin used in conventional designs, with reduction in parts, increase in rigidity, and possible weight saving. Several experimental applications are described below.

Aileron tab. An experimental tab was constructed which was identical in function and strength to the aileron tab of the Constellation airplane. The surfaces

Fig. 7 Typical sections suitable for production by rolling.

and beam of this tab were formed in one piece from simulated rolled-ribbed sheet stock as shown in Fig. 8. This skin section replaced the 0.016-in. outer skin, the 0.016-in. beaded inner skins, and the 0.032-in. beam web in the conventional design; enough weight was saved here with rolled-ribbed sheet to permit use of a casting for the central actuator fitting. Thus, considerable simplification was achieved, and it was estimated that, even if the rolled-ribbed sheet were to be *machined* for production, a net saving could be achieved. Parts and cost comparisons are indicated in Table 2.

Wing aft structure. The trailing-edge surfaces of the Constellation wing carry no wing bending loads, but only the direct air loads of low magnitude, which are transferred forward immediately to the main wing structure. Consequently, the aft surfaces are made as light as possible, the skin gages and stiffening being determined chiefly by handling requirements, manufacturing problems, and oil-canning tendencies (Fig. 6).

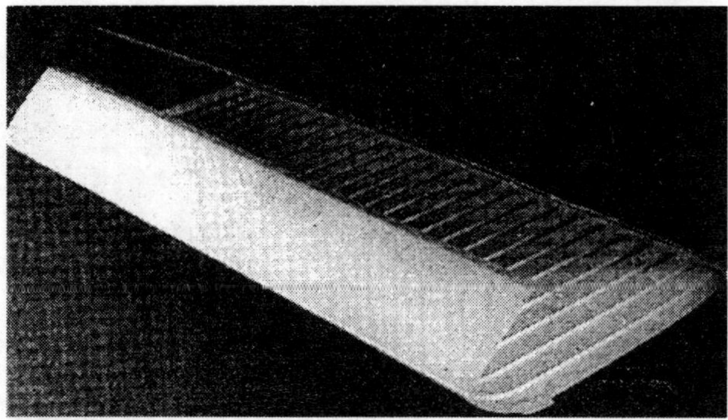

Fig. 8 One-piece tab skin of rolled-ribbed sheet: formed and trimmed, before assembly.

An equivalent structure made with rolled-ribbed sheet is illustrated in Fig. 9. The improvement in smoothness is evident from the photographs. Only half as many ribs were required, and the necessity for contour-formed chords was eliminated. The aft beam or "false spar" was also constructed of rolled-ribbed sheet. Comparative data and cost estimates for the 60-in. section in conventional structure and as designed with rolled-ribbed skin are presented in Table 2.

Aileron trailing edge. On lightly loaded trailing-edge structure, the use of rolled-ribbed sheet with stiffening elements oriented chordwise often provides

Table 2 Cost comparisons: rolled-ribbed structures

	Aileron tab		Wing aft structure	
	Integrally stiffened construction	Conventional construction	Integrally stiffened construction	Conventional construction
Weight (total weight per unit, lb)	4.65	4.80	28	28
Number of parts				
Total number detail parts per airplane	16	64	116	152
Total number different parts per airplane	7	25	47	61
Number of attachments				
Rivets per assembly	220	490	1310	1380
Spotwelds per assembly	——	560	——	2260
Cost (based on production of 50 airplanes)				
Material	$ 25.02[a]	$ 4.57	$120.30[a]	$ 17.63
Labor	46.16	84.57	102.40	227.90
Tooling	58.05	112.85	79.60	123.10
	$129.23	$201.99	$302.30	$368.63

[a]Includes total cost of machining the integrally stiffened sheet from 0.125-in. plate stock.

Fig. 9 Wing aft structure with rolled-ribbed skin, view of lower surface.

sufficient rigidity and strength so that all internal framework can be eliminated. For example, the aileron trailing edge of the Constellation airplane is constructed of 0.020-in. skin supported by many light ribs spaced 4 in. apart. An alternate design was made using rolled-ribbed sheet; all ribs were eliminated, and a slight weight advantage was also obtained.

Miscellaneous. Several other advantageous applications for rolled-ribbed sheet have been suggested: ducts, tanks, shell structures, leading-edge skins, rib webs, fairing, equipment and furnishings, instrument cases, cabinets, junction boxes, cabin partitions, doors, etc. Many suggested uses are outside the aircraft field—for example, roofing and siding, household furnishings, containers, and architectural and interior-decorating purposes.

Fabrication Techniques

No new tools or special techniques are required to use rolled-ribbed sheet. It may be treated much like ordinary sheet stock, but sufficient bend radius must be allowed for bends across the stiffener sections. Samples of formed parts made on conventional tools are illustrated in Fig. 10.

Rolled-ribbed stock may be cut and trimmed with conventional power shears if the stiffeners are protected from the clamping rams by a piece of heavy sheet used as a buffer plate. The portion to be trimmed away is placed under the blade

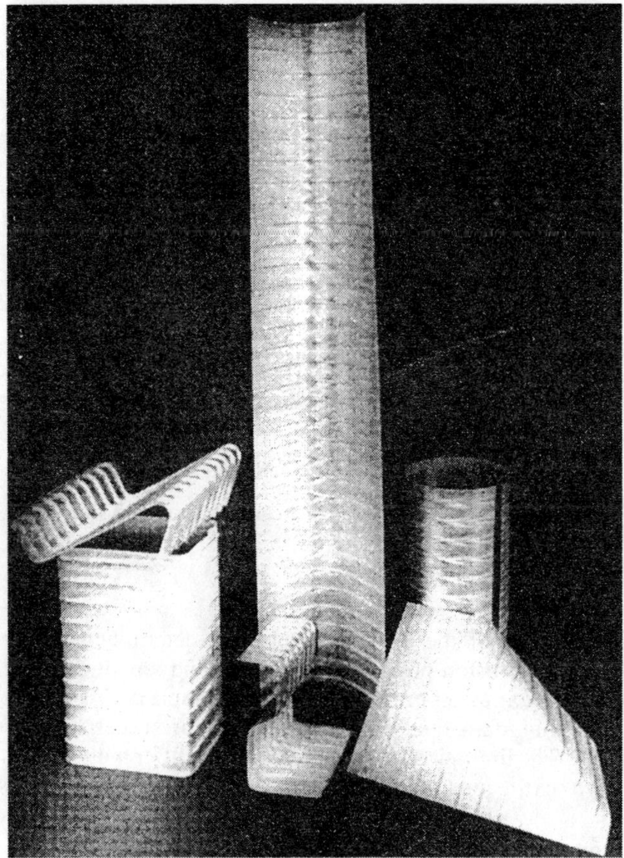

Fig. 10 Sample forming in rolled-ribbed sheet: conventional sheet-metal tools were used.

since this becomes irregularly bent. Gentle curvatures may be handled on ordinary rolls; sharp bends may be made with the power brake using the conventional punch and dies. Excellent joggles have been formed on the power brake using shims and plates.

In some cases it is desirable to remove the stiffener elements over a portion of the rolled-ribbed sheet to permit sharp bends (such as for a tab trailing edge) or to allow attachment to a plain flange. This has been accomplished satisfactorily as a routing operation, using a flat-bottom cutter turning at high speed and clamping plates adjacent to the area to be worked.

Extruded Integrally Stiffened Sheet

Previous Work

Several shapes of practicable dimensions for integrally stiffened aircraft surfaces have already been extruded although as far as is known none have as yet been used for this purpose.

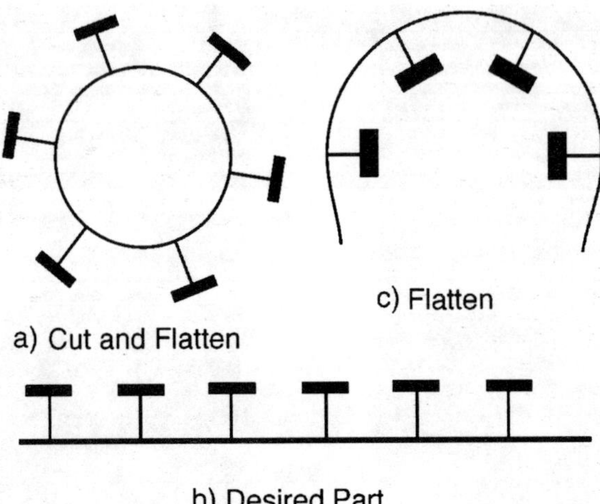

Fig. 11 Various methods of producing desired shape using extrusion.

The maximum size of the extruded section is determined primarily by the dimensions of the extrusion press (die diameter), and length is limited by billet size. The available extrusion pressure limits the complexity of the section shape and determines minimum thicknesses and fillets for successful extrusion of a particular shape. The limitations are interrelated and differ with various materials; the larger size extrusions in the harder alloys must have correspondingly thicker webs.

To take maximum advantage of press capacities, Dow Chemical Company several years ago produced an "octopus"-type section similar to that shown in Fig. 11a. This circular section was then split, heat treated, and flattened to form a 17-in. wide sheet with integral bulb stiffeners. Dow Chemical Company has also produced a 9-in. wide T-stiffened section, similar to Fig. 11b, by extruding in the flat. Both of these sections were made in magnesium alloy. Difficulty was experienced in obtaining satisfactory flatness in the surface after heat treatment.

Current Availability

Section shape. A logical configuration is provided by regularly spaced T-stiffeners, the spacing being determined by skin-buckling requirements and the stiffener flanges providing high crippling strength and high section moment of inertia. An unsymmetrical flange may be advantageous for attachment but is not quite as efficient (tends to roll) in bending or column action, particularly if the web and skin are thin.

Flat extrusion. In Fig. 11b, a section is shown which might be used in the upper-surface structure of the wing of a large transport airplane. This section is naturally limited in width by the extrusion press die diameter.

Circular section extrusion. Under a contract with the U.S. Air Force, Reynolds Metals Company is proceeding with the development of tools and processes to extrude the section shown in Fig. 11a and split, heat treat, and straighten the product. Successful results have been obtained in extruding the hollow shape in 61S, 24S, and 75S alloys, and it is believed that parts of widths and thicknesses comparable to standard sheet sizes can be produced on present equipment. Considerable variation in section details will also be permissible. This method of extrusion makes possible as a future development the production of a tapered-skin integrally extruded section.

Semicircular sections. To obtain sections of width larger than available die diameters and yet avoid some of the extrusion and straightening problems encountered in Fig. 11a, the shapes shown in Fig. 11c have been proposed by Alcoa. The thickness limitations in this case are approximately the same as for flat extrusions.

Application

For surface structure such as wing panels, fuselage, and empennage surfaces, where loads are high and these are few irregularities in contour or structural continuity, integrally stiffened extrusion shows exceptional possibilities. Extruded sections have high physical properties, and, because of the excellent detail configuration, they will generally carry unusually high compressive stresses even with widely spaced supports. Thus, both weight advantage and simplification of internal structure can be obtained.

The T-stiffener configuration makes for simplicity and efficiency where attachments, splices, and joints are required. Because extrusion will be obtainable in long lengths, reduction in the number of chordwise splices is generally possible. For some applications it may be worthwhile to machine the extruded section to obtain taper in skin thickness and section depth and to make integral reinforcements and attachment pads.

Press Forging

Production of thin-skinned surface panels with deep-section integral stiffening by press forging has become more feasible with recent advancements in technique and press capacity. During the war, Germany had a press of 30,000 tons capacity; some large slab-shaped parts were made in quantity on this press, but none approached stiffened-skin surface structure in mininum thickness or complexity of detail. The largest press in this country has a capacity of 18,000 tons and is operated by Wyman-Gordon Company in Worcester, Massachusetts.

Current Development

Under a U.S. Air Force contract, Lockheed is now working with the Wyman-Gordon Company to determine if integrally stiffened structure, such as that shown in Figs. 1 and 2, can be profitably produced by press forging. Considerable progress has been made in development of dies and in forging technique. Results

Fig. 12 Integrally stiffened test section produced by press forging.

of an early phase of the program are illustrated in Fig. 12. This panel is 25 in. square with 1-in. stiffeners spaced at 1½-in. intervals. A few years ago, the forging of such specimens was considered impossible; results to date definitely prove the feasibility and practicability of making large surface panels by this method although more development work remains to be done.

Forging to Contour

The question of forging to finished contour or forging flat is still to be answered. Forging dies that will produce a part to finished contour may be more expensive than dies producing only a flat blank. However, if the dimension can be held within required limits, it would eliminate forming. If dimension cannot be held or surface smoothness is insufficient, other additional machining or straightening operations are required. If the forging could be made to required dimension and surface smoothness was satisfactory, the problem of machining for desired accuracy at joints and other attachment areas would be more difficult in a formed part than in a flat.

Applications

Press forging would practically eliminate fabrication and assembly operations in the aircraft plant. In certain cases, considerable weight advantage may also be found. The wing panel illustrated in Fig. 1 weighs 76 lb in conventional construction; as an aluminum-alloy forging, its weight is only 56 lb; if forged from magnesium alloy, 46 lb.

Figure 13 illustrates the design of a wing-surface panel for press forging. This panel must carry both end compression and normal air pressure; in this case, this is best accomplished with two sets of stiffening elements. In the forged design, these are obtained by the waffle-grid arrangement. The weight is the

Fig. 13 Press-forging design simulated by machining: proposed one-piece surface panel for P-80 wing (splice was necessitated by limited material size).

same as that of the conventional design, but the part is made of one piece instead of 49.

Another example of a part suitable to the press-forging method is the one-piece bulkhead rib shown in Fig. 2.

Casting

Casting as a production method has inherent advantages because it eliminates many operations in preparation of the material as well as in fabrication and assembly of the part. It also permits the complexity of detail which is usually necessary for high structural efficiency. That it is not used more widely is due to the following disadvantages:

1) Material properties. Strength of cast material is generally only about half as much as that for the wrought alloy.

2) Technological limitations. As the size of castings is increased, the problems of warpage, shrinkage, homogeneity, and porosity become more severe. If minimum thicknesses are too large, cast surface structure will not be able to compete on a weight basis.

Studies were made of the possibilities of both aluminum- and magnesium-alloy castings for wing-surface and tail-surface panels for the F-80 airplane. Weight penalties were indicated in every case, but casting should be given consideration in connection with new design problems, not only because of the possible effect of special requirements or conditions, but also because of recent advancements in the technique of casting.

Conclusions

The major cost item in present-day aircraft construction is assembly and assembly tooling. Through integral stiffening design—combining skin and skin stiffeners in one unit—assembly costs may be greatly reduced.

Additional advantages are found in the inherent high rigidity and surface smoothness of integrally stiffened construction.

Many different methods may be used to achieve integrally stiffened construction. Some show disadvantages in parts fabrication and material cost which offset the savings in assembly. For example, machining from thick plate affords a good method of experimental or prototype construction, and, as a production method, it is itself worthy of serious consideration. However, fabrication costs are high, and the large material wastage incurred is basically unsound.

Material costs are much lower for parts produced by extrusion, but, because this method results in a constant section, machining will generally be required to gain the additional advantages of thickness taper, integral doublers, and integral pads. Meanwhile, the present high cost of machining (fabrication cost) can be reduced by the use of suitable high-speed machining equipment and, if warranted by quantity production, the development and use of automatic controls.

For lightly loaded portions of the aircraft structure, a light-gage rolled-ribbed sheet offers great possibilities for simplification and reduction in cost and the additional advantage of external smoothness.

Press-forging will be especially advantageous for large quantity production; in such cases, it offers great flexibility of detail design and maximum simplification and reduction in number of parts. It will probably have size-vs-minimum-thickness limitations determined by press capacities.

It appears that, by the combined use of several of these methods, the overall cost of aircraft construction can be materially reduced. Much more development work must be done before the advantages can be fully realized.

Bibliography

Altwicker, H., *Large Forgings in Germany*, Industrial Planning Division, Directorate, Procurement and Industrial Planning, Air Material Command, U.S. Air Force, Wright–Patterson Air Force Base, Dayton, OH.

Boyajian, A. Z., "The Design of Simplified Structures for Low-Cost Airplane," *Aeronautical Engineering Review,* Vol. 5, No. 7, July 1946, pp. 18–24, 29.

Dow, N. F., and Hickman, W. A., *Structural Evaluation of an Extruded Magnesium-Alloy T-Stiffened Panel*, NACATN No. 1518, Feb. 1946.

Mohler, O. E., *Saving Time and Money,* Pegasus, Fairchild Engine and Airplane Corporation, Jan. 1949.

Sandwich Construction

N. J. Hoff
Polytechnic Institute of Brooklyn and *Skydyne, Inc.*

S. E. Mautner
Skydyne, Inc.

What Is Sandwich Construction?

The characteristic feature of the sandwich construction is the use of a multilayer skin consisting of one or more high-strength outer layers (faces) and one or more low-density inner layers (core). In particular, airplane structures built according to the sandwich principle are true monocoques—that is, they are hollow bodies having a comparatively thin skin of the sandwich type which carries all the loads imposed upon the structure and at the same time provides an aerodynamically clean, smooth, streamlined surface. In sandwich monocoques inner bracing elements are needed only where concentrated loads are applied so that elsewhere the entire interior of the structure is available for crew, passengers, cargo, and equipment.

The use of a low-weight spacer between the high-strength load-carrying layers of the sandwich skin results in a light, yet stable, skin construction that does not buckle under the loads that occur in flight and at landing. Materials at present available for the faces are plywood, wood pulp fibers, and aluminum alloy; for the core, cork, balsa wood, and synthetic materials such as cellulose acetate.

Historical Development

True monocoque airplane structures consisting only of a strong, thin, lightweight, smooth shell with no external or internal bracing elements have long been the dream of the airplane designer. They enclose the motor, pilot, passengers, and cargo with the least amount of waste space, and, thus, their cross-sectional area is held to a minimum. Moreover, their surface is smooth and their shape streamlined. For these reasons, monocoques offer the smallest possible resistance to flight and are ideal for economic air transportation.

When in the 1920s the First-World-War-type biplane was abandoned in favor of the streamlined monoplane, efforts were made to build the monoplane as a true monocoque. It turned out, however, that the monocoque skin buckled under small compressive or shearing stresses, and the structure collapsed prematurely unless it was reinforced internally. In some of the wooden monocoque fuselages

Aeronautical Engineering Review, Vol. 3, No. 8, Aug. 1944, pp. 29–43, 63.

built in those days, the reinforcing elements were only transverse rings, also known as frames or ribs. The great weight of this construction, the heterogeneity of wood and its low resistance to humidity, as well as the general dislike of engineers for wood soon resulted in the replacement of wooden monocoques by aluminum structures.

Unfortunately, aluminum-alloy sheet is even more liable to buckle than commercial plywood is. Thus, it was necessary to add longitudinal reinforcing elements to the transverse rings to keep aluminum-alloy aircraft from failing at low values of the stress. The longitudinal reinforcing elements are known as longitudinals, stringers, or cap strips.

The design and manufacture of the aluminum-alloy-reinforced monocoque, or semimonocoque, was developed to great perfection in the 1930s. As is well known, practically all the modern fighter and bomber planes are built as semimonocoques. According to present-day practice, greatest stability and strength coupled with smallest weight are attained if the sheet metal used for covering, as well as for bent up or drawn reinforcing sections, is extremely thin and the number of reinforcing elements large.

The drawbacks of this type of construction are rather obvious. It contains so many small, thin-walled elements riveted together that its manufacture is expensive. Jenkins stated in a paper[1] that there are 3,000,000 rivets in a large-sized aircraft and that approximately 50% of the total cost of the airframe is made up of the cost of fastening together the many pieces that constitute the structure. In the authors' opinion, the modern aluminum-alloy semimonocoque attained its present form mainly because the development was brought about by big orders of the armed forces who are not deterred by high production cost. Under present conditions, the only way to build airplanes even halfway economically is to build them in series greater than presumably ever needed in peacetime production. With smaller orders, the cost of tooling up for series production is prohibitive.

Since the reason for the complexity and high cost of the structure is the low stability—that is, the liability of the aluminum-alloy skin to buckle—the obvious thing is to look for another skin that does not have this drawback. It is known that the stability of the skin depends primarily on its thickness. Therefore, low-density materials that can be used in thick gages without increasing the structural weight are at a premium for monocoques. It appears, however, that no single material is now available that combines the required light weight with the necessary strength. It is logical, therefore, to make use of a thick layer of ultralightweight material sandwiched between two layers of thin, but strong, sheet.

The authors are unable to present a complete history of the idea of the sandwich construction in aircraft. To their knowledge, the earliest description of a sandwich wall is contained in a patent granted in 1924 to von Kármán and Stock.[2] S. E. Mautner began experimenting with a plywood-cork sandwich system in 1934 in the airplane plants of the Schneider-Creusot concern in France. In 1938, he showed at the Salon d'Aeronautique a low-priced cabin monoplane having a sandwich wing. In Great Britain, de Bruyne[3] has been working for many years on the utilization of reinforced plastics in aircraft. It appears that the sandwich system was first adopted in England in the fuselage and in the leading edge of the aileron of the DeHavilland Albatross. The system attained worldwide publicity when it was disclosed that the almost legendary DeHavilland Mosquito bomber had a sandwich-type fuselage.[4] As far as is known, the first sandwich airplane part manufactured in this country was an aileron built by Skydyne in 1939.[5]

Strength and Stiffness of the Sandwich Skin

Experimental Data Available

There is little information available in technical literature on the strength and stiffness of the sandwich construction. Some data are contained in an article by Gough, Elam, and de Bruyne[3] published in 1940 and in a more recent one by Gordon[6] published in 1943. In the past year, experiments have been made with sandwich-type test specimens at the Polytechnic Institute of Brooklyn by Albert J. Culleu and Joseph Kempner under the direction of N. J. Hoff. A short report is presented here on three test series on tension, torsion, and bending, respectively, carried out with sandwich material having a balsa core and spruce plywood faces. These tests were considered preliminary to a more intensive research program, and no attempt was made in them to control the moisture content of the wood or to establish averages of great numbers of individual tests. Consequently, there was a considerable amount of scatter in the test results. Nevertheless, it was possible to observe typical trends, which are reported here.

Tension Tests

In the tension test series altogether 20 specimens were tested. All specimens had spruce plywood faces consisting of two mutually perpendicular veneers $1/64$ in. thick arranged at 45 deg to the direction of tension. The thickness of the balsa core varied from $1/16$ to $1/2$ in.; the specific weight of the balsa, from 10.4 to 15.0 lb per cu ft; and the fibers of the balsa wood were parallel to the direction of tension.

The tests were carried out on a Fairbanks lever-type testing machine of 2000-lb capacity. The test specimens were attached to the machine by means of pins and lugs to insure absence of end moments. Figure 1 shows the test set-up; Fig. 2, broken specimens. In about half the tests, strains were measured by Huggenberger extensometers; in the other half, by Baldwin-Southwark metalectric strain gages. Separate strain readings were made on the two faces of each specimen, since the presence of a bending moment could be detected from the occurrence of differences in the strain readings. When such differences were noticed, the specimen was shifted until the two gages gave identical readings. This procedure was considered necessary because of the variations in the elastic properties of wood which may give rise to bending moments even though the load is applied along the geometric centerline of the specimen. The accuracy of the load readings was ± 1 lb. The Huggenberger extensometers had a magnification of approximately 320 and could be read with an accuracy of about 10^{-4} in. per in., while the accuracy of the electric strain readings was of the order of 10^{-6} in. per in. For each specimen, strains were plotted against loads, and, in each case, the curve obtained was straight up to failure. A typical plot is shown in Fig. 3. An average ultimate tensile stress F_{tu} was obtained from the test data by dividing the load by the total cross-sectional area of the specimen, and an apparent modulus E_{ap} was calculated by dividing the difference between two suitably chosen average

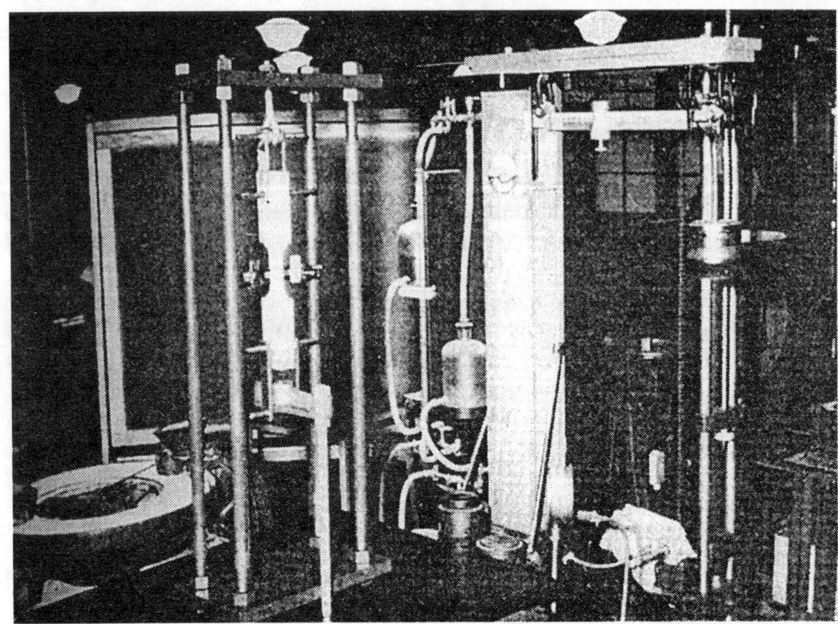

Fig. 1 Tension test set-up with Huggenberger extensometers.

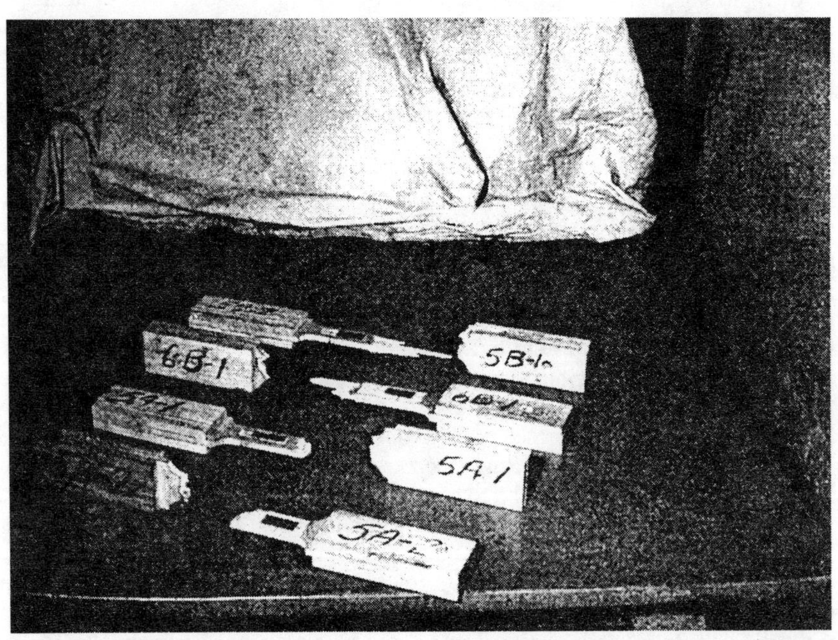

Fig. 2 Tension test specimens after failure.

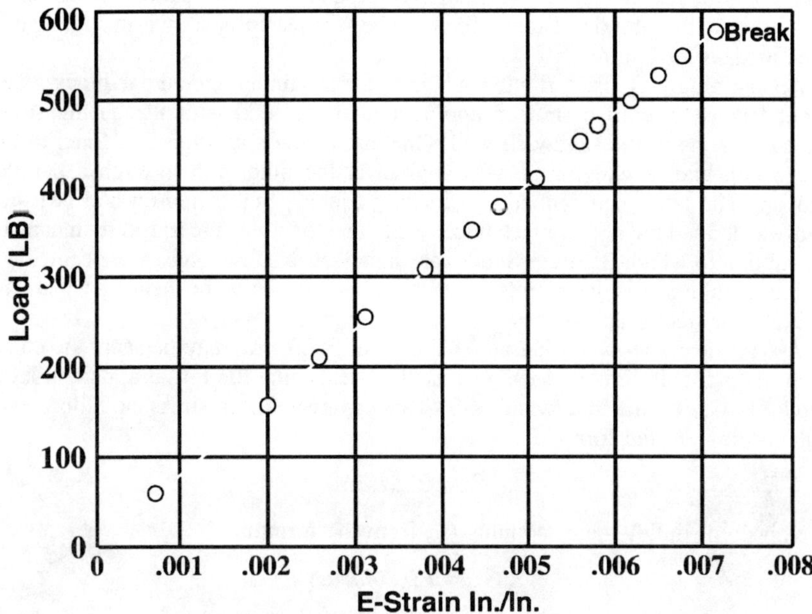

Fig. 3 Typical load-strain diagram in tension.

stress values by the corresponding difference in strain. The averages of the 20 test values are:

$$F_{tu} = 3500 \text{ lb per sq in.}$$

$$E_{ap} = 715{,}000 \text{ lb per sq in.}$$

No definite variation was observed in the values of the failing stress with increasing thickness of the balsa core or with increasing density of the balsa wood. However, a slight trend toward increasing values of the apparent modulus with increasing thickness of the core and increasing density of the balsa wood was noticed.

It is well known that plywood is little suited to carry tension applied at an angle of 45 deg to the direction of the fibers. It must be concluded, therefore, that the tensile strength of this type of sandwich skin is derived almost entirely from the balsa core. The strength-to-weight ratio was not unfavorable, since the average weight of the sandwich skin was only 16 lb per cu ft.

Torsion Tests

In the second test series, 17 specimens were tested in torsion. The basic group in this series consisted of three circular cylindrical specimens of $3\frac{5}{8}$ in. overall diameter and 9 in. clear length. The sandwich wall contained a balsa core of $\frac{1}{4}$-in. thickness and two plywood faces each consisting of two $\frac{1}{64}$-in. spruce veneers.

The fibers of the veneers were mutually perpendicular and subtended an angle of 45 deg with the axis of the cylinder. The test set-up is shown in Fig. 4; two specimens after failure, in Fig. 5.

An oak plug was inserted into each end of the cylinder and an oak block glued around it in order to permit an application of the load without causing stress concentrations in the sandwich wall. One end of the specimen was fixed to the test stand, and a pure torque was applied at the other end through a parallel linkage. The latter was pinned to a loading triangle that rested on ball bearings and was loaded through a mechanical jack. The load was measured by means of a load link to which Baldwin-Southwark metalectric strain gages were cemented. Relative angular displacements of the two end blocks were measured with the aid of long steel scales and hands.

A typical torque-twist diagram is shown in Fig. 6. As may be seen, the curve is not straight. Its initial slope was used to determine the initial apparent shear modulus G_{ap} of the sandwich skin. The apparent shear stress at failure was calculated from the formula

$$F_{su} = T_{max}/(2At)$$

the apparent initial shear modulus G_{ap} from the formula

$$G_{ap} = TSL/(4A^2t\alpha)$$

where

T = the applied torque in in. lb
T_{max} = the maximum applied torque in in. lb
A = the area included by the median line of the wall of the section in sq in.
t = the total wall thickness in in.
S = the developed length of the median line in in.
L = the free length of the specimen in in.
α = the angle of twist in radians

The average values obtained in the tests are:

$$F_{su} = 1116 \text{ lb per sq in.}$$

$$G_{ap} = 85{,}666 \text{ lb per sq in.}$$

It is known that the best arrangement for plywood to carry torsion is that in which the grain subtends an angle of approximately 45 deg with the plane of the torque. On the other hand, if balsa wood alone is tested in shear, it fails under small loads. It was expected, therefore, that in the torsion test specimens the balsa core would play little part in resisting torsion. This anticipation, however, was refuted by a group of tests in which one end of the specimen had only an internal oak plug and the other had only an external oak block. Obviously, in this test arrangement, the torque had to be transmitted through the balsa core. Nevertheless, the values of the apparent failing shear stress and the apparent initial shear modulus were only 28 and 15% smaller, respectively, than the values obtained with the specimens of the basic group. It is believed that the effect of the plywood in restricting the deformations of the balsa core may have been responsible for this unexpectedly favorable result.

Fig. 4 Torsion test set-up.

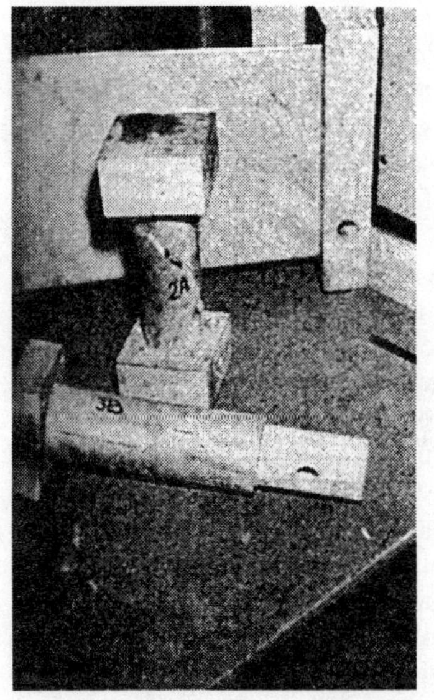

Fig. 5 Torsion test specimens after failure.

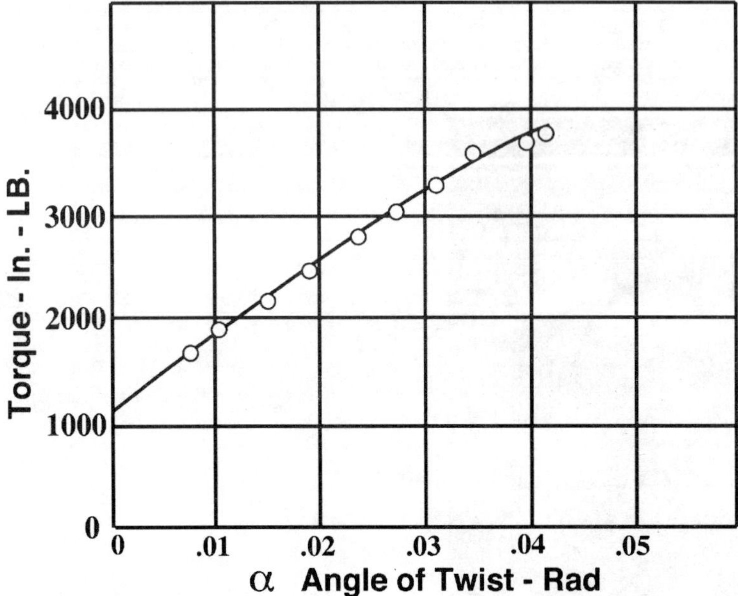

Fig. 6 Typical torsion torque-twist diagram.

Bending Tests

In the third series of tests, 25 specimens were tested in bending. The specimens were 15 in. long and 2 in. wide and had on both sides two $\frac{1}{64}$-in. thick layers of spruce veneer with mutually perpendicular grain arranged at an angle of 45 deg with the lengthwise direction. The thickness of the balsa core between the plywood faces varied from $\frac{1}{16}$ to $\frac{1}{2}$ in.; the specific weight of the balsa wood, from 10.4 to 15.0 lb per cu ft; and the fibers of the balsa ran in the lengthwise direction.

The tests were carried out on a Fairbanks lever-type testing machine of 2000-lb capacity. Figure 7 shows the test set-up in which the specimen was supported by two $\frac{1}{4}$-in.-diam dowels. One of the dowels was held in its supports by two ball bearings in order to facilitate movement of the specimen when loaded. To avoid crushing of the specimen under the concentrated load at its midpoint, the force was applied through a smooth wooden block having a convex surface in contact with the specimen. Arrangement for two different distances between the dowels may be noticed in the photograph. This was necessary since the thinner specimens exhibited large deflections under loads.

Loads were measured with an error of ± 1 lb. The order of magnitude of the error in the deflection readings was 10^{-4} in. Deflection readings were obtained through the use of a calibration constant correlating deflections to the revolutions of the loading wheel of the testing machine.

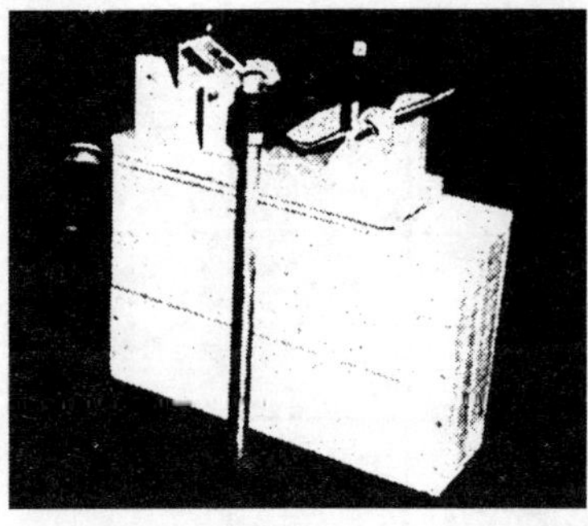

Fig. 7 Bending test set-up.

It was observed that, after the maximum load had been reached, the specimen continued to hold and exhibited extremely large deflections before actual rupture occurred. Practically all specimens failed through fracture of the fibers on the tension side at the middle of the specimen. Definite anticlastic curvature of the specimens was noticed in all the tests.

A typical load-deflection diagram is shown in Fig. 8. The modulus of rupture F_b is calculated from the formula

$$F_b = M_{\max} c/I$$

the apparent modulus of elasticity in bending from the formula

$$E_{bap} = PL^3/(48I\delta)$$

where

$$I = bt^3/12$$

and

$M_{\max}$ = the maximum bending moment in in. lb.
P/δ = the slope of the tangent to the initial portion of the load-deflection curve in lb per in.
b = the width of the specimen in in.
t = the thickness of the specimen in in.
c = one-half the thickness of the specimen in in.
L = the distance between supports in in.

The average modulus of rupture was

$$F_b = 5815 \text{ lb per sq in.}$$

the average apparent initial modulus of elasticity in bending

$$E_{bap} = 472{,}000 \text{ lb per sq in.}$$

No systematic variation of these values with thickness or density of the balsa-wood core was observed.

Comparison with Aluminum Alloy

The comparison of the merits of a material by means of a strength-to-weight ratio based on some arbitrarily chosen formula is misleading. The actual value

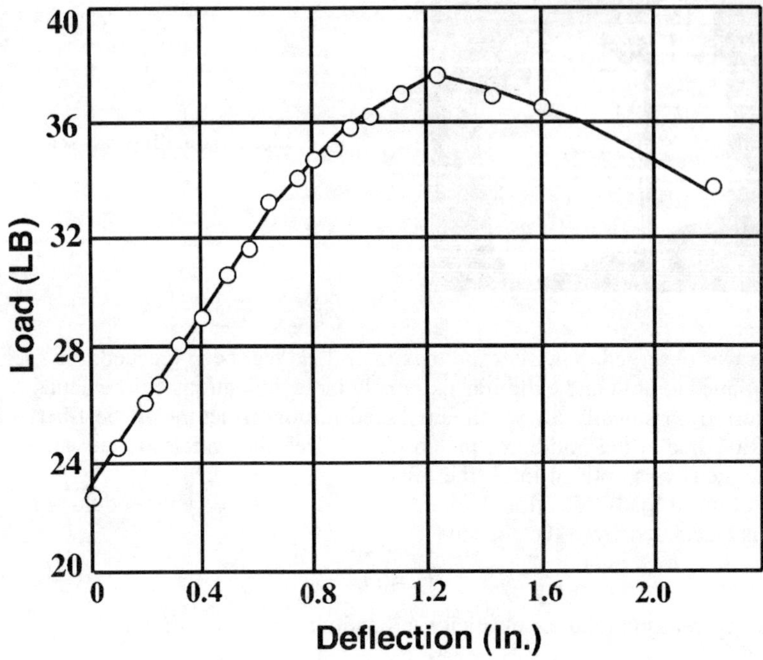

Fig. 8 Typical bending load-deflection diagram.

of a structure depends upon its behavior under many different kinds of loading, as well as upon considerations of design, manufacture, availability, price, etc. It is not intended, therefore, to establish a definite rating for the sandwich construction in the comparison that follows, but rather to give an indication of its advantages or disadvantages regarding the particular types of loading investigated in the experiments just described.

Since design considerations do not restrict the thickness of the skin of a monocoque, all comparison will be based upon the assumption of a thickness ratio of aluminum to sandwich skin corresponding to an equal weight of the two. With an average value of 16 lb per cu ft for the sandwich skin and 172 lb per cu ft for aluminum alloys, the thickness of the latter must be 0.093 times the thickness of the former. If the thickness of the sandwich skin is assumed to be $^5/_{16}$ in., that of the aluminum skin must be 0.029 in.

The comparison of the tensile properties is straightforward. A strip of sandwich skin 1 in. wide and $^5/_{16}$ in. thick carries $(^5/_{16}) \times 3500 = 1092$ lb. A strip of 24ST aluminum-alloy skin 1 in. wide and 0.029 in. thick fails under $0.029 \times 62{,}000 = 1800$ lb. Obviously, 24ST aluminum alloy is better for this kind of loading than is the type of sandwich construction investigated. The suitability of the sandwich system can be improved, however, through the use of spruce veneers with the grain arranged parallel to the direction of tension.

A similar consideration of the shearing strength as established in the torsion tests gives the following results. The sandwich strip fails under $(5/16) \times 1116 = 349$ lb. With an ultimate shear stress of 37,000 lb per sq in. the 24ST aluminum-alloy strip carries $0.029 \times 37{,}000 = 1070$ lb. Thus, aluminum alloys appear to be preferable in this case of loading as in the preceding one. The conclusion is not correct, however, since such a thin gage aluminum-alloy sheet buckles long before the ultimate shear stress is reached, unless it is reinforced, while all the twisted sandwich cylinders showed a clear tension break across the diagonal grain. If, however, the aluminum-alloy cylinder is reinforced, the weight of the reinforcing elements must be accounted for in the comparison. For a circular cylinder of 1.8-in. radius, 9-in. length, and 0.029-in. wall thickness, Donnell's formula[6] gives a critical stress in torsion $f_{cr} = 21{,}000$ lb per sq in. The actual critical stress is about two-thirds of the theoretical—that is, 14,000 lb per sq in. The use of this value in the place of 37,000 lb per sq in. results in a total failing load of 405 lb. This is still slightly higher than the load carried by the sandwich wall, but a radius of curvature of 1.8 in. is exceptionally small in aircraft structures. With increasing radius, however, the buckling stress of the aluminum-alloy cylinder decreases, and the advantage of the stable sandwich system becomes pronounced.

The section modulus of a strip 1 in. wide and $5/16$ in. thick is 0.0162 cu in. The bending moment carried by such a strip of sandwich material is $0.0162 \times 5815 = 94$ in. lb. The section modulus of a strip 1 in. wide and 0.029 in. thick is 0.00014 cu in. If the 24ST aluminum-alloy strip is assumed to have a modulus of rupture in bending of 100,000 lb per sq in., the bending moment carried by it becomes $0.00014 \times 100{,}000 = 14$ in. lb. From these data the advantage of the sandwich system is obvious for this kind of loading.

In conclusion it may be stated that, from the strength-weight standpoint, the advantages of the sandwich skin lie in its local bending strength and rigidity and in its stability which makes the danger of buckling remote. Preliminary tests of specimens having aluminum-alloy faces and a balsa or a synthetic core show even more favorable results, but no report on these can be made available at present.

Design and Manufacture

Figure 9 shows an aileron manufactured according to current practice in aluminum-alloy airplane construction and the Skydyne sandwich aileron designed to replace it. Figure 10 is a photograph of the cross section of the sandwich aileron

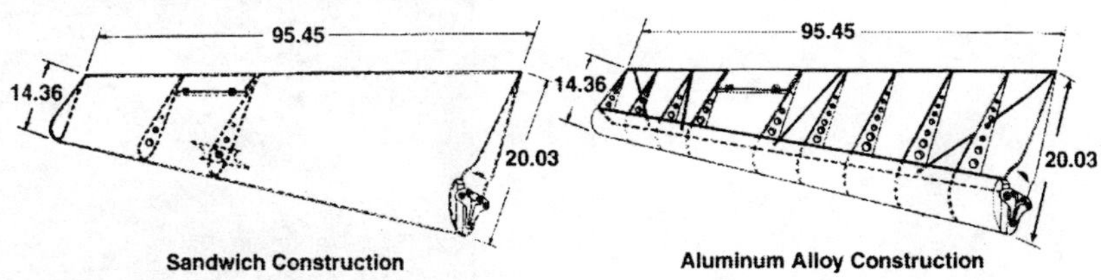

Fig. 9 Comparison of aluminum-alloy and sandwich ailerons.

and illustrates the arrangement of the balance lead. The original aileron weighed 23.75 lb including 11.75 lb of balance lead; the sandwich aileron weighed 22 lb, including 12 lb of balance lead. The simplicity of the design of, and the small number of parts in, the sandwich aileron are to be noted. The next few photographs illustrate different phases of the production procedure as developed for this particular job.

Manufacture of the aileron begins with cutting balsa wood on the band saw or rip saw and edge gluing it to panels of the required size. A panel is then bent to the shape of the leading edge. The balance lead is inserted and fixed temporarily by gluing it in place with rubber cement. A phenolic glue is applied by a spray gun to the surfaces of the leading-edge piece. Two other panels are planed to a tapering thickness, sent through the glue-spread machine, and edge glued to the leading-edge piece.

Two panels of mahogany plywood are prepared for gluing to the inner and outer surfaces, respectively, of the balsa-wood core. Each panel consists of two veneers $\frac{1}{64}$ in. thick. The grain in one veneer is perpendicular to that in the other, both being arranged at 45 deg to the spanwise direction.

The assembled plywood and balsa core are placed in a hardwood mold that is reinforced with heavy steel channels (Fig. 11). A rubber bag is inserted into the assembly, and an internal air pressure of about 40 lb per sq in. is applied. A heating pad is attached to the surface of the rubber bag, and the hardwood mold is provided with metal heating elements. The assembly is kept in the mold for 10 min at a temperature of about 220°F during which the gluing operation is completed.

The edges of the glued product are usually not quite satisfactory; therefore, they are trimmed about 1 in. to final size in a pattern. Figure 12 is a photograph of the completed aileron shell.

Figure 13 shows all the reinforcing elements needed for the aileron. They are the prefabricated ribs that close the two ends of the aileron and the tab cutout, the reinforcements at the points of application of concentrated loads, and the trailing-edge insert. After the application of phenolic glue to the surfaces of these elements, they are inserted in the aileron shell, and the entire assembly is placed into an assembly jig that locates the elements precisely. The assembly jig also serves as a press, its upper part being tightened to the lower one through eight bolts by which means pressure is exerted upon the glue lines of the reinforcing elements of the aileron. The assembly jig press is shown closed in Fig. 14 and open in Fig. 15. Heat is applied to the glue lines through metal strips provided

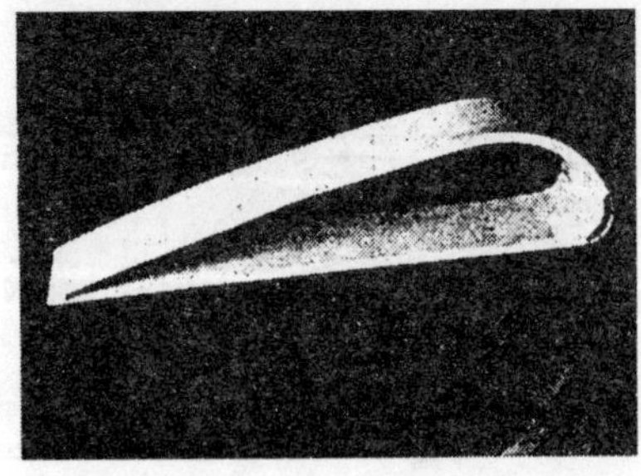

Fig. 10 Sandwich aileron shell section.

Fig. 11 Aileron inside pressure mold open.

Fig. 12 Aileron shell.

Fig. 13 All components of aileron.

Fig. 14 Aileron assembly jig press closed.

Fig. 15 Aileron assembly jig press open.

in the jig. A temperature of about 300°F is needed in this operation since the heat is supplied to the glue line from only one side. The trailing-edge insert is held firmly in place by a number of horizontally arranged screws which are clearly visible in the photographs.

When the aileron is taken out of the assembly jig press, a normal dipping operation is undertaken in order to seal it. After the few external attachments have been screwed on and a finish coat of enamel is applied, the aileron is ready for use.

The manufacturing procedure just described is a typical example and applies in all detail to the particular aileron only. It can be modified to suit special demands, and various improvements on it are possible. The use of molds cast in plastics and heating by high-frequency electric current may be mentioned as examples.

Another airplane part built according to the sandwich principle is shown in Fig. 16. It is a helicopter boom, or tail cone, molded all in one piece and weighing 25 lb. The overall length of the boom is 175 in.; its diameter at the front end, 20 in.; at the rear end, 6 in. The required ultimate bending moment at the front end was 84,000 in. lb, the permissible total deflection at the rear end 1½ in. The manufacturing procedure used in this case differed from the one just described inasmuch as the internal rubber bag was replaced by an external rubber bag from which the air was sucked out by an ejector. An interesting detail of the design is the application of an aluminum-alloy reinforcing collar to the front end of the boom. It is cyclewelded to the wood and serves the purpose of reducing the

Fig. 16 Helicopter boom.

bearing pressure and the shear stress in the wood at the points where high concentrated loads are introduced into the tail cone through the attaching bolts.

Conclusions

It was shown in the section on the history of the sandwich construction that the incentive for its development was given by the desire to build a true monocoque airplane. The true monocoque has long been recognized as the ideal in aircraft construction because of its excellent space utilization, its suitability for streamlining, and its inherent smoothness of surface.

The same properties are also present in the modern aluminum-alloy semimonocoques, but they are achieved at the cost of great complexity of the structure caused by the longitudinal and transverse reinforcing elements necessary for the stabilization of the thin aluminum-alloy skin. Because of the great number of small reinforcing elements, the complexity of their fitting, and the labor involved in their riveting, aluminum-alloy semimonocoques are inherently expensive. The cost can be reduced only in mass production in which, however, the expense of tooling up is great.

The armed forces of the world are willing to pay the cost of this expensive production, but the private owner will certainly look for a more reasonable design. The aluminum-alloy semimonocoque does not have a good chance, therefore, in the postwar light plane field, particularly since the numbers of each type produced are likely to be small and, thus, no advantage could be gained from mass production.

The drawbacks of aluminum-alloy semimonocoque construction are not found in sandwich construction. It combines the advantages of monocoque construction with inexpensiveness in manufacture. It is well suited to small series production, since the small number of molds and jigs needed can be built of wood at a low cost.

The structural strength and efficiency of the sandwich construction has advanced rapidly in recent years. Phenolic glues increased the strength, weathering properties, and mold resistance of the skin; the development of the cycleweld procedure made possible the use of metal for the load-carrying outer layers. This last item should be particularly emphasized, since it opens up a field of entirely new possibilities for the efficient design of fittings which formerly was the weak point in wood construction.

A few additional merits of the sandwich skin are the damping of vibrations and the good insulation against noise and heat. Its attractiveness for the light cabin plane is enhanced by great local strength and rigidity. The pilot and the passenger need not be afraid of leaning against a sandwich wall or stepping on any part of a sandwich floor. It can carry such loads safely. Indeed, a sandwich-type fuselage is itself a natural double-walled cabin which combines functional structural design with luxury. It will be possible to construct a small cabin plane on the sandwich principle which is just as streamlined and racy as the fighter or bomber plane flown at present by the potential postwar purchaser of the light plane at a cost he can afford.

References

[1] Jenkins, E. S., "Rational Design of Fastenings," War Engineering Annual Meeting of the Society of Automotive Engineers, Detroit, MI, Jan. 10–14, 1944.

[2] von Kármán, T., and Stock, P., British Patent No. 235,884, June 21, 1924.

[3] Gough, G. S., Elam, C. F., and de Bruyne, N. A., "The Stabilization of a Thin Sheet by a Continuous Supporting Medium," *Journal of The Royal Aeronautical Society,* Vol. 44, No. 349, Jan. 1940, p. 12.

[4] Hoff, N. J., "It's the Skin!" *Air Progress,* Vol. 3, No. 3, Sept. 1943, p. 13.

[5] Klemin, A., "Moulded Shell Construction in Plywood," *Aero Digest,* Vol. 42, No. 1, Jan. 1943, p. 148.

[6] Gordon, H. R., "Aircraft Design Considerations," *Modern Plastics,* Sept. 1943.

THERMAL-PROTECTION SYSTEMS

The Shuttle Tile Story

Paul A. Cooper and Paul F. Holloway
NASA Langley Research Center

Much has been written in both general and technical aerospace publications about the delay of the first launch of the Space Shuttle because of structural problems with the orbiter's ceramic reusable thermal-protection system (TPS). This article explains the problems and their resolution.

The orbiter has basically a conventional skin-stringer aluminum-aircraft structure. The properties of aluminum dictate keeping the maximum temperature of this structure below 350°F in operations. Aerothermal heating during ascent and re-entry will create surface equilibrium temperatures well above this figure and, in many places, will push them above the melting point of aluminum (1220°F).

This heating necessitates some form of insulation. The Shuttle design goal of 100-mission reusability with minimum turnaround time between flights ruled out ablative heat shields (except for some use between elevons). The need for reusability in effect dictated a lightweight, nonablative TPS able to protect the aluminum substructure from high surface temperatures and, at the same time, withstand the thermal cycles and environmental loads of spaceflight. Developing such a system greatly challenged the Space Shuttle's designers.

In the early 1970s, NASA and the Space Division of Rockwell International, the prime contractor for the orbiter, agreed to use a newly developed TPS ceramic material formulated by and manufactured by Lockheed Missiles & Space. It acts as an excellent insulation where surface equilibrium temperatures range between 750 and 2300°F. This ceramic, being highly brittle, has strain-to-failure performance considerably below the yield strain of aluminum and a low coefficient of linear thermal expansion. Thermal and mechanical expansion and contraction of the aluminum skin would cause ceramic material bonded to it to crack.

The designers decided to protect the reusable surface insulation (RSI) from excessive strain in two ways. First, the ceramic insulation was placed on the aluminum in the form of individual tiles with side dimensions of the order of 6 in. or less. About 30,000 tiles of various sizes and shapes cover slightly over 70% of the orbiter's exterior. Gaps between tiles permit them relative motion as the aluminum skin expands or contracts and as the substructure deforms causing skin-surfacc curvatures. The allowance for relative motion alone does not prove sufficient to protect the integrity of the ceramic material. A tile bonded directly to the structure will still crack as the aluminum skin under it expands locally or undergoes local lateral deformations. Some means must isolate the strain of the aluminum substructure from the tile. To do this, the tile was first bonded to a felt pad using an elastomeric, room-temperature-vulcanizing (RTV) silicone

Astronautics & Aeronautics, Vol. 19, No. 1, Jan. 1981, pp. 24–34, 36.

adhesive, and then the tile and pad combination was bonded to the aluminum skin with the same adhesive. This strain-isolation pad (SIP), having very low shear and extensional moduli, protects the brittle ceramic material from deformations of the aluminum structure.

Initially, the loads expected on the TPS came well within the strength of RSI and SIP, but as the design of the orbiter progressed, mission requirements became firmer, and load predictions became refined, it became obvious that the TPS would have to withstand loads higher than anticipated. Although they reduced margins of safety, the loads in most cases caused stresses within the strength of the RSI and the SIP as considered independently. However, tests of the RSI/RTV/SIP as a *system*—made for the first time about a year and a half ago—revealed the group tensile strength to be less than the tensile strength of the individual components. This caused negative margins of safety over large areas of the orbiter, and the TPS in certain areas did not have sufficient strength to survive the tensile loads of a single mission. That was the basic structural problem which contributed to the delay of the first flight.

This situation forced an intense effort—involving several NASA centers, industrial concerns, and universities—to understand thoroughly the TPS as a structural system and solve the problems associated with the high-load areas. As a part of this thrust, in the fall of 1979, NASA Langley Research Center began investigating static and dynamic structural mechanics of the TPS and its individual components. Using results of this investigation, we will attempt to give the reader some insight into the material and structural characteristics of the TPS, the problems in its design that contributed to the delay of the first launch, and the techniques used to resolve these.

TPS Description

The isotherm plot in Fig. 1 gives typical expected maximum surface temperatures for the orbiter in a nominal trajectory—a range of maximum surface equilibrium temperatures from a low of 600°F on the upper surface aft of the cockpit to a high of almost 2700°F at the fuselage nosecap. The orbiter's skin (mainly 2024, 2219, or 2124 aluminum, with graphite/epoxy used for the cargo-bay door) has, as mentioned, a designed maximum-use temperature of 350°F.

The various insulation procedures used for temperature control of the structural skin (Fig. 2) include reinforced carbon-carbon (RCC), two types of ceramic reusable surface insulation tiles, and a limited amount of nonreusable ablative material. RCC—a carbon cloth impregnated with additional carbon, heat-treated, and then coated with silicon carbide—has a reuse temperature of 2900°F. The two types of ceramic RSI tiles, one of which is made in two different densities, cover regions experiencing surface temperatures between 700 and 2300°F. A blanket of felt reusable surface insulation (FRSI) coated with room-temperature curing silicon covers areas where surface temperatures do not exceed 700°F during entry or 750°F during ascent. Other thermal barriers include thermal windows, aerothermal seals to restrict hot-gas flow into control-surface cavities, and tile filler bars (strips of coated felt bonded to the structural surface between tiles to prevent direct heat-radiation or convection to the skin). Tile gap fillers—pads made from an alumina mat covered with a ceramic fabric—bonded to filler

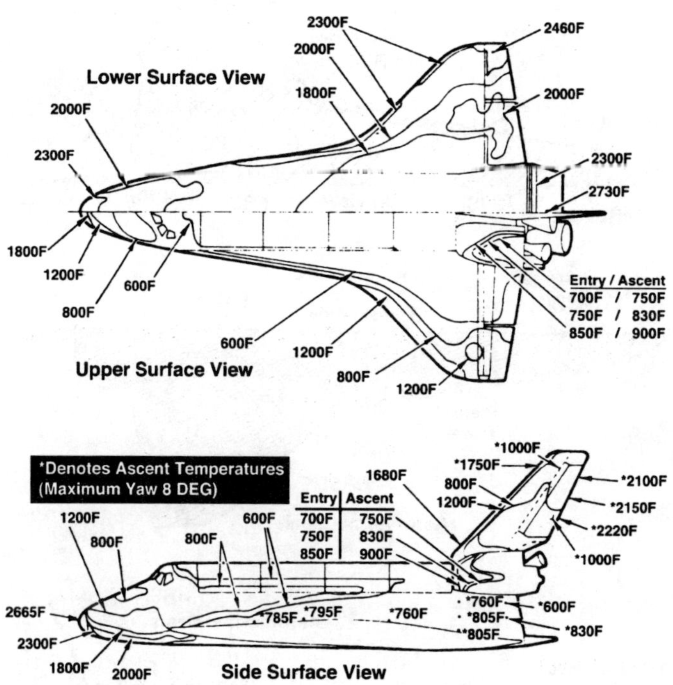

Fig. 1 Orbiter isotherms for a normal flight. Maximum surface temperatures expected for trajectory 14414.1C.

bars between tiles in high-pressure-gradient areas restrict the flow of hot gas between adjacent tiles. Pads of this type are also used as thermal barriers around structural penetrations such as landing-gear doors. Internal blankets insulate regions where RCC protects the structure from internal surface heat radiation and are used extensively throughout for thermal management.

There are three types of ceramic RSI: Class I, Class II, and LI-2200 tiles.

Class I tiles cover areas where the maximum surface temperature should run between 700 and 1200°F. They have a white ceramic coating with a low solar absorptance to help maintain low temperature in orbit by reflecting solar radiation. These tiles are designated low-temperature reusable surface insulation (LRSI).

Class II tiles cover areas where the maximum surface temperature should run between 1200 and 2300°F. Their black ceramic coating, with a high surface emittance, will efficiently radiate heat during re-entry. These tiles are designated high-temperature reusable surface insulation (HRSI).

Lockheed manufactures both the Class I and Class II tiles from a 9-lb/ft^3 ceramic RSI designated LI-900. The third type of ceramic RSI, designated LI-2200, has

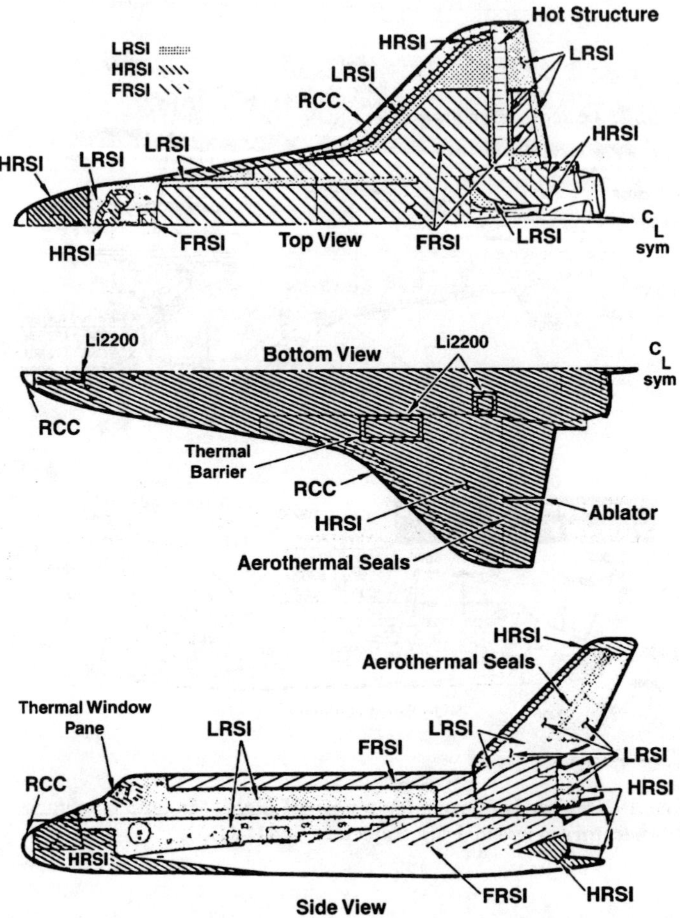

Fig. 2 TPS distribution.

the same coating as Class II tiles, but a higher density (22-lb/ft^3) and strength. The orbiter uses only a small number of LI-2200 tiles in areas of high concentrated loads, usually around penetrations such as landing-gear doors or in the forward fuselage area near RCC interfaces needing its higher heat resistance to handle surface temperatures which might reach 2600°F.

Table 1 shows the weight and surface area covered by the different types of external insulation. Tile thickness varies according to heat load and requirements for maintaining the aerodynamic outer moldline. The tiles run less than ½ in. thick for LRSI at the upper midfuselage region and as thick as 4–6 in. for HRSI on the body flap's lower surface. Both the LI-900 and LI-2200 tiles are cut and shaped from larger ceramic blocks to fit specific orbiter areas. The blocks are composed of compacted 1.5-micron-diam silica fibers bound together by colloidal

Table 1 Distribution of types of TPS[a]

TPS[b]	Area, sq ft	Weight, lb
FRSI	3,581	1,164
LRSI	2,741	2,376
HRSI	5,164	11,046
RCC	409	3,094
Miscellaneous	——	3,094
Total:	11,895	21,815

[a]Orbiter 102 configuration. [b]Includes bulk insulation, thermal barriers, and closeouts.

silica fused during a 4-h sintering process in which temperatures reach 2400°F. Voids comprise over 90% of the resultant tile. The tiles are then coated on five sides with reaction-cured glass (RCG) consisting of silica, boron oxide, and silicon tetraboride and glazed at 2000°F. A silicon polymer waterproofs the uncoated side.

Direct bonding of the tile to the aluminum substructure would excessively strain the highly brittle RSI and RCG coating. An isolation pad reduces the strain. This pad consists of a felt with a low shear modulus bonded to the uncoated side of the tile and the tile-felt assemblage bonded to the aluminum substructure.

Nylon fibers (trade name Nomex) form the pad. Repeatedly passing a barbed needle through the pad in a sewing-like procedure compacts the fibers oriented transversely to the pad.

Figure 3 depicts the complete TPS assemblage in schematic form. The majority of the strips of filler bar bond only to the aluminum substructure and not to the RSI, thus providing an SIP vent path during ascent. In selected areas, the tiles also bond to the filler bars to distribute loads over a larger area, reducing the stress on highly loaded tiles. The RCG coating on the sides of the tile does not extend to the filler bar. Thus, some uncoated surface allows the porous tile to vent.

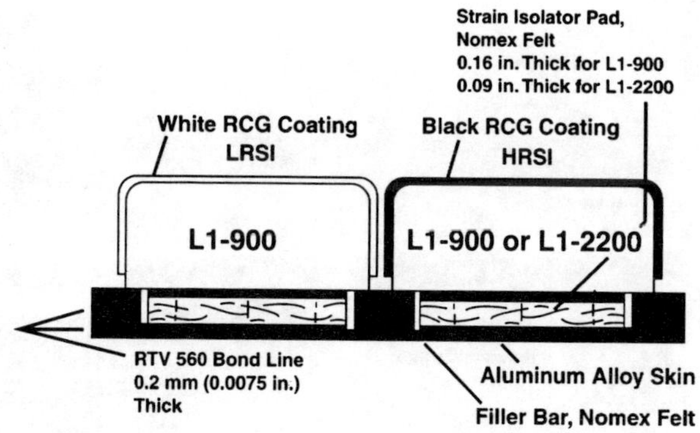

Fig. 3 Schematic of TPS assemblage.

TPS Flight Loads

Figure 4 describes a typical mission profile. The orbiter experiences a variety of loads during ascent. These must be considered in determining the limit loads on the TPS.

Ignition of the Shuttle's main engines creates an oscillatory pressure wave that loads tiles in the aft region of the orbiter, but this wave should damp rapidly. Engine noise can cause acoustic pressure loads directly on the tile and can excite aluminum substructure motion that in turn excites the TPS, causing inertial loads. These inertial loads vary for a given noise spectrum according to the local response characteristics of the aluminum substructure. Engine noise levels as high as 165 dB are expected in the vertical-tail region during liftoff.

During ascent, the tile experiences aerodynamic loads—aerodynamic pressure gradients and shocks, buffet and gust loads, acoustic pressure loads caused by boundary-layer noise, and concomitant substructure motion and unsteady loads resulting from vortex shedding from the connecting structure to the external tank.

Figure 5 describes the typical acoustic environment during ascent. Noise levels should exceed 160 dB over much of the orbiter's surface either at liftoff or during ascent. (Local load conditions during ascent, such as occur near attach points of the external tank during separation, require special TPS designs for local situations and will not be discussed further here.)

The TPS configuration will cause some additional loads. Since the tile and the SIP are highly porous and, thus, have an internal pressure which tends to

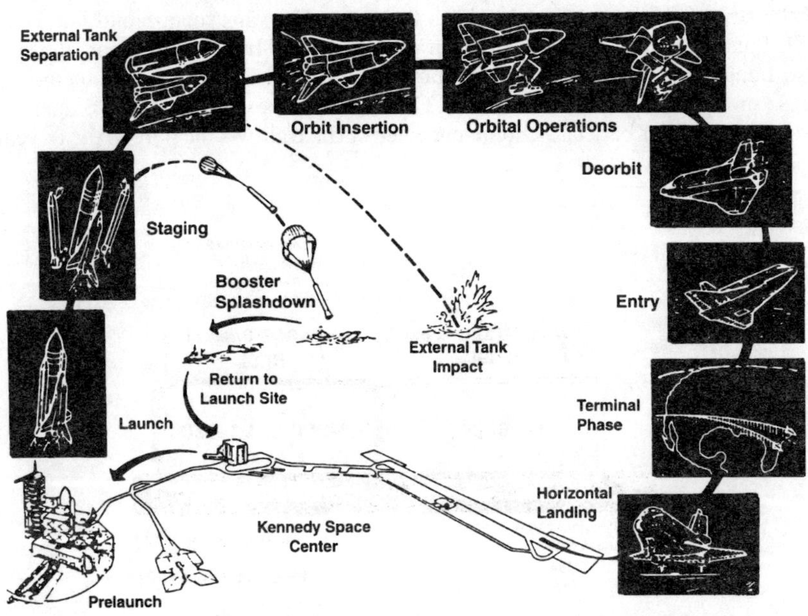

Fig. 4 Typical mission profile.

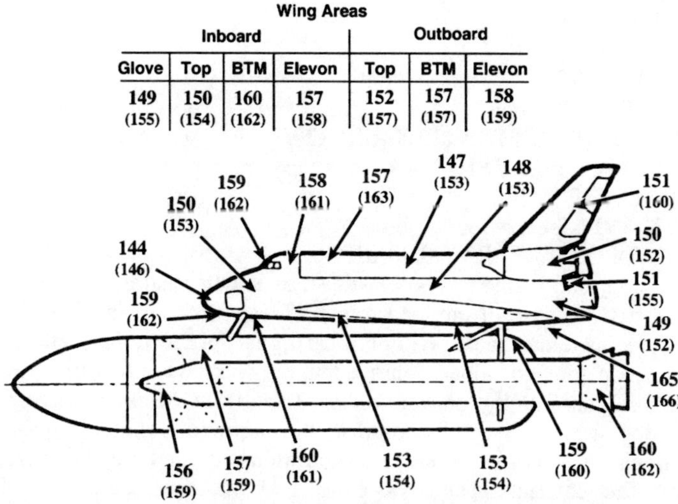

Wing Areas						
Inboard				Outboard		
Glove	Top	BTM	Elevon	Top	BTM	Elevon
149 (155)	150 (154)	160 (162)	157 (158)	152 (157)	157 (157)	158 (159)

Fig. 5 Ascent aerodynamic noise levels. Maximum space average levels for nominal and wind-dispersed (XXX) vehicle attitudes.

equalize to the external environment, the tiles will experience a varying internal pressure during ascent as the tiles and SIP vent. Tensile forces can develop in the TPS due to the differential pressure caused by the vent lag.

Besides stresses caused by flight loads and consequent out-of-plane substructure deformation, the TPS experiences small residual tile stresses as a consequence of coating/tile thermal-expansion differences created during the firing of the coating. Moreover, the SIP has mismatch stresses due to normal installation tolerances allowed between the tile bonding surface and the aluminum skin.

During re-entry, the orbiter experiences out-of-plane substructure deformation, boundary-layer acoustic noise, steady aerodynamic loads, unsteady aerodynamic loads caused by boundary-layer separation and vortices, loads from aerodynamic maneuvers in the landing approach, and landing loads. All must be considered in assessing TPS life.

After first flight, a large amount of data from development flight instrumentation (including continuous readouts from over 300 thermocouples, calorimeters, and radiometers and over 200 pressure and acoustic sensors) will help refine load and heating predictions, and actual flight data will improve estimates of TPS life.

What Reduced TPS Strength

Tensile loads applied normal to the SIP are transmitted across it at discrete regions along the *transverse* fiber bundles. Stress concentrations from this local load transfer cause a reduction of about 50% in *system* tensile strength for both the LI-900 and the LI-2200 TPS, as follows (values in psi, from internal Rockwell International documentation):

LI-900 ceramic RSI	24.0
0.160-in. SIP	41.0
RTV-560 adhesive	480
RTV/LI-900 RSI/RTV/0.160 SIP/RTV	11.7
LI-2200 ceramic RSI	60.0
0.90-in. SIP	68.0
RTV-560 adhesive	480
RTV/LI-2200 RSI/RTV/0.90 SIP/RTV	30.2

NASA Langley conducted a photoelastic study to demonstrate that the reduced strength occurred because of the load-transfer mechanisms of the SIP. A highly sensitive photoelastic material (giving optical signals proportional to the internal stress level when viewed under polarized light) was bonded to the SIP using RTV adhesive. The SIP/photoelastic-material interface exhibits discrete stress risers all along the interface. In contrast, the aluminum/photoelastic-material interface shows only corner stress concentrations, caused by the differential stiffnesses of the materials. The stress concentrations measured at the SIP/photoelastic-material interface (as high as 1.9) account for the reduced strength of the SIP/RSI assemblage.

Densification of Bonding Surface

Rockwell International and NASA Johnson and Ames investigated several procedures to strengthen this RSI/SIP interface. The most efficient of these, a densification of the bonding surface of the RSI, was selected in October 1979. In the densification procedure, the voids between fibers at the bonding surface is filled by a ceramic slurry—a mixture of DuPont's Ludox (a colloidal silica) and a silica slip consisting of a mixture of small particles of silica and water. A controlled amount of this mixture, pigmented to give an identifying light-gray color, is brushed on the surface to be densified. The mixture is air dried for 24 h and oven dried at 150°F for 2 h and then waterproofed by exposing the tile to vapors of methyltrimethoxy silane (Dow-Corning Z-6070) and acetic acid at 350°F.

This mixture provides a hard, strong, nearly continuous densified layer approximately 0.02 in. thick. The density decreases gradually toward the interior of the tile. Most of the densification material remains within 0.11 in. from the bonding surface for the LI-900 RSI.

Densification of the tile surface strengthens the SIP/tile interface sufficiently so that failure under a static tensile loading occurs in the parent undensified RSI below the densified region for the LI-900 RSI and either the RSI below the densified region or the SIP for the LI-2200 RSI.

Densification increases the average static tensile strength almost 100% for the LI-900 RSI and over 60% for the LI-2200 RSI, as shown here in psi:

0.160 SIP/undensified LI-900	11.8
0.160 SIP/densified LI-900	22.6
0.090 SIP/undensified LI-2200	30.2
0.090 SIP/densified LI-2200	46.3

Proof Test

To obtain allowable-strength predictions, a large number of tensile tests of individual material components (the RSI, SIP, and RTV) preceded full-scale tile production. When it was discovered that the system had considerably less strength than the *weakest* of the individual components, these "allowable" values could no longer be used in evaluating the margin of safety of the tile at various places on the orbiter. Indeed, new statistical interpretations of a small number of complete system tests (RSI/RTV/SIP/RTV) gave a large number of tiles already installed on the orbiter a negative margin.

Rather than rely on such predictions, a proof tensile load equal to 1.25 times the maximum stress expected during flight (limit stress) was applied to tiles already installed on the orbiter. If the tile survived the proof load, presumably it had sufficient static strength to give a positive margin of safety. If it failed (as determined by either separation from the substructure or exceeding measured noise counts in an acoustic emission test taken during the loading process), it was removed and densified.

In addition, any tile whose predicted limit stress dictated a proof stress higher than 10 psi was automatically removed, densified, and reinstalled. And every tile installed after October 31, 1979, has been densified.

Moreover, a nondestructive test has been introduced that assures using only high-strength RSI tiles. On the basis of several hundred tensile tests, Rockwell has found a strong correlation between the velocity of sound through the tile material and its tensile strength.

The combination of tile-material acceptance, based on this sonic test, and verification of the strength of the bond from the tension tests of tiles in place assures the integrity of the densified tiles under static loading.

SIP Static Response

Problems attend the design of the TPS other than the SIP/RSI interface strength treated by densification. A tensile load applied normal to the plane of the SIP and slowly increased will deform the strain-isolation pad a considerable amount at very low loads. As transverse fibers straighten and begin to carry load, the stiffness of the pad increases. The pad shows a decreasing rate of deformation for the same increase in load. To put it a third way, the SIP material exhibits a nonlinear stress-strain behavior (see the first load cycle of Fig. 6, which represents a proof-test load sequence).

To add to this complexity, during this moderate loading process, the transverse fibers realign themselves so that, after a complete tension-compression cycle, as would occur during a TPS proof test, the material not only has a permanent set, but also has a different and *even more nonlinear response* as loads are again applied (Fig. 6).

Indeed, the form of the static response of the SIP to load continuously changes as more load cycles are applied. The material develops a continuously increasing region of low stiffness. The SIP material behavior, thus, is highly nonlinear, dependent on its prior load history, and after "load conditioning" has a sizeable low-stiffness region.

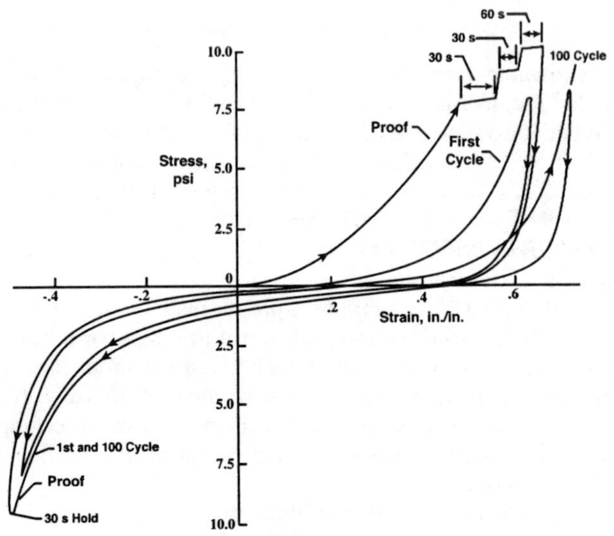

Fig. 6 Typical nonlinear stress-strain response of SIP under initial loading and continued load conditioning.

Tests have also shown that under constant load the SIP can exhibit a large amount of nonrecoverable creep and that the ultimate strength is a function of the rate of loading. At higher load rates, the SIP exhibits higher strengths.

An analysis developed by NASA Langley takes into account this nonlinear behavior in calculating TPS limit stresses. A series of laboratory tests on single TPS specimens checked this analysis. Figure 7 shows a square tile first proof tested and then subjected to combined tension and moment loading. Tile displacement was measured at one point. The graph in Fig. 7 compares the analytically predicted displacements at the same point. The prediction is very good. Although such a test does not totally substantiate the stress analysis, it does increase confidence in the nonlinear approach.

In brief, the usual analytical methods depending on linear superposition to add stresses due to individual load conditions can yield misleading results. The various load conditions and substructure displacements should be applied simultaneously using the nonlinear stress analysis. In general, the nonlinear analysis predicts higher stresses in the SIP than a linear analysis unless the loading is primarily caused by substructure deflection. Figure 8 indicates the complexity of stress distributions that can be expected in a tile. It shows the interface stress distribution from an assumed shock passage across an outboard-elevon tile.

TPS Dynamic Response

The SIP's nonlinear displacement under load causes an unusual response of the TPS to sinusoidal dynamic acceleration. The resulting wave form, shown in

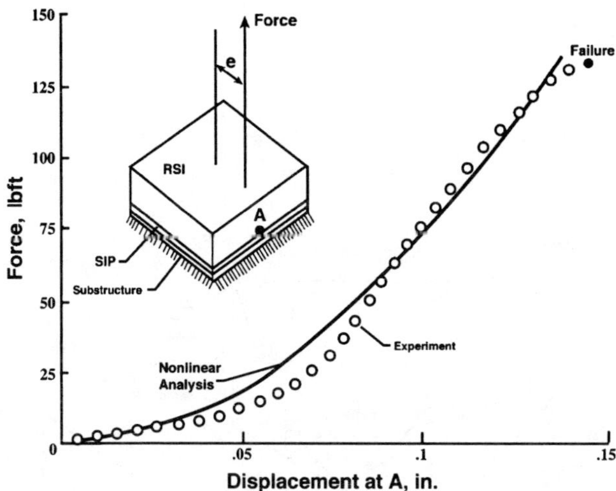

Fig. 7 Combined static tension and moment test of thermal-protection system.

Fig. 9, has a sharp peak acceleration. This causes a stress amplification of the maximum input acceleration of up to a factor of 4 at resonance. The orbiter, however, should experience only random vibration input during a mission. Random-vibration tests predict the maximum response acceleration to be no greater than 2.5 times the maximum input acceleration.

Nonlinear characteristics of TPS material and its resultant variable damping create a very complex dynamic response. For example, the natural response frequency of the SIP is a function of the applied acceleration and, thus, varies with the drifting dynamic loads. Also, inspection of the transverse fibers in the SIP shows that they do not run completely normal to the plane of the pad; so the in-plane and normal motions of the SIP should be coupled.

A dynamic analysis accounting for these complexities has been developed that contains both material and viscous damping effects evaluated empirically from vibration tests. The analysis has been checked by comparing the measured dynamic responses in an experimental random-input dynamic-loads test. Figure 10 shows results. The analysis accurately predicts the general response and the output acceleration amplification.

Another interesting dynamic response has been observed in controlled tests and has been predicted with the nonlinear dynamic analysis. At a driving frequency different from the natural frequency of the TPS, a large lateral parametric response occurs at half the driving frequency. This dynamic instability could reach a critical level under sinusoidal dynamic inputs. However, we expect only randomly varying dynamic inputs during the orbiter's flight, and, even if the input centered about the parametric response frequency, we doubt that large lateral motions can occur.

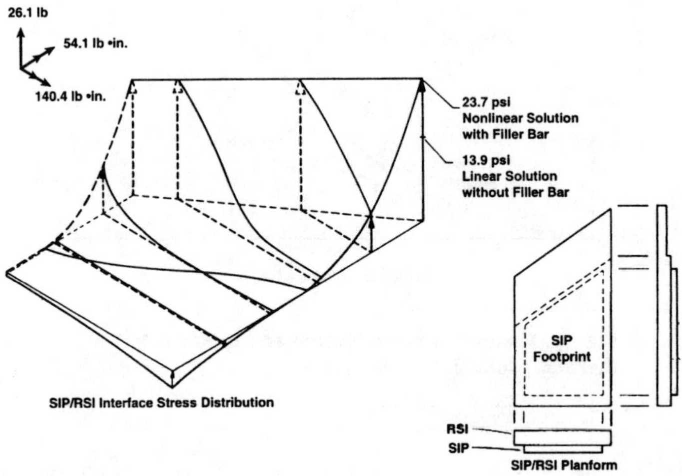

Fig. 8 SIP/RSI interface stress distribution for an elevon tile.

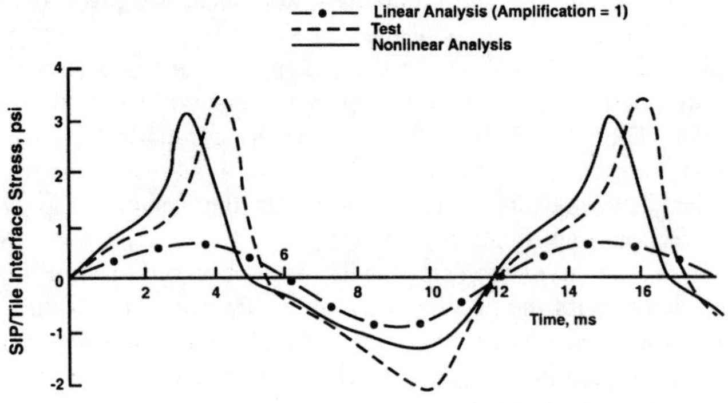

Fig. 9 Dynamic waveform of tile response to sinusoidal base shake.

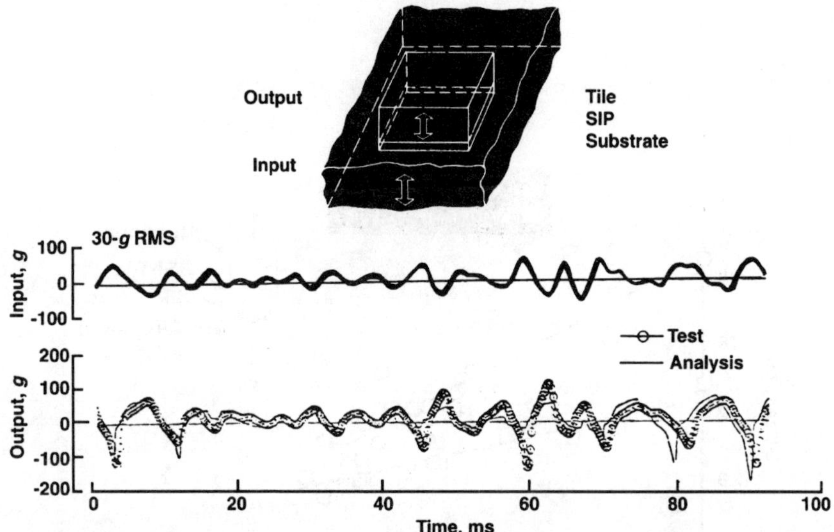

Fig. 10 Dynamic analysis predicts tile response to random substrate input motion.

TPS Fatigue

The dynamic motion of the TPS poses another concern for its structural integrity. Repetitive loadings due to liftoff and high-speed boundary-layer aerodynamic noise, plus oscillating shocks repeatedly traversing the tiles, could cause fatigue damage.

As can be seen in the *S–N* fatigue curves of Fig. 11, the greater static strength achieved by densification of the bonding surface of the RSI does not fully translate into an equivalent increase in fatigue strength because the mode of fatigue failure differs for the undensified and densified TPS. Results plotted for the undensified TPS represent failure due to complete separation at the SIP/RSI interface. Results plotted for the densified TPS, on the other hand, represent failure defined as a total out-of-plane movement of 0.25 in.

Figure 12 shows displacement history of the densified TPS during a typical fatigue test for a constant-amplitude, fully reversed sinusoidal load applied at 1 Hz. As the number of cycles increases, the SIP continues to unravel with a continual increase in total specimen travel and a continual expansion of the low-stiffness region. Eventually, the SIP will completely separate, but before this occurs a tile would be loose enough so that a small exciting force during flight could cause it to move into the airstream. The relative motion of one tile to a neighboring tile during entry could disrupt airflow, trip the boundary layer prematurely, and increase heating downstream. The amount of relative tile motion required to trip the flow depends on the local boundary-layer thickness, and, thus, the measure of acceptable motion increases toward the aft end of the Shuttle as the boundary layer thickens.

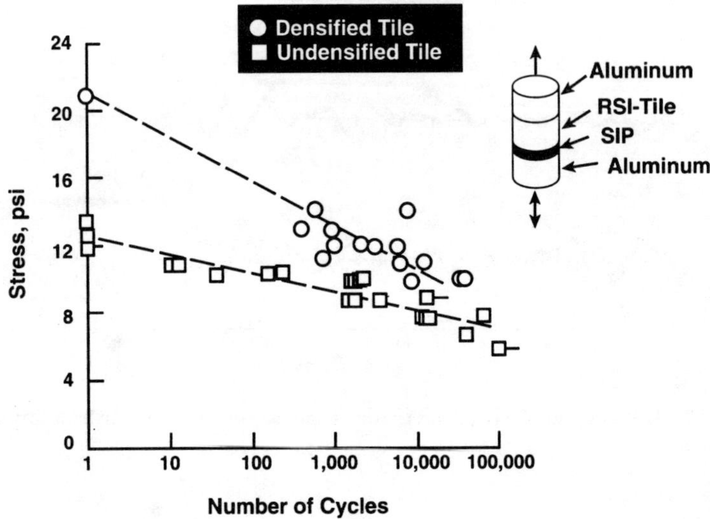

Fig. 11 Fatigue characteristics of LI-900 tile/0.160 SIP TPS.

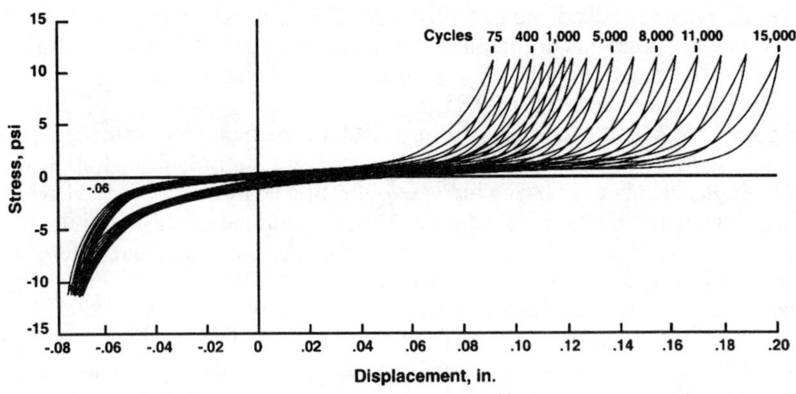

Fig. 12 SIP extensional behavior under repeated load for densified TPS toward the loosening of tile.

The fatigue results of Fig. 11 show that cyclic loading causes a relatively large reduction in the stress levels that both the densified and undensified TPS can withstand for a small number of cycles. Since the fatigue failure of densified TPS takes the form of excessive elongation of the SIP, rather than separation in the parent RSI, a further increase in strength of the RSI ceramic by chemical reformulations or changes in manufacturing processes would not improve the TPS lifetime. In fatigue, SIP represents the weak link in the densified TPS.

Based on current calculations of cumulative damage from fatigue, some tiles, undensified as of October last year, will have limited life.

Data from development flight instrumentation will be used after the first few flights to refine our knowledge of expected loads and to refine tile-lifetime predictions based on updated stress histograms. Since undensified tiles will have less life expectancy than the densified, NASA plans to remove and densify the remaining undensified tiles during the normal refurbishment periods between developmental flights.

TPS-Lifetime Assessment

Results to date have greatly increased our knowledge of the basic properties and response characteristics of the orbiter's RSI/RTV/SIP TPS. However, there are still several questions to be answered before flight, among them a projection of the single- and multiple-mission life of tiles subject to the combined flight loadings.

To make a proper assessment of TPS life, two series of flight-simulation tests are planned at Langley Research Center by NASA and Rockwell, and several equivalent lifetimes will be completed before the flight-readiness review scheduled next month.

The first series (known as DI-19), a multienvironment fatigue test, subjects a forward fuselage test structure with densified HRSI thermal protection to a mission simulation with eight different environmental sequences—including rain, liftoff acoustics, ascent heating, on-orbit cold soak and vacuum, entry heating, terminal-area acoustical loads plus, finally, landing loads. Several of these conditions will be combined with structural loading of the test component where appropriate.

The second test series, known as the Shuttle TPS CLOT (Combined Loads Orbiter Test), will be performed in Langley's 8-ft Transonic Pressure Tunnel. Tile arrays placed on three different representative orbiter structures will be tested under the appropriate ascent dynamic pressure for Mach numbers between 0.7 and 1.4. Controlled shock passages and high aerodynamic acoustic loadings will be applied to the heavily instrumented tiles, and tile responses noted.

Many other tests are being conducted concurrently at NASA centers and by Rockwell. These include fatigue tests that, when coupled with analysis and cumulative fatigue predictions, will yield realistic testing of TPS life.

Concluding Remarks

Recent experiments and stress analyses have shown the TPS to have marginal strength in many areas of the orbiter. Modifications of the TPS, such as tile densification, have been made to correct these strength deficiencies. Densification of the tile surface brings the system static strength up to the strength of individual components. The failure level in a static tension can, thus, be predicted accurately using nondestructive sonic testing of the RSI before bonding.

Densified TPS is sufficiently strong in all areas to withstand expected static loads during a mission. In areas where dynamic loads predominate, however, the TPS may have limited life because of excessive SIP extension.

A modification to this characteristic is desired to make the TPS an operational system. For example, the SIP could be modified to decrease its prior-load-history dependency. Bonding the tile to the filler bar can also reduce the SIP peak stresses. A means of mechanically attaching the tile to the skin—such as an auger concept developed by NASA Johnson Space Center—is under evaluation for flight qualification. Developmental research for both the improved strain-isolator pad and variations of the auger concept has begun both in industry and within the NASA centers.

The fragile nature of the brittle ceramic tile and its coating has renewed interest in finding alternative reusable thermal-protection systems. For example, NASA Ames and Lockheed have developed a stronger but lighter ceramic insulation, called fiber-reinforced ceramic insulation (FRCI). But the material still must undergo a series of thermal and mechanical characterizations and qualification tests, and only a small amount of FRCI can be considered for possible use on the second orbiter to replace the heavier LI-2200 RSI. Additional emphasis is also being placed on metallic TPS on both standoff wall and new multiwall concepts. Vought is developing an advanced carbon-carbon material with greater strength and reusability than the current RCC.

It will take intense R&D to create and flight-qualify any new TPS with the long-term reliability required for reusable space-transportation systems.

For the foreseeable future, RSI represents the only reusable TPS ready for flight, and even it requires additional effort to insure that it will give the full 100-mission lifetime.

All these points notwithstanding, the ceramic RSI is one of the best lightweight thermal insulators ever developed, and, in addition to its key role on the Shuttle orbiter, it should find wide use in future high-technology applications.

Refractory Metals for Thermal-Protection Systems

Eldon E. Mathauser
NASA Langley Research Center

Within the past decade, considerable effort has been directed at studies of refractory metals with the objective of utilizing them in high-temperature environments. Substantial progress has been made in understanding the behavior and use of the refractories, but many problems still remain, and to date relatively little use of these metals has been made in other than specialized applications (e.g., rocket-motor nozzles, furnace structures, and lamp filaments).

This discussion concerns the status of refractory-metal sheet for thermal-protection systems and draws attention to factors that have a major influence on utilization of refractories, such as availability, material properties, fabrication, protective coatings.

This assessment is based partly on experience with refractory metals at the NASA Langley Research Center and partly on a review of current literature. The Langley studies have been directed at structural applications of sheet and have included investigation of mechanical properties, coating behavior, fabrication, joining, and structural performance. A test specimen fabricated at Langley to investigate application of refractories to thermal-protection systems is a corrugated molybdenum-alloy heat shield panel approximately 20 in. square. Support clips of molybdenum alloy are riveted to the corrugated shield and pinned to an inner nickel-alloy structural panel. The panel carries high-temperature fibrous insulation. Panels of molybdenum alloy have been subjected to cyclic high-temperature tests in a radiant-heating facility, and similar panels of columbium alloy have been subjected to flutter tests in a supersonic wind tunnel. The performance of the panels was found to be generally satisfactory in the flutter tests and marginal with respect to coating performance in the radiant-heating tests.

Availability of Refractory Metals

The availability of good-quality alloys of molybdenum, columbium, tantalum, and unalloyed tungsten has improved markedly in the past five years. This development has had its impetus largely in the refractory-metal sheet-rolling programs sponsored by the government[1] and in the requirements of lifting re-entry vehicles such as Asset and the recently canceled X-20. Sheets of adequate size and range of thicknesses can be obtained in all of the common refractory metals. Quality and uniformity are generally high.

Astronautics & Aeronautics, Vol. 2, No. 7, July 1964, pp. 66–69.

Some of the other refractory metals, including vanadium and chromium alloys, are not developed sufficiently to be considered for vehicle application in the near future.[2] Another material not classified as a refractory metal has been receiving considerable attention lately for possible aerospace-vehicle applications. This material, nickel strengthened by refractory-oxide particle dispersion, appears to have considerable promise for thermal-protection systems at temperatures below approximately 2400°F.[3]

Mechanical and Physical Properties

The strength of refractory metals is generally not of primary importance in determining their usefulness in thermal-protection systems. They must have adequate strength at elevated temperatures, of course, to preclude the possibility of creep. Not enough data are available to define this problem clearly, but it appears that molybdenum- and tantalum-base alloys and tungsten will probably encounter creep problems at different temperatures above 2500°F. Many of the currently available columbium alloys have low strength at high temperatures and retain little resistance to creep deformation, as shown in Fig. 1.[4-6] This graph presents typical tensile strengths, 1-h rupture stresses, and stresses that produce 0.2% creep strain in 1 h in the temperature range from 2000–3000°F.

Note that stresses less than 1 ksi (ksi = 10^3 psi) will produce 0.2% creep strain in 1 h at 2500°F. Because of this lack of resistance to creep, allowable stresses for columbium alloys in structural applications would be extremely

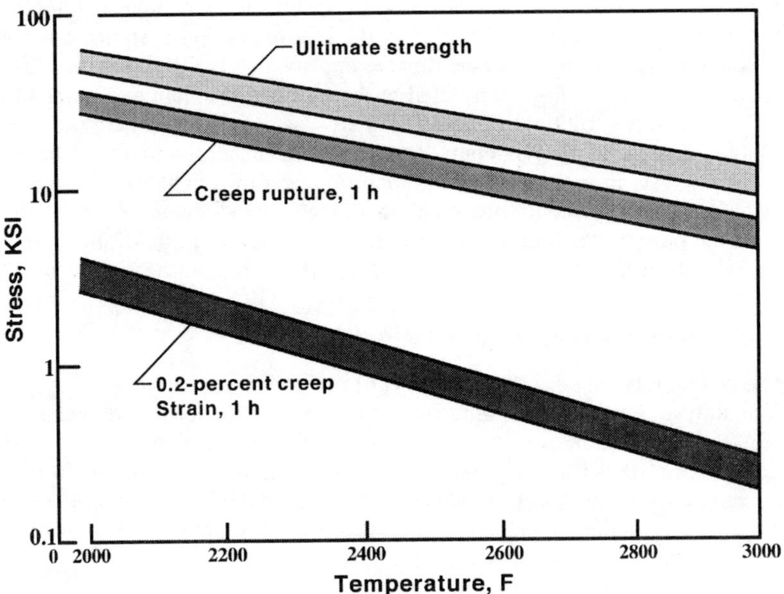

Fig. 1 Characteristics of columbium alloys. Representative properties for several recently developed.

low, particularly for temperatures above 2500°F. The 0.2% creep strain indicated here corresponds in magnitude to the permanent strain associated with the yield stress.

The majority of the columbium- and tantalum-base alloys are ductile at room temperature, and brittleness is not expected to be a factor in forming them.[7] Moreover, they do not undergo embrittlement from recrystallization through exposure at elevated temperatures. On the other hand, some of the molybdenum alloys and tungsten are brittle at room temperature and require special care in fabrication to prevent cracking and possible delamination, particularly in thin sheet. Furthermore, molybdenum and tungsten sheet, generally used in a cold-worked condition, become recrystallized when exposed to high temperatures. Recrystallization usually causes a pronounced loss in ductility. For this reason it may not be desirable or practicable to reuse molybdenum or tungsten components in thermal-protection system components that have been subjected to high temperatures.

Fabrication

Columbium and tantalum alloys in general can be formed at room temperature. For present-day molybdenum alloys, many simple forming operations can be done satisfactorily at room temperature. Some complex forming operations may require moderate temperatures in the 300–600°F range. Successful forming of tungsten generally requires temperatures of 1000°F or higher.

Conventional machining approaches are generally acceptable for the refractory metals except tungsten. This material is very difficult to machine, and specialized methods such as electrical-discharge machining may be desirable. Established procedures such as chemical milling and newly developed methods such as electrochemical machining are also feasible. They are primarily of interest for molybdenum and tungsten.

Refractory metals can be joined most successfully with mechanical fasteners, such as rivets, screws, or bolts. The joining method is strongly influenced by the protective coatings necessary with refractory metals. Welding may be considered as a joining method, although some concern regarding reliability of welded joints in refractory-metal structures exists. Brazing is also of interest. Several developmental efforts are underway to establish its potential for refractories.

Continuing efforts are being made to investigate conventional approaches such as fusion, resistance, and arc spotwelding and less conventional methods including electron-beam welding and diffusion bonding.[8,9] Severe embrittlement and cracks occur in weld areas of columbium- and tantalum-base alloys if gaseous contamination occurs. Molybdenum and tungsten welds are not sensitive to embrittlement by gaseous contamination, but they do become severely embrittled from recrystallization in the weld area. For this reason welding of molybdenum and tungsten cannot, so far, be considered a satisfactory joining method. Welding of columbium and tantalum under carefully controlled conditions may be acceptable for some of the currently available alloys.

Protective Coatings

The availability of a suitable protective coating is perhaps the most important single factor that must be considered in the utilization of refractory metals in thermal-protection systems. Protective coatings are required to prevent oxidation at elevated temperatures. Moreover, columbium and tantalum alloys require them to prevent embrittlement by gaseous contamination. Many types of coatings have been investigated for the refractory metals, and steady development and gradual improvement in coating performance and reliability are being made.[10–12]

Among various coatings either commercially available or in an advanced stage of development, the disilicide-base one is perhaps the most successful, particularly for molybdenum alloys. For columbium alloys, a commercial coating containing chromium, titanium, and silicon is one of the outstanding types available now. For tantalum-base alloys, a tin-aluminum coating is probably among the best available. No commercially available coating with proved performance exists for tungsten, but several promising ones are in an advanced stage of development.

Although they vary considerably in performance, depending on many factors, the commercially available coatings described in Fig. 2 should provide approximately 1 h of protection at 3000°F, 50–100-h protection at 2500°F, and 500–1000-h protection at 2000°F. These values are representative of data from several types of coatings and apply to alloys of columbium, molybdenum, and tantalum and to unalloyed tungsten.

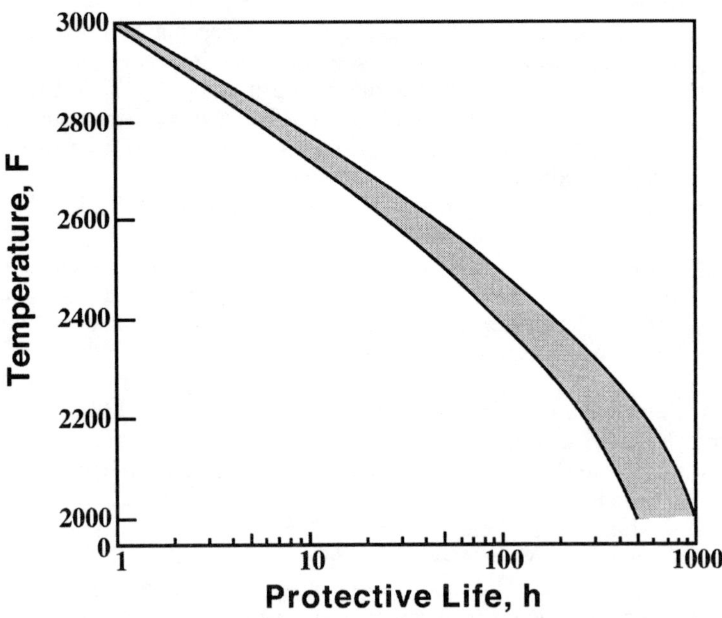

Fig. 2 Oxidation protection provided by typical present-day coatings for alloys of columbium, molybdenum, tantalum, and unalloyed tungsten.

Considerable degradation of the performance represented in this graph can be brought about by many factors, most significant perhaps is reduced or low pressure. Low pressure associated with high temperatures may lead to evaporation of the protective coating. Another important factor is thermal cycling or cyclic exposure to elevated temperatures.[13] Coating life may be reduced by an order of magnitude under intermittent heating, compared with the life of continuously heated coated specimens. Another item of importance in establishing coating performance, particularly for thin sheet, is diffusion at the coating-metal interface, with a resultant loss of substrate material.[14] With this effect, the decrease in thickness of the molybdenum-alloy sheet of approximately 17% is attributed to diffusion. The coated sheet was exposed at 2500°F in air for times up to 29 h. The diffusion at the coating-metal interface produces a reduction in sheet thickness that makes use of foil material, such as honeycomb core, impractical when coating is required.

Another significant aspect of protective coatings concerns variation in performance between large and small specimens. Because of possible variations in coating, the life of large coated components will generally be less than the life shown by small test-coupons. Then, too, distortions may occur in large components during the coating operation, and it is often not possible to make the final assembly of components distorted, and possibly embrittled, during coating.[15]

Experience with coated molybdenum heat shield panels indicates that embrittlement of the coated component and overall growth and distortion from the coating application are potentially serious problems. Some of the molybdenum-alloy support clips, for instance, were broken during the final assembly of shield onto load-carrying structural panel. This occurred largely because of growth and distortions of the corrugations. Smaller heat shield components and improved coating techniques would be expected to alleviate this problem.

Other considerations bearing on coating performance include damage by micrometeoroid impact, need for predictable and uniform high emissivity, and accelerated deterioration or pest failure of coatings.[16] The pest phenomenon may be described as an unusually rapid failure of the coating at intermediate temperatures even though protection is obtained at higher temperatures.

Recent development efforts in coatings have been directed at improving performance at reduced pressures and high temperatures, identification of new coatings for the 3000–4000°F range, study of basic behavior of coatings to define protection and failure mechanisms, and development of multicomponent graded coatings.

Concluding Statement

The status of refractory metals for thermal-protection systems might be summarized as follows:

1) Generally good-quality refractory metals are available. Most interest appears to be directed at recently developed columbium- and tantalum-base alloys that exhibit moderate strength, relative ease of fabrication, and no embrittlement from recrystallization.

2) Creep problems may be encountered with some alloys, particularly if higher temperatures of use are achieved through improved coatings.

3) Mechanical fastening should probably be preferred for joining. Welding under carefully controlled conditions may be feasible for some alloys. Brazing and diffusion bonding may also prove of interest for some applications.

4) Availability of suitable protective coatings is the most important factor determining whether refractory metals can be utilized in thermal-protection systems.

5) Major improvements in quality and uniformity of coatings have been made in the past five years. Commercially available coatings are expected to provide approximately 1-h protection at 3000°F, 50–100 h at 2500°F, and 500–1000 h at 2000°F.

6) Coating performance is seriously degraded by many factors, including reduced pressures, thermal cycling, and diffusion at the coating-metal interface.

7) Current experience indicates that thermal-protection systems should preferably be composed of several small, rather than a few large, components. This approach is suggested because distortions, nonuniform application of coatings, and assembly difficulties are considerably more severe with large components.

References

[1] Ogden, H. R., "Status Report No. 2 on Department of Defense Refractory Metals Sheet-Rolling Program," Battelle Memorial Institute, DMIC Rept. No. 176, Oct. 15, 1962.

[2] Manning, C. R., Jr., and Mathauser, E. E., "Investigation of Some New Materials for Aerospace Vehicle Applications," Sixth National SAMPE Symposium, Seattle, WA, Nov. 18–20, 1963.

[3] Manning, C. R., Jr., Royster, D. M., and Braski, D. N., "An Investigation of a New Nickel Alloy Strengthened by Dispersed Thoria," NASA TN D-1944, July 1963.

[4] Schmidt, F. F., and Ogden, H. R., "The Engineering Properties of Columbium and Columbium Alloys," Battelle Memorial Institute, DMIC Rept. No. 188, Sept. 6, 1963.

[5] Bartlett, E. S., and Van Echo, J. A., "Creep of Columbium Alloys," Battelle Memorial Institute, DMIC Memorandum No. 170, June 24, 1963.

[6] Bartlett, E. S., et al., "The Current Status and 1970 Potential of Selected Defense Metals," Battelle Memorial Institute, DMIC Memorandum No. 183, Oct. 31, 1963.

[7] Schwartzberg, F. R., Ogden, H. R., and Jaffee, R. J., "Ductile-Brittle Transition in the Refractory Metals," Battelle Memorial Institute, DMIC Rept. No. 114, June 25, 1959.

[8] Mishler, H. W., and Monroe, R. E., "Electron-Beam Processes," Battelle Memorial Institute, DMIC Rept. No. 174, Sept. 15, 1962.

[9] Young, W. R., and Jones, E. S., "Joining of Refractory Metals by Brazing and Diffusion Bonding," ASD-TRD-63-88, Jan. 1963.

[10] Krier, C. A., "Coatings for the Protection of Refractory Metals from Oxidation," Battelle Memorial Institute, DMIC Rept. No. 162, Nov. 24, 1961.

[11] Sama, L., "Oxidation Resistant Coatings for Tantalum Alloys and Other Metals," ASD-TDR-63-160, Feb. 1963.

[12] Nolting, H. J., and Jefferys, R. A., "Oxidation Resistant High Temperature Protective Coatings for Tungsten," ASD-TDR-63-459, May 1963.

[13] Rummler, D. R., Stein, B. A., and Pride, R. A., "A Study of Several Oxidation-Resistant Coatings on Mo-0.5 Ti Alloy Sheet at 2500 F," NASA TN D-2040.

[14] Stein, B. A., and Lisagor, W. B., "Diffussion Studies of Several Oxidation Resistant Coatings on Mo-0.5 Ti Alloy at 2500 F," NASA TN D-2039.

[15] Bartlett, E. S., and Gibeaut, W. A., "Production Problems Associated with Coating Refractory Metal Hardware for Aerospace Vehicles," Battelle Memorial Institute, DMIC Memorandum No. 172, July 26, 1963.

[16] Bartlett, R. W., "Investigation of Mechanisms for Oxidation Protection and Failure of Intermetallic Coatings for Refractory Metals," ASD-TDR-63-753, Part 1, July 1963.

Thermal Protection of Space Vehicles

Peter E. Glaser
Arthur D. Little, Inc.

The idea of scaling the thermal barrier, so often discussed publicly, conjures up visions of a brilliantly glowing nose cone successfully resisting aerodynamic heating while re-entering the Earth's atmosphere. Although space vehicles encounter their most severe problems during this brief period, they must, in addition, overcome the less severe, but also less obvious, problems created by aerodynamic heating in extended flight through the upper atmosphere, by solar radiation, or by internal heat generation. Consequently, most of the approaches favored by nose-cone technology are not applicable.

The problems generated by these nearly steady heating conditions have sparked the development of an integrated system designed to protect various parts of a space vehicle. This system must combine maximum insulating effectiveness with the lowest possible weight. Unlike most present insulations, a thermal-protection system is usually a combination of materials which, besides providing maximum insulating effectiveness, gives structural strength without imposing too great a weight penalty. The advantage of a thermal-protection system is that the insulating function and the load-carrying elements can be designed separately. The insulation can be designed to withstand the predicted heating conditions, and the load-carrying elements can be designed to withstand outside forces, such as aerodynamic loads, vibration, acceleration, thermal shock, and particle erosion. The insulation and load-carrying elements can then be combined to form the vehicle structure.

For Various Heating Conditions

Thermal-protection systems can be used under different heating conditions, as, for example:

1) Prior to flight, to reduce the boiloff rates of cryogenic propellants.

2) During flight, to protect the vehicle structure, propellants, occupants, and instruments from the effects of aerodynamic heating or solar radiation.

3) While in orbit or deep space, to provide thermal control over the space-capsule environment.

Heat flows from the warmer to the colder surface of an insulating material by 1) conduction through the solid portions, 2) conduction or convection of the gas filling the spaces between the particles, or 3) radiation across these spaces and through the solid matrix. Although these heat-flow mechanisms are well understood individually, their combined effects are still being investigated. The

Astronautics, Vol. 5, No. 4, April 1960, pp. 40, 41, 105.

individual conductivities of each mechanism, however, can be minimized at various temperatures and pressures.

Heat conduction through the solid insulating material can be reduced if the conduction paths are broken up with finely divided materials such as fibers or powders. As long as the individual particles do not sinter or agglomerate under external pressures or when exposed to high temperatures, resistance to heat flow will be formed at the surface of each particle. In fibers or powders, the gas filling the interstices between particles is usually the dominant contributor to overall heat flow for moderate temperature differences. In general, large gas molecules have low thermal conductivity. Therefore, some benefits could be derived from the use of a gas of high molecular weight. This technique, however, limits the improvement in insulating value to approximately the difference in free-gas conductivity between the gas and the air. A far greater improvement can be obtained by removal of the gas filling the spaces between particles—that is, by "vacuum insulation."

Previous Work

Vacuum insulation is certainly not a new development. In 1898, James Dewar demonstrated that an evacuated space between two suitably silvered glass vessels forms an effective heat barrier. For quite some time it was widely held that air had the lowest thermal conductivity of any combination of materials. In 1910, however, Smoluchowski showed that thermal insulators more effective than air could be produced from fine powders subjected to a moderate vacuum. The low values of thermal conductivity thus obtained can be explained by the following effects.

1) As the pressure is lowered and gas molecules are removed, less heat is transferred after the mean free path of the molecules exceeds the average particle spacing.

2) Thermal conductivity at that point decreases substantially, until a lower limit is obtained when most of the gas molecules have been removed. The larger the particle spacing, the lower the pressure required for the thermal conductivity to approach its limiting value.

3) At this lower limit, the thermal conductivity still has a finite value because heat can be transferred by radiation, solid conduction between the particles, and residual gas conduction.

Although the amount of radiant heat passing through an insulation in cryogenic applications is far less than in high-temperature applications, heat so transferred plays a major role in the overall performance of the system. Maintaining the insulating effectiveness obtained by the removal of environmental air and by the breaking up of conduction paths requires the attenuation of radiant energy that otherwise might pass through the material and nullify the benefits derived from reduced heat transfer by other mechanisms.

To assure the structural integrity of a thermal-protection system, two approaches can be used. In the first, the insulation is placed in a space between two pressure vessels, with the inside vessel containing the material to be insulated, and the outside vessel designed so that it will withstand external pressures and other outside loads. Suitable supports should be provided to tie the inner vessel to the outside vessel, because the insulation need not be designed to carry loads. Despite the high insulating effectiveness thus produced, a considerable weight penalty makes this design almost impractical for space vehicles. However, it has been used widely in the transportation and storage of cryogenic fluids.

In the second approach, the insulation is designed to be load-bearing. Thus, supports and heavier pressure vessels can be eliminated. Moreover, the thermal-protection system can be shaped more easily and made to conform to the contours of the surfaces to be insulated. Thin-skinned vacuum insulation—a term used to describe a thermal-protection system's ability to carry and transmit loads imposed upon it after it has been evacuated—has been developed to confine the insulation and hold a desired vacuum without appreciable diffusion for extended periods. This type of insulation, however, must still rely on an underlying load-bearing structure or vessel to which loads can be transmitted. Because a load-bearing insulation has a higher density and, hence, increased weight, its thermal conductivity should be lower than that of a lighter insulation to be equally effective. An index of the product of thermal conductivity and density allows us to compare different insulating materials. The choice of design must be governed by the magnitude of the index applied to the complete system.

Storable Cryogenics

Thermal-protection systems for cryogenic fluids such as liquid oxygen, liquid nitrogen, or liquid hydrogen are being developed so that such fluids can be stored aboard a space vehicle for an extended period. Thus, the complexity of propellant fuel-handling systems is reduced, and cryogenic fluids can be carried during space missions without incurring substantial losses caused by boiloff of the gases in the storage container.

To obtain the most efficient cryogenic insulation, we must mount the greatest possible number of radiation shields of low emissivity in series and separate them from each other and the boundaries by spacers. These shields must have the lowest possible thermal conductivity, have a minimum area of contact, and be under a high vacuum. Although a number of material combinations have been investigated for the attainment of these objectives, there is no theoretical limit to the very low thermal conductivities of such multiple-layer radiation shield insulation, since it is theoretically possible to reduce both the thickness of the radiation shields and the thermal conductivity of the spacers and obtain still higher insulation efficiencies.

A protection system for cryogenic fluids using multiple-layer radiation shield insulation may be composed of alternate layers of thin reflecting foils of aluminum and spacers of low conductivity materials with only a few thousandths of an inch thickness assembled in sandwich form.

The open space between the vessel to be insulated and the outside shell is filled to the required thickness with this type of insulation and evacuated to a pressure of the order of 10^{-6} mm of mercury, depending upon the particle size and the density of the spacer materials. A 1-in.-thick assembly of 60 layers of aluminum foil with alternate submicron-diameter glass-fiber spacers evacuated to 10^{-6} mm of mercury and insulating a tank of liquid nitrogen can provide up to 500 times greater insulating effectiveness than the insulation commonly used in household refrigerators. Because of this very high insulating effectiveness, both the thickness and, therefore, the weight of such a thermal-protection system can be minimized.

Attaining High Vacuum

A flexible outside shell covering the multiple-layer radiation shield insulation acts as a vacuum barrier which needs to be only thick enough to withstand diffusion, while the insulation is designed to withstand the compressive forces of atmospheric pressure or aerodynamic loads. The attainment of a high vacuum in the insulated space is not so difficult as it first appears because, with cryogenic fluids such as liquid hydrogen, any gas remaining in the space would tend to diffuse to the cold surface and either condense or freeze.

Once the vehicle has reached high altitudes, the thin shell no longer will have to be a good vacuum barrier, since the vacuum existing in space will be quite adequate to assure satisfactory performance of the insulation. Thus, even occasional penetration by micrometeorites will not tend to deteriorate the insulating effectiveness and, provided that such penetrations are relatively small and few in number, the diffusion of air as the vehicle re-enters the atmosphere and experiences aerodynamic heating would not greatly reduce its effectiveness, particularly if the system is compartmentalized.

We can overcome the effects of aerodynamic heating on a structure either by letting the structural materials withstand the high temperatures generated on the exposed surfaces (hot structure) or by protecting the structure with insulation (cooled structure). The hot structure requires refractory materials that will not deteriorate in the high-temperature environment and that can be designed to carry the loads imposed on them without creating a weight penalty. Difficult problems confronting materials research and development of suitable refractory materials and necessary joining techniques have to be solved to make this approach feasible.

The cooled structure uses a thermal-protection system and an underlying load-bearing aluminum structure designed in accordance with known criteria. This structure may be provided with an auxiliary cooling system.

The arrangement of a thermal-protection system applied to the outside surface of a space vehicle protects the internal structure from the effects of aerodynamic heating. The design principles are the same as those employed in a protection system for low-temperature applications. The primary requirements are 1) materials that can withstand temperatures of 2000–4000°F; 2) the use of radiation cooling at the outside surface, which is coated so that it will reject a large portion of the heat generated by radiation to the surrounding space; and 3) attenuation of radiant energy passing through the insulation by suitable radiation absorbing and scattering media.

The development of thermal-protection systems for hypersonic-speed vehicles required new high-temperature insulations. Figure 1 shows the improvements that can be obtained in an insulating material if the radiant energy passing through it is attenuated. If this insulation is made to be load bearing and is protected with an outside shell, aerodynamic loads can be transmitted to the underlying structure. A suitable design of the outside shell can compensate for thermal expansion caused by differentials in the temperature profile.

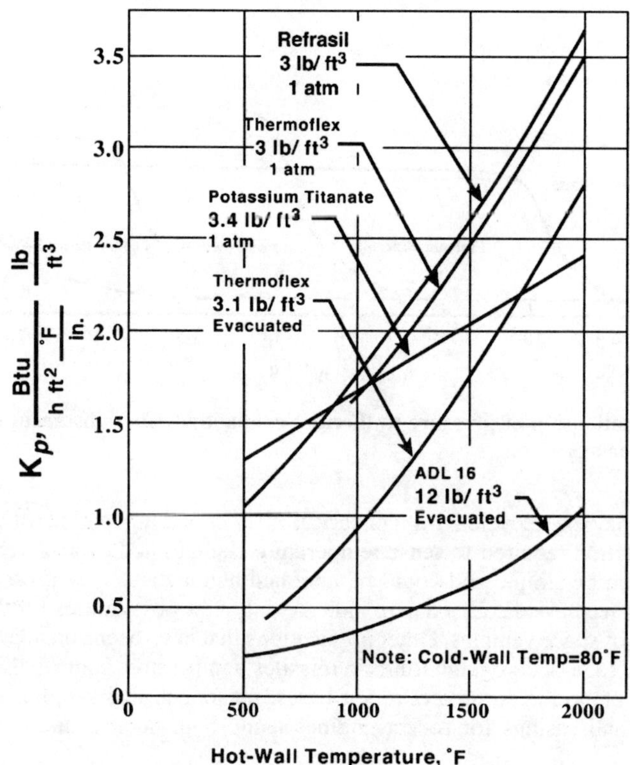

Fig. 1 K_p of various insulating materials.

Space Vehicle Requirements

Manned or instrumented space vehicles or stations may require a thermal-protection system that can control the environment temperature of the capsule and either withstand solar radiation or conserve heat energy on a deep-space mission. For vehicle thermal control, systems of the type that we have discussed can be used. Since the system's effectiveness depends largely upon the gas pressure in the enclosed space occupied by the fibers or powders, a fairly simple control system can be devised. This system can vary the gas pressures in compartmented sections of the thermal-protection system and, thus, increase or decrease insulating effectiveness. The variations in gas pressure can cause substantial changes in heat transfer through the thermal-protection system as shown in Fig. 2. Thus, the side being heated by solar radiation may be so arranged that insulating effectiveness will be either increased or reduced to provide a comfortable temperature in the environment of the space vehicle.

Such a thermal-protection system, combined with surface coating on the capsule, could meet most of the conditions expected in space travel. Since very small pressure changes are needed, only a very small quantity of gas is required.

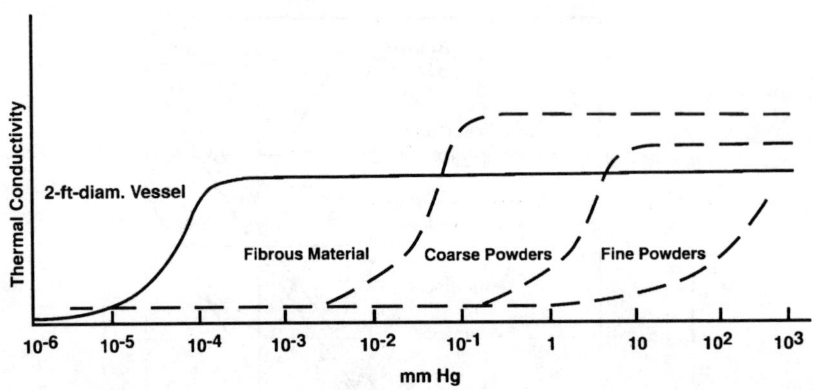

Fig. 2 Relationship of pressure to thermal conductivity for substances of various particle spacings.

The gas may be recycled for conservation over a long period of time. The instrumentation required to sense temperature changes and control vehicle environment can be simple and could be operated automatically by preset controls.

Systems we have discussed here indicate only new possibilities for the thermal protection of space vehicles. Other applications that have been considered include onboard storage of cryogenic fluids in missiles; temperature control of instrument packages; umbrellas to protect moon bases, personnel, and supplies from solar radiation; and casings for rocket engines using both nuclear and conventional fuels.

The successful development of thermal-protection systems implies that some of the requirements for spaceflight can be met without undue sacrifice in vehicle performance and that such systems can increase our capabilities for space exploration through the use of materials and techniques now available to us.

STRUCTURES FOR AIRCRAFT

Very Large Vehicles—to Be or . . . ?

Winfield H. Arata Jr.
Northrop Corporation

Ever since the successful development of the airplane by the Wright Brothers in 1903, technological breakthroughs have permitted unforeseen expansion in the capabilities and applications of aircraft. In this vein, there appears to be a continuing, measurable trend of greater efficiencies associated with scale both in air- and water-based vehicles. This and the more-detailed article by Larry Noggle, which follows, highlight the trend toward larger vehicles, indicate advanced technologies that back the trend, and cite commercial and defense applications benefiting from vehicles of very large size.

Much effort would be required, of course, to bring any concept for a large new vehicle into service. The field, as a matter of fact, must yet sort out the best concepts, establish priorities, develop the needed technologies and facilities, and establish funding needs and sources. But it should begin doing this because the following advanced technologies will mature in the next decade:

- Fuel-efficient gas turbine engines.
- Alternative fuels.
- Digital avionics.
- Composite materials and structures.
- Advanced aluminum alloys.
- Aerodynamic techniques—laminar flow control, winglets, wing-in-ground effect, and active controls.
- More-accurate navigation systems for en route and terminal operations.
- Meteorological intruments.

As has happened so often in the past, such technical advances should improve economies of operation in many ways.

Initial developments should favor military applications in the following areas: logistics airlift, including troop transfer; in-flight refueling; surveillance and early warning; airborne command; airborne missile launch, both strategic and tactical; and sea control. Obviously, large vehicles could reduce reliance on overseas bases.

A similar list can be made for seaborne large vehicles incorporating advance technologies: logistics and troop transfer, missile launch, launch of tactical and surveillance aircraft, submarine tending, and sea control.

The speed of such vehicles should give them much greater survivability than conventional surface ships in transit between continents—a survivability not

Astronautics & Aeronautics, Vol. 17, No. 4, April 1979, pp. 20–25, 33.

normally associated with past military operations. This factor looms large when time does not permit rapid replacement of lost vehicles. Speed may also reduce the number of craft needed and, thus, manpower.

A military development would plausibly foster civil cargo and passenger transports.

Large size would permit movement of commodities not normally being carried today and, thus, should broaden the market and infuse it with new ideas. Certainly, it is possible to foresee shipment of oversize spare parts, construction equipment, raw materials such as forest products, medical equipment, and large multimode containers.

Igor Sikorsky's large aircraft developments in Russia during World War I may be thought of as the beginning of a trend. Table 1 lists some large aircraft developed since 1929. It includes one rigid airship, the *USS Macon*. Putting these large vehicles in perspective, the semilog plot of their weights (Fig. 1) shows a linear trend that may forecast developments in the next decades. Study projects of major U.S. airframe contractors and NASA can be related to the trend line shown.[1,2]

Over the past few years, manufacturers, the U.S. Air Force, and NASA have been studying and publishing their concepts for very large conventional takeoff and landing (CTOL) aircraft, such as a 12-engine supertanker.[3] That aircraft concept, the Boeing RC-1, had a planned payload of 2,320,000 lb and a takeoff gross weight of 3,550,000 lb. Its wing spanned 478 ft.

Mikolowsky, Noggle, and Stanley have discussed very large conventional CTOL transports and considered alternate fuels.[4] They compare the design impact of JP, liquid methane, liquid hydrogen, and nuclear fuel. Payload for the comparison was 350,000 lb and mission radius, 3600 n mile.

In October 1977, NASA held midterm briefings at the Langley Research Center for contractual studies, entitled "Cargo/Logistic Airlift Systems Studies" by

Table 1 Large aircraft

Year	Company	Model	Gross weight, lb
1929	Junkers	G-38	66,000
1929	Dornier	DO-X	123,200
1933	Goodyear	Macon	403,000
1934	ANT	20	92,595
1938	Boeing	314	84,000
1942	Messerschmitt	323E-2	99,225
1943	Junkers	JU 390	160,930
1943	Blohm & Voss	BV238V-1	208,000
1946	Lockheed	Constitution	184,000
1946	Bristol	Brabazon	290,000
1946	Saunders–Roe	Princess	330,000
1946	Hughes	Hercules	400,000
1946	Northrop	B-35	209,000
1955	Convair	XC-99	400,000
1965	Lockheed	C-5A	769,000
1969	Boeing	747-200B	775,000

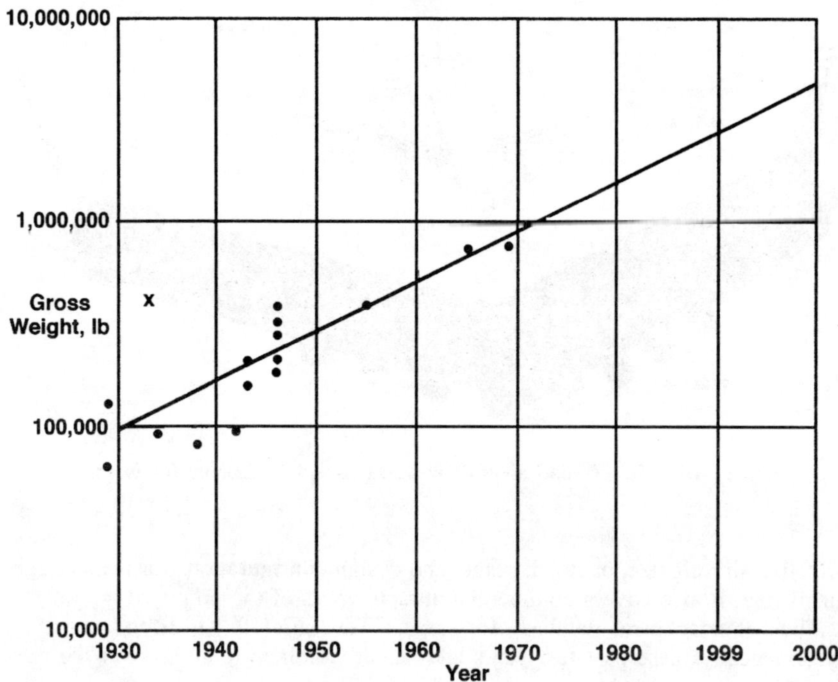

Fig. 1 Trend in aircraft weight, based on Table 1.

prime contractors Douglas Aircraft and Lockheed-Georgia.[5] The parties gave oral presentations in Washington, D.C., in June 1978. The objectives of the studies were:

- To estimate the air-cargo growth potential for both the international and U.S. domestic market systems projected for operation by 1990.
- To identify likely commodity and distribution-system characteristics.
- To determine air "eligibility" criteria.
- To estimate the market demand for large vehicles.

Ensuing studies were to explore questions of marketing timing for a new aircraft before the year 2000, market growth stimulated by new aircraft, a common civil/military aircraft, design criteria, and cost benefits for an entire air-cargo system.

Meanwhile, late in February 1978, NASA had convened a conference on "CTOL Transport Technology—1978" at Langley Research Center. From it came two volumes of papers covering propulsion, structures and materials, laminar flow control, advanced aerodynamics and active controls, operations and safety, and advanced systems.[6] And following the conference, D. William Conner of Langley published a summary of results, "CTOL Concepts and Technology."[7]

In the summer of 1978, the U.S. Air Force and NASA announced a two-year contract to Analytic Services, Inc., for studying next-generation large cargo aircraft. This explores technical and cost data necessary to identify the most cost-

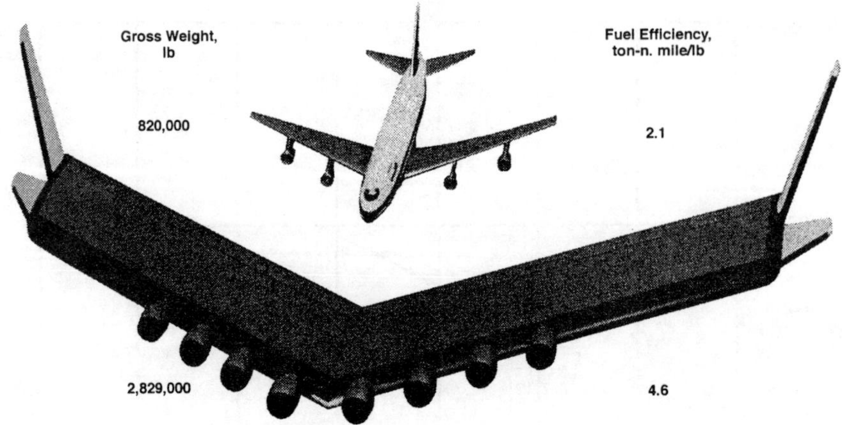

Fig. 2 Distributed-load aircraft design concept by Boeing for NASA.

effective aircraft size, payload, range, and design configurations for the strategic airlift role. It also covers commercial attractiveness of the proposed aircraft.

Thus, the technical database for large CTOL aircraft has been expanding systematically these past few years and should continue to do so over the next several.

As a follow-on to large conventional CTOL aircraft, new advanced concepts are being proposed and studied.[1,2] F. A. Cleveland treated large aircraft, including a large flying wing, in his 1970 Wright Brothers Lecture.[8] H. W. Withington of Boeing has given a general review of future possibilities, both conventional and advanced,[9] such as flying wing, or span-loader, with a capacity for 1700 passengers plus cargo. And at the 1978 AIAA Aircraft Systems and Technology Meeting, Jobe of the U.S. Air Force Flight Dynamics Lab and Kaufman and Vachal of Boeing gave a paper on wing planforms for large military transports.[10]

In another analysis of distributed-load-aircraft concepts, Whitener of Boeing describes one model (Model 759-211) with a takeoff gross weight of 2,829,000 lb and a payload of 1,300,000 lb, which takes the form of 104 containers 8 by 8 by 20 ft each.[11] Figure 2 compares this model to a version of the Boeing 747. A slightly smaller model (Fig. 3) has six turbine engines and a gross weight of 2,350,000 lb. Its planned payload would be 900,000 lb. Note the loading of cargo through a wing tip.

Wing-in-ground effect (WIGE) concepts have been discussed for a number of years, both in the United States and Europe. Moore of Lockheed-Georgia presented a conceptual design study of a power-augmented ram-WIGE aircraft at the 1978 AIAA Aircraft Systems and Technology Meeting.[12] A Soviet eight-engine seaplane concept was depicted in 1974.[13] In 1976, at the Third AIAA/SNAME Advanced Marine Vehicles Conference, Gallington et al. reviewed WIGE work.[14]

Multiple fuselages represent another design concept being considered in some quarters. From NASA, Fig. 4 compares in scale a C-5A with a twin-fuselage

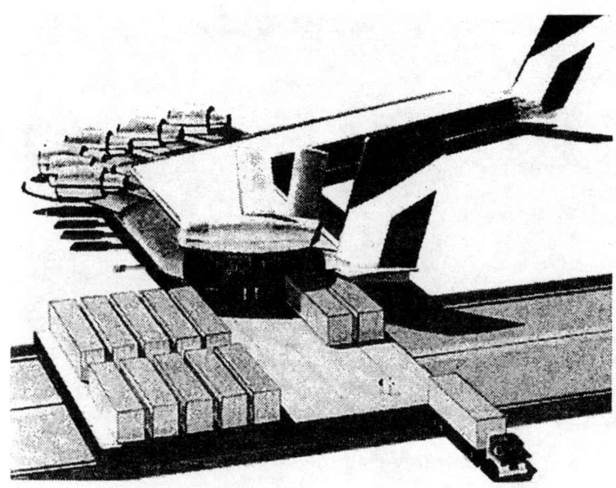

Fig. 3 Slightly smaller span-loader studied by Boeing for NASA. Courtesy NASA Langley.

cargo model representing a takeoff gross weight of 2,160,000 lb and payload of 1,200,000 lb. The cockpit in the center of the model would be of DC-9 size.

Werner Pfenninger, now at NASA Langley, has advanced a laminar flow control (LFC) concept (Fig. 5) for a 1,000,000-lb-takeoff-gross-weight transport. It would have a payload of 265,000 lb and, under the conditions assumed, a range of 11,000 n mile. It would have the following drag-minimization features: supercritical wing, laminar flow control, lift-drag ratio of 48, aspect ratio of 16.3, and controllable split wing tips.

Nuclear-powered concepts include the interesting NASA Langley plan shown in Fig. 6—a nuclear-powered tug able to tow two fully loaded C-5As. The tug has a design gross weight of 1,943,000 lb.

Fig. 4 Twin fuselage cargo concept. Courtesy NASA Langley.

Fig. 5 Pfenninger's now familiar LFC concept. Courtesy NASA.

In 1977, the U.S. Air Force and Canada completed a joint air-cushion-landing-system (ACLS) program with a converted XC-8A aircraft and made this statement: "ACLS could handle aircraft as heavy as 3,000,000 lb—and be competitive in weight with conventional landing systems." Figure 7 shows an illustration of an ACLS adapted to a 1,000,000-lb-gross-weight transport concept.

The NASA Aircraft Energy Efficiency (ACEE) program includes the E^3 (Energy Efficient Engine) propulsion project. Propulsion systems for very large aircraft should benefit from NASA's E^3 work as well as the current efforts of engine manufacturers.

While nuclear propulsion has its advocates, for a number of reasons conventional gas turbines will probably be used with very large aircraft. A change in weight and operational safety factors might alter this conclusion.

Fig. 6 NASA concept for a nuclear-powered "tug" aircraft. Courtesy NASA.

Fig. 7 Air-cushion landing system (ACLS). Courtesy Bell Aerospace/Textron.

Numerous articles have been written on alternative fuels.[4,6] The pros and cons reflect the waxing or waning of proven petroleum reserves. In the near term, it is certain the JP-fuel specifications will be broadened to accept alternate hydrocarbon sources, such as coal, oil shale, and tar sands.[15] Long-term alternative sources may include liquid methane and liquid hydrogen.[4,6,8,16]

This makes a good point to comment on lighter-than-air (LTA) craft. Ever since the large zeppelins, ending with the German LZ-129 and LZ-130 and the United States' *Akron* and *Macon* developments in the mid-1930s, LTA-vehicle resurgence has been a possibility, but not until recent studies sponsored by NASA, the Navy, and DOT have LTA concepts received serious consideration. New missions involve radical departure from previous types. They include offshore unloading, heavy cargo handling, surveillance patrol, and operations from the airborne platform.

Figure 8 shows a concept for short-range, heavy-lift operations. It combines features of blimp and helicopter. The design depicted has a payload of 75 tons; the concept might be pushable to 300 tons.

Concerning sea-based large vehicles, the U.S. Navy has completed a two-year survey of new technologies for ships and aircraft. The broad study investigated air- and sea-loiter aircraft, LTA craft, WIGE aircraft, air-cushion vehicles (ACVs), surface effect ships (SESs), hydrofoil and planing craft, and small waterplane twin-hull craft (SWATH).

Progress along the course of this Advanced Naval Vehicles Concepts Evaluations (ANVCE) has been reported at the AIAA/SNAME Advanced Marine Vehicles Conferences.[17,18] (A very brief summary of each concept has taken the form of interviews of the ANVCE staff.[19])

The Navy has been operating a 190-ton SWATH vehicle in the Hawaiian islands as a helicopter platform (Fig. 9); it has a 20-kn cruise speed. Figure 10 depicts the concept for a 4000-ton SWATH with a 30-kn cruise speed operating as a small aircraft carrier. Concepts up to 30,000 tons are under study.

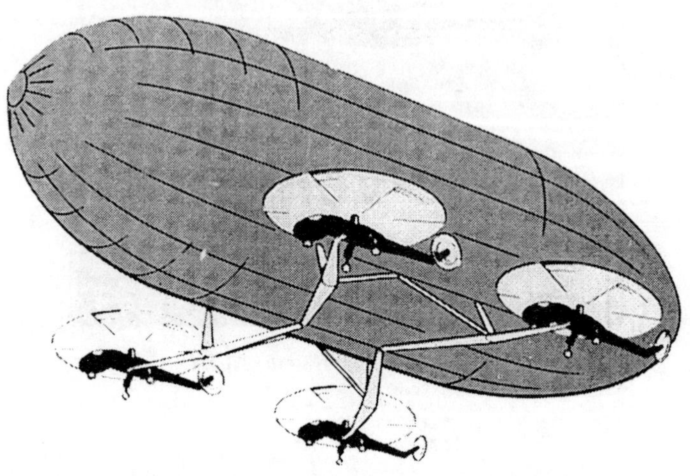

Fig. 8 Combination of blimp and helicopter. Courtesy NASA.

Fig. 9 The 190-ton SWATH vehicle *USS Kaimalino.* Courtesy USN.

Fig. 10 Navy illustration of a 4000-ton SWATH concept. Courtesy USN.

Rohr has a 3600-ton SES under construction for the Navy (Fig. 11). It should have a cruise speed of 80–100 kn. Figure 12 depicts a larger 10,000-ton SES cruising at 50–60 kn. SESs up to 20,000 tons are envisioned.

The ANVCE office had point designs made by McDonnell Douglas and Lockheed-Georgia on WIGEs weighing up to 1,000,000 lb, capable of performing both combat and logistics missions. Sea-keeping abilities and maneuvering just above the water surface are expressed concerns. However, its operational potential warrants further development.[12–14,17,18]

Relative to LTA, the ANVCE office looked at air-buoyant and semiair buoyant point designs. One air-buoyant design was 783 ft long and capable of a dash speed of 80 kn.[19]

This month AIAA holds the first Very Large Vehicle Conference. Its four sessions should answer these key questions:

- What are the requirements?
- What are the most promising concepts?
- What technological developments are still needed?
- Where do we go from here?

Fig. 11 Rohr 3600-ton SES. Courtesy USN.

Fig. 12 The larger 10,000-ton SES. Courtesy USN.

References

[1] Whitehead, A. H., Jr., "Technical and Economic Evaluation of Advanced Air Cargo Systems," NASA TM-78672, Feb. 1978.

[2] Nagel, A. L., "Studies of Advanced Transport Aircraft," NASA TM-78697, May 1978.

[3] Taylor, M. D., "The RC-1, A Conceptual Arctic-Resource Air Transport," ICAO Bulletin, April 1973.

[4] Mikolowsky, W. T., Noggle, L. W., and Stanley, W. L., "The Military Utility of Very Large Airplanes and Alternative Fuels," *Astronautics & Aeronautics,* Sept. 1977, p. 46.

[5] Whitehead, A. H., Jr., "Characteristics of Future Air Cargo Demand and Impact on Aircraft Development," AIAA Paper 78-1524, Aug. 1978.

[6] NASA Conference, "CTOL Transport Technology—1978," CP-2036, Parts I and II.

[7] Conner, D. W., "CTOL Concepts and Technology Development," *Astronautics & Aeronautics,* July/Aug. 1978, p. 29.

[8] Cleveland, F. A., "Size Effects in Conventional Aircraft Design," *Journal of Aircraft,* Vol. 7, No. 6, Nov./Dec. 1970, pp. 438–512.

[9] Withington, H. W., "Passenger-Aeroplane Developments—The Next Three Decades," ICAO Bulletin, Feb. 1978.

[10] Jobe, C. E., Kulfan, R. M., and Vachal, J. D., "Wing Planform Geometry Effects on Large Subsonic Military Transport Airplanes," AIAA Paper 78-1470, Aug. 1978.

[11] Whitener, P. C., "Distributed Load Aircraft Concepts," AIAA Paper 78-100, Jan. 1978

[12] Moore, J. W., "Conceptual Design Study of Power Augmented Ram Wing-In-Ground-Effect Aircraft,"AIAA Paper 78-1466, Aug. 1978.

[13] “New Soviet Hybrid Aircraft Depicted,” *Aviation Week,* Jan. 28, 1974.

[14] Gallington, R. W., et al., “Recent Advances in Wing-In-Ground-Effect Vehicle Technology,” AIAA Paper 76-874, Sept. 1976.

[15] Churchill, A. V., et al., Future Aviation Turbine Fuels,” *Journal of Aircraft,* Nov. 1978.

[16] Brewer, G. D., “Liquid Hydrogen—A Logical Choice to Fuel Future Commercial Aeroplanes,” *ICAO Bulletin,* Feb. 1976.

[17] Mantle, P., and Meeks, T. L., “The Advanced Naval Vehicle Concept Evalution,” AIAA Paper 76-846, Sept. 1976.

[18] Mantle, P., and Meeks, T. L., “An Evaluation of Advanced Naval Vehicles,” AIAA Paper 78-748, April 1978.

[19] Rhea, J., “A Compass for the Crystal Ball,” *Sea Power,* Jan. 1978.

Large Vehicle Concepts

Larry W. Noggle
U.S. Air Force Aeronautical Systems Division

Charles E. Jobe
U.S. Air Force Flight Dynamics Laboratory

In 1973, two events combined to give new impetus to interest in very large vehicles. First, the oil embargo emphasized the need to investigate alternative energy sources and less-energy-intensive transportation concepts to reduce the nation's dependence on foreign petroleum supplies. Second, the Mid-East War dramatized the military need for a rapid long-range, heavy-lift airlift not dependent upon foreign bases or overflight rights. Since very large vehicles inherently possess the volume needed to accommodate "alternative" low-density fuels, such as hydrogen or methane, and also provide the payload and fuel capacity required for long-range independent military operations, they became an object of considerable study within DOD.

In the civil transportation sector, very large vehicles were likewise studied because of their suitability for using alternative fuels. They received additional support from studies that projected an air-cargo market that could support a large dedicated air freighter by the end of the century.

This article briefly surveys most of the very large vehicle concepts examined by the Air Force, Navy, NASA, and industry in recent study efforts. Because of similarities in many design concepts and in the interest of brevity, we have chosen to discuss types of vehicles and present representative examples, rather than treat specific concepts in detail. By this means we hope to convey the major characteristics of each and their major technological implications. We will be leading to the conclusion that, for aircraft, conventional configurations will serve known needs.

Conventional Large Aircraft

A number of studies have examined large conventional aircraft configurations that capitalize on the weight and fuel savings made possible by advanced technology.[1–4] These efforts have examined designs with payloads from 110,000 to 600,000 lb and gross weights from 326,000 to 1,700,000 lb.

A few of these studies have been devoted to exploring the feasibility of integrating design requirements for civil and military transports—to produce a common aircraft suitable for military airlift and economically competitive as a dedicated commercial air freighter.

Astronautics & Aeronautics, Vol. 17, No. 4, April 1979, pp. 26–32.

Fig. 1 Conventional aircraft concept designed for civil/military cargo commonality.

Figure 1 depicts a large aircraft in the military airlift role. It features a double-arc fuselage (radii of upper and lower portions of fuselage different) that can hold four-abreast 8-by-8-ft commercial cargo containers and outsize military cargo. It has a wing span of 330 ft, a length of 280 ft, a gross weight of 1,480,000 lb, and a range of 6200 n mile with a payload of 400,000 lb. The configuration depicted is optimized for minimum gross weight. Configurations optimized for minimum fuel consumption usually have less sweep, greater aspect ratio, large wing area, and substantially greater operating weight.

Figure 2 shows a smaller aircraft that has potential for serving the passenger as well as the air-freight market. It employs over-the-wing engines to give a low-wing planform while achieving a low cargo deck. It has a payload of 115,000 lb and a takeoff weight of 349,000 lb for a range of 3600 n mile.

Fig. 2 Smaller conventional aircraft concept for passengers as well as cargo.

Laminar Flow Control (LFC) Aircraft

A large aircraft (Fig. 3) could employ laminar flow control (LFC) to conserve fuel while yielding very long ranges. The Fig. 3 design concept has wing and tail surfaces slotted to provide laminar flow to 70% chord. The reduction in the profile drag of the wings and empennage due to LFC increases the cruise lift-to-drag (*L/D*) ratio from approximately 27:1 to 40:1. Suction for the LFC subsystem is created by four ram-air turboshaft engine/compressor units on the trailing edge of the wings and two units on the empennage. The design calls for a wing span of 455 ft, length of 282 ft, takeoff gross weight of 1,550,000 lb, and aspect ratio of 14. It would carry 350,000 lb of payload 10,000 n mile.

Nuclear-Powered Aircraft

Some of the study efforts have investigated aircraft concepts that employ cryogenic and nuclear fuels.[1,2,5] For the most part, the optimum configurations for these aircraft resemble conventionally fueled (JP) aircraft in appearance. The only significant variations occur in the fuselage shape of cryogenically fueled designs, which house fuel tanks in the fuselage, generally in an upper lobe.

Figure 4 depicts a nuclear-powered-aircraft exception to this trend. The canard configuration suits nuclear propulsion because it allows the large concentrated load of the reactor and integral shielding to be positioned near the aircraft's center of gravity (CG) without obstructing access to the cargo area. The plane shown has a payload of 400,000 lb at a gross weight of 1,560,000 lb. It has a wing span of 308 ft and a length of 275 ft. The nuclear subsystem weighs 391,000 lb and features a containment vessel designed to survive high-speed impacts without rupture; it delivers 230 Mw.

The design of a nuclear-powered aircraft poses unique considerations in structures, landing gear, and controls. Structural support for the very heavy nuclear subsystem must be carefully integrated into the airframe to avoid excessive weight penalties. Landing gear must be designed for repeated landings at nearly maximum gross weight, since there is little weight change due to fuel consumption. Moreover, since the payload's CG is far forward of the wings, the aircraft experiences relatively large shifts in CG between the fully loaded and the unloaded condition. This places exceptional emphasis on design of the stability and control subsystem.

Nuclear-powered aircraft do not exhibit the familiar payload/range characteristics of fossil-fueled aircraft, since they offer virtually unlimited range at maximum payload. Because of this, the designer must give greater attention to the ride quality and vibration environment, since mission duration will usually be limited by crew comfort and tolerance to noise and vibration.

Benefits from Technology Improvement

All of the aircraft concepts investigated benefit from present-day aerodynamic computational techniques that include the effects of viscosity and allow the design of optimum wing sections for a given application. They also benefit from the introduction of active controls for flutter suppression and alleviation of gust or maneuver loads; active controls greatly reduce the weight penalty for the high-aspect-ratio wings. Automatic control subsystems reduce vertical-tail size by

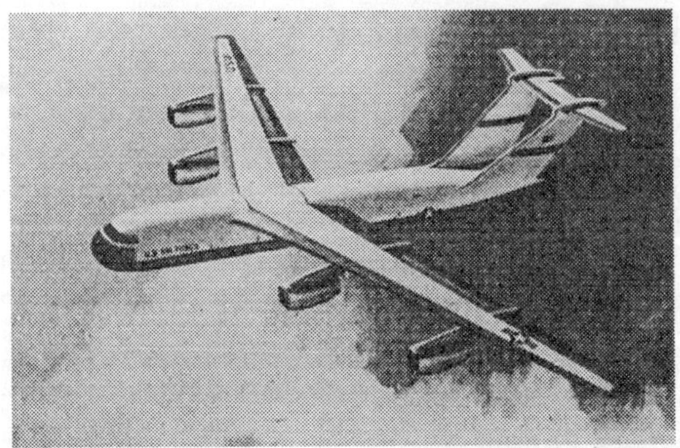

Fig. 3 Laminar Flow Control (LFC) concept.

countering the effects of engine failure at takeoff and by permitting relaxation of the longitudinal static-stability margins.

Improvements in low-speed performance result from elimination of flap cutouts, larger flap span, and increased flap efficiency through the addition of nonplanar features such as vertical fins at the extremes of flap segments.

Careful integration of the engine nacelle with the wing design produces higher critical-Mach-number contours and better stiffness and flutter characteristics than those of the wing alone.

Composite primary structures employed run 15–20% lighter than the aluminum structures they replace. High-purity aluminum alloys having higher strength and toughness reduce the weight of critical structural members.

Lightweight carbon-composite brakes save considerable weight, improve wear life, and reduce production costs. Advanced skid-control subsystems reduce tire wear by as much as 30% while improving safety.

The technology-improvement factors employed in the design process might be summarized as shown in Table 1.

Fig. 4 Nuclear-powered aircraft concept.

Table 1 Technology improvement factors

Factors	Weight savings, %
Composite structures	
Wing	22
Empennage	27
Body	18
Landing gear	8
Nacelles	17
Propulsion (engines)	6
Flight controls (fly by wire)	30
Hydraulics (high pressure)	30
Electrical (adv. generators)	10
Active controls (countering reduced static stability and, for wing, alleviating maneuver loads)	
Wing	14
Body	2
Horizontal stabilizer	12

Propfan Outlook

Energy efficiency could be substantially improved by propfan propulsion. Operating at Mach 0.8, the propfan offers a 10–15% improvement in specific fuel consumption (SFC) over advanced turbofans and provides an increase in aircraft range of 6–8%. Propfan propulsion systems may become available by the 1990s for vehicles of modest size, but much development effort will be required before they can be considered for aircraft in the million-pound-gross-weight class. Improvements will be needed in lightweight high-power gearboxes, propellers, noise attenuation, and maintenance and support to make the propfan fully competitive for very large vehicles.

Nonconventional Aircraft

The designer of very large vehicles has an opportunity to exploit fully an old idea that has yet to make it—the flying wing, or span-distributed-load aircraft.[6,7] Figure 5 illustrates a flying-wing configuration that combines the aerodynamic improvements of supercritical airfoils with the structural efficiency (weight fractions on the order of 25%) of span-distributed loading to carry payloads from 400,000 to several million pounds. Such a plane employs active flight controls to good effect. Its winglets double as vertical stabilizers. The commercial air freighter shown here has a wing span of 407 ft at a sweep of 30 deg. It weighs 2,350,000 lb at takeoff and can deliver a payload of 900,000 lb some 3600 n mile.

Sweptwing designs exhibit a critical low-speed instability during takeoff. Since the aircraft does not rotate at takeoff, the lift coefficient must be generated at ground-roll attitude. Another difficulty, characteristic of all large aircraft and particularly a span-loader, concerns the aircraft/airport interface. The wing spans generally exceed existing taxiway clearance limits, and the optimum arrangement of landing gear would exceed existing runway and taxiway pavement widths.

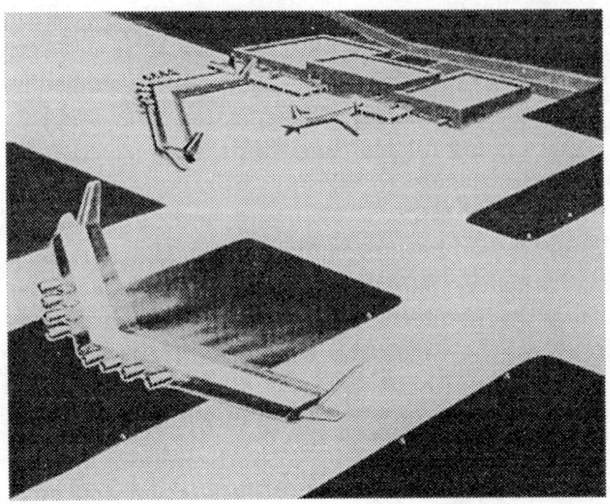

Fig. 5 Very large sweptwing span-distributed-load aircraft.

Wing-in-Ground-Effect Vehicles

Near the ground, wings experience an increase in lift and decrease in drag due to reduced downwash (downward motion of the air aft of the wing). The increase in *L/D* from proximity due to the ground depends inversely on the ratio of height above the ground to wing chord. Thus, wing-in-ground-effect (WIGE) vehicles require a large wing chord and relatively large, flat expanses to achieve efficiency at safe heights above the ground. For this reason many WIGE designs have been sea-based aircraft.

The history of WIGE vehicles, dating back to the turn of this century, was recently reviewed by Ollila[8] at the U.S. Air Force Flight Dynamics Laboratory/Foreign Technology Division WIGE Technology Workshop.[9] WIGE concepts have primarily taken the form of large cargo transports. Other potential applications include search and rescue, surveillance, utility transportation, and recreation. Our remarks will concern two current designs:

- A Navy concept utilizing the power-augmented-ram (PAR) effect.[10]
- A hybrid PAR-WIGE design with wing additions for flight at altitude, by Douglas.[11]

In most respects the conventional WIGE vehicle has poorer performance than the PAR-WIGE due to takeoff and landing requirements.

The Navy PAR-WIGE design (Fig. 6) uses exhaust from the forward-mounted engines, along with tip plates and trailing edge flaps, to raise static pressure under the wing and lift the vehicle out of the water. This eliminates the long takeoff and landing runs characteristic of conventional WIGE vehicles. Tests at the David Taylor Naval Ship R&D Center demonstrated that the PAR cushion can be generated at high loadings and in relatively rough sea conditions. Since a PAR-WIGE vehicle does not cruise on the PAR cushion, it has a relatively high transit speed (200–400 kn) and can loiter for long periods on the ocean surface. It has been considered for the tactical-sea-control, transoceanic-transport,

Fig. 6 PAR-WIGE vehicle.

and strategic-deterrence missions in the 1990–2000 period. Payloads range from 520,000 to 750,000 lb for these missions and gross weights exceed 2,000,000 lb.

Figure 7 shows a variant of the PAR-WIGE studied by Douglas. It retains the augmented ram wing as inboard wing panels and adds high-aspect-ratio outboard panels to permit efficient flight as a conventional aircraft. Estimates indicate that this variant would be approximately 20% more efficient as a transport than a basic PAR-WIGE vehicle. But it cannot operate from air fields at design takeoff gross weight, since the benefit derived from the PAR concept results in a low thrust-to-weight ratio and the thrust is insufficient to satisfy rate-of-climb requirements for land operations. Adding landing gear would further penalize the design for land operations.

WIGE and PAR-WIGE vehicles offer the basing flexibility of seaplanes and high aerodynamic efficiency when operated at minimum ground or sea clearances.

Fig. 7 Hybrid ram-wing vehicle.

The PAR feature is a new technology that can reduce the power mismatch between cruise and takeoff conditions for seaplanes. Moreover, the higher wing loading and lower thrust-to-weight ratios permitted with this concept increase structural efficiency. Considerable wind-tunnel test data support WIGE design work. Nevertheless, it will take still more analysis to understand the nonlinear aerodynamics (including hysteresis) and the stability and control characteristics exhibited by WIGE vehicles.

Lighter-Than-Air (LTA) Vehicles

Airships can be broadly divided into two categories: the fully buoyant conventional airship and the partially buoyant hybrid—based upon the buoyancy parameter (Beta), the ratio of static lift (buoyancy) to total lift or gross weight. Conventional airships are fully buoyant ($\beta = 1$). They can be further categorized as rigid, nonrigid, and semirigid.[12]

Nonrigid airships use a flexible fabric envelope filled with slightly pressurized lifting gas. Rigid airships usually have an aluminum ring-and-girder frame with an attached outer covering of neoprene-coated cotton and several internal gas cells arranged longitudinally. Semirigid designs employ a gas envelope affixed to a metal keel. A pressurized rigid airship combines the features of both the rigid and nonrigid by using an aluminum hull. Historically, the principal lifting gas has been hydrogen; but now designers prefer helium for safety.

Partially buoyant hybrid airships ($\beta < 1$) depend upon additional dynamic lift from rotors, wings, hull, or combinations of these to achieve flight. A NASA-sponsored study evaluated the technical feasibility of such airships and their potential for fuel-conserving transportation.[13] The Navy has also sponsored airship concept studies related to future missions.

The primary attributes of long range, ability to lift large and heavy payloads at low fuel consumption, and low noise and other pollution especially suit the airship to several missions—surveillance, transport of heavy voluminous payloads, coastal patrol, sea control, observation, and moving goods in remote areas. Airships can haul large objects well because their lifting ability increases with the cube of linear dimension while structural weight increases in virtually direct proportion to surface area—a reverse of the familiar "square-cube" law of airplanes.

The NASA study chose three types of vehicles for detailed study: a passenger feeder airship for regional airports, a heavy-lift short-haul cargo mover, and a large rigid craft for long-endurance naval missions.[14]

Figure 8 illustrates an 11,200,000-cu-ft rigid airship designed to perform naval missions independent of support from other than home base. Typical missions include stalking submarines and other antisubmarine warfare, ocean (plus sound) surveillance, and convoy escort. The vehicle shown could stay on station 20–30 days and cruise at speeds of 75 kn at an altitude of 5000 ft.[15]

The heavy-lift airship concept illustrated in Fig. 9, a "Helistat" has a payload of 75 tons and a range of 100 n mile. Proposed by Piasecki Aircraft, the concept consists of four CH-54B helicopters and a 2,500,000-cu-ft nonrigid airship hull. Analysis indicates it can haul payloads up to 250 tons.

Mayer has surveyed airship studies and recent developments.[16] The last large rigid airships were retired in the late 1930s. The Navy disestablished its LTA

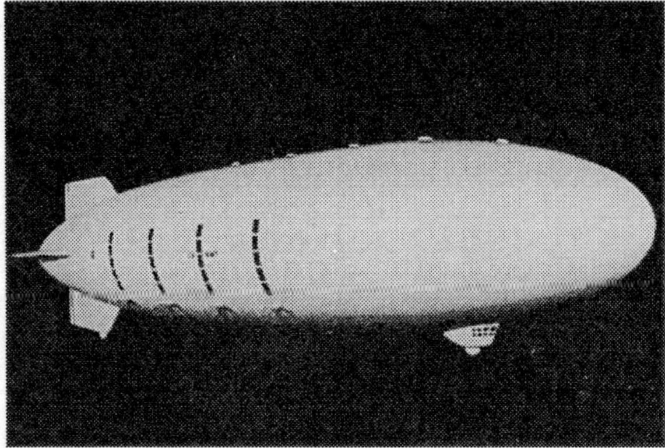

Fig. 8 Concept for a 11,200,000-cu-ft rigid airship.

commands and decommissioned its nonrigid fleet. Thus, several decades of technology have passed since any serious investigation of airships. Over that period aeronautical engineering has seen dramatic improvements in structural and covering materials, propulsion systems, and flight controls that could increase airship performance and economics. And hybrid concepts such as the Helistat have evolved that resolve many former problems. Thus, the future of the airship has not become a closed case.

Surface Effect Ships

Not confined to flying concepts, interest in very large vehicles spans a spectrum of air-cushion vehicles (ACVs) configured for land and water applications. The

Fig. 9 This 2,500,000-cu-ft ''Helistat'' airship concept could haul 250 tons of payload.

Navy has been investigating the potential of high-speed air-cushion vehicles for more than two decades.

An illustration in the preceding article (Arata, Fig. 11) shows an artist's rendering of the 3600-ton surface effect ship (SES) the Navy has under development.[17] Two partially submerged rigid sidewalls and two flexible seals, bow and stern, in contact with the water contain the SES's supporting cushion. This arrangement permits using either waterscrew or waterjet propulsors and provides stability at speeds in excess of 80 kn.

Current technology limits the maximum size of SESs to a displacement of 4000 tons. However, this is not an intrinsic limit, and further advancements in the technologies listed below could yield SESs of several thousand tons displacement:

- Power production: requires low weight and high-power output with low specific fuel consumption.
- Power transmission: requires lightweight reduction gears with ratings of 50,000 hp or greater, high-power clutches, and flexible design concepts for coupling various combinations of engines, lift fans, and propulsors.
- Power conversion: requires high-power, supercavitating propellers with ratings of 50,000 hp or greater.
- Structures: requires design and construction techniques for lightweight, high-strength hulls plus sidewall construction with flexibility for multiple propulsors.

We note that the developments identified in the areas of power generation and transmission would also profit propfans and WIGE vehicles.

Concluding Remarks

Several of the foregoing concepts and their merits have been known to designers for many years. Nevertheless, the conventional-large-aircraft configuration has evolved as the most efficient for current missions and fuel prices. Moreover, recent studies of conceptual aircraft designs for the 1980s and beyond indicate that the changes necessary to increase fuel efficiency and decrease noise and pollution will not greatly change general conventional configuration, arrangement, or size. To win further development support, very large aircraft of unconventional configuration must satisfy, we believe, new or unique mission requirements or provide the commercial capabilities to generate new markets. With precisely this intent, this month experts will be discussing many large-aircraft concepts at the AIAA Very Large Vehicle Conference. Perhaps they will reveal a way to change this prevailing conclusion.

References

[1] Mikolowsky, W. T., Noggle, L. W., and Stanley, W. L., "The Military Utility of Very Large Airplanes and Alternative Fuels," *Astronautics & Aeronautics,* Sept. 1977, pp. 46–56.

[2] Barber, E. A., et al., "Innovative Aircraft Design Study, Task II," The Boeing Aerospace Company, Vol. 1, June 1977.

[3] Barber, E. A., et al., "Innovative Aircraft Design Study, 1977," The Boeing Aerospace Company, Vol. 1, D180-24713-1, July 1978.

[4] Kulfan, R. W., and Vachal, J. D., ''Application of Laminar Flow Control to Large Subsonic Military Transport Airplanes,'' Air Force Flight Dynamics Laboratory, AFFDL-TR-77-65, July 1977.

[5] Muehlbauer, J. C., et al., ''Innovative Aircraft Design Study, Task II, Nuclear Aircraft Concepts,'' Lockheed-Georgia Company, LG77ER0008, April 1977.

[6] ''Technical and Economic Assessment of Span-Distributed Loading Cargo Aircraft Concepts,'' NASA CR-144963, June 1976.

[7] ''Technical and Economic Assessment of Swept-Wing Span-Distributed Load Concepts for Civil and Military Air Cargo Transports,'' NASA CR-145229, Oct. 1977.

[8] Ollila, R. G., ''An Historical Review of WIGE Vehicles,'' presented at AFFDL/FTD Wing-in-Ground Effect Technology Workshop, Wright–Patterson Air Force Base, OH, Sept. 1978.

[9] Osburn, R. F., and Galloway, C. E., *Proceedings of the AFFDL/FTD Wing-in-Ground Effect Technology Workshop,* to be published as FTD TM, Jan. 1989.

[10] Papadales, B. S., ''The Performance of a Conceptual Multimission Power-Augmented-Ram Wing-in-Ground Effect Vehicle,'' David Taylor Naval Ship Research and Development Center, ASED-395, Sept. 1977.

[11] Douglas Aircraft Company, ''Wing-in-Ground Effect Vehicles,'' pamphlet, June 1977.

[12] Strother, E. F., ''Lighter-Than-Air Concepts and Recent Developments,'' *AIAA Student Journal,* Vol. 16, No. 2, Summer 1978.

[13] Ardema, M. D., ''Feasibility of Modern Airships, Preliminary Assessment,'' *Journal of Aircraft,* Vol. 14, No. 11, Nov. 1977.

[14] Goodyear Aerospace Corp., ''Feasibility Study of Modern Airships, Phase II-Executive Summary,'' NASA CR2922, Nov. 1977.

[15] Marcy, W. L., ''Conceptual Design of a Rigid Airship for U.S. Navy Operations in the 1990s,'' AIAA Paper 77-1195, Aug. 1977.

[16] Mayer, N. J., ''LTA: Recent Developments,'' *Astronautics & Aeronautics,* Vol. 15, No. 1, Jan. 1977.

[17] Fee, J. J., and Handler, E. H., ''The Navy's Large Surface Effect Ship,'' Navy Sea Systems Command, pamphlet, May 1977.

Development of Optimum Structure for Large Aircraft

L. M. Hitchcock
The Boeing Airplane Company

Introduction

There are available today at least two major techniques or methods for the design of aircraft structure. One of these is the so-called *stretch method.* In this method, the structure is initially designed understrength and the deficiencies corrected by means of a static-test development program. In other words, the static test becomes a design tool, and it is apparent that lightweight structure can result. The second method involves designing the structure to support full design loads. It employs rigid weight control, reliable stress analysis methods, knowledge from past experience, and developmental testing for allowable stresses and loads to obtain light structure. It is this second method that will be discussed and demonstrated to be of overall advantage in the development of large aircraft. For convenience, this method shall be referred to as *100% design.*

The philosophy of 100% design is considered beneficial in the development of large aircraft because it can:

1) Result in lightweight structure.

2) Assure that the static-test program will be completed in a reasonable time period.

3) Provide maximum continuity for the production program.

4) Permit early flight test without excessive load factor restrictions limiting evaluation of performance, stability, and control of the airplane.

These are recognized as important items in all aircraft production development, but they are especially important for large aircraft. It is concluded that 100% design satisfies all four of these needs and that, in this sense, the method can yield essentially ''optimum'' structure.

Lightweight Structure

To achieve low structural weight by the method of 100% design (or by any design method), it is necessary to follow a rigid weight-control program. We have found that one of the requisites in the weight-control program is the utilization of analytical methods for the preliminary weight estimates. These methods account for critical loads, geometric relationships, the method of construction, and the

Aeronautical Engineering Review, Vol. 13, No. 11, Nov. 1954, pp. 50–55.

allowable stress levels of the materials. The refinements in method permit the evaluation of even small changes in the design parameters and improve the ability to determine efficient structural arrangements in the preliminary design stage. Once the airplane configuration is selected, it is our policy to establish optimistic target weights with minimum allowances for extra features and contingencies. This emphasizes the overweight effects of extra features and special design problems and, thereby, provides a constant challenge to offset such increases by design refinements. During the structural layout stages, a continuous weight evaluation is made of this design development. The current weights are compared with the target weight allowances and periodically reported to all responsible design groups. Reasons for weight differences are carefully analyzed, and steps are taken not only to eliminate such overweights, but also to better the target weights wherever possible. It follows, of course, that reliable stress analysis is necessary to achieve the desired goals.

It has frequently been held that practical analysis methods are not satisfactory for obtaining light structure. However, when analysis and experiment differ widely, it is probable that analysis has been applied without due regard to the limitations of the ever-present assumptions. Inasmuch as the magnitude and complexity of modern structural design problems make analysis indispensable, it has been our practice to modify analysis methods on the basis of the results of past tests for application to future designs. (The modifications are, of course, applied in the simplest manner consistent with good engineering practices.) With a continuous improvement in experience and judgment on this basis, it becomes possible to exploit analysis with the expectation of better correlation with experiment.

Correlation of analysis with test results enabled us to use the experience gained on the B-47 airplane more efficiently in the design of the B-52. Furthermore, the prototype jet transport we have built has a structure that is designed almost exclusively on the basis of judgment gained in analyzing and testing the C-97, B-47, and B-52 airplanes. By a process of review, fundamental conservatisms have been consistently reduced so that the analysis methods used are more applicable now than in the past. To collect significant facts from static-test programs for review, we have always liberally applied strain gages to well define stress paths and intensities. For example, on the B-52 program, a total of 2500 strain gage locations are used; 600 of these are recorded on one critical test condition.

Another aid in the design of light structure is the extensive testing of joints, panels, composite structures, and representative components to find true allowable stresses and loads in the structure. In so doing, we attempt to exploit the strength of the structure in the developmental test phase instead of in the static-test phase as for the stretch method. For example, the B-52 program has involved over 130 test programs to determine the stress allowables to be permitted in all questionable locations. These tests included a true half-size model of the body in the region of the bomb bay and wheel wells. The results of the tests on the model indicated the manner in which the torsion loads were transmitted, the allowable shear flows on skins of different thicknesses and radii, the influence of the support spacing, and the optimum cross sections of the body frames. These tests were extended beyond the elastic range to the ultimate strength of the model so the results were fully applicable to the ultimate design conditions of the airplane.

Our experience has confirmed that the methods just described will give light structure with fairly definable strength limits if the policies are pursued with sufficient vigor. Also, the method of 100% design is found to be conducive to successful static-test programs.

The Static-Test Program

To be successful, a static-test program must be completed within a reasonable minimum time period. It is important, then, to avoid excessive premature failures with their attendant delays. The bugaboo of premature failure is hard to avoid.

We may divide premature failures into two types, the local and the catastrophic. Local failures are those that are only moderately extensive so that the test specimen may be repaired and modified for subsequent retesting in a short time. Catastrophic failures are those that are impractical to repair and retest without excessive delays and substitution of major sections or components built in the construction jigs.

Local failures can be troublesome even though they do permit the reuse of the test article.

Catastrophic failure is well illustrated by the experience of testing the B-50 wing. The first failure in the program occurred at 92% of load. The article is a complete washout. For subsequent testing, it was necessary to replace the right outboard wing and the entire left wing and rework the right inboard wing. A second catastrophic failure occurred at 95% of load, and it was once more necessary to replace entirely the left wing. Upon further testing the right wing failed locally at 99.5% of load. This level was considered evidence of structural integrity for the condition tested with suitable minor changes incorporated. The reworked wing was subsequently tested to destruction in another condition at 101.5% of load. The final cost of the program was approximately three times what it would have been without the premature catastrophic failures. In regard to the effects of delay, the program was even more unsatisfactory.

The XB-29 program had many premature local failures, and, as a result, the actual testing time was more than 13 months instead of the three months or so scheduled. On the other hand, the B-47 test program was completed with but few and minor premature failures so that the time schedule was exceeded by only two months. Further, the B-47 static-test program was enlarged in scope after testing began. The satisfactory experience of the B-47 program is attributed to a more successful application of the method of 100% design.

Our past experiences have proved to us the need for completing static-test programs without excessive delays. It is felt that this need is even more important today than previously, for with the increase in aircraft design complexity there has been an accompanying increase in the number of tests required in the static test program. For the future, we must anticipate airplanes of still greater complexity with difficult environmental test problems so that delays will become increasingly undesirable.

The concern regarding static-test delays is so great that, in spite of the advantages inherent in the method of 100% design, we have considered it necessary to test initially to only 95% of ultimate load in the critical design conditions.

This procedure lessens damage in any one test, which could precipitate premature failures in subsequent tests, and generally reduces the probability of catastrophic failure and the attendant delays. Proof of 100% strength is established by correlation between the 95% tests, the stress analysis, and the ultimate destruction tests. The destruction tests are conducted for the condition most critical for each major component of the airplane.

It was previously stated that premature failures are difficult to avoid, but our experiences to date with the method of 100% design and the policy of testing to 95% of load have shown that premature failures in lightweight structure may be minimized and static-test delays reduced to a level that can be tolerated.

Production Continuity

In addition to the problems of designing light efficient structure and completing the static-test program without excessive delays, there is the related need of maintaining production schedules with a minimum of disruption. From our experience on large airplanes, the production model design usually begins before the completion of the X-airplane, and production starts before the static tests. Frequently, it will be found that the prototype is so different in concept from the production airplane that its structure is not representative of the production series. This was the case on the B-47 series, in which the X-model was not static tested, while the B-model, with a stronger structure, was. The last large airplane prototype that we static tested was the XB-29. However, even in this instance, the B-29 airplane was in production at the time the static test was being conducted on the XB-29.

The static-test airplane is usually the first of the production series, and production is concurrent with the static-test program. Obviously, then, the design of the structure must be reliably finalized before extensive production tooling can be constructed. Production items, such as forgings, extrusions, and so on, require considerable lead time so it is necessary that the static test be completed, insofar as possible, without turning up major deficiencies in these items. While premature failure in static test may indicate only superficial structural changes, these may have a more serious influence on production parts and tooling.

We have also found that the size of the airplane can have an influence on the relations of the static-test and production programs. For example, because of the size of the B-52 airplane, it has been necessary to use a part of the production area as the test facility. The area used is about one-fourth that of a special hangar used for final production and flight testing of the production airplanes. As new designs are tested only about every four years, it is not economically feasible to provide a facility of the size required, suitable for static testing only. Here, again, it is important to minimize the delays of the static-test program because of disruptions that could result to production schedules.

Aside from the direct influence of premature static-test failures on production tooling, the changes to be made to items already produced must be considered. In the case of the B-50 wing, it was necessary to modify 20 existing wings because of the deficiencies discovered in static test. This incident, however, was insignificant in comparison to the difficulties experienced on the B-29 production program at Wichita. The delays resulting from the premature failures on the XB-29 prevented completion of the static test in the scheduled time so that much early production had to be modified to incorporate the changes determined as a result of the static tests. One of the more undesirable features of the situation was the necessary diversion of effort from production to static-test fixes in time of war. Since these early unsatisfactory experiences, we have been especially careful in designing structures that will be of satisfactory weight and will still come through the static test without excessive premature failures or serious structural faults that cause excessive delays and disrupt production. On the whole, and in spite of some past difficulties, the method of 100% design has enabled us to conduct production and static-test programs concurrently, with good results. In fact, such a situation exists for the B-52 airplane today.

Early Flight Testing

On most Boeing designs, flight testing has commenced prior to the static-test program. For safety, the airplane is speed and load-factor restricted below design limit values until the static-test program is completed. By designing to 100% of load, the load-factor limits permitted are reasonably adequate for evaluating the performance, stability, and control characteristics of the airplane. It is particularly important that the limits be set as high as possible on low load-factor airplanes, such as heavy bombers and transports, so that a proper evaluation can be made. For these airplane types, structure design to less than 100% of load, as in the stretch method, could be so marginal that complete flight evaluation would not be practical. We have found that the method of 100% design has provided adequate strength for safe flight, providing even for the inadvertent exceedings of the restrictions which sometimes occur.

Additional Considerations

The method of 100% design does give lightweight structure that is compatible with the needs of the overall airplane development program. It will not give the lightest possible structure for a given airplane but, for that matter, neither will any other method that is economically feasible. Accordingly, it is generally expected that the 100% design may give strength slightly in excess of that needed for the immediate airplane loads. This excess is not waste for it may be utilized in enlarging the concept of the airplane.

That successful airplane designs experience growth has been historically demonstrated many times. All Boeing production designs have experienced growth. A notable example of this is the B-50 airplane, which is largely identical to the B-29 in geometry but vastly improved in gross weight and payload limits.

It is in anticipation of just such growth that we have exploited our static-test programs by failing the structure in as many parts, and for as many conditions, as is practicable. For example, the ultimate strength of the B-47B wing was fairly well established for the entire span by failing the inboard and left-hand outboard wings in positive bending and the right-hand outboard wing in negative bending. Applied to the whole airplane, this procedure fully discloses the growth that may be immediately available and minimizes the structural changes that may be required for future growth. It is apparent from previous statements that the limited excess strength of the B-47B was used. This permitted an increase in the ratio of maximum gross weight to weight empty, with the range performance benefiting.

It is not claimed that growth should be considered in the original airplane design, but it is believed that the method of 100% design holds definite advantages over the stretch method when it is accepted that, historically, growth may be expected. Based on the past experience of premature failures in static test, and remembering that the stretch method involves initially underdesigning the structure, we might hypothesize accumulated failure records for several designs and compare the two methods. In the long run, a positive advantage must accrue to the 100% design method.

Conclusions

It is apparent from what has been said that inordinate delays are what must be avoided. We have not used the stretch method of design, but our experiences on the B-29 and B-50 airplanes are believed indicative of what may be expected in testing stretch designs and lead us to the conclusion that the delay potential of this method is too great for large airplanes. On the other hand, the philosophy of 100% design has proved practical in reducing delays in the static-test and production programs and in permitting early flight test evaluation of the design and has given acceptable lightweight structure. Furthermore, aside from satisfying the direct needs of the production program, it is obvious that 100% design offers better growth possibilities than the stretch method. In view of all the benefits obtainable by the method of 100% design in the airplane production development program, it is proper to classify as optimum the light, efficient structures that can be designed by this method.

It is agreed that in favorable circumstances the stretch method can be extremely satisfactory. However, it must be understood that it is not applicable to all types of airplanes in all circumstances. Somewhere there must be a limit, and we are convinced that it is far below the modern heavy bomber or transport. Those who accept this premise with reluctance may contemplate the possibility of determining the so-called optimum least weight structure of the B-52 on a cut-and-dry basis.

Design of Airframes for Nuclear Power

Clarence L. Johnson and F. A. Cleveland
Lockheed Aircraft Corporation

Introduction

With the increasing activity in the field of nuclear power for aircraft, it seems in order to discuss some of the technical features and problems connected with aircraft so powered. Naturally, security limits will prevent extremely detailed presentations on many aspects of design features of nuclear-powered aircraft, although every year makes the feasibility of the type more and more a certainty. Many problems which seemed insurmountable a few years ago have proved to be of a rather minor nature or nonexistent; others, unexpected, have appeared.

The use of nuclear power is of great interest to the aircraft designer because it provides a means for giving the airplane nearly infinite range, to all intents and purposes. After a half century of striving to make aircraft carry reasonable loads farther and farther, the advent of a type of powerplant that will solve the range problem is of the utmost importance. To be able to base military aircraft in the United States and still have a striking power any place in the world is a goal worth striving for. The tremendous implications in the elimination of foreign bases, with the great savings in cost to the taxpayer, are a great factor in encouraging work on the nuclear aircraft. We are reaching the limit of range with our conventional powerplants and are having to resort to air-to-air refueling procedures to obtain reasonably acceptable radii of operation.

The majority of the papers which have been written to date on the application of nuclear power to aircraft have dealt almost exclusively with the powerplant itself, only cursorily with the airframe. Those of Kalitinsky, Leverett, and Silverstein were focused on the inherent problems involved in perfecting an airborne reactor-engine propulsion unit; the first two of these were somewhat concerned with the feasibility of creating such a powerplant *at all* in a practical form. Hibbard's address to the Air Force Association in the spring of 1954 dealt more with integrated airframe—powerplant aspects of nuclear-powered airplanes but, of necessity, fairly briefly.

It is the intent of this paper to cxplore the nuclear-powered airplane almost exclusively from the viewpoint of the airframe designer. It is hoped that such an examination will bring to light the real impact of the new propulsion concept by revealing the differences occasioned in the airframe design and by casting them in terms of advantages, or disadvantages, in the design of a useful airframe. It appears that the strategic bomber, by requiring both high speed and great

Aeronautical Engineering Review, Vol. 16, No. 6, June 1957, pp. 48–57.

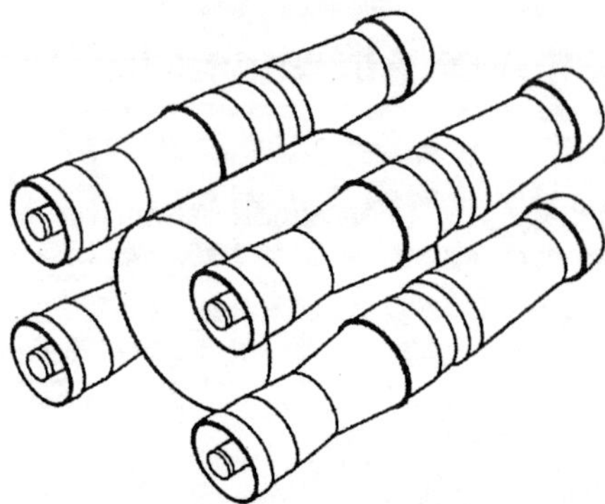

Fig. 1 Typical powerplant.

endurance and because of the inherent low-altitude potential advantages over similar chemical airplanes, will be the first candidate for a nuclear powerplant.

We will eliminate chemical augmentation as a significant feature of the airplane operation mainly to keep the discussion more straightforward. We are, thus, discussing only the all-nuclear airplane with the exception that some chemical fuel is carried for take-off and landing go-arounds.

As currently proposed (Fig. 1), most of the nuclear powerplants are large and clumsy, even without considering the shielding required for personnel safety. The nuclear powerplant in its early form is inefficient primarily because of the low temperature limits of the operating cycles. These limits are set by the ability of current materials to stand high temperatures. When improved materials are available, we would expect the nuclear powerplant to advance rapidly in its overall efficiency, with a consequent improvement in ability to install such powerplants in airplanes of smaller size than those currently contemplated.

Basic Powerplant Characteristics Are Threefold

In order to clarify and categorize the airframe reactions to the presence of the nuclear power package, let us first determine just what the basic characteristics of the powerplant are which dictate their recognition in the airframe design. In their most elementary form, they are only three:

1) A highly concentrated weight in the reactor.
2) A powerful radiation field emanating from the reactor.
3) No consequential flight endurance limitations by the powerplant fuel.

We believe that the first of these characteristics is quite familiar. The reactor with the shielding attached to it, generally called the "reactor shield assembly," is a very dense affair. It may weigh from 25,000 to 100,000 lb, but, instead of having a density of 20–25 lb per cu ft (as have modern turbojets) or 50 lb per cu ft (as has typical chemical fuel), the reactor shield assembly may have a density of from 100 to 200 lb per cu ft. We will explore the implications of this concentration of mass later.

The radiation field from the reactor proper is also well known. If it were possible to put as much shielding on the reactor as is done on ground reactors, we could reduce the radiation therefrom to a negligible amount. But the total weight of shielding required to do this would be prohibitive; in fact, we are forced to the so-called divided shield concept in order to reduce the total shield weight to an acceptable amount. Divided shielding (Fig. 2) is, of course, simply a division of the shielding between the reactor and the crew compartment in such a fashion as to result in near-minimum total shielding weight. As is apparent from the illustration, it is possible to go too far in either direction, and, as we will discuss later, the optimum dividedness is not a function of total shield weight alone but must also be compromised in recognition of several other factors. Divided shielding, then, while permitting lower radiation levels in the crew compartment for a given total shield weight, actually results in *higher* radiation levels every place else in and around the airframe.

The third powerplant characteristic is the most familiar and is the fundamental reason that we are in this business at all. Its significance in a major sense is quite obvious; in many minor senses, however, it leads to less familiar, but interesting, effects which we will discuss.

Now we have established the assumptions under which we will operate in our discussion, and we have classified the important powerplant characteristics which will bear, singly and together, on our airplane design. We propose first to examine the disadvantages of having the nuclear propulsion system in the airplane and, second, to investigate the benefits to be expected.

Probably the first difficulty encountered by the designer in his first layout of a nuclear-powered airplane is the tremendous congestion at the center of gravity (CG) of the airplane (Fig. 3). It appears to be generally desirable to have the

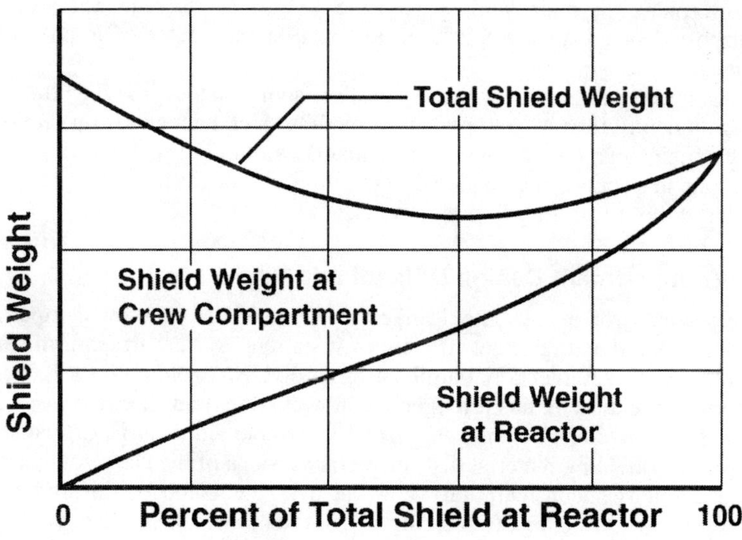

Fig. 2 Shield dividedness effect on total shield weight.

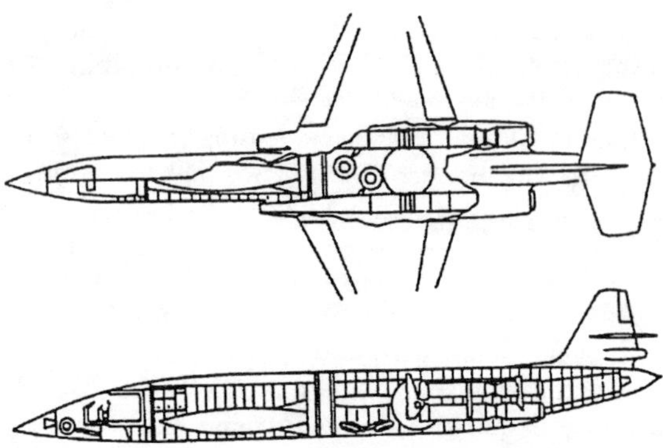

Fig. 3 Typical inboard profile.

engines in close proximity to the reactor to reduce the many problems of transferring the reactor heat output to the engine air. The combination of the reactor shield assembly and the turbojets accounts for something in excess of 50% of the airplane equipped weight empty. The wing, if unswept, certainly passes through the fuselage in the CG area. The center of gravity of the fuselage structure inherently lies close .to the airplane CG. The main landing gear must lie very close to the airplane CG. Aircraft fixed equipment tends to be spread out in the fuselage and cannot be very helpful in rebalancing. About all that is left to aid the designer in pulling the CG away from the powerplant are the crew compartment shielding and the empennage. The latter of course operates so as to cancel part of the effectiveness of the crew shield in this respect. It is apparent, then, that the powerplant, the main landing gear, the wing, and—to reduce CG travel—the disposable load all vie for fuselage space at the center of gravity of the airplane.

Wing sweep can alleviate the congestion somewhat by moving the wing-fuselage juncture forward and perhaps providing thick enough airfoil sections to stow the main gear. Our studies indicate that this compromise does not generally pay off and that sound, imaginative design can overcome the important congestion problem without resorting to wing sweep.

Crew Compartment Design Difficult and Vital

The second problem we should like to discuss is that of crew compartment design. Layout of comfortable, efficient crew stations is a real art requiring expert attention, even in conventional airplanes. The divided shield concept focuses so much more attention on this arrangement, however, that our former efforts seem almost trivial by comparison. The crux of the whole situation is, of course, the thickness of shielding material distributed on all sides of the crew compartment. Simple geometrical considerations show that every cubic foot of shielded volume costs from 50 to 500 lb of shielding. This situation makes it mandatory that every possible effort be made to save every bit of volume not essential. Human engineering to a degree never before necessary is demanded, and equipment refinement and miniaturization of all items which must remain with the crew is of paramount importance. When crew station volumes, including equipment, get down into the range of 50–75 cu ft per man, apparently simple problems of movement within the compartment and of space to stretch the body to full length are no longer easy. In addition, provision of safe and rapid emergency crew escape is more difficult because of the thick shielding, as well as the restricted space. It is of vital importance that design in this area be done in acute consciousness of the relative value of the pound saved to the unit of human efficiency sacrificed.

One other aspect of crew station design should be pointed out. Nuclear radiation, whether neutrons or gamma rays, is basically directional in nature—that is, if the nuclear airplane were in a vacuum, only a rear face shield would be required at the crew compartment. But, as shown in Fig. 4, the radiation is scattered, or deflected, through angles as high as 180 deg by hitting atoms in the air or aircraft structure so that considerable radiation enters the crew compartment from the sides and from the front. Because the radiation energy decreases as the scattering angle increases, the side radiation is weaker than the direct beam on the rear face, and that hitting the front face of the compartment is weakest of all. Now (Fig. 5) the lower the energy of a given radiation flux, the less the material thickness required to attenuate it by a given factor. In addition, scattered radiation has to travel a greater distance, from a simple geometrical viewpoint, which leads to a flux decrease inversely proportional to the square of the distance traveled. This situation leads to two interesting conclusions. The crew compart-

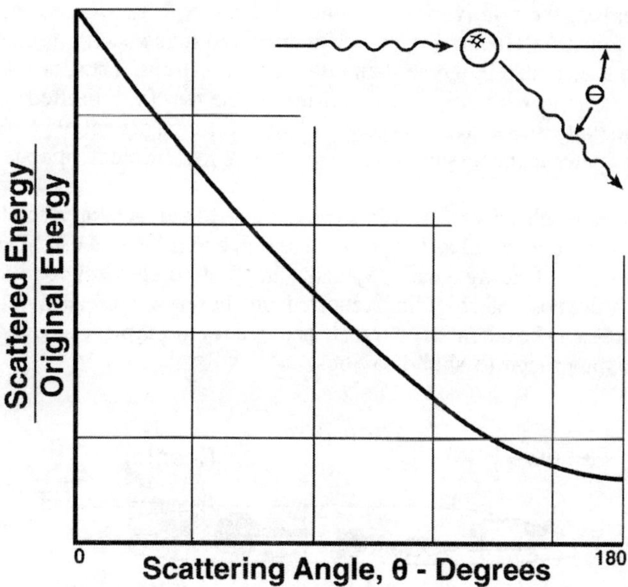

Fig. 4 Radiation scattering.

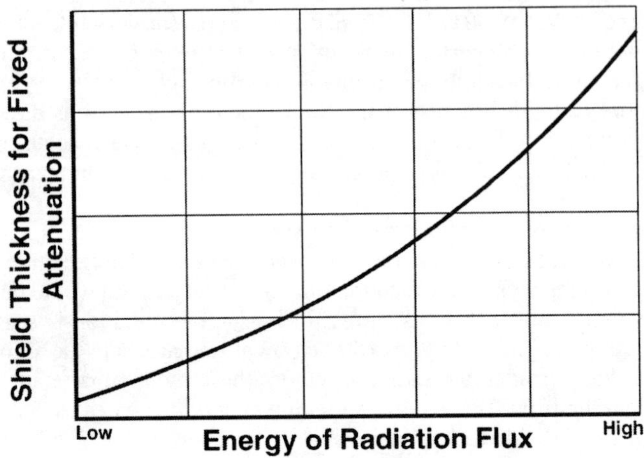

Fig. 5 Shield change with flux energy.

ment shielding becomes progressively thinner as one progresses from rear to sides to front, which means that not just cubic feet but the geometrical disposition of the cubic feet is important in optimizing crew station design. Also, it is much easier to replace the relatively thin front shielding with transparent materials of equivalent radiation attenuating power than it is to make such a replacement on the sides; the rear face is worse than either sides or front. Thus, windows in the crew compartment will likely be nonexistent in the rear face, limited in the sides, and only difficult in the front because of the still-consequential thicknesses of transparent material necessary to replace the more efficient opaque shielding removed.

The directional nature of the radiation leads also to the fact that aircraft structure and components are useful as shielding material, and judicious use of such things as the wing box, landing gear, payload, and fuel for landing go-arounds can reduce the thickness of shielding required on the crew compartment rear face (Fig. 6). Care must be taken, of course, to account for the expenditure of disposable items in this approach to shield savings.

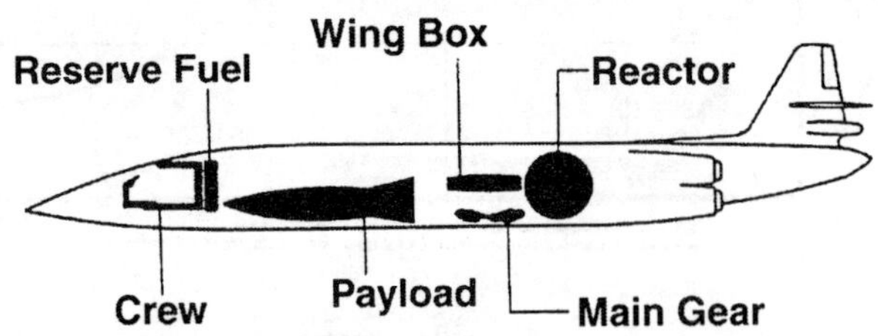

Fig. 6 Equipment and structure as shielding.

Structural Relief Not Available

The divided shield concept leads directly to another design situation which is not ideal. We normally think of the fuel and equipment weights carried in a fuselage (Fig. 7) as being rather well distributed along its length. This permits relatively efficient and light fuselage structure and results in reasonable moments of inertia in pitch and yaw and in fairly high frequency fuselage vibration modes. Installation of a heavily shielded crew compartment near the nose of the airplane changes this situation profoundly. A heavier fuselage structure is required for simple static loads, at least. It may be that the dynamic loads arising from the much lower natural frequency in vertical or side bending will lead to further structural increases. The effect of increased moments of inertia in pitch and yaw on the several dynamic stabilities of the airplane will probably not be great (we have not explored this in detail), but they must be considered.

In conventional aircraft, considerable primary wing structural weight is saved by so-called inertia relief—that is, any weight items which are installed outboard of the wing root decrease the wing structure necessary to carry a given total airplane gross weight. Virtually all airplanes enjoy some of this saving because of fuel carried in the wing; many have the further advantage of tip or pylon-mounted fuel tanks; wing-mounted powerplants, either partially submerged or in strut-mounted nacelles, afford considerable relief; landing gear hung from the wing provides some help; and internal or external wing-mounted armament affords savings. The inertia relief of all of these items is somewhat tempered, of course, by the additional down-bending inertia loads they cause in landing or taxi conditions.

The all-nuclear airplane will probably benefit from few, if any, of these bending reliefs (Fig. 8). The lack of any consequential amount of chemical fuel certainly

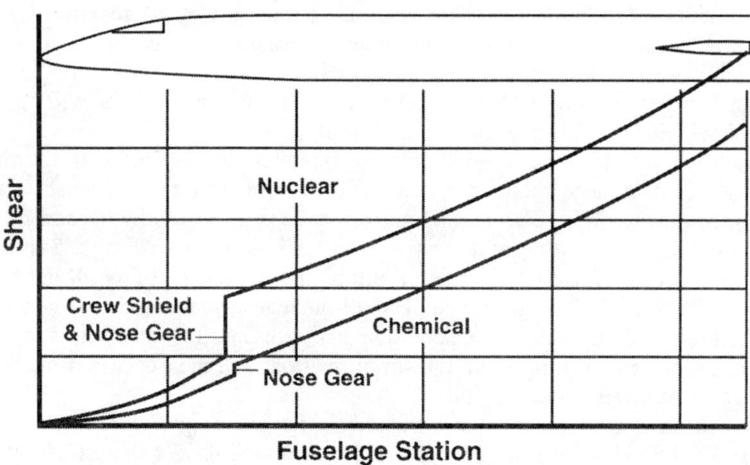

Fig. 7 Comparison of fuselage flight shear loads.

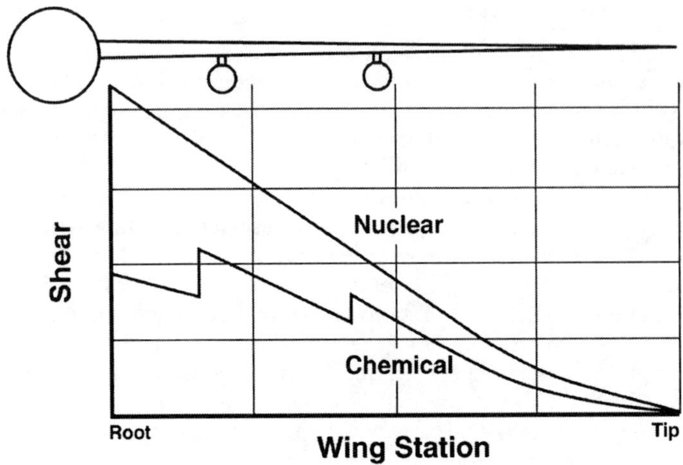

Fig. 8 Comparison of wing flight shear loads.

eliminates aid in that respect. The overriding desirability of having the engines close to the reactor to reduce the heat-transfer problem will most likely keep the engines in the fuselage. If thin, straight wings are used, gear stowage therein is not too attractive. The only bending relief item remaining is armament; some aid may be obtained in this direction.

Another type (Fig. 9) of structural relief generally exploited in conventional aircraft is that due to the difference in the maximum gross weight and the landing weight, due mainly to fuel consumed during the mission or, if necessary, dumped prior to landing. Typically, the landing gear, after satisfying takeoff and taxi load requirements, need be designed for something like half of the maximum gross weight at landing impact. Since the nuclear airplane we are considering uses essentially no fuel during its mission and cannot even be counted on to have disposed of its payload, no such relief is in sight; the landing gear will have to be designed for the full airplane gross weight.

The last type of weight relief normally exploited in conventional aircraft is the flight performance increase as the gross weight decreases because of fuel consumed during flight. The landing performance is improved by the same gross weight reduction which permitted savings in the landing gear design. Similarly, maximum speed, cruise speed, rate of climb, and maneuverability all get better during a flight as fuel is consumed. The all-nuclear airplane does not benefit in this fashion and, though the calculation and presentation of nuclear aircraft performance is much simplified, the simplification is at the expense of performance gains at lower gross weights.

Though several of the points discussed so far are outgrowths, to a varying degree, of the radiation from the reactor, let us now consider the effects of this radiation more directly. We shall examine four major problem areas in which the radiation field is the prime factor in reaching certain development decisions:

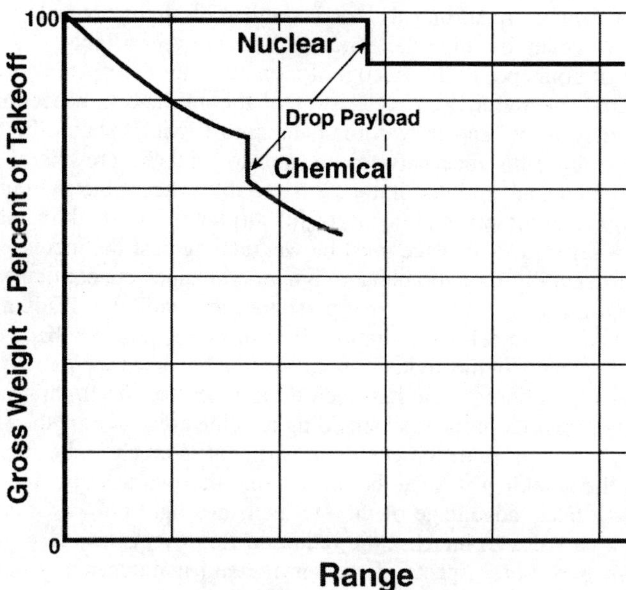

Fig. 9 Gross weight variation with range.

1) Crew personnel radiation exposure.
2) Maintenance personnel radiation exposure.
3) Airframe materials damage.
4) Airborne equipment integrity of operation.

We believe that we may state without reservation that the flight crew is the single most sensitive element in the airframe from a radiation point of view. Unfortunately, the determination of the tolerable human radiation dose is very difficult and may well prove to be one of the most complex research areas in the attainment of operational aircraft. Gamma rays and neutrons of the same intensity damage the body to different degrees, to begin with, and the effects of each of these two types of radiation also vary with the energy of the radiation. As a further complication, various organs of the body are sensitive in varying degrees to radiation, and the effects on these organs become maximum at various times after exposure. Finally, all of the biological reactions are to some degree functions of the dosage rate, the total dose, and the time between exposures. The difficulties faced by the radiobiological people in endeavoring to set up exposure criteria for crew personnel are evident.

Once such criteria are established, however, the shielding engineer and the airframe designer are just beginning their analysis in order to translate the biological specifications into an operational airplane shield. The combined shielding at the reactor and at the crew station must provide an optimal ratio of neutron-to-gamma flux inside the crew compartment. Decisions are necessary by the eventual user of the airplane regarding the total number of flight hours desired for an individual crewman, defining the expected time between flights, and stipulating the number of hours per flight. Recognition in all of these respects must be given to the different operational characteristics of the airplane in peacetime training and proficiency flying and in wartime campaign flight. It should be noted here that the separation distance between the reactor and the crew is an important factor in shield design and that it, too, is a variable susceptible to limited control by the designer in the original layout of his airplane. The shielding savings due to increased separation distance must be weighed against the increased fuselage weight and generally unfavorable aerodynamic changes concomitant.

The foregoing discussion of *crew* personnel and radiation leads naturally to that of *ground* personnel and radiation. It is in this regard, we believe, that the largest single penalty of the divided shield concept becomes apparent. It is obvious (Fig. 10) that, if a dosage rate has been established for the flight crew and if a consequential fraction of the total shielding is at the crew station, then the dosage rate everywhere *outside* of the crew compartment is much higher than it would be if *all* of the shielding were at the reactor. In other words, any aspect of shield design which takes advantage of the fact that the flight crew is in a restricted, delineated area tends to increase the radiation flux everyplace else.

There are two more aspects of the service and maintenance problem which we will mention before turning to the favorable factors. Because a large percentage

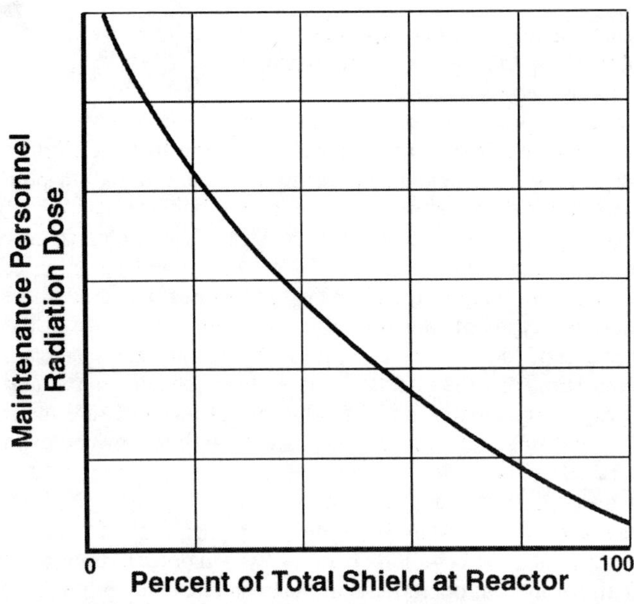

Fig. 10 **Shield dividedness effect on ground crew radiation dose.**

of aircraft service and maintenance work is on the powerplant, this work will have to be done in an area closer to the reactor than the crew compartment. Also, it will probably not be practical to limit the work of maintenance personnel in accord with a high permissible dosage rate at widely spaced time intervals; they should be able to work at least a half day, five days a week, on nuclear airplanes. This means that they will probably be limited to a much smaller dosage rate than will be flight crew, very probably near the present Atomic Energy Commission maximum laboratory dosage rate of 7.5 milliroentgens per hour. Crew dosage rates may be 10–100 times this value.

"Dark" Maintenance Situation Lighted by Improvement Factors

The maintenance situation looks very dark, indeed, at this point. Let us examine the factors which improve it.

First, the reactor will not be at full power during service. It will probably be completely shut down. Unfortunately, a good deal of decaying radioactivity remains even in this state, but the decrease compared to full power is greater than one order of magnitude.

Second, it is probably not desirable or necessary to initiate service immediately after shutdown; since the radioactivity remaining is decaying with time, a wait of a few hours decreases the dosage rate. Along these same general lines, the flux after shutdown is almost exclusively gamma rays, which simplifies the problem somewhat.

A third aid in the maintenance picture is the possibility of removing from the reactor some of the radioactive materials after shutdown. The degree to which this may be accomplished is unalterably defined by the structure and design of the specific reactor, but some cases hold considerable promise of flux reductions by one or two orders of magnitude.

The last recognized means of achieving normal, manual powerplant service is through the use of a technique known as shield augmentation. Since neutron emission is not a postshutdown problem and since the reactor neutron shield might conceivably be a liquid, it is possible to consider draining or removing the neutron shield and replacing it with a high density liquid or slurry having much better gamma attenuation characteristics. Flux reduction by better than an order of magnitude is certainly possible if the mechanical difficulties can be overcome.

We have so far carefully avoided mentioning the use of remotely controlled handling devices for service and maintenance. We believe that some simple operations can be performed by manipulators of one kind or another—for example, accomplishment of several of the flux reduction means we have discussed would require such an approach. But for plain, everyday, operationally practical upkeep of airframes, powerplants, and equipment, it is next to impossible to replace the two hands and the Mark I eyeball of a mechanic with steel claws and a TV tube. We have endeavored to quantify this statement, and it appears that something between 5–10 times the number of manhours is required to accomplish typical service work by manipulators *with direct vision* as is required with a conventional manual contact approach. Let us qualify these remarks to a limited extent; there will certainly be a number of areas where adjustments and checks must be made either before or after shutdown and ground safing. It will be of paramount importance for the airframe designer to provide for these procedures in such a fashion that manipulators can be used as efficiently as possible. In the event that the reactor can in no way be rendered ground safe, it will be necessary to remove the powerplant in toto to special maintenance shops for its service, at the same time making the rest of the airplane accessible to normal service operations. In this case, the design of the powerplant for manipulator maintenance, even in special shop areas, will be a supreme test of the powerplant designer's skill.

We have discussed the biological damage of nuclear radiation and its impact on many aspects of aircraft design. A simpler, more easily tested, and probably not quite so profoundly damaging characteristic of radiation is in the materials of which airplanes are made. A really thorough assessment of the problem is beyond the scope of this paper; many of the basic mechanisms of radiation damage are not yet fully understood by even the metallurgists and physicists. In addition, material damage *outside* of the reactor is serious, in general, to the degree to which the divided shield concept is exploited. Thus, the trend toward more shielding at the reactor, suggested by service and maintenance considerations, aids radiation damage reduction similarly.

Material Radiation Damage May Prove Significant

The fields in which radiation damage may prove significant are at least eight. They are not listed in any particular order because their relative occurrence, and the degree to which the individual effects are important, vary with the flux level and type, the material being considered, and the use intended for the material in the airplane. As we will see, only a few of these should prove important, and only a couple will be vital.

The first field, *dimensional changes*, refers to changes occurring without regard to stress on the subject part. The total integrated dosage required to cause significant change is such that only materials inside the reactor shield itself could conceivably be affected. It is, thus, very improbable that the airframe designer need consider this.

The second item, *creep acceleration*, is, in the purest sense, of interest primarily as a dimensional change. As in the first item we mentioned, creep acceleration will be most critical inside the reactor, but it may also prove to be somewhat troublesome in some instances in the airframe. It appears that some materials undergo considerable creep acceleration at high stress levels and/or high temperatures due to radiation. Certainly in supersonic aircraft with significant aerodynamic heating of structures and even in subsonic airplanes having prestressed connectors, it will be necessary for the designer to account carefully for potential radiation-accelerated creep characteristics.

Our third category, *hardness, strength, embrittlement increases*, may prove quite important. Fortunately, work-hardenable alloys in the hard condition are not further hardened appreciably. Nevertheless, for fittings and connectors near the reactor, the possibilities of increases in stress concentration and notch sensitivity and decrease in impact strength will require consideration, as will the effects on fatigue life.

Our fourth and fifth terms, *accelerated diffusion at faying surfaces* and *accelerated stress corrosion*, spring from a single characteristic of radiation—it accelerates diffusion of atoms in a metal. Plated pistons and riveted or bolted connections where dissimilar metals rub may, through microdiffusion across the plane of contact, lead to embrittlement or other important metallurgical changes. Accelerated stress corrosion is probably much more serious a problem in the reactor, but reactor attachment fittings and support structure may require some consideration in this respect.

The sixth item, *transmutation,* is included mainly for completeness. It is certainly true that, under slow neutron irradiation, most metals gradually transmute into others. For example, aluminum becomes silicon, and copper becomes nickel and zinc. Some of the metals become isotopes of the original element, but none of the common structural materials is transmuted into liquids or gases, and, even more important, the percentage of the original metal converted in an airplane lifetime is probably less than the percentage of unknown impurities in the original material. Thus, transmutation is not expected to be recognized in the airframe design.

Induced radioactivity, the seventh category, appears to be one of the most important in the entire list. In general, the material being bombarded by neutrons must have four characteristics for it to be critical in this regard.

1) It must exhibit a fairly high tendency to capture neutrons within its nucleus.

2) The isotope resulting from this capture must be unstable—that is, radioactive.

3) The radiation emitted from the isotope must include a consequential amount of medium- or high-energy gamma rays.

4) The half-life of the isotope must be significantly long.

Speaking loosely, it is the product of these factors which is important; a material deficient in one respect may be so high in another that it remains on the critical list. Fortunately, such materials as aluminum, iron, titanium, and magnesium arc markedly deficient in at least one of the criteria. Cobalt, zinc, tungsten, manganese, chromium, and molybdenum, on the other hand, are somewhat discouraging, to various degrees. In general, it appears that it is not the fundamental element of common materials, but their alloying constituents, which are most troublesome. No matter how the lists of good, intermediate, and bad materials finally end up in detail, it is obvious that accurate material quality control, not only of alloying elements but also of impurities, will be of sharply increased significance. The designer will have to become fairly conversant with induced radioactivity characteristics and will have to keep his material specifications within the bounds prescribed to keep the airframe from becoming too hot to handle.

We have saved *molecular structure changes* until last because they are characteristic of mainly organic materials, and organics, in turn, respond to irradiation in mainly this fashion. Peculiarly enough, either of two opposite effects predominates in organics in this regard. In most cases, the basic molecular chains are broken resulting usually in poorer physical characteristics often in gas evolution, and eventually in destruction of the material as a usable component. In a few cases, the opposite occurs: The basic molecular chains cross-link themselves without cleaving, and the result is greater strength and density and generally improved physical characteristics, particularly heat stability. Most of our typical aircraft organics—such as rubber, lubricating oils, fuels, hydraulic fluids, electric insulations and dielectrics, and the transparent butyrates and polyesters—are hurt to one degree or another by radiation. Rubbers lose elasticity; lubricants, fuels, and hydraulic fluids evolve gas and tend to gum; and the transparent plastics weaken and discolor. Fortunately, the majority of these materials will probably last through one or more overhaul periods. In the particular case of organic fluids, some latitude of choice is possible in selecting materials inherently better than others, and the use of additives can be helpful in promoting radiation stability. But again, as with metals and alloys, a whole new additional criterion of material selection must be learned and applied if sound aircraft design is to accrue.

Electronic Equipment May Suffer Serious Effects

We have hinted at changes in dielectric properties of materials previously. If we view electronic equipment operating in a radiation field, we find that even more serious effects may be expected. Perhaps the most serious is the possibility that the ionized sphere of air surrounding the airplane and centered at the reactor may attenuate electromagnetic propagation. If this proves to be true, then radio transmission and reception in the airplane would be adversely affected, and, even more critical, radar ranges would be reduced. These phenomena would occur, of course, only to the degree that the flux level outside the reactor is high, and tendencies toward greater reactor shielding would reduce this type of deterioration of equipment performance. It is still not known whether or not this attenuation occurs at all frequencies, however, or, if it does, what the threshold ionization level for detectable attenuation is at the various frequencies.

The effect of the radiation on electronic components is better known today. Four known types which become essentially unusable in high radiation fluxes are germanium transistors, electrolytic condensers, phototubes, and bias cells. These are all replaceable by other types which appear to be quite satisfactory even under long, fairly intense irradiation. But induced radioactivity again assumes considerable importance in that, for example, capacitors, relay and switch contacts, and vacuum tubes become radioactive fairly readily, and a cooling-off period of several weeks before handling such components would create a pretty difficult service problem in the aircraft electronic systems. Considerable relief in all these respects is potentially available to the aircraft designer through placement of the electronics gear far from the reactor and by locating it in the radiation shadow of some mass in the airplane. Some of the reactor and engine controls, however, will undoubtedly incorporate electronic elements of necessity close to the reactor, and these will not be so easily rendered satisfactory.

Three other equipment problems should be mentioned here. Relay and switch contacts wear faster in the ionized air of a radiation field, and photographic film fogs very quickly in even modest fluxes. Ozone, a product of air ionization, hastens the deterioration of rubber.

Viewing the equipment problem in general, it is seen that a majority of the work to be done in improving components is outside of the control of the airframe designer, but he can, by efficient arrangement of his layouts, ease the radiation level to which much of the equipment is subjected by as much as two to four orders of magnitude; his potential contribution in this respect should certainly be recognized early in the equipment development program.

We mentioned earlier the likelihood that the nuclear engines would be kept close to the reactor and, therefore, that wing-mounted engines would probably not be available for wing bending relief. For the sake of completeness, the other implications of not hanging the engines on the wing should be mentioned. By scrutinizing the reputed and real advantages of wing-mounted engines, we can quickly see what design problems may accrue if wing mounting is not feasible. The oldest reasons for wing engines are the requirements of propeller clearance and some reduction in fuselage vibration. Neither of these is particularly applicable to a nuclear-powered turbojet airplane, though the clearance problem would be if nuclear turboprops are considered. The attainment of good turbojet inlet ram recovery is always simpler with nacelle inlets, and the difficulties of jet exhaust washing the aft fuselage are automatically eliminated. Both of these problems can be solved with fuselage installations, but more design attention is required.

Main landing gear stowage often works out very nicely in engine nacelles but can be handled in the fuselage or wing root. The absence of cutouts in the wing for gear stowage is of consequential advantage in attaining high torsional stiffness. Maintenance accessibility of wing nacelles is *inherently* better than that of engines submerged in a fuselage, but the differences can be made slight by good design. Engine fires in the air or explosive failure of a turbojet are *intrinsically* somewhat better isolated in individual nacelles, but there are many sound design approaches to equal or better the situation with submerged powerplants. The newest advantage of wing engines is the use of the engine mass to aid in increasing critical wing flutter speeds. This is mostly applicable to swept wings of high aspect ratio and is, thus, not necessarily appropriate to most of the configurations likely in the nuclear bomber.

Some of the problems created by locating the engines in the fuselage are real; some are academic. In general, we are confident that the advantages of such an installation far outweigh the advantages of wing mounting, particularly in the case of the nuclear airplane.

Challenge to Designers Seen in Need for Cooling Apparatus

In the design of today's turbojet aircraft, we have gotten far away from the cooling radiator problems of the old liquid-cooled reciprocating engine. It appears that, with nuclear power, we will have to brush up on this skill. Some of the potential nuclear cycle candidates have radiators in the turbojets, but these are outside of the direct responsibility of the airframe designer. We refer, instead, to the fact that at least the shield, and possibly some of the other reactor components, may require cooling through the use of a relatively conventional liquid-to-air heat exchanger. The amount of heat to be handled may approach 10% of the total reactor power. Incorporation of a cooling system of this capacity in a high-speed airplane with only nominal weight cost and reasonable drag penalties is a great challenge to the designer and thermodynamicist and requires extreme integration of the airframe and powerplant design effort.

Another cooling problem is inherent in reactors of all varieties. When a reactor is shut down, though the fission process stops, considerable heat continues to be generated by the decay of the fission products. This phenomenon, known as ''afterheat,'' decays with time, but the amount of heat during the first week or two is great enough to require forced cooling to avoid having the core melt itself and destroy the reactor. The situation leads to one easy and one difficult problem. Following a normal landing, either an engine must be run continuously to remove the afterheat, an auxiliary air supply must be attached, or the reactor fuel must be removed. In the event of a crash or an accident during takeoff or landing which stops all engines, the afterheat is certainly going to damage the core.

The last design problem created by the nuclear powerplant is one of the potential, not certain, type. It is possible that the response characteristics of the reactor to increased or decreased power demand, either inherently or because of limitations built in for safety, will be slow compared to those desired during takeoff or landing. Solution to such a difficulty intimately involves both the powerplant and the airframe manufacturer and should be thoroughly analyzed by them quite early in the development of the powerplant and airframe.

We have now completed a discussion of the design problems expected to result inherently from the application of nuclear powerplants to useful aircraft. We are sure that many of those apparent difficulties looming as most serious now will prove to be most tractable to handle; we are equally sure that some of the most innocent will prove to be real backbreakers. In any event, let us proceed to a survey of the airframe design *advantages* which we may reasonably expect to accrue with our nuclear powerplant. Because discussion of *new* problems is inherently more interesting and stimulating than discussion of *ex*-problems, we will treat the advantages very briefly.

We shall only mention the salient advantage of increased endurance, or range, since it has been so frequently extolled in every discussion of nuclear-powered aircraft. We will see a little later, however, that this lack of endurance sensitivity implies some very definite secondary attributes to be enjoyed in actual airplane design and construction.

One of the immediate enjoyable prospects following removal of chemical fuel from an airplane is the reduced center-of-gravity travel to be expected (Fig. 11). Though some CG movement will remain, associated with the armament, payload, gear position, and crew, the net result should be a reduction of travel from the typical 12 or 16% to half that or less. This has one significant effect—the tail volume may be reduced consequentially. This is brought about, of course, because all of the critical longitudinal trim moment changes are reduced, and the overall tail moment envelope required becomes accordingly smaller.

Another very real advantage having to do with center of gravity and trim problems arises from the divided shielding concept. Even if the most rearward CG limit is held closely during design, it usually tends to drift aft as soon as the airplane gets into service and modifications commence; often, the compromises necessary to handle the problem are very undesirable. In the nuclear aircraft, a new means of compensating for this is available. If the shield dividedness is such that small changes in dividedness cause very small changes in total shielding weight, as is expected to be the case, then it will be possible to increase slightly the shield on the rear crew compartment face and decrease the reactor shadow shield thickness. This is a powerful tool for changing balance, particularly in the terminal phase of project design, and is almost free weightwise. We might add that, if the accuracy of the science of shielding computation does not improve

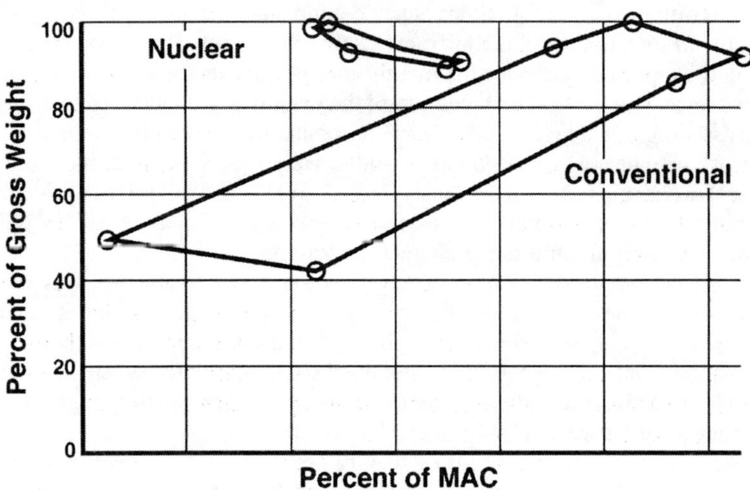

Fig. 11 Comparison of balance diagrams.

considerably before we design the first airplane, we are going to be in real unknown weight and balance troubles in even the design stage!

You will recall that we discussed the lack of wing inertia relief at some length when we considered the disadvantages imposed by the nuclear powerplant. The removal of the inertia relief items from the wing provides, however, modest compensation. In essence, the wing structure becomes much simpler, cleaner, and more efficient because the compromises in structural arrangement and design for fuel stowage and engine mounting need not be made. Whereas it has been in the past fairly easy to provide access doors in the multirib structural box of thick, low-speed wings, it cannot be done cheaply, weightwise, in thin multispar, high-speed wings. Thus, a consequential weight saving accrues because the nuclear airplane does not need the internal wing access.

We mentioned earlier that, if wing sweep is not used, the area around the airplane center of gravity will likely be somewhat congested by the necessity of locating the wing box, landing gear, and nuclear powerplant close to the CG. Again the congestion problem is not all bad in that the major load paths from the gear to the powerplant and from the wing to the powerplant will be very short. This leads to greater structural efficiency and reduces the weight of heavy structure required to carry the powerplant load to the wing and main landing gear.

Cost of an Extra Pound Not So Great

If one views the conventional chemical airplane as being built up of completely linear components—that is, if engines, airframe equipment, and airframes are all rubber so that they can grow or shrink to match perfectly any change in gross weight and size to maintain airplane *performance* constant—then the addition or subtraction of a pound of weight causes a substantially greater change in the total aircraft gross weight. Since such component flexibility is not generally available *in fact*, except in the airframe itself—wing area, for example—the cost of added weight normally results mainly in a penalty in some phases of aircraft performance. But the best real measure of the penalty is still obtained by assessing the gross weight increase on a ''rubber'' airplane. In chemically powered aircraft the gross weight cost for a pound of added weight may be anything from 3 to 10 lb. This is obviously a seriously large ratio, and a major portion of it is generated by the extra fuel and associated system provisions necessary to keep speed, rate-of-climb, and range characteristics constant.

The all-nuclear-powered aircraft enjoys a marked advantage in this respect. Though its engines and reactor, its landing gear and equipment, and its airframe structure must go up somewhat in weight, as do the same elements of the chemical airplane, the net increase is considerably less, particularly when the mission incorporates a sizeable endurance requirement. It appears that the weight-increase ratio may vary between 1.5–4, rather than 3–10.

If the situation is cast in terms of real, rather than rubber, airplanes, a similar advantage accrues. The chemical version will suffer a modest decrease in speed and climb performance and a more serious decrease in range. The nuclear airplane will lose speed and climb performance only. This entire reassessment of the cost of a pound will no doubt have a marked effect on the criteria applied to deciding the wisdom of equipment addition to nuclear aircraft.

We should mention, before leaving this particular subject completely, that the real cost of a pound of additional drag is susceptible to the same analysis and gives the same results. This is of particular importance 1) when equipment protruding into the airstream is added, or when the aerodynamic airplane is modified for new equipment, or 2) if the original estimates of airplane drag prove to be lower than the drag actually encountered.

By far the predominant factor in the vulnerability of chemically powered aircraft to nonexplosive bullets, to shrapnel and debris, and to small explosive shells is the fuel system. It achieves this dubious eminence because not only is a square foot of fuel more sensitive to projectiles than all other aircraft elements, excepting personnel, but also there are many times more square feet of fuel than of any other component or system. In addition, the fuel system has the further unfortunate characteristic of spreading damage by fire quickly and extensively; it, thus, generates by its inherent qualities a continuing, multiplying source of damage to other systems in the airplane.

The all-nuclear airplane has an obvious and tremendous superiority in this respect. The reactor itself may prove to be somewhat vulnerable, but its vulnerable area is relatively small, and the net gain from elimination of large fuel capacity will be enormous. It should be noted here that engine fire hazards in noncombat operation of conventional aircraft are great because of the fuel-loaded plumbing, pumps, and controls adjacent to the engine. If chemical fuel is used only for takeoff and landing in the nuclear airplane, the entire system near the engines can be purged and rendered inert during nearly all of the mission.

In addition to the reduction of fuel system and powerplant vulnerability is that associated with the crew. The rather thick layers of shielding surrounding the crew compartment will likely protect the personnel almost completely from fragments and small projectiles, particularly those approaching from the sides, top and bottom, and aft.

For the sake of fairness we should mention that the nuclear airplane does not have any particular advantage against *large* explosive warheads. These are designed to cripple airframe structure *wherever* they hit, and no outstanding differences may be expected between chemical and nuclear airplanes, with the possible exception of the crew station area where the shielding may be of some help.

Endurance Advantage Outweighs All New Problems

Viewing the comments we have made, both good and bad, now in retrospect, it would appear that the aircraft designer is to be faced with many more new problems than he is to be relieved of old ones. This is most assuredly the case if we count only the *numbers* of problems on each side. If we consider *both* quantity and quality of the advantages and disadvantages, ignoring the flight endurance advantage, we will again come out on the short end as designers. But the importance—the quality—of the endurance characteristic is such as to dominate completely this apparent negative balance and render it unimportant. And this unique characteristic is one to be greeted enthusiastically, not by the designer alone, but also by the military analyst, the military operating command, and the aircraft crew. Clever, creative aircraft design will remain the sine qua non, as it has been in the past, if we are to exploit the nuclear powerplant application to the fullest.

Structures for Supersonic and Hypersonic Vehicles

Turning Up the Heat on Aircraft Structures

Alan Dobyns
Sikorsky Aircraft

Charles Saff
McDonnell Aircraft Company

Robert Johns
NASA Lewis Research Center

The structures of helicopters and subsonic, supersonic, and hypersonic fixed-wing aircraft use many similar design and analysis methods, with significant variations based on their different operating environments. Hypersonic aircraft incorporate the most exotic materials and structural designs because of the harsh conditions encountered. High-frequency fatigue loading influences helicopter rotor component design, while requirements for economy and low weight help to shape fixed-wing aircraft.

If yesterday's thrust in aircraft structures called for best performance at low cost, today's seeks lowest cost for good performance. This shift is creating new thrusts in four areas: integrated design, in which cost weighs as heavily as any other parameter; improved design analysis tools; low-cost fabrication techniques; and more-sophisticated test methods. The primary emphasis of all this effort is to produce low-cost structures in minimum time while assuring performance, integrity, and supportability.

Integrated product development teams are the mechanism used to ensure that functional and cost requirements are met. No longer are designs developed in one area, checked in several others, then "thrown over the wall" to a manufacturing group that must do the best it can. Manufacturing is now a primary member of the team.

CAD/CAM systems are the integrated product team's primary tool. These evolving systems are now sophisticated enough to look at form, fit, and function; moreover, they are integrated with analysis packages that permit material selection, sizing, and dimensional tolerances to be checked as well. Three-dimensional CAD/CAM data are becoming the language through which design, manufacturing, and engineering communicate.

Work continues on linking CAD/CAM tools with finite element analyses so that load paths can be defined early and both global and local sizing accomplished

Aerospace America, Vol. 30, No. 9, Sept. 1992, pp. 34–37.

quickly. While design engineers long for the day when finite element models will be developed automatically, the current systems already do much of the work.

The use of CAD/CAM early in the preliminary design cycle has enabled engineers to know sooner what the parts are, what they look like, what can be sized, and what is dictated by fit and minimum gage concerns. We now have the tools to optimize composite laminates and, thus, omit the time-consuming hand iterations. But work remains to be done in analysis methodology. No generally accepted failure criterion exists for composites. We still rely heavily on test data to design composite structures and to link the design to its certification. This dependency adds to the cost of selecting composite materials.

Modeling of fabrication processes is one area in which three-dimensional CAD/CAM and finite element analyses have reduced costs. Simple modeling of metal-forming and composite-forming operations has been so successful that first time quality is now routine, and process verification parts are used only for the most complex formed parts. This modeling has accelerated acceptance of advanced manufacturing methods such as superplastic forming/diffusion bonding and thermoforming of some composite structures.

But complex composite parts have proven costly. Comparisons of metals vs composites often favor metallic structures. Weldable metals and cocured composite parts are being examined to lower costs and limit the number of fasteners and other parts. However, large-scale fabrication of these parts will require improved nondestructive evaluation techniques.

Prototyping is another key ingredient in low-cost manufacturing. Inexpensive materials can be used for checking the design and fabrication to ensure that the first production parts in high-cost materials are usable.

A more recent advance in structural development is the coupling of sensor technology to fabrication and testing. Sensors for temperature, pressure, and other parameters related to resin curing are now used to assure part quality and minimize variations from nominal design requirements. These same sensors can be used to monitor temperatures and pressures after fabrication.

Although initial work focused on embedded sensor systems, these pose a myriad of ingress/egress problems, and the latest technologies are nonintrusive. Computer-enhanced video technology is used to record and reduce test data and to highlight results for analysis. The ability to rerun a failure in slow motion, zoom in, and computer-enhance the image has enabled greater understanding of how materials deform and fail.

Only through these low-cost techniques could structures as complex as the National Aerospace Plane center tank be successfully developed. This tank includes composite panels, superplastic-forming/diffusion-bonding stiffeners, and complex joining. Tested under extreme heat and cryogenic temperatures, the tank repeatedly withstood some of the most severe thermal gradients ever imposed on an airframe structure.

Design conditions for helicopters differ markedly from those of fixed-wing aircraft. Crashworthiness considerations, as well as static and dynamic loads, heavily influence a helicopter's fuselage design, while the rotor components are affected by fatigue and reliability concerns. In the RAH-66 and Boeing 360, these parts employ graphite and Kevlar advanced composites to save weight.

High-frequency fatigue loadings affect helicopters more strongly than fixed-wing aircraft, primarily because of rotor-blade effects. The variation of load on each blade as it revolves results in cyclic loads on the rotor hub, transmission, and airframe at multiples of the revolution speed. In addition, the wake from each rotor impacts the fuselage and tail separately, causing vibratory loads on each. Simulation programs are now the principal means used for analyzing main rotor blades. Graphite and fiberglass composites are very popular now for rotor blades because of their excellent fatigue performance.

There is more variability in measured loads for a given maneuver in a helicopter than in a fixed-wing aircraft. Thus, it is difficult to develop a representative loads spectrum for calculating component replacement times and inspection intervals. As a result, flight-test data are recorded for each maneuver with a number of pilots, and conservative loads are used in fatigue life calculations.

Flight data recorders now being developed for the helicopter fleet will allow data on the usage and/or loads experienced by each vehicle to be used for calculating component fatigue lives. Current recorders emphasize usage monitoring. Loads monitoring in the dynamic system, however, presents problems because slip rings on the main rotor mast as used in flight tests represent a maintenance burden and are unreliable for service applications. Continuing pursuits in this area involve the possibility that airframe system measurements can be related to dynamic system loadings.

The recognition of a given flight regime, such as left turn or rolling pull-out, from monitored angles, velocities, accelerations, and control positions has been demonstrated. Thus, a flight data recorder could provide the usage history of each aircraft. In the future, these data will enable more accuracy in defining inspection intervals and replacement times, thereby increasing safety.

The Army is requiring that the dynamic components for the RAH-66 helicopter be substantiated for "six nines" (0.999999) reliability, or one structural failure in a million flight hours. The airframe design must be damage tolerant, with depot-level inspection intervals of 3000 flight hours. The RAH-66 will be constructed mainly of advanced composites, which generally have good fatigue and damage-tolerant performance. In addition, much of the airframe structure is designed with crash loads. Thus, operating stress levels remain well below the endurance limit.

Helicopter designers tend to place more emphasis on crash survivability than do fixed-wing aircraft designers, possibly because of differences in the vehicles' environments. From a structures standpoint, the key is to reduce both accelerations on passengers and encroachments into the cabin. Designers tend to place the engines and main rotor transmission above the passenger cabin in order to put the lift vector over the center of gravity. Thus, the frames that support the engines must be designed to resist crash loads.

Of the several crash analysis computer codes available, KRASH is the most popular. Key parts of the structure are modeled with beam elements to simulate their plastic behavior during a crash. The Euler equations of motion are integrated to produce a time history of the loads and deformations of critical members during the crash event.

In a typical crash, the landing gear initially absorbs energy with air-oil shock absorbers and with crushing of the wheels and tires. At the end of shock absorber travel, the landing gear attachment deforms, absorbing more energy. The fuselage underside then contacts the ground and begins crushing. Much effort has gone into designing floor structure that crushes in a controlled manner.

As the final link in crash survivability, passenger seats are designed to stroke up to 10 in. in a crash to reduce vertical accelerations on the occupants. Bell and Sikorsky Advanced Composite Airframe Program helicopters have been crash tested at NASA Langley to demonstrate that the analytical techniques used accurately predict the effects of a full-scale crash.

Designers of the hot structures used in supersonic-cruise and hypersonic vehicles start with the same design and analysis technology base used for subsonic aircraft but must also factor in the thermal effects on material properties, creep, and fatigue.

From aviation's beginnings, the challenge has been to fly higher and faster. The sonic barrier loomed large for many years, and the first supersonic military aircraft could fly faster than the speed of sound for only a few minutes. Eventually, military vehicles such as the SR-71 reconnaissance aircraft could cruise at more than three times the speed of sound. The British-French Concorde has been flying at nearly twice the speed of sound for several years. The sonic barrier of earlier years has been replaced more recently by the thermal barrier.

Aluminum can be used for external skins at speeds up to only about Mach 2.4. In that range, the strength of the aluminum begins to decrease rapidly because of the high aerodynamic heating and skin friction. The SR-71's titanium structure allows it to fly in excess of Mach 3. The Space Shuttle reaches speeds greater than Mach 25 to achieve low-Earth orbit; its aluminum airframe is allowed only because it has a nonload-bearing thermal-protection system that keeps the aluminum relatively cool.

The United States embarked on a national effort to produce a High Speed Civil Transport (HSCT) by the year 2005. Early studies indicate that the design speed will be somewhat higher than Mach 2 to allow for an economical aluminum and composite airframe.

Because new advanced composite materials are critical to HSCT performance, NASA initiated the Enabling Propulsion Materials program. New high-temperature ceramic matrix composites are being developed for the combustor liner, where very high temperatures are needed to get the NO_x levels down and where environmental resistance and long-term durability are vital.

High-temperature intermetallic/metal-matrix composites are being developed for the exhaust nozzle structure, where strength-to-weight ratio and long-term durability are critical (Fig. 1). Nozzles made with today's conventional materials could weigh as much as 2–3 tons per engine; so these efforts are vital to a successful HSCT.

Unlike subsonic engines, the HSCT engine flies at or near its maximum thermal-mechanical loading conditions throughout the cruise period (perhaps 80% of total flight time). In addition, acoustic fatigue at high temperatures is a much more severe condition for the HSCT nozzle because of the far higher noise levels it must suppress to make the aircraft environmentally acceptable.

Another advanced project with extremely challenging thermal-structural requirements is the National Aerospace Plane (NASP) or X-30. This single-stage-

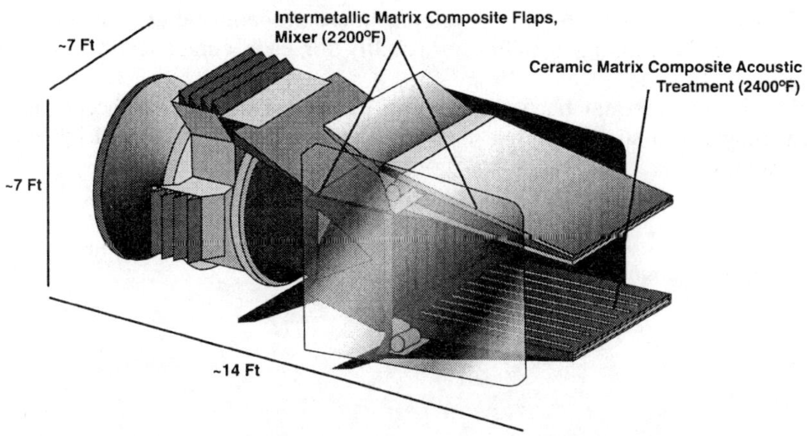

Fig. 1 Exhaust nozzle structure design for HSCT.

to-orbit vehicle will take off horizontally from a runway and fly into low-Earth orbit.

Unlike the Shuttle, the X-30 will have a hot structure made largely of advanced composites. Critical areas of both the airframe and engine will be actively cooled with high-pressure gaseous hydrogen that will then be burned as fuel in the propulsion system. Air is compressed by the forebody of the aircraft before entering the engine inlet. The airframe's aft portion is shaped as an external nozzle to provide lift and thrust. Fuel is also burned in the exhaust stream just aft of the engine to reduce drag.

The X-30's liquid hydrogen fuel boils at −423°F. Some parts of the engine and airframe have temperatures approaching 3000°F. This thermal mismatch and resulting differential expansions require special structural considerations. The fuel tank is made of graphite/epoxy composite, which has lightweight as well as low thermal conductivity and expansion coefficients. Coated carbon-carbon composite serves as a thermal-protection system in the airframe's most intensely heated areas. The primary airframe material is silicon carbide fiber/titanium-matrix composite, chosen for its high temperature capability and relatively light weight.

The enormous heat flux in the NASP engine's combustor region challenges the ability of currently available materials to carry the thermal-mechanical loads, even with active cooling. Hydrogen gas passing through channels within the combustor walls is an extremely effective coolant. But the high temperatures and thermal strains generate highly nonlinear material behavior, cyclic plasticity, and creep. These conditions present formidable difficulties for designers and analysts and for their computer hardware.

Advanced nonlinear three-dimensional anisotropic finite element codes are necessary for solving these problems. The solutions also require constitutive material models that relate stress and strain to basic material properties as a function of temperature and time. This area is in its infancy, and the need for such models is critical. Also in development are probabilistic analysis methods that enable determining a failure probability for each component for various loading conditions.

Clearly, the thermal barrier still exists. The past decade has been one of great progress in structures, but many problems remain to be solved by future generations.

Getting Up to Speed in Hypersonic Structures

Michael W. Kehoe
NASA Ames–Dryden Flight Research Facility

Rodney H. Ricketts
NASA Langley Research Center

This decade once again offers an opportunity for exploring the hypersonic flight regime. As the 21st century approaches, hypersonic technology will become the baseline for more-advanced commercial aerospace systems that significantly reduce the cost and time required for transporting passengers and cargo around the world. This technology will also create new military transportation systems for carrying astronauts and equipment into space.

Previous development of hypersonic technology culminated with the X-15 aeronautical research program in the late 1950s and early 1960s. The X-15 flew 199 test flights with altitude excursions greater than 300,000 ft and speeds above Mach 6. Late in the program, a white ablative coating was applied to protect the structure from the high temperatures encountered during hypersonic flight.

Following completion of the X-15 effort, a few hypersonic-flight programs were considered, but none was funded. The X-20 DYNA-SOAR, a manned boost-glider vehicle, was funded initially but later canceled.

Breakthroughs in propulsion, materials, and vehicle-system integration now make sustained hypersonic flight possible. Present development of hypersonic technology centers on the X-30 National Aerospace Plane (NASP) program. The NASP's objective is to demonstrate an aerospace vehicle that can takeoff from a runway and achieve a single-stage-to-orbit capability. As the NASP flies through the atmosphere from subsonic speeds to orbital velocities, its structure will be subjected to much greater surface heating and temperature gradients than have previously been encountered. These extreme temperatures result from the need to fly at very high dynamic pressures to provide the air-breathing propulsion system with enough oxygen.

Such temperatures can seriously affect the structural integrity of a hypersonic vehicle. Thermal effects such as extremely hot surface temperatures and large temperature gradients will cause variations in the structure's elastic characteristics. The effects of these structural variations must be known during the design phase of such vehicles.

For today's subsonic and supersonic aircraft, structural dynamic stability considerations are primarily aeroelastic and aeroservoelastic. Aeroelasticity is an interaction between the vehicle's inertia, stiffness, and unsteady aerodynamic forces. Dynamic aeroelastic phenomena, such as flutter, can produce divergent

Aerospace America, Vol. 30, No. 9, Sept. 1992, pp. 18–20, 29.

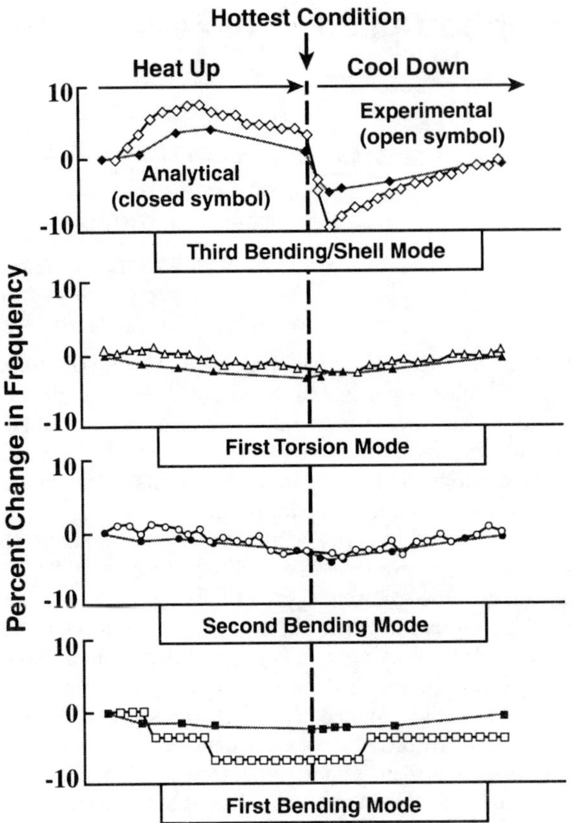

Fig. 1 Thermal gradients affect the vibration frequencies of a builtup wing box.

oscillations of the aircraft structure. Aeroservoelasticity is an interaction between the aircraft structure, the flight-control system, and the unsteady aerodynamic forces. Aeroservoelastic instabilities are also known as control-structure interactions. These unfavorable couplings can be avoided through prudent design. The effects of aerodynamic heating on these two instability mechanisms must be considered in finite element model (FEM) analysis methods.

Hypersonic-vehicle designers are relying increasingly on FEM analysis methods to ensure that there is sufficient vehicle structural dynamic stability throughout the desired flight envelope. These analytical methods account for both thermal gradients and material property changes for purposes of determining the stiffness and vibration characteristics of heated structures (Fig. 1).

These analytical methods are being validated through a series of systematic correlations with experimental data acquired during vibration tests and flutter model tests. The ability to conduct tests in a hostile environment presents new challenges for the discipline of structural dynamics.

Experimental measurement of a structure's modal characteristics is required for accurate characterization of dynamic models used in the FEM analytical methods. Modal testing is conducted to measure the structure's resonant frequencies, dampings, and mode shapes. These test results form a foundation of experimental data that permits correlation with computations and enable verification of finite element analysis methods.

Vibration tests of heated flat panels made of either metallic or composite materials have been conducted at NASA's Ames–Dryden Flight Research Facility. Several different types of structures have been tested, and the maximum temperature at which data were acquired is 1400°F (Table 1).

These tests took place in an enclosed oven that had feedback temperature control. The oven, which used quartz heating lamps, was divided into three separate zones so that each test article could be subjected to uniform, nonuniform, and transient heating.

Each test article was instrumented with thermocouples and accelerometers. The test article's vibratory responses were measured with accelerometers at temperatures of up to 500°F. Above that temperature, a technique employing a laser vibrometer was developed to measure the responses.

Several vibration tests of a heated box beam have been conducted at NASA Langley. The box beam, constructed of metallic spars, ribs, and skins, was equipped with accelerometers and thermocouples. Heat was applied via a quartz lamp, and the effect on vibration frequency was measured during the transient heating up and cooling down of the structure. The data underline the importance of including in the finite element analysis both material property changes originating from elevated temperatures and internal stresses originating from thermal gradients.

NASA conducted a flutter test of a solid aluminum wing model at its Wallops Island Facility in about 1960. The model, which was cantilevered at the root, was placed in the flow of a Mach 2 hot air jet. The model underwent transient heating by the jet and reached a maximum temperature of 800°F. During the test, the model went into and out of flutter as the thermal gradients first developed and then subsided. The thermal gradients were a result of the uneven aerodynamic heating, which reduced the model's torsional stiffness.

Recently, finite element analysis methods were developed to model these torsional stiffness changes (Fig. 2). Results were obtained for both the hot wing model and the cold wing model to determine the effects of the property change only, the prestress only, and the combined property and prestress effects (Fig. 3). The combined predicted effects on torsional stiffness compared well with the

Table 1 Vibration test data were acquired at various maximum temperatures

Test article	Maximum temperature, °F
Flat aluminum plate	700
Flat composite plate	375
Builtup aluminum structure	800
Titanium honeycomb panel	1400

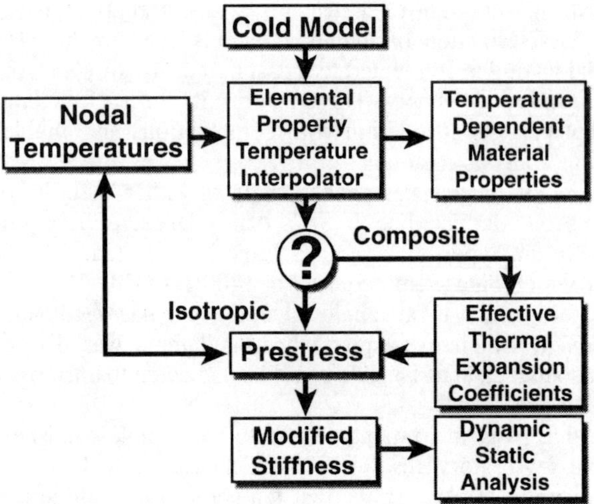

Fig. 2 Finite element analyses now include prestress conditions.

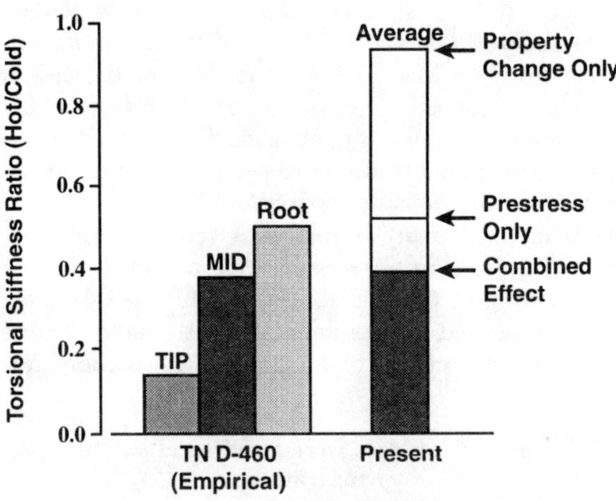

Fig. 3 Reduction of wing torsional stiffness is due primarily to prestress conditions.

experimental results, which showed a reduction of as much as 60%. Most of this reduction was caused by the prestress condition.

The hot stiffness model developed from this test was used in a flutter analysis. The analysis predicted this flutter occurrence to within 8% of the measured flutter speed. Detailed analysis of the results shows that the primary degradation in torsional stiffness, which caused the flutter instability, can be attributed to the thermal stresses created by the large thermal gradients in the wing.

Tests being considered for further validation of analytical methods include high-temperature, high-speed wind-tunnel testing and in-flight measurement of steady and unsteady pressures at Mach 3 and higher.

Aerothermoelastic tests are being considered for the NASA Langley combustion-driven wind tunnel. This facility can test models at up to Mach 7 and provides realistic flight environments from 55,000 ft to 125,000 ft. Data gathered from such tests would validate finite element analyses at realistic flight conditions, where temperature will significantly affect structural stiffness.

Tests are also being proposed for the SR-71 aircraft located at NASA's Ames–Dryden facility. Their objective is to establish a comprehensive database of steady and unsteady pressure measurements at speeds up to Mach 3. This database will be used to validate emerging computational fluid dynamics analysis methods.

An experimental database is essential for validating the finite element analysis procedures used to account for both thermal gradients and material property changes experienced by hypersonic-vehicle structures. Analyses conducted so far have shown that the structural dynamic stability of such vehicles can be degraded by heat-induced reduction of material properties. These analyses also show that the thermal stresses may be either beneficial or adverse in this regard. It is, therefore, important to have accurate analysis methods for assessing the structural dynamic stability of hypersonic vehicles. This validation can come only through a reliable experimental database established by comprehensive hostile environment testing.

Structures for Hypervelocity Flight

Peter K. Shih, Allen D. Zwan, Michael N. Kelley, and Jack Prunty
General Dynamics

In recent years, design and analysis of hypervelocity vehicles has taken a great deal of engineering. Aerodynamic heating at hypersonic speeds will drive structural temperatures higher than on conventional aerospace vehicles. Hypervelocity vehicles will fly in the atmosphere at higher altitudes where exotic phenomena such as nonequilibrium flow chemistry are the rule rather than the exception. Cryogenic fuel will sharpen thermal gradients within the vehicle itself. Above all, structure must be lightweight and highly efficient for maximum performance. So thermophysical analysis has become a full partner in structural design, including that with advanced materials.

Air Force Wright Aeronautical Labs sponsored a study using a cryogenically fueled, rocket-powered transatmospheric vehicle for thermo-structural concept development and weight optimization. The vehicle's hypersonic lift-to-drag ratio was high, for maximum glide range. Analysis started from an altitude of 250,000 ft at Mach 25 and took the vehicle through an equilibrium glide to a synergetic plane change that permitted a 45-deg shift in orbital inclination during a once-around, suborbital flight. After identifying a suitable configuration and trajectories, researchers next defined aerodynamic heating loads using standard codes such as the lab's AEROHEAT and ASHAB from General Dynamics. Both the aerodynamic heat flux and radiation equilibrium temperature were calculated at several locations on the vehicle surface. Analysis of the thermal-protection system used the heat fluxes, and the temperatures were part of material selection criteria. Study results showed that peak temperatures on windward surfaces vary between 2600°F and 3000°F; so thermal protection must use carbon-carbon.

Noncatalytic effects first became widely known during review of flight-test heating data from the Shuttle. Around most vehicles flying hypersonically high in the atmosphere, diatomic air molecules dissociate into atomic oxygen and nitrogen inside the shock layer. These atoms recombining in the presence of a catalyst such as a metallic aircraft skin, for example, transfer additional heat to the skin. But on the Shuttle, silica-coated tiles and silicon carbide-coated carbon-carbon were noncatalytic; so heating was less than predicted. Study of noncatalytic surface effect on the hypervelocity vehicle used a nonequilibrium viscous shock layer code. The effect tended to lessen as the vehicle nose radius decreased and on surfaces well downstream of the nose. So noncatalytic surface effects on overall aerodynamic heating are expected to be minimal for high lift-to-drag vehicles with small nose radii.

Aerospace America, Vol. 27, No. 5, May 1989, pp. 28–31.

In addition to that of thermal loads, definition of structural loads must precede detailed thermo-structural design. Critical load conditions included maximum side load and axial acceleration during boost, hypersonic maneuvers with both zero and full fuel loads, gusts, and landing loads. Limit load factors based on the Shuttle and other manned vehicles were 3 *g* axial and 2 *g* normal with full fuel and about 5 *g* axial and 3 *g* normal for the low-fuel case.

The study examined several basic types of thermo-structural concepts. The integral tank designs use cryogenic tankage as the primary load-carrying structure, whereas the nonintegral designs have a load-carrying fuselage structure supporting the tankage. Three variations of each concept are differentiated by the use or absence of thermal protection on the windward and leeward surfaces.

Within each of the basic concepts were variants using different materials and allowing investigation of structural temperature variation. Conventional structural materials ranged from 2219 aluminum with a peak structural temperature of about 350°F to Rene 41 with 1600°F. Tankage material also varied, and thermal-protection alternatives included tailorable advanced blanket insulation; metallic multiwall tiles; and panels made of superalloy, refractory metal, and carbon-carbon. Thermal analysis sized thermal-protection and insulation requirements by imposing heat fluxes calculated previously on a thermo-structural concept. Researchers varied insulation thickness to attain the desired structural temperature.

Maximum angle of attack and hypersonic maneuver were the prime cases for wing-body bending. Ascent acceleration affected the design in parts of the tankage, and the landing loads affected local body panel shear. Design load was considered in conjunction with other phenomena such as resistance to panel flutter, sonic fatigue, and minimum gage.

After an initial screening of the concepts, load intensities were used for detailed structural sizing and weight estimates for the remaining concepts. Calculated weights included body shells and lower surfaces, cryogenic tankage and insulation, and wing spars and skins. Other structures such as nosecaps, leading edges, the crew compartment, and control surfaces were not considered in detail.

Weight comparison led to selection of a nonintegral aluminum-lithium alloy tank concept with hybrid thermal protection as the lowest weight configuration. Wing structure unit weights are about 6.9 lb/ft^2 of planform area, nearly 30% lighter than the Shuttle wing, which benefits from a more benign trajectory, and competitive with a future space-transportation concept proposed by Langley. Body structure weights, at 4.9 lb/ft^2 without tankage and 6.1 lb/ft^2 with, are 20–40% lighter than the Shuttle.

The selected thermo-structural concept is a frame-supported monocoque structure with the base of the body supported by the center section of the wing and a forward lower panel. The leeward and side structure in the forward crew/equipment area, the leeward wing structure, and the leeward wing-body junction use Inconel 718 nickel superalloy. The balance of the leeward structure, which in the hybrid concept requires no thermal protection, uses titanium 6Al-2Sn-4Zr-2Mo, a high strength-to-weight ratio alloy with a peak temperature capability of 1050°F. The windward structure consists of Rene 41 superalloy, allowing a structural temperature of 1600°F. The wing is a multispar arrangement in which the leeward surface skin panels and multiple discrete spar caps on the windward

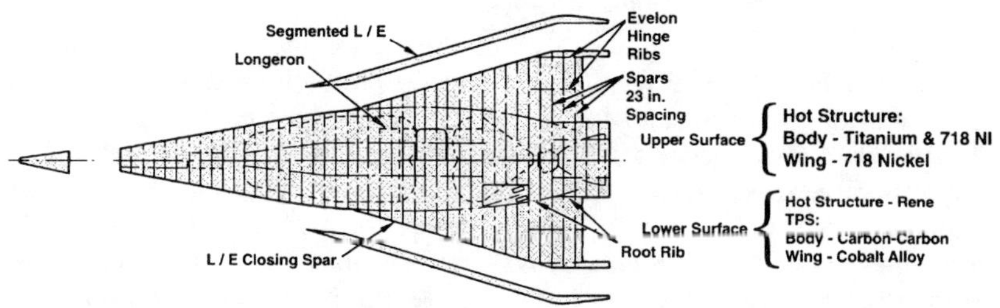

Fig. 1 Optimized hypervelocity vehicle thermo-structural concept features thermal-protection system on the windward surface only.

surface transmit the bending loads. Upper skin panels carry the drag loads; differential bending of the spars resists torsion. Cross radiation from the lower skin panels through the wing structure reduces internal thermal gradients. The wing spars feature sinusoidal corrugated webs of Rene 41, while the leeward wing structure consists of sandwich panels of Inconel 718.

The hybrid design limits thermal protection to the windward surface only (Fig. 1). Slip-jointed carbon-carbon cover panels over an encapsulated fibrous insulation blanket protect the body. On the wing, similar panels of cobalt super-alloy overlay a minimal insulation layer around each spar cap. The vehicle has hot load-carrying structure, except for the windward surface. So most of the vehicle surface is a rugged, airplane-like metallic skin, and the balance is relatively durable, highly reliable, radiative thermal protection designed for inspection and maintenance.

Cryogenic tankage of lightweight aluminum-lithium alloy has a double-lobe design for maximum propellant capacity inside an aerodynamically efficient fuselage shape. The liquid hydrogen tank, located forward, is supported by a kinematic system of two linkages forward and three trunnions aft. The liquid oxygen tank, located aft, is similarly supported by three trunnions and one linkage. The unusual liquid oxygen tank adapts the thrust cone structure of the Atlas missile, in which the inertia of the liquid oxygen directly bears 40% of the thrust load. Covering the tankage is a layer of fibrous insulation purged with gaseous helium to prevent air from liquefying on the tank surface, a process called cryopumping. The payload compartment and main landing gear bay are between the two tanks.

The design concept's prime advantage over competing integral tank concepts is highly efficient integration of the wing center section with lower body structure. An uninterrupted wing carrythrough is more efficient than carrying the wing bending moment around several fuselage frames. Also, safety and reliability of the simpler nonintegral tank should be much better than the integral design for several reasons. The simple, well-defined stress distribution of the nonintegral tank aids analysis of fracture mechanics, while failure modes and points of potential failure are much fewer. And the simple kinematic tank attachment is far less trouble prone than the multiple interfaces of an integral tank under thermal stress.

The concept has several innovations. Eliminating weighty acreage of minimum-gage lower-surface skin in favor of discrete spar caps heightens bending efficiency while decreasing the potential for thermal stress. Unobstructed cross radiation within the wing lowers peak temperatures and resultant thermal gradients. Thermally protective cover panels have fail-safe attachments and are structurally efficient, durable, rugged, and can be removed for maintenance or inspection. The complete tankage and propulsion system can be installed on the wing structure platform and checked out prior to fuselage assembly, a feature that aids field maintenance.

After thermo-structural optimization techniques yielded a promising concept, application of advanced technology began. Considerations included materials expected to be available commercially by 1995, such as silicon carbide/aluminum, silicon carbide/titanium, and silicon carbide/superalloy metal-matrix composites; rapid solidification rate alloys; intermetallic compounds; advances in organic-matrix composites; and improved carbon-carbon coating systems.

Specific locations where advanced materials were applicable included body and wing primary structure as well as thermal protection on the windward surface. Rapid solidification aluminum and titanium replaced titanium and superalloy, respectively, in the body. Wing structure was rapid solidification titanium and silicon carbide/superalloy, while silicon carbide/superalloy replaced cobalt alloy in thermal-protection panels on the wing. Improvements in carbon-carbon coating technology promised weight savings in thermal-protection panels on the windward body.

Study results predicted that weight savings on the vehicle structure studied will amount to 18%. More would derive from vehicle designs better suited to advanced materials, but close cost examination should precede such a commitment.

Hypersonic Structures and Materials: A Progress Report

L. Robert Jackson, Sidney C. Dixon, and Darrel R. Tenney
NASA Langley Research Center

Alan L. Carter
NASA Ames Research Center

Joseph R. Stephens
NASA Lewis Research Center

Building on discoveries already made in structures and materials applicable to hypersonic vehicles depends on upgrading long neglected test facilities, techniques, and instrumentation.

Ongoing research has provided detailed and accurate information about the aerothermal, aerothermoelastic, and acoustic loads a hypersonic vehicle must withstand. Materials research over the past decade has barely scratched the surface of high-payoff technology but has provided structural designers with several promising schemes for saving weight while maintaining strength and the benefits of favorable thermal effects. Key materials being analyzed for major airframe structures and engine components include metallic alloys, intermetallic compounds, and metal-matrix composites. Significant advances have also been made in advanced carbon-carbon composites, ceramics, and ceramic-matrix composites. Finally, considerable effort has gone into the development of geometrically efficient structural concepts and fabrication techniques for high-temperature applications.

In each area, however, many technical challenges remain. Weight of a hypersonic, air-breathing, single-stage-to-orbit (SSTO) vehicle is probably more critical than for any previous vehicle design, and selection of structures and materials must be especially sensitive to thermal effects.

Unfortunately, much of this nation's high-temperature test capability has been unused, unimproved, and, in some cases, dismantled since the early 1970s. NASA, the U.S. Air Force, and the aerospace industry are looking at the feasibility of upgrading heating and cryogenic capabilities; adding vacuum; and improving instrumentation for measurement of strain, temperature, pressure, and deflection. High-temperature facilities must be upgraded and expanded, especially for cryogenic tank and actively cooled structure, as well as for life cycle testing of

Aerospace America, Vol. 25, No. 10, Oct. 1987, pp. 24–30.

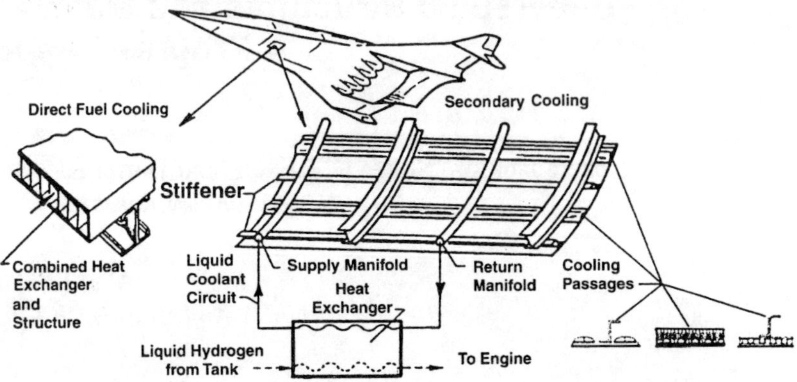

Fig. 1 Designers of hypersonic vehicle structure can choose between direct fuel cooling, where cryogenic fuel flows through the structure, and secondary cooling, where another liquid coolant flows through the structure to a separate heat exchanger where heat is dumped into the fuel. Direct cooling shows the most promise for engine structure, and secondary cooling elsewhere. Alternative designs also are shown for cooling passages.

airframe and engine structure (Fig. 1). A significant challenge is securing the capability to test with combined loads, including dynamic loads, and with liquid hydrogen fuel in the test hardware. Such testing capability is essential if hypersonic vehicles are to become a reality.

Structural designers cannot meet the stringent weight requirements of future high-speed vehicles unless they can accurately predict the loads. Key aerothermal, aerothermoelastic, and acoustic loads are high in part because an SSTO vehicle with air-breathing propulsion follows a high dynamic pressure trajectory to achieve adequate propulsive efficiency. Unlike conventional rocket-propelled vehicles, an air-breathing hypersonic vehicle must deal with severe heating due to its configuration and relatively long exposure to high dynamic pressure.

A current experimental and analytical research program at NASA Langley is concentrating on generic local aerothermal loads. The program will provide a database for use in preliminary design and will be used for validation of computational fluid dynamics (CFD) codes.

Part of this program deals with the interaction of an oblique shock with the bow shock of a cylinder. More specifically, a generic representation of shock-on-lip for a hypersonic vehicle with air-breathing propulsion shows that optimum propulsion efficiency occurs when the precompression shock impinges on the engine cowl lip so that all compressed air is captured by the engine. The test models consist of single cylinders mounted with the axis of each parallel to a sharp leading-edge shock generator. The shock generator can be articulated at various angles to the free stream flow. Tests have been conducted in the Langley 8-ft High Temperature Tunnel and the Calspan 48-in. and 96-in. shock tunnels.

A schlieren photograph of the jet interaction flow field shows the oblique shock generated by a 12.5-deg wedge, the displaced bow shock, the reflected shock, and the continuation of the bow shock. Results include a supersonic jet in the subsonic region that can impinge nearly normal to the surface.

The highest heating rate amplification factor obtained for Mach 8 flow in one experiment was about 10 times the undisturbed stagnation point heating rate.

These data help fill a void in the database for two-dimensional leading edges. However, to determine other effects, such as nonequilibrium chemistry effects, experiments are needed at high Mach and Reynolds numbers.

Recent developments have demonstrated that the inviscid features of the interference pattern and the surface pressure can be accurately determined. These developments include finite element CFD procedures using shock-capturing techniques and a gridding technique that adaptively refines and re-refines the grid based on flow field gradients. This refining and re-refining is achieved by placing more grids in areas of high gradients and removing elements from areas of low gradients. The localized and intense nature of the heating doubles the requirements of the viscous solution to this problem. Implicit time-marching techniques to resolve the boundary-layer heat transfer and skin friction are under development at several NASA research centers.

The effects of aeroacoustic loads on the structure will be an important consideration in the design, verification testing, and, ultimately, the operation of future hypersonic vehicles. Large areas of the vehicle will be exposed to the high temperature and acoustic loads of the engine exhaust. Technology challenges include:

- Estimation of the intensity and frequency of acoustic loads associated with hypersonic air-breathing propulsion and aerodynamic flow over the vehicle.
- Prediction of response and sonic fatigue of advanced materials and structures operating at elevated temperatures.
- Specification of combined environmental testing to qualify the structure for the anticipated operating life cycle.

Material developments over the past 10 years have provided structural designers with several promising systems for saving weight (Fig. 2). Structural weight limits require the use of thin-gage, low-density alloys and composites. Metallic alloys, intermetallic compounds, and metal-matrix composites for major airframe structures and engine components offer advantageous characteristics. These include light weight, high stiffness, high thermal conductivity, high temperature capability, and good fabricability. Rapid-solidification-rate (RSR) processing is producing a new family of alloys, the elevated temperature properties of which will be a 50% improvement. The new alloys will also retain these properties for much longer periods of exposure because of better stabilization of the microstructure.

Titanium alloys and intermetallic compounds are attractive for temperature applications up to 1000°F. Several NASA research centers, the U.S. Air Force, and the aerospace industry have recently emphasized alloys of titanium intermetallic compounds, specifically the aluminides. Research aims at controlling microstructure using alloy chemistry and processing techniques such as RSR to improve the properties and use temperature. Studies are complete on powder making and subsequent consolidation. Fabrication to thin-gage sheet as well as technology for forming and joining are under development. Major hurdles that must be overcome

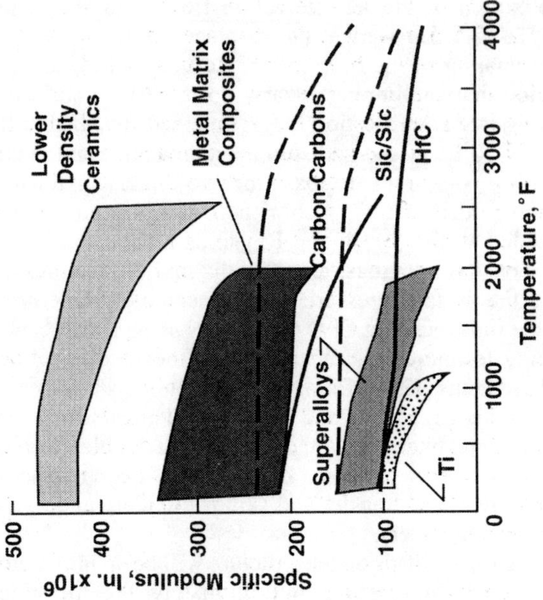

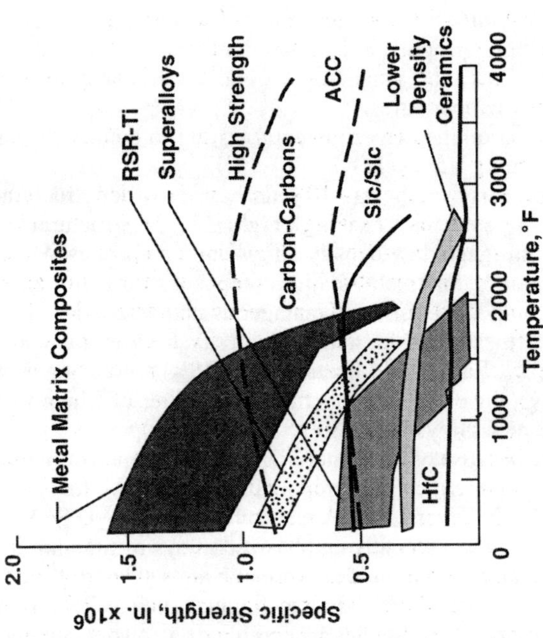

Fig. 2 Advanced materials provide the structural designer with a variety of properties to choose from, but the choices narrow considerably above 2500°F.

include the environmental effects of the Earth's atmosphere, embrittlement from hydrogen exposure, permeability, and catalytic effects. Another is limited room temperature ductility, which is characteristic of the titanium aluminides.

Under development are intermetallic-matrix composites that have the potential to withstand 1500–1800°F. NASA Lewis and several other laboratories are concentrating their research on composite fabrication techniques. They are exploring foil rolling, plasma spraying, arc spraying, and powder metallurgy processes to optimize the consolidation of the aluminides. A major effort by Avco Corporation is currently directed toward smaller diameter silicon-carbide fibers and possible new fibers such as titanium diboride. These will allow fabrication of thin foil-gage sheet. Fiber-matrix interaction at the higher temperatures for some metal-matrix composites means coating the fibers to minimize reaction rates.

For some structural components such as the engine cowl lip, strength and thermal conductivity greater than that of nickel or copper are essential. Metal-matrix composites such as tungsten or graphite-reinforced copper can provide the required strength and thermal conductivity. Some prototype components are currently undergoing evaluation. More-efficient metal-matrix composites are under development by the major aerospace companies.

Carbon-carbon composites are attractive candidate materials for hot structures and thermal-protection systems because of their strength retention at high temperatures. The Space Shuttle is a point of reference as the only existing reusable hypersonic vehicle. Reinforced carbon-carbon (RCC) was developed for the Shuttle orbiter's nosecap and wing leading edges, baselined in 1973. Even though this material was made from low-strength rayon-based carbon fibers, its strength efficiency is superior to both superalloys and ceramic composites at temperatures higher than 1800°F. Development of advanced carbon-carbon (ACC) under the sponsorship of NASA Langley has resulted in a 100% increase of in-plane strength over that of RCC. The ACC material currently being evaluated uses woven carbon cloth made with poly-acro-nitro-base graphite fibers. Matrix-dominated out-of-plane properties of ACC are about one-half those for RCC.

Promising approaches to improving interlaminar properties include alternate fiber and matrix materials, enhancement of fiber-matrix interface bonding, variations in composition and process to impart higher through-the-thickness strength, and use of three-directional reinforcement. Current investigations are aimed at improving out-of-plane tension and interlaminar shear strengths.

Carbon-carbon parts must be protected from oxidation by a high-temperature oxidation-stable coating. An oxidation-protection system for reinforced carbon-carbon on the Shuttle's wing leading edges consisted of a silicon-carbide conversion coating, a crack sealer, and an overcoat sealer. It has performed well in flight at temperatures to 2500°F. Shuttle coating technology for carbon-carbon should be adequate for SSTO missions. For vehicles with multiple missions such as SSTO and hypersonic cruise, however, significant technical progress will be required to develop coatings for use over a broad range of operating temperatures.

Significant advancements have also been made in ceramics and ceramic-matrix composites over the past 10 years. They offer the advantages of high temperature strength, high strength-to-weight, and outstanding environmental resistance without dependence on coatings. The matrix and fibers are inherently oxidation resistant to about 3000°F. NASA Lewis has recently developed the processing of silicon-carbide fiber reinforced, reaction-bonded silicon nitride to yield a strong tough composite. Investigators have established that this reaction-bonded silicon nitride is capable of strength retention for up to 100 h at 2550°F. Monolithic ceramic may be required for nosecap and leading-edge inserts. Ceramic composites could also find applications in airframe, control surfaces, and engine structure. Greater across-ply and interlaminar strengths and more-efficient use of the fibers for reinforcement are possible. These will improve the strength-to-density ratio.

Advanced fabrication techniques, such as superplastic forming and diffusion bonding, permit fabrication of low-cost, high-geometric-efficiency metallic structures that were not previously available. Under NASA contract, industry has conducted design, fabrication, and testing studies of uncooled and uninsulated structures, insulated structures, and actively cooled structures. Actively cooled structures must be used when radiation equilibrium temperatures, which are dependent on configuration and trajectory, exceed the use temperature of structural or thermal-protection system materials. These are about 2000°F and 3000°F, respectively. Moreover, an actively cooled structure may offer a lower weight vehicle than either thermally protected or hot structures.

Considerable effort has gone into the development of geometrically efficient structures for high-temperature applications. One type uses curvature to provide axial stiffness while alleviating transverse thermal stresses. In addition to improving design and developing fabrication techniques for such structure, NASA Ames has tested for structural allowables and tested large components under combined mechanical and thermal loads to the maximum design conditions and temperatures up to 1340°F (Fig. 3). Analysis of the data by NASA Ames and NASA Langley indicates that curved-element hot structures represent a mature technology ready for flight demonstration.

The cryogenic fuel used by the Shuttle is carried in the expendable external tank. Future systems designed for full reusability, however, will undoubtedly

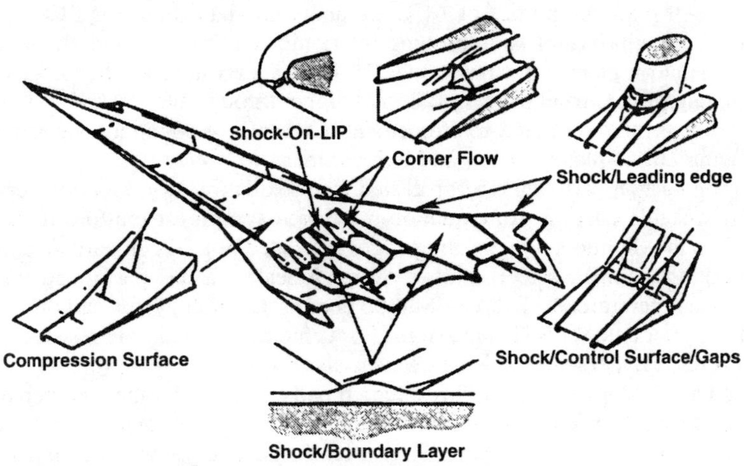

Fig. 3 Complexity of structure and aerothermal loading is a challenge to designers of hypersonic vehicles that must be tested and certified for manned flight.

carry their own cryogenic fuels internally, becoming virtual flying cryogenic tanks. As a result, structural designers must go beyond conventional considerations such as thermal protection and support of vehicle mechanical and thermal-structural loads. Design must also address cryogenic containment problems.

The primary problem that distinguishes hydrogen tanks from other cryogenic tanks is the proclivity of liquid hydrogen to condense other gases because of its extreme temperature, as low as −423°F. Air, or any purge gas other than helium, condenses on the outside surface of the tank, and the dramatic drop in volume due to the phase change produces a partial vacuum. The vacuum pumps additional gas to the surface, where it condenses. This cryopumping transmits large quantities of heat to the fuel, causing hydrogen boiloff. If the gas is air, it also produces a potential safety hazard because the initial liquefaction of air is oxygen rich.

A metallic thermal-protection system with fibrous insulation for reduced heating to the tank and structure is proposed for the insulated fuselage and tank walls of reusable hypersonic vehicles. However, the amount of fibrous insulation required varies, depending on the temperature limit of the exterior tank surface. Tank walls are welded to provide leak-free containment of the cryogenic propellants.

Work on hydrogen-cooled engine structures at NASA Langley began with the Hypersonic Research Engine (HRE) program in the 1960s. That program culminated, from a thermal-structural standpoint, in tests of a complete flight weight hydrogen-cooled structural assembly model in the Langley 8-ft tunnel. These tests confirmed the suitability of the basic hydrogen-cooling approach.

Two major thermal-structural problems were uncovered that must be solved before a hydrogen-cooled scramjet can become a practical reality. First, the coolant requirements must be reduced. The HRE required almost three times as much hydrogen for coolant as for fuel. Second, the thermal fatigue life must be increased. HRE had an anticipated fatigue life of only 135 operational cycles.

Studies in the 1970s of airframe-integrated scramjets with high potential performance led to the sweptback, fixed-geometry, hydrogen-fueled, rectangular scramjet concept. This has the several structural advantages of fixed geometry as well as reduced wetted surface area and heating rates. All engine surfaces wetted by the airstream are regeneratively cooled by circulating the hydrogen fuel through a cooling jacket, which is adequate to about Mach 10.

The fabrication and material technology required to obtain reasonable thermal fatigue life for the cooling jacket has been developed and experimentally validated. A major manufacturer is currently fabricating a fuel injection strut for testing in NASA Langley's combustion and mixing research apparatus. Tests are expected to be complete in 1988. Contractors are now studying low-density, high-conductivity metal-matrix composite materials for hypersonic engine structure.

Significant progress has been made in structures and materials technology for reusable launch vehicles during the last 20 years. One of the most exciting developments is the use of metal-matrix composite materials and advanced carbon-carbon materials for structural and acreage thermal-protection system applications. But research has just begun on key high-payoff technology such as that for hydrogen-cooled structures and for composite structure and insulation of cryogenic tanks.

Developing HST Structural Technology

W. R. Laidlaw and E. W. Johnston
North American Aviation, Inc.

Prospects for our country developing a hypersonic transport (HST) have been viewed with various enthusiasm in the light of the fact that the supersonic transport is about to be accelerated into a major engineering, fabrication, and flight-testing effort. Skeptics have already evaluated an HST project with negative attitudes and conclusions. Countless examples exist, however, illustrating how the scientific community proves too optimistic in near-term evaluations and too pessimistic in long-range projections.

For instance, many engineers predicted that jet engines would never replace piston engines—"Their fuel consumption is too great!" they said. They also scoffed at the introduction of the subsonic jet transport. "Who is going to pay the extra fare?" they said—"Who needs to get there that fast?" Meanwhile, the subsonic jet transport market exceeded all predictions and continues to expand. Today, we are hearing cliches like these in reference to the SST, and it can be reliably predicted that the skeptics will have to revise their thinking about this transportation speed regime, too.

So, looking to these past experiences for guidance of human reactions to projected progress in modes of transportation, the hypersonic-transport system may arrive sooner than current projections indicate. We should apply our newly acquired wisdom as we consider the hypersonic transport, keeping in mind that the SST has some major milestones to achieve.

The primary objective of the HST would now appear to be in the definition of the technology and its acquisition in a systematic and appropriate manner. Technological progress should be paced by an agreed-upon time schedule and consist of appropriate analytical design studies, followed by exercising test hardware specimens in ground facilities and then flight testing of subscale vehicles.

Focusing hypersonic technology directly toward an HST system would achieve three things. The technology would have a direct, unequivocal application; it could be pursued in a continuous, logical fashion; and, by virtue of this environment, the personnel would be well motivated.

Let us turn to HST concepts that have been propounded and problems associated with structures. Conceptual studies of the hypersonic transport have been conducted for a number of years.[1–5] Studies and analyses performed by NASA, the U.S. Air Force, and industry indicate that HST systems fall into general categories:

1) Mach 6–8 design cruise speed; liquid hydrogen or methane fuel; multiple-mode propulsion with subsonic-combustion ramjet propulsion; and 80,000–100,000-ft cruise altitude.

Astronautics & Aeronautics, Vol. 4, No. 10, Oct. 1966, pp. 40–43.

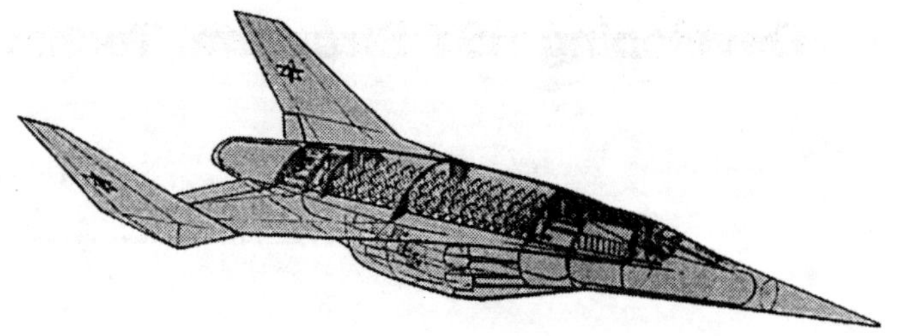

Fig. 1 Mach-12 HST concept.

2) Mach 10–14 design cruise speed; liquid hydrogen fuel; multiple-mode propulsion with supersonic-combustion ramjets employed at the design speed; 110,000–140,000-ft cruise altitude.

Needless to say, as time proceeds, the characteristics of systems will see greater refinement.

At the present time, two general categories of HST can be described. The first has a nominal cruising speed of Mach 6. The main structural components are for crew, 150 passengers, fuel compartments, wings, empennage surfaces, and a subsonic-combustion turboramjet propulsion system. The concept described would employ methane fuel. It would be about 300 ft long. A similar configuration employing liquid hydrogen fuel would have larger fuel volume.

Conceptual arrangement for a Mach-12 cruising speed is shown in Fig. 1. A very high level of integration would be needed for the inlet (forward fuselage), inlet-cowl-combustion section, exhaust nozzle (aft fuselage), liquid hydrogen tankage, and passenger (150) compartments. This vehicle might be 200 ft long, but it would have a much wider fuselage.

In either case, the number of passengers and fuel volume would exert strong influences on the design arrangements.

Structural Temperatures

Large portions of the HST will experience increases in temperature on the external surfaces from the SST temperature of 600°F to around 1800–2400°F. Fuel temperatures would be –423°F for hydrogen or at least –259°F for methane. With the HST, severe thermal gradients in the structure clearly become a major concern.

Figure 2, depicting environment, will help describe significant predicted temperature levels of the two concepts.

The Mach-6 HST would accelerate out of the ''sonic boom'' permissible flight zone at approximately 1500-psf flight dynamic pressure to a duct-pressure-limit restraint and eventually climb to Mach 6 at approximately 100,000 ft. Index external skin temperatures would be approximately 1400°F. With liquid hydrogen (–423°F), an index thermal gradient of 1823°F could be experienced. With methane (–259°F), the thermal gradient would be 1659°F. The lower temperature gradient could have a significant influence on system selection.

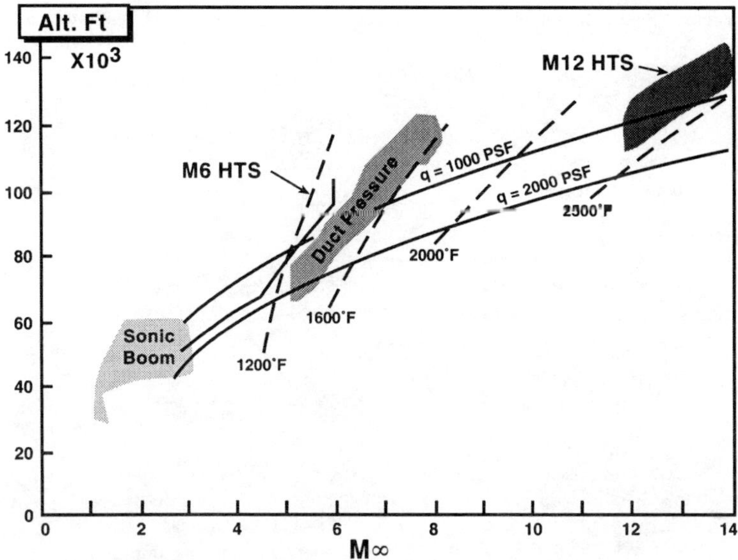

Fig. 2 HST environments.

The Mach-12 HST cruising regime would cover altitudes of 110,000–140,000 ft at flight dynamic pressures of 1500 Q to perhaps as low as 500 Q. The index external temperature would be approximately 2400°F with a temperature differential of 2823°F for liquid hydrogen fuel—the only reasonable fuel choice for this flight regime.

It is obvious that designing lightweight, reliable, long-life structures for such vehicles presents a great engineering challenge.

Referring to the Mach-6 concept, the structural concepts shown in Fig. 3 were developed through various study efforts. From studies such as these, basepoint designs can be evaluated in greater depth first analytically and then through laboratory specimen testing and static-load heat testing of small specimens and later large specimens and then on up the line of higher-confidence-level tests of large assemblies.

It is important to re-emphasize that sufficient concepts have been generated. It is timely and appropriate that direct and systematic structural test programs be defined and implemented and that significant engineering test results be obtained, documented, and disseminated in structural-efficiency terms of weight, reliability, flutter limits, fatigue life, fabrication costs, etc.

Materials

The SST materials program deals with the more promising titanium alloys and stainless steels for the basic airframe and with the superalloys for the hotter areas. In the HST, the superalloys will become the primary external and wing internal structure. Preferably, refactory alloys will not be used for the hotter,

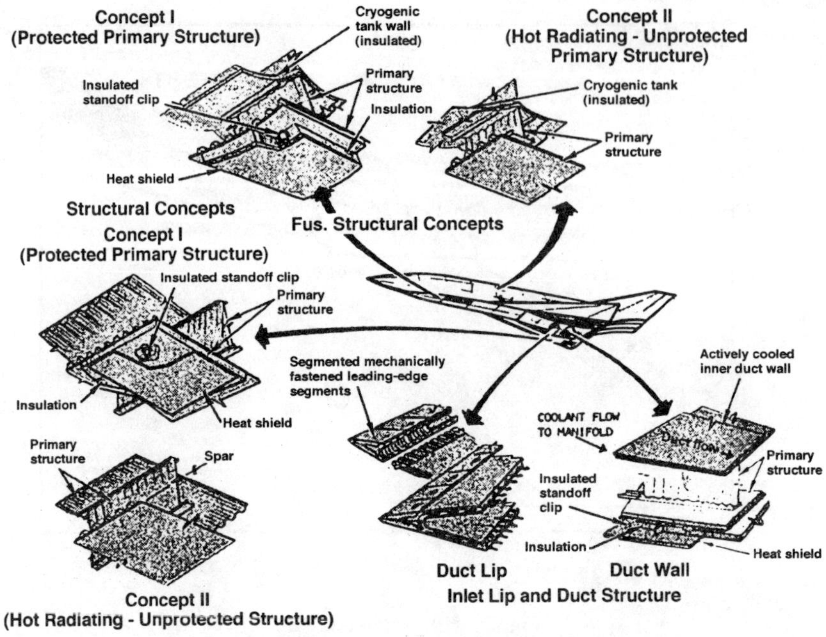

Fig. 3 HST structural concepts.

external leading-edge components. Engine components will be regeneratively cooled. In view of lessons of the past, now is the time to acquire significant test data and knowledge of the superalloys on strength throughout the entire temperature range, welded strength, fatigue, stiffness, thermal stress, toughness, stability, cost, availability, producibility, formability, corrosion, weldability, and brazability. Advances in diffusion bonding of titanium have also been applied to the superalloy, Rene 41. We recommend technology programs in this area.

Aerothermoelasticity

For guidance in the important subject of aerothermoelasticity, we might look to "A Survey of Aerothermoelasticity" by I. E. Garrick in the January 1963 *Aerospace Engineering.* Garrick discusses rigid-body aerodynamics, aerothermodynamics, thermoelasticity, static aeroelasticity, dynamic aeroelasticity, vibration, and dynamic aerothermoelasticity, encompassing 10 interacting areas which he surveys.

The field has produced other experiences that must also be applied in formulating structural technology programs. The XB-70A, for instance, represents a complete flying laboratory that has experienced modest temperature aerothermoelasticity of these interacting factors.

Similarly, the X-15 research airplane provides the only true-environment flying laboratory for an HST system. Examples of important findings already obtained through the X-15 program include:

1) Control-surface torsional rigidity being designed by thermal strain during heating, rather than only at high temperatures, as had been originally anticipated.

2) Panel flutter influenced significantly by Mach number; additional real-environment flight tests could be made for panel length-width ratio, skin thickness-length, and modulus of elasticity effects.

3) Panel flutter influenced by temperature ratio, as shown by wind-tunnel tests.

Structural-Testing Approach

The projected size of the HST probably precludes it from being tested as a whole in any single high-temperature structural test facility. But both the facilities at Wright–Patterson Air Force Base and at NASA Flight Research Center could be expanded for an HST test program.

Small components would and should be tested first, and the components tested in the final programs should be the largest possible. Although less than ideal, this approach has been successfully used in the XB-70A and X-15 programs.

Since the X-15 has been acknowledged to be "the most successful research airplane of all time" and since little, if any, structural damage was experienced in flight testing, the structural-testing approach of that program deserves review and detailed study in planning HST work. Major components tested included wing box, wing-leading-edge segments and panels, wing-fuselage A-frame, front spar, horizontal-tail box, curved panels, and fuselage specimens.

Following this approach, the HST vehicle could also be broken down, in a systems engineering fashion, into major components such as fuselage, wing, controls, vertical surfaces, and landing gears. The fuselage would be broken down further, for example, into forward fuselage and crew compartment, passenger compartments, fuel compartments, rear fuselage, and interconnecting structure for rear fuselage and vertical tail. The breakdown would proceed to as many tiers as the structural designers have required for component and subassembly fabrication and also to the depth necessary to perform the necessary structural tests.

At this point, the structural designer faces difficult questions. How many components, subassemblies, should be tested? In what facilities? How can we compress the testing? This suggests the key role of some central management.

Moving the technology forward in a concerted manner requires, we believe, program management through some appropriate agency. Without a central management, advances in structural technology, and other technologies as well, probably would come randomly and would lack pacing and force.

Concluding Remarks

HST speeds and altitudes mean exceptional reliability requirements for the structure. The final structural-testing procedures must have the unequivocal confidence of the nation's decisionmakers. This suggests that research vehicles will have to test full-scale structural components in the real environment for representative cruising times. The road to that goal requires program planning now.

References

[1] Pietrangeli, G. J., and Nice, E. V., "The Feasibility of a Mach 7 Transport Employing Airbreathing Propulsion Systems," Society of Automotive Engineers, 427C, Oct. 9–13, 1961.

[2] Weber, R. J., "Propulsion for Hypersonic Transport Aircraft," NASA TM X-52025.

[3] Gregory, T. J., Petersen, R. H., and Wyss, J. A., "Performance Tradeoffs and Research Problems for Hypersonic Transports," AIAA Transport Aircraft Design and Operations Meeting, Seattle, WA, Aug. 10, 1964.

[4] Petersen, R. H., Gregory, T., and Smith, C., "Some Comparisons of Turboramjet-Powered Hypersonic Aircraft for Cruise and Boost Missions," AIAA/RAeS/JSASS Aircraft Design and Technology Meeting, Nov. 15–18, 1965.

[5] Weber, R. J., and Dugan, J. F., Jr., "Methane Fueled Propulsion Systems," AIAA Preprint 66-685, AIAA Second Propulsion Joint Specialist Conference, June 13–17, 1966.

Engineering for the High-Temperature Age of Flight

Frederick L. Bagby and Associates
Battelle Memorial Institute

Elements involved in the high-temperature engineering problem represent a degree of complexity seldom encountered in modern-day technology. And increased complexity generally means increased uncertainty in the detail engineering of components and systems. This, in turn, leads to heavy reliance on experimental methods to assure that desired design objectives are achieved.

Although research is gradually reducing unknown areas of high-temperature environments, much still remains to be determined. High temperatures and high rates of heat transfer, the possibility of chemical surface reactions, and complex stress patterns, all occurring simultaneously, represent service conditions difficult to duplicate in the laboratory. Laboratory studies frequently involve modeling procedures, a technique not always applicable for the study of the response of materials to thermal environments because properties such as specific heat and coefficients of expansion are constants.

Instrumentation Is Always a Problem

Another limitation to laboratory studies is the always present problem of instrumentation. Some of the temperatures and transient conditions involved cannot be measured with the usual types of instruments. Full-scale operational testing, thus, remains the only completely positive means of evaluating designs and for determining the exact nature of the high-temperature environments of interest. Unfortunately, full-scale testing also represents a slow and costly procedure for obtaining technical data, especially when flight testing is involved.

Evaluation of materials for high-temperature service requires simulation of at least four environmental conditions: temperature, rate of heat transfer, chemical conditions at the surface, and imposed forces such as viscous shearing forces in a boundary layer. Among the high-temperature testing devices in use or contemplated for investigating the reaction of materials to one or more of those conditions are the following:

- Solar furnace. Can provide temperatures up to 6000°F and heat fluxes approaching 2000 Btu/ft^2 s. It is useful for studying small material samples where heat flux and surface emissivity characteristics are of primary interest but is not adaptable to tests requiring high-velocity gas flows.

Astronautics, Vol. 2, No. 9, Sept. 1957, pp. 46–51.

• Electric resistance and induction heating. Can heat materials to melting points with ease and rapidity but does not simulate conditions where material is heated from an external source, such as high-temperature gas flows.

• Rocket combustors. Service conditions in combustor and nozzle of all types of operational rockets can be simulated closely in small-scale units. Also, throat areas of rocket nozzles are adaptable for simulating erosion and oxidation conditions for materials intended to withstand aerodynamic heating rates up to and exceeding 3000 Btu/ft^2 s. Heating rates can be varied to simulate variations in aerodynamic heating rates of flight vehicles having specified mission profiles.

• Combustion jets and tunnels. Useful for testing materials in high-temperature, high-velocity gas streams. While it provides lower heat fluxes than those obtainable in a rocket combustor throat, it can simulate oxidation environments of free flight. Specific heat ratio of air can be simulated at any given temperature.

• Electric arcs. Show promise as a dynamic high-temperature testing device. Temperatures of over 10,000°F are obtainable with high heat fluxes. However, they cannot simulate the chemical or gas velocity environment of aerodynamic heating conditions in their present stage of development.

• Plasma jets. Provide extremely high-temperature flows of ionized gases. Heat fluxes of at least 2000 Btu/ft^2 s can be obtained, and temperatures of over 20,000°F are theoretically obtainable. Jets are limited for aerodynamic-heating studies in the present state of development because of the low kinetic energy of the discharge gas flow.

• Shock tubes and tunnels. Can simulate hypersonic gas flows, including temperatures and chemical characteristics, very closely. Their deficiency from the standpoint of material testing is the extremely short duration of test periods, which last only microseconds. Also, instrumentation requirements are extremely difficult to meet.

• Electromagnetic accelerators. Small material test specimens can be suspended in an electric field within a tube containing air or other gases. Accelerated through the long tube, a particle can reach hypersonic velocities. Accelerators provide close approximation to free-flight conditions, but considerable development is required to make such devices practical for material studies.

• Pressurized ballistic ranges. Free-flight conditions and Reynolds numbers can be simulated exactly by firing projectiles from special hypervelocity guns down a pressurized ballistic range. Adaptability to material studies is limited.

• Heated air jets. Ceramic storage bed heaters provide pressurized air jets with temperatures as high as 4000°F. This device is useful for studying material properties in high-temperature air, although heat fluxes are moderate.

• Wind tunnels. Most expensive of all test facilities, the wind tunnel is the only type that can simulate free flight closely. Large hypersonic blowdown tunnels, using ultrahigh-temperature preheated air and capable of simulating temperature and velocity variations of a vehicle flight profile, appear feasible.

Brief mention must be made of a special high-temperature testing situation. This is the use of atomic fireballs as high-intensity heat sources. Such research has been carried out and should provide a unique insight into material properties under the peculiar transient conditions of an atomic blast.

New Measuring Technique Needed

As previously mentioned, measurement of temperatures is a problem common to all types of high-temperature research. Thermocouples and radiation pyrometers are suitable for measuring material temperatures in many situations. But thermocouples are limited to temperatures under 3000°F. Another technique is needed when higher temperatures are involved.

The use of radiation pyrometers is acceptable where geometrical considerations of the test set-ups do not handicap use of the technique. Spectroscopic methods can be used with good accuracy to measure temperatures above 3000°F, but the technique is cumbersome, and extensive calibration procedures are required. A calorimeter technique, where a high-temperature gas sample is mixed with a known low-temperature sample, serves well in many conditions.

The list of devices for high-temperature, high-velocity gas flow research is lengthy. Most of the devices, however, are unable to simulate exactly all aerodynamic-heating conditions. Even wind tunnels and shock tubes are limited in their adaptability to the study of rarefied gas dynamics problems typical of ultrahigh-altitude flight.

Properties Need Careful Study

In the final analysis, material properties will play a major role in the era of high-temperature aeronautical engineering. A thorough understanding of the characteristics and limitations of a wide variety of materials will be required on the part of engineers if they are to achieve reliable, lightweight powerplants and airframes. Ingenuity and skill in exploiting particular characteristics of a material will be extremely important. Where material properties, such as melting points, are fixed, the problem is mainly one of learning to live within the basic physical limitations of a given material.

The standard materials ranging from aluminum alloys through high-temperature alloys continue to be important, with increased interest now being shown in the properties of these materials over their entire temperature range. New structural forms of some of these materials, such as sandwich construction, are arousing interest because of advantages offered under certain high-temperature conditions.

Development Is Under Way

Significantly, there is a growing interest in previously neglected metals and in metallic protective coatings. Alloys of columbium, tantalum, and tungsten show promise for the temperature range just above 2000°F but must be protected against oxidation. Intensive development of many of these new materials and coatings is just beginning.

Ceramics and cermets show promise as structural materials with working temperatures considerably higher than those possible with metals. The crystalline ceramics hold promise both as structural materials and as insulative or ablation coatings for the intense heating conditions experienced by re-entry bodies. Amorphous or glassy ceramics appear to have logical applications as radomes and transparencies when plastics prove inadequate. Cermets are the transition between metals and ceramics with brittleness limitations.

Improvement Appears Likely

Although used in small quantities, rubbers and plastics represent a limiting factor on high-temperature progress. Substantial improvement in the working temperature levels of some of the plastic and rubber materials appears likely, however, particularly if fundamental data are obtained on the oxidation characteristics of these materials. In this category of materials, the short-time, high-temperature exposure characteristics are substantially better than the long-time exposure characteristics. And, despite their temperature characteristics, plastics do have potential value as insulation and as ablating coatings for resistance to intense aerodynamic heating.

A look at the future of high-temperature technology as related to the field of aeronautics reveals the vital need for a new type of engineering technology. The aeronautical engineer can no longer get by on a nodding acquaintance with science. Instead, the chemist, physicist, metallurgist, and engineer will have to work together intimately to solve the high-temperature problems that now exist, as well as those that can be foreseen.

Research in this area is moving at a rapid pace, and some results, in terms of new material concepts, are now becoming apparent. One example is the substitution of metals for carbon in polymers, giving rise to a family of organo-metallic compounds potentially capable of withstanding much higher temperatures than present fluids, rubbers, and plastics. Another example is metal laminates made of layers of different metals using powder metallurgy techniques and ultimately worked into forms and shapes in a manner similar to current metal-working techniques.

Finally, basic research into the physics of the solid state promises to provide ways of improving properties of materials at all temperatures. Understanding of the ways in which atoms arrange themselves in crystals and of how to control these arrangements may provide answers to the present inconsistencies in material properties. Such knowledge can also pave the way for the development of custom-made materials. Cermets and materials compounded of rubber and plastic are the first attempts to develop materials of mixed characteristics.

But this new concept of materials technology is still in its early stages. Its full potential, far from fulfillment, can scarcely be estimated at present.

It is already providing us with new materials and new techniques which help us come closer to a solution of many of the problems associated with the high-temperature environments that will be encountered by tomorrow's vehicles.

There can be little doubt that further development of this new kind of engineering technology will, in the not too distant future, provide us with components and systems for vehicles with high-temperature capabilities well beyond those that exist today.

High-temperature materials for aeronautical applications

Type of material	Temperature range, °F	Applications	Manufacturing processes and availability	Comments and limitations
		Metals and alloys		
Aluminum alloys				
7075 T6	<250	General airframe structure	Available as sheet, forgings, extrusions	Little chance of increasing temperature range except in special case of sintered aluminum powders
2024 T4	<300	General airframe structure	Available as sheet, forgings, extrusions	
Magnesium alloys				
HZ32, EZ33, EK41	<500	Wheels, miscellaneous parts	Available as castings	New alloys may be satisfactory up to 800°F
HK31, HM31	<600	General airframe structure	Available as sheet, forgings, extrusions	
Titanium alloys				
6Al-4V	<500	General airframe and engine structure	Available as sheet, forgings, bar	Titanium presently in development stage; future alloys may be useful up to temperatures of 1200°F
7Al-3Mo	<700	General airframe and engine structure	Available as sheet, forgings, bar	
5Al-2.5Sn	<800	General airframe and engine structure	Available as sheet, forgings, bar	
Stainless steels				
Precipitation hardening	<700	Airframe skins and structure	Available as sheet and plate	Future alloys may be useful to 900°F
Martensitic	Up to 900–1100	Airframe structure	Available as sheet and plate	Future alloys may be useful to 1200°F
Austenitic	Up to 1200–1400	Powerplant structure and components	Available as sheet and plate	Development to 1500°F appears as limit
High-strength steels				
AISI 4130 and modified 4000 series	<400	Fittings, machine elements	Available as plate, bar, forgings, castings	Future alloys may be useful to 700°F
Die and tool steels	<1100	Fittings, machine elements	Available as plate, bar, forgings, castings	Future alloys may be useful to 1400°F
High-temperature alloys				
Wrought nickel base Cast nickel base	<1600	Combustion liners, turbine blades	Available as sheet, forgings, castings	May be useful to 1800°F
Wrought cobalt base Cast cobalt base	<1650	Combustion liners, turbine blades, airframe structure	Available as sheet, forgings, castings	May be useful to 1800°F
Molybdenum	<2200	Combustion liners, turbine blades, airframe structure	Available as plate, bar, and sheet	Subject to rapid oxidation; no commercial protective coatings available
Special purpose metals				
Chromium alloys	<2000	Engines and airframes	Experimental alloys not commercially available	Lack room temperature ductility; no coatings required
Beryllium	Up to 1500 (without protective coating)	General airframe structure	In laboratory development stage only	Raw materials extremely expensive

(*cont.*)

High-temperature materials for aeronautical applications (*continued*)

Type of material	Temperature range, °F	Applications	Manufacturing processes and availability	Comments and limitations
		Metals and alloys (continued)		
Columbium	(see applications)	These metals in alloy form all show promise for general structural application in the temperature range above 2000°F	Has been fabricated as bar, rod, slab, and strip; raw material reserves improved in recent years	Development just beginning; requires protective coatings
Tantalum			In development stage	Requires protective coatings
Tungsten			In development stage	May ultimately be usable at over 3000°F; requires protective coatings
Copper	<1900	Heat sink structures	Available in any form	—
Metallic protective coatings				
Aluminum	<1800	Coating of gas turbine buckets	Applied as a thin, diffused coating	—
Chromium	<1200	Protection of oxidizing metals	Electro or vapor deposit	Resists erosion and oxidation
Nickel	<1200	Protection of oxidizing metals	Electro or vapor deposit	Resists erosion and oxidation
Diffused Cr and Ni layers	(See comments)	Protection of oxidizing metals	Electroplated coating	Excellent oxidation resistance; 1000 h at 1800°F; 500 h at 2000°F; 100–200 h at 2400°F
Cobalt tungsten alloys	<2000	Protection of molybdenum	Electroplated coating	High hardness and good oxidation resistance in slightly oxidizing atmospheres
Nickel tungsten alloys	<2000	Protection of molybdenum	Electroplated coating	
		Ceramics and cermets		
Crystalline ceramics				
Conventional porcelains	<2200	Protective coating on metals	Sintered shapes	Generally have lower strength than oxides but are cheaper to produce
Highly crystalline porcelains	2200+	Protective coating on metals	Sintered shapes	Use less binder and are more refractory than conventional porcelains
Ceramic oxides	<4500	Protective coatings and thermal insulation of high-temperature structures	Sintered shapes, extrusions, flamesprayed coatings, preformed shapes, moldings	Classification includes oxides of zirconium, aluminum, magnesium, etc.; may be useful as ablation-type coatings
Graphite	<6000	Structural shapes	Molded or extruded shapes subsequently baked, machinable	Strength increases with temperature to about 4600°F; promising structural material if protected from oxidation

(*cont.*)

High-temperature materials for aeronautical applications (*continued*)

Type of material	Temperature range, °F	Applications	Manufacturing processes and availability	Comments and limitations
		Ceramics and cermets (continued)		
Reduced ceramics	<7000	Structural shapes	Bonded or sintered shapes, hot pressings	Classification includes carbides, borides, nitrides, etc., all subject to oxidation; SiC is best and cheapest where resistance to high-temperature oxidation is required
Amorphous or glassy ceramics				
Vitreous enamels	<1500	Protective coatings	Spray coatings	—
Glass-bonded ceramics	<1500	Radomes, structural shapes	Preformed shapes	These glass-bonded ceramic materials are limited primarily by the softening temperature of the binder used
High silica, low-expansion glasses	1500–2200	Radomes	Preformed shapes	
Fused silica	2000–3000	Radomes, windows	Moldings	
Cermets				
Metal-bonded ceramics	1400–2700	Protective coating, turbine blades, valves	Powder techniques, sintered shapes	Brittleness intermediate between metals and ceramics; may extend temperature range of superalloys; many require protection from oxidation
		Plastics and Rubbers		
Plastics				
Thermoplastics				
Polyacrylics	Up to 215–275	Windows, canopies, dials, luminated panels	Parts fabricated by injection molding or extrusion; availability good	—
Polyamides (Nylon)	Up to 325	Gears, bearings, cams, bushings, coatings for metal parts, fibers	Injection molding; availability good	—
Polyethylene	Up to 170–250 (depending on type)	Cable insulation, pipe, valves	Injection molding or extrusion; availability good	—
Polystyrene	Up to 180–225 (depending on type)	Housings, containers, insulation (foam)	Injection molding or extrusion; availability good	—
Polyvinylchloride	Up to 125–150	Pipe, cable insulation, tubing	Injection molding or extrusion; availability good	—
Fluorocarbons (Teflon, Kel F)	Up to 450–550	Electrical insulation, seals, gaskets, hose, high-temperature applications	Injection molding and extrusion; availability good, but expensive	—

(*cont.*)

High-temperature materials for aeronautical applications (*continued*)

Type of material	Temperature range, °F	Applications	Manufacturing processes and availability	Comments and limitations
		Plastics and rubbers (continued)		
Thermosetting				
Amino plastics	275–400	Electrical outlets, housings, heat-resistant knobs, handles	Molding, impregnating; availability good	—
Epoxy	250–300 (for long-time service) up to 1000 (for short-time exposure)	Electrical components, aircraft structural parts, wing sections (reinforced with glass fiber), high-temperature adhesives for fabricating honeycomb and other aircraft structures	Molding, casting, laminating; availability good	New catalyst systems for improved high-temperature use are under development
Polyesters-polyurethanes	400–500 for long-time service	Radomes, high-strength structural parts (same as for epoxy)	Casting, impregnating; availability good for polyesters, fair for polyurethanes	Development for high-strength, high-temperature use
Phenolics	250–350 (for long-time service) up to 1000 (for short-time service)	Laminates and asbestos-filled gaskets for high-temperature service, stabilizers	Laminating, molding; availability good	—
Silicones	−90–480, up to 1000 (for short-time exposure)	High-temperature insulating varnishes, ozone-resistant coatings, gaskets, seals	Laminating, coatings; availability fair (high price)	Rapid curing of resins
Elastomers				
General purpose				
Natural rubber		Tires, insulation, and miscellaneous noncritical parts	Extrusion, compression molding; availability good	Synthetic natural rubber
GR-S	−65–200 350 (for very short service)	Tires, insulation	—	—

(*cont.*)

High-temperature materials for aeronautical applications (*continued*)

Type of material	Temperature range, °F	Applications	Manufacturing processes and availability	Comments and limitations
		Plastics and rubbers (continued)		
Polybutadiene rubber		Tires, insulation	—	High temperature
Butyl rubber		Tires, tubes, insulation	—	Butyl tires
Oil- and fuel-resistant rubber				
Nitrile	–65–250 (for extended use)	Gaskets, seals, hose	Extrusion, compression molding; availability good	Improved oil-resistant types
Neoprene	350 (for short service)	Gaskets, seals, hose	—	—
High-temperature rubbers				
Fluorocarbons	–50–450 (extended service) to 600 (for short service)	Fuel-resistant gaskets and seals, chemical resistance for peroxides, nitric acid	—	Improved types having higher tensile properties
Silicones	–100–500 (extended service) 800 (for short service)	Seals, low- and high-temperature gaskets, insulation	—	—
Polyacrylates	–50–350 (for extended service) 400–450 (for short service)	Seals, gaskets	Extrusion, compression molding; availability poor	—
Other special purpose types				
Hypalon	Up to 300	Ozone resistance	Availability fair	—
Polyurethanes	200–250	High-abrasion resistance	Molding, casting; availability poor	Oil, fuel, and high-temperature resistance

Unique Structural Problems in Supersonic Aircraft Design

F. P. Mitchell
Avro Aircraft Limited

Introduction

The extremely rigid geometric requirements demanded by supersonic aircraft have greatly increased the complexity of our future aircraft. It has resulted in a large increase in technical manpower required, with a resulting high cost. Very thin wings, low-aspect-ratio wings, smaller diameter fuselages, high control surface loads, and high pressures and temperatures form part of the aerodynamic requirements which must be met. The job facing the structures engineer is to produce a structure meeting all the rigid aerodynamic requirements but with a weight lower than that of preceding aircraft structures. The task is both difficult and challenging, requiring large groups of engineers working in close cooperation. The work is characterized by rational analysis, rather than by broad assumptions which in the past have been satisfactory. Let us examine the major structural problems separately.

Load Analysis

The problem of analyzing complex low-aspect-ratio wings and fuselage combination must be solved for on a rational basis. The fuselage or center section can no longer be assumed to be a rigid support for the wing but bends under load, thereby influencing the load distribution in the wing to a large extent. The fuselage is then an elastic support. The displacement of the elastic support is dependent on the magnitude of load in the fuselage which is, in turn, tied into the load in the wing (Fig. 1).

It is evident that the fuselage-wing combination must be analyzed as a complete complex redundant structure. The days when we could analyze the wing separately from the center section are past. Then, the problem facing the structures engineer was clear, the solution to the problem not so clear.

There are different theoretical methods of approach to the problem—for example, the method advocated by D. Williams in RAE *Structures Report 156* which concentrates on solving for deflections. This method is certainly satisfactory for aeroelastic problems, but it does not satisfactorily produce stress distributions; to work from deflections to stress is at best inaccurate and at the worst impossible. Stress concentrations of a local nature cannot be sensed by deflections; however,

Aeronautical Engineering Review, Vol. 15, No. 1, Jan. 1956, pp. 34–39.

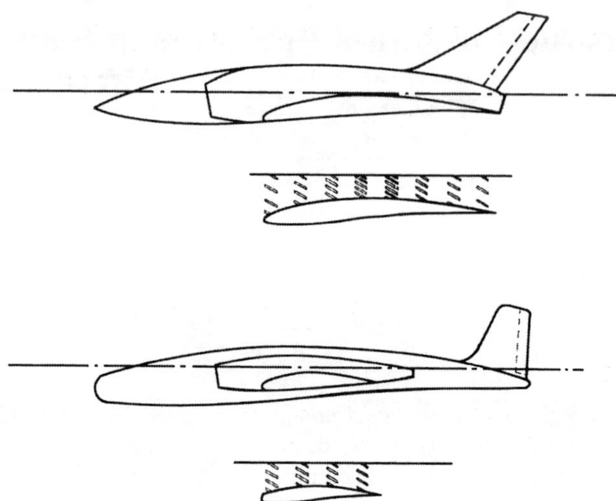

Fig. 1 Aircraft body distortion.

it is common knowledge that quite crude assumptions concerning the stress distribution yield displacements which favorably compare with experiment.

Hence, it was decided to calculate directly for stresses and to compute displacements from known stresses. The wing was considered to act as a plate (a two-dimensional problem) rather than as a beam (a one-dimensional problem) as is customary for straight or swept wings with relatively small taper. In brief, the method of analysis assumes an approximate moment distribution so as to satisfy the equilibrium conditions exactly while leaving strains still incompatible. This assumed distribution was corrected by several properly chosen self-equilibrating groups of internal loads akin to the stress functions of the Southwell theory of plates, their magnitude being determined by the Castigliano theorem. By solving a larger number of linear equations, the stresses were obtained and the deflections computed. The problem was set up in a matrix form. Air loads were applied as a large number of concentrated loads. Due effect of taper, oblique panels, and Poisson's ratio was included. The use of high-speed digital computers was essential; even then, it was not apparent that the computation facilities were reliable enough to allow the handling of matrices of the order 173 by 173.

Versatility was absolutely essential in such a large analysis; it must be a tool for the structures engineer to use (too often such an analysis can become a burden).

The first step to meet this requirement was to solve for the unknowns on a unit-load basis. With this type of solution and with the use of high-speed digital computers, it was possible to obtain the stress distributions and shapes with extremely little effort for a multitude of aerodynamic cases.

The characteristic of a matrix solution and a unit-load solution also allows the use of self-equilibrating loads to represent changes of structure stiffness. There are obvious limitations to this type of repair, and complete versatility in relation to design changes is not possible. This weak point had to be remedied in some way. The only solution was to carry on an independent analysis, using proved methods of analysis but with broad assumptions to set closely the main structural members. This is where general engineering skill and experience must play important parts.

It has been found that the two types of analysis complement each other and provide a good check. This approximate analysis was also necessary to produce detail stress distributions within the framework of the matrix.

Supersonic aircraft employing thin wings are subject to large distortions under flight loads. This structural distortion under load has been recognized as an additional hazard to structural integrity and has necessitated its acceptance as a major design requirement.

There are several aspects concerning shapes which must be considered—for example, secondary or induced stresses into fuselage frames due to bending and twisting of the wing.

Because of compression stresses in the upper surface and tension stresses in the lower surface, the wing takes a well-defined shape. A fuselage former attached to this wing will distort in some manner. Since the former is extremely light compared to wing skins, the restraining loads on the wing are extremely light, but the stresses induced into the frame can be rather large. It is possible to reduce the magnitude of the induced stresses by careful design.

Figure 2a shows the worst possible arrangement. This type of arrangement could cause failure of the former at quite low wing stresses. Figure 2b shows an important design improvement. The induced stresses are still significant, however. Figure 2c shows an optimum arrangement.

Assuming that all possible efforts have been made to reduce the induced stresses, the structures engineer is faced with the problem of making a frame good for static loads as well as induced loads. Since the induced stresses are a direct function of the stiffness of the frame, there is no easy solution. The objective is clear, however; the stiffness must be kept as low as possible, and the allowable stresses must be kept high. A trial-and-error method was found to be most satisfactory. A frame was sized for static loads and the stiffness calculated. The induced stresses were then calculated and superimposed on the statically determined stresses. This operation was repeated until a satisfactory set of stresses was obtained. Accurate stiffness calculations and control of stiffness are absolutely essential. Increased stiffness of the frames due to poor analysis or manufacturing tolerances would cause an increase of loads into the frame which would cause premature failure of the attaching structure.

A more difficult analytical problem regarding shapes is the transition problem between a deflecting structure as shown in Fig. 2 and a more rigid type of former that one would find forward of the wing. Since it was not possible to make the solid-type of fuselage former flexible enough to absorb the distortion, it was necessary to use specially designed frame or frames to reduce distortions to an acceptable value. These frames were designed almost entirely from the stiffness point of view, and by use of them it was possible to control the distorted shapes.

One thing is clear: The structures engineer must have a complete grasp of stiffness and distorted shapes as he has in the past controlled strength. This has increased by several times the technical man-hours required.

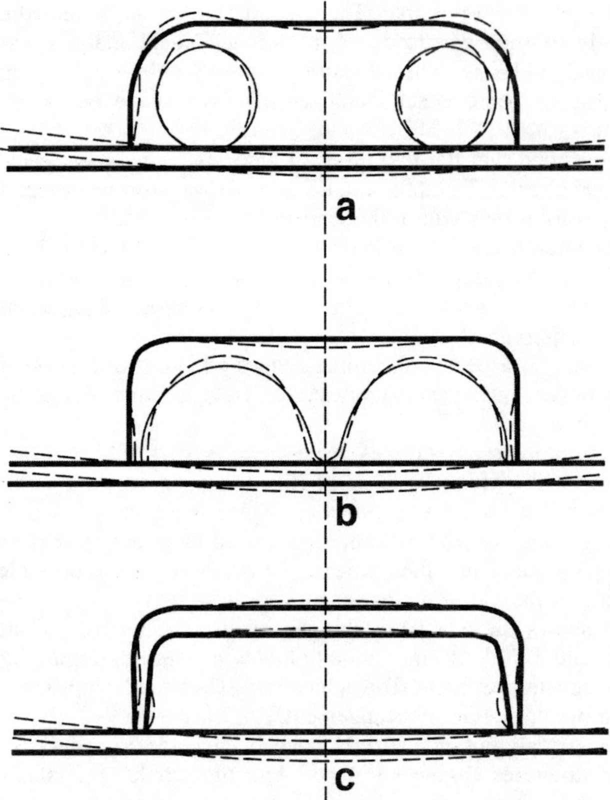

Fig. 2 Distortion of frames attached to a deflecting wing.

Fatigue

Fatigue problems will continue to receive more attention by the structures engineer. It is generally agreed that one cannot afford a weight penalty to improve the fatigue life of a fighter aircraft. It is, however, important that the best fatigue-life-to-weight ratio be obtained and that the structure have a uniformity of fatigue life—i.e., there should be no point where the stress concentration factors greatly exceed the average factor.

This latter requirement has caused some interesting problems in relation to thick wing skins on supersonic aircraft. A particular typical spanwise lap joint has been proved to be particularly bad from a fatigue point of view. The offset of load from the skin to the joint plate causes extremely high local stresses in both plates. Fatigue testing was considered necessary earlier in the design stages of the aircraft in order to obtain some design allowables for these joints. In a joint that has been high loaded, the distortion of the plates and the elongation of the bolt in the holes are apparent.

As was suspected, the joints were proved to be unsatisfactory in the highly loaded areas of the wing, and it became necessary to investigate alternative types of joints.

Figure 3 shows the result of this investigation. Type A is a normal lap joint. Type B uses a high heat-treat steel strap just below the skin. This method did not improve the offset but did put the steel in the high stressed area. Since steel has a much higher endurance limit than aluminum, there was a significant improvement in the fatigue life of the joint.

Type C is, however, the optimum arrangement. A thin steel strap is used externally by machining a step in the skin. This arrangement reduces the offset to a minimum and provides satisfactory fatigue characteristics.

The necessary use of large high-strength forgings in the thin wings and fuselages has produced a serious fatigue problem. The characteristics of this material differ greatly in the longitudinal grain, transverse grain, and short transverse grain. Careful control of stresses, especially in the short transverse grain direction, is essential in providing a satisfactory fatigue life.

Careful detail design and detail stress analysis are probably the most important single factors in providing a satisfactory fatigue life. Detail stress analysis of complicated fittings is difficult and time consuming, and although a large amount of work may not be justified to provide static strength, it is necessary to provide satisfactory fatigue life.

However, analytical work will not replace fatigue testing. Too often, the results of fatigue testing do not agree with theory. Relatively insignificant parameters may greatly affect the fatigue life of an aircraft component.

Early and continued fatigue testing and development is a major aspect in providing structural integrity.

Internal Pressures

Supersonic aircraft are literally flying pressure vessels. In the past, the only significant pressures from a structural point of view have been in the cockpit and fuel tanks. Now, the complete aircraft structure is subject to variations of pressure and depressions of quite high magnitude.

This fact has influenced the structural design greatly, and a great deal of ingenuity has been required in order to handle the problem with a minimum of weight penalty.

The analytical questions have not been severe, although the problem has proved to be particularly interesting in several respects. It has opened a completely new field in structural research aimed at providing the optimum strength-to-weight ratio. A considerable amount of research has been done on optimum structures using varying parameters of end load, shear, and torque. The addition of pressures into the set of equations has a pronounced effect on the final configuration of the optimum arrangement. Since the amount of weight involved is large, obtaining the optimum arrangement was absolutely necessary.

Analytically, the stability of distorted depression vessels proved to be the most significant problem.

The problem is primarily a detail one, and for that reason it is impossible to cover it in this paper in a satisfactory way. It has been mentioned primarily

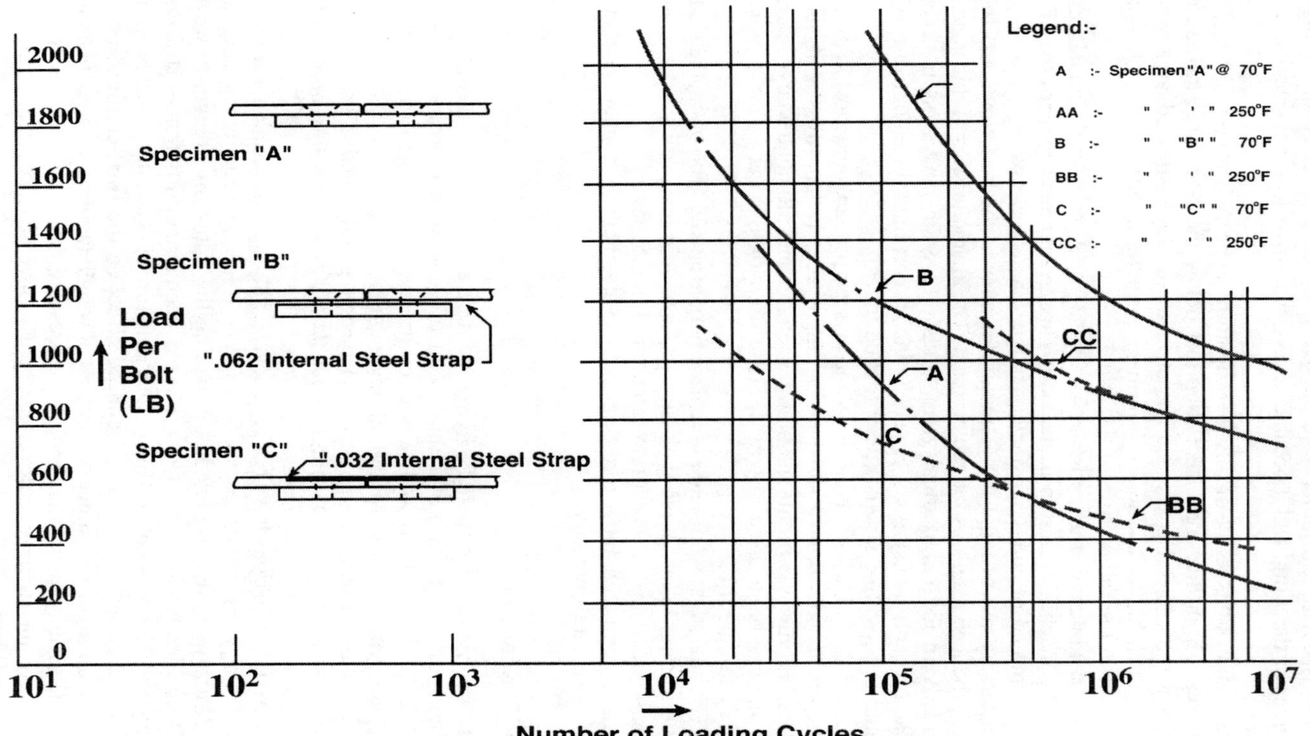

Fig. 3 Repeated loading of $^1/_4$-in.-diam NAS C'SK bolted joints.

because it is an additional aspect that has greatly influenced the structural design of supersonic aircraft.

Landing Gear

It would be reasonable to suspect that landing gear would not present any unusual problem in supersonic aircraft. This is, however, far from the truth. The exacting geometric requirements complicate the undercarriage problem tremendously—not only in mechanical design, but also in load analysis.

The gear must be stored in a limited space envelope; it must cut the very minimum of prime structure, and, therefore, it cannot be properly braced. It must be long enough to allow high angles of attack for takeoff and landing of the aircraft, and, yet, it must be capable of bringing to rest heavy, high-speed aircraft on normal runways with the minimum of weight penalty.

The dynamic condition of a landing gear during wheel spin-up has been a well-known phenomenon for many years. However, in the past, it has been satisfactory to assume the gear as a cantilever beam fixed at the aircraft structure. Recommended magnification factors of ANC-2 are based on this type of gear.

The problem of the dynamics of an elastic gear mounted on an elastic structure is an enormously more complicated one. The method for calculation of natural frequencies, stiffnesses of *U/C* strut, and wing has been developed. An equivalent vibrating system of *U/C* strut-wing has been introduced with six degrees of freedom. The frequency equations for the system have been solved by interation method for the two lowest frequencies. The structural aerodynamic damping has been taken into account by multiplying the undamped responses by a calculated decay factor.

Design parameters also had to be altered to derive the optimum dynamic characteristics. Metering pins, spring-controlled valves, and geometric arrangement of fore and aft bogie can be used to alter the forcing function due to spin-up and, thereby, significantly alter the dynamic characteristics. Figures 4 and 5 show a plot of the displacement function with a corresponding forcing function. The variation of forcing function alters the displacement function considerably. Steps in the forcing function curve are characteristic of a fore and aft bogie where one wheel contacts the ground first and proceeds to spin-up before the second wheel hits. This has the advantageous effect of extending the time of spin-up. The analysis included dynamics of the gear-wing in the vertical direction, as well as in the horizontal.

The amount of technical man-hours required for this program is approximately 15 times that required for a more conventional gear in a low-speed aircraft.

Heat

Aerodynamic heating has been much publicized of late as a major structural problem in supersonic aircraft. A brief discussion of this problem, then, should be sufficient for this paper.

There are three main aspects of heat that must be considered.

Strength Properties

Elevated temperatures have the effect of reducing the allowable stresses and elasticity of all metals. This is not a difficult problem to cope with; however,

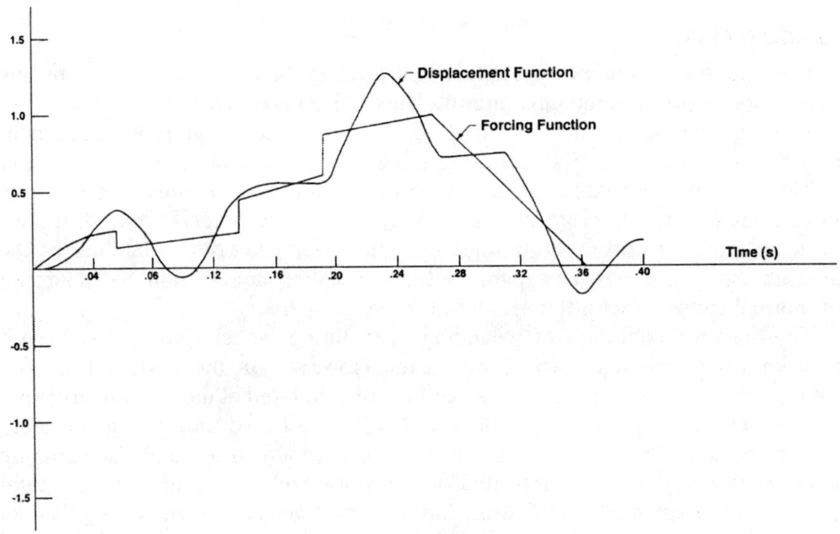

Fig. 4 Landing gear displacement function.

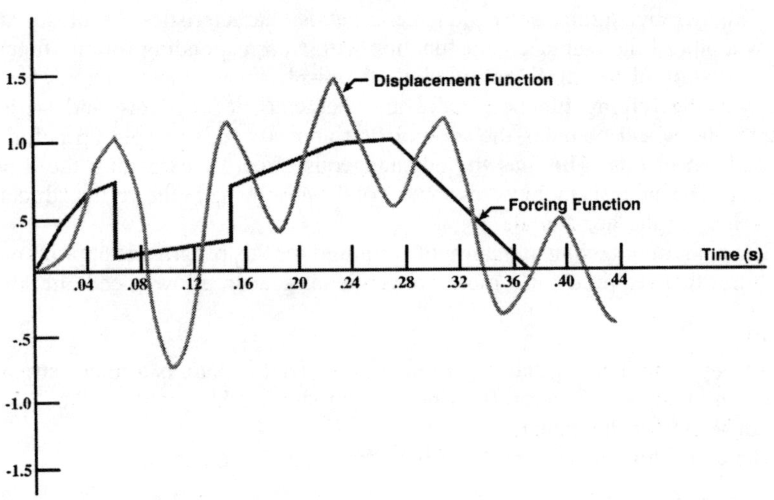

Fig. 5 Landing gear displacement function.

considerable laboratory testing is necessary to produce full sets of material properties at various elevated temperatures.

Transient Temperatures

The transient temperature problem is by far the most difficult that will have to be tackled. The flow of heat through a structure causes a heat gradient. This gradient causes induced stresses, as well as a change of shape of the structure. Careful design is necessary to minimize the adverse effects of this temperature gradient. Thermal insulation, conductors, composite materials, and unconventional structural arrangements can be used by the designer to improve or impair the flow of heat, depending on which is the most desirable.

A large volume of experimental data will be necessary, especially in relation to thermal conductance of joints. Special techniques and methods will have to be developed to apply large amounts of heat to the test specimen under extremely close time control.

Thermocouples and strain gages will be required to measure heat and strains on a continuous trace throughout the structure.

Creep

Creep has been listed separately here because of its importance in the elevated-temperature problem. This is particularly critical with high-speed aircraft where time at temperature is likely to be a significant feature.

New Materials

The great strides being made aerodynamically must be paralleled by advances in the metallurgical field. New and better aircraft structural materials are very much in demand. New alloys are being developed which appear promising, but the development of these materials is being viewed with considerable impatience by the structures engineer.

Superhigh heat-treat steels, high-strength magnesium, new high-strength forgeable aluminum alloys, and high-strength titanium alloys appear to be promising developments. If these materials can be made available in production quantities, the success and performance of supersonic aircraft will be improved considerably.

Aircraft production departments must also take active steps in the development of new materials. Each new material presents a multitude of special problems. Development of the necessary production techniques is an important phase in the complete acceptance of a new alloy.

Conclusion

This paper covers a few of the major aspects of supersonic-aircraft-structure problems—and in a very broad manner. The multitude of more detail problems in itself poses as a major problem that can be solved only by large numbers of well-trained technical personnel. The job of planning, coordinating, and supervising such large numbers of technical personnel is an imposing task. Last but far from least, it requires a great deal of money, not only for the analytical work, but also for the testing and development associated with it.

Structures for Space Systems

Structures That Adapt to Space

Ben K. Wada
Jet Propulsion Laboratory, California Institute of Technology

John Tracy
McDonnell Douglas Space Systems

Current and future spacecraft structures will require improvements in reliability, performance, testability, and cost-effectiveness. Unless a structural system meets these challenges, it probably will never be implemented in an operational system.

Large precision systems embody all these challenges. Dimensions may range from 10 to 100 m, which is larger than the launch vehicle shroud; thus, such a structure must be either deployed or constructed in space. The precision requirements within which a structure's dimensions must be maintained can range from 10 nm to tens of microns. These requirements call for maintaining precision in the space environment at various mission-unique temperatures; these range from sub-Kelvin to near room temperature for a 10–30-year mission duration. Composite materials allow tailoring of a structure's passive characteristics to meet specific requirements such as coefficient of thermal expansion, stiffness, strength, damping, and weight. But a totally passive system cannot meet the challenges of the future.

The recent incorporation of adaptive structures into the design of spacecraft may enable adjustment of the geometry and structural characteristics during operation. This adjustment is made either through remote commands or automatically in response to external adverse stimuli. Adaptive structures hold great promise for improving the design of spacecraft structures.

Rapidly advancing computational capabilities for predicting the response of structural systems are allowing more-effective development of design and hardware through concurrent engineering, optimization studies, and computer-aided design and manufacturing. These capabilities help to increase performance and reduce total development costs. The computational results' accuracy depends upon specific input parameters; examples include material properties, their variation with the space environment and with time, identification and definition of nonlinear parameters, and damping.

Because of the degree of uncertainty in input data that will always exist, analytical predictions can only be used to aid in the design and verification process.

There is little development of new test approaches for meeting future challenges. Very little experimental data are available on the submicron response of

Aerospace America, Vol. 30, No. 9, Sept. 1992, pp. 38–41.

a structure subjected for 10–20 years to the space environment or on dynamic response in the submicron range for validation of computational predictions.

Ground validation tests to establish the characteristics of structures in the micron range appear improbable because the Earth's gravitational field masks parameters of significance in space. The use of flight tests for obtaining data is expensive and time consuming and is often valid only for the particular flight configuration. Moreover, data from one structural system are difficult to apply to another even though the two may appear similar.

In the past four years, interest in adaptive structures has grown exponentially. Their use can be applied equally to both precision structures and other large systems; the difference lies in the characteristics of the actuators and sensors needed by each. Recent problems associated with Galileo and the Hubble Space Telescope reveal the difficulties associated with structures this large, and this new approach holds great promise for improving designs and reducing costs for such systems.

Adaptive structures became feasible with the development of actuators and sensors that had sensitivity on the order of the structural strains obtained during operation. These components, which are directly incorporated into precision structures, have nanometer resolution. Without power, such a structure functions passively, and, with power, the actuators impart a displacement that changes the structure's geometry or characteristics.

At present, ceramic piezoelectric, polymer piezoelectric, electrostrictive, and magnetostrictive materials have the resolution and bandwidth (0 to a few kHz) for actuation and sensing in precision systems. Noncontact displacement sensors have the resolution required for measuring displacements. Development of control circuitry on a small "chip" that can be directly integrated into the structure is currently progressing. Power supplies are also being miniaturized and space qualified to allow their integration on a spacecraft.

During the past few years, the use of active members integrated within laboratory structures proved the feasibility of adapting the structure at submicron levels. Truss-type structures have received emphasis because they are the configuration of choice for both deployable and constructable designs. Both the Jet Propulsion Laboratory (JPL) and MIT are working on adaptive structures for precision systems.

Currently undergoing tests at JPL are the Phase B structure, Precision Segmented Reflector structure, Focus Mission Interferometer structure, and MIT test structure. Experimental results show the feasibility of path length control of 10 nm, submicron control of static displacement, operational passive dampers at a few nanometer displacements, active damping at micron-level displacement, and use of active members for system identification. Both local collocated controls and distributed controls added damping to the structure. Special microdynamic test equipment at JPL enables static or dynamic testing of the actuators, joints, or other mechanical parts with resolution in the 2-nm range.

Martin Marietta completed its work on the Passive and Active Control of Space Structures program to develop technology in design and predictive capabilities for passive damping.

The ASTREX test facility at the Phillips Lab at Edwards Air Force Base will include embedded piezoelectric, in-line piezoelectric actuator, and magnetostrictive active members. This testbed rapidly slews the spacecraft through large angles and then points it quickly and accurately. Most of the structures described above are about 5 m in dimension.

Research is in progress at NASA Langley to develop teleoperated robotics systems for assembling truss structures one element at a time.

Another area of structures at JPL involves the adjustment of errors in two-dimensional surfaces used in optical benches or mirror surfaces. The purpose here is to maintain the critical dimensions of optical elements or to maintain surface tolerances. The feasibility of placing many small actuators behind a mirror to correct for small wavelength errors has been proven in the past three years. More recently, piezoelectric actuators bonded to the back of composite mirror structures have successfully corrected long wavelength displacement errors.

NASA and the SDI Organization have recently sponsored programs for using adaptive structures in space experiments and applications. Without this step, spacecraft designers will not develop confidence in the viability of this technology. Wright–Patterson Air Force Base and JPL jointly conducted a KC135 flight experiment on a 12-m truss boom with active members. This exercise demonstrated the simplicity of using piezoelectric actuators to add active damping without precise knowledge of the structural characteristics.

SDIO is developing small experiments and plans to fly them within the next few years. They include TRW's Adaptive Control Technology Experiment, which will attempt to show effective vibration suppression using members with embedded piezoelectric actuators and sensors. JPL's Cryocooler Vibration Suppression experiment will seek to suppress motion at the tip of the cold finger where the sensors are located. Two vibration suppression approaches on the cryocooler experiment include piezoelectric applique for canceling the motion of the cold finger and adding a piezoelectric system to control the rigid body motion of the sector of the cooler. The experiment will be flown on the UK Space Technology Research Vehicle.

NASA plans to fly two small experiments. In one, MIT's Mid-deck Active Control Experiment, a beam structure with two simulated instruments must be precisely pointed. Active members with piezoelectric actuators will provide active damping. In the other experiment, Jitter Suppression For Precision Space Structures, McDonnell Douglas will attempt to demonstrate the use of active and passive damping for suppressing vibration to meet precision pointing requirements. The reflight Wide Field Planetary Camera under development at JPL will seek to correct the prescription of the Hubble Telescope; within the Active Fold Mirror are electrostrictive actuators to actively adjust its angular position during its operational lifetime.

NASA and SDIO are planning other flight experiments. During several past attempts to develop a stand-alone structures flight experiment, growing cost estimates and flight schedules resulted in cancellation.

JPL, the University of Colorado, and Duke University are performing research on deployable/constructable systems, which require greater reliability. In many deployment concepts, length of members must be precisely controlled, and gaps in the joints must be eliminated to assure precise locations of the preselected nodes. Errors or changes resulting from environmental inputs can induce internal member loads during deployment and may lead to binding during that process.

Similarly, during assembly of indeterminate trusses, mismatch of member lengths can make installation of the member impossible.

During deployment or assembly, analysis establishes areas of maximum strain energy. By placement of a simple active member that extends with tensile load and contracts with compressive load, the degree of binding or internal structural loading can be eliminated or precisely controlled. The ability to include in the design additional active members as a backup enhances reliability.

Currently, a structure's performance depends on the uncertainty of the fixed dimensions of the structural members (or surface) and joint gaps during the system's operational lifetime. Incorporating adaptive structures into the design gives the ability to change the geometry, preload joints, and modify structural characteristics during that lifetime. This capability enhances the structure's performance.

The added expense of actuators, sensors, and local controllers makes it difficult to justify incorporating adaptive structures in terms of potential cost savings. But such savings do result from the lower costs associated with characterization of materials, fabrication, thermal control, analysis, and testing. Adaptive structures do not require the same initial degree of precision as do passive systems because the ability to make adjustments while the system is operating can compensate for uncertainties. Also, because the structure itself will place fewer demands on the control system, the requirements for controls can decrease significantly. The structural characteristics can also be modified to better fit the controller.

Possibly the most important feature of adaptive structures is that they provide a means for validating a design through ground tests. Current and projected limitations in ground test technology make many of today's structural systems difficult to validate on the ground, and future requirements will be still more demanding. The major current problem is that the Earth's gravitational field overwhelms significant parameters that are present in 0 *g*. For example, small joint gaps are preloaded because of gravity and behave as a linear structure on Earth. In space, however, these gaps can cause the system to respond randomly when subjected to a deterministic input. Controlling such a system is almost impossible.

Determining the static position of large precise structures on the ground is difficult; adaptive structures can relax the ground test requirements by 2–4 orders of magnitude. The ground test objectives are to establish the uncertainty of the structure, confirm that the performance requirements in space are within the uncertainty, and show that the range of active members covers these structural uncertainties. Unless the hardware can be validated by ground test, the system will never be adopted for flight because of the large uncertainties about its performance. In addition, the active members can be directly used as a force excitation source to establish the state of the structural system. Research at JPL is addressing the problems associated with ground test approaches controlling joint gaps, static position adjustment, and the need for structural redundancy.

Future spacecraft missions will require large precision structures for observing the planet Earth, other planets in our solar system, and objects in outer space. Still, without reliable, cost-effective structural systems that can be ground tested and meet performance requirements, the objectives will be descoped, the projects rejected because of risk and cost, and less desirable alternatives selected instead. Recent developments in adaptive structures promise to answer these future challenges.

New Economic Structures for Space in the 1980s

Klaus P. Heiss
Econ, Inc.

Over the past decade *Astronautics & Aeronautics* has presented many columns and articles expressing deep concern about U.S. R&D issues: on the erosion of R&D funding by government and industry, on education and training of manpower to conduct R&D, the uses of this limited national pool on large government programs vs industry efforts, and the increasing gap between the *awareness* of aerospace opportunities and the *pursuit* of them by government and industry.

Whereas the 1950s and 1960s represented an era of enterprise, expansion, risk taking, and growth, the 1970s devolved as a decade of retrenchment, cutbacks, monetary illusion (created by inflation), limited risk taking, and management by "stretch-out" rather than accomplishment. Funding of R&D has been cut back from over 3% of GNP in recent years, the reduction of 1% coming from cutbacks in *federal* funding of R&D and, thus, principally cutbacks in aerospace R&D.

In short, whether by accident or design, we experienced a period of *disinterested* policy, the consequences of which now show up in the strategic balance and changing patterns in aerospace world trade.

At the heart of this retrenchment was NASA and the U.S. space program. The vast successes of the 1960s—in manned spaceflights, space communications programs, and space science—created a foundation for the decades to follow, but the 1970s brought singular failures—by NASA and others—to capitalize on it. From an era of challenging goals, NASA wandered into indecision, confusion, lack of purpose, or *imposed* inactivity, mostly due to budgetary and fiscal constraints. Also, the notion was held by technological Luddites (for want of a better term) that the federal budget deficit and all other societal issues can be solved by cutting back aerospace programs.

Any other nation with the space-technology base of the United States of the late 1960s would have proceeded into the 1970s by making its space program a principal component of its technological, economic, and foreign policies. For two principal reasons the United States failed on this score:

- NASA was not given a challenging new space goal (space-transportation systems are means and not goals).
- The United States did not create a plan to put its vast market forces in pursuit of opportunities to exploit space on a worldwide basis. Both of these causes of failure have to be corrected.

Astronautics & Aeronautics, Vol. 19, No. 1, Jan. 1981, pp. 19–21.

No economic studies, however skillfully done, can break through this morass of institutionalized inactivity and passivity, this collective withdrawal. The U.S. space effort needs institutional restructuring.

First, NASA should be formally charged with creating a strong, large-scale technology base in geosynchronous orbit (GEO) by 1990. This goal should encompass scientific, technological, and applications objectives. The United States can meet this goal with an operational Space Shuttle. No other nation or group of nations can.

The Large Technology Base (LTB)—we will dignify the goal with capital letters—would be comparable in its technological challenge and achievement to the landing of men on the moon in the 1960s. It holds massive economic, technological, and scientific implications. It will sweep in an economic revolution of space and ground communications systems, switching capabilities in space (rather than on the ground), applications in both industrial and developing countries, information processing, evaluation and storage (''libraries'' in space), and operational support for missions by spacecraft in many other orbits (e.g., Earth-observing systems—Landsat, Seasat, weather satellites, etc.). Support to handle the massive flow of data from such activities can be done in space *at great economy* to individual space missions.

Establishing a new U.S. space goal, especially for NASA, will not in itself bring about the full development of U.S. space-program opportunities. We must restructure the very basis and mode of space operations and applications. We must bring about accountability in space programs, government and industry cooperation to the benefit of the U.S. economy, and the institutional setting to allow full private as well as government initiatives in pursuit of space opportunities.

Such a restructuring of NASA for the 1980s would re-establish it as a research and technology (R&T) development, rather than a space applications and operations, agency. Many people familiar with NASA consider the conflict between R&T and applications/operations a debilitating influence on the agency—a factor which has inhibited rather than spurred progressive thinking and programs in space.

NASA has uniquely been suited as an organization in pursuit of national technology goals. This capability the nation should use to its best advantage in the 1980s. However, major *system operations*, such as space transportation, space applications, and transfer of uses of space technology, have to be done in some other, more-accountable, and more-productive way. I think we need two new institutions—a *Space Transportation Company* and a *Space Bank* for space applications and technology-transfer projects.

With the advent of the Space Shuttle, the role of space-transportation services should be separated out of NASA, the Department of Defense, or *any other* governmental agency and transferred to a Space Transportation Company. The example of Comsat in space communications may furnish a useful reference on how to go about such a transfer. Although any such transfer is complex and, initially, will mean heavy government involvement and cooperation, at some future time an entity for space transportation should be spun off to industry or the investment public.

The roles of a Space Transportation Company may encompass operations of the Space Shuttle fleet; operations of the launch sites (ETR and WTR); space-transportation support functions; acquisition and operation of upper stages, including advanced OTVs; acquisition of additional Space Shuttle vehicles (orbiters 6 and 7); specification of technology requirements in space transportation to NASA, launch services, and operation of expendable vehicles not performed by industry; and, most important, marketing of STS services at home and abroad in a business rather than government environment.

Some of these functions could be transferred immediately (e.g., marketing). Others will have to await the Space Shuttle's reaching full operational capability. The acquisition of additional orbiters could be transferred quite soon (with the full support and cooperation of NASA).

It is not beyond reason to "spin off" one of the current NASA centers to become the base of such a Space Transportation Company. However, the new entity has to have complete freedom to draw upon the best talents in industry and government to organize its team and its operations.

Separate from the Space Transportation Company (a support services or operations-oriented entity), a Space Bank should be formed in lieu of the current NASA applications and technology-transfer programs.

The LTB presents challenging *technology* goals at a subsystem level: building and controlling large structures; power supplies (to support the many diverse missions of LTBs); switching stations; energy generation, storage, and distribution networks; information processing, storage, and retrieval systems; and defense systems to make LTBs secure against terrorists and other nations.

The *scientific* uses of LTB span observations of the Earth and other solar system bodies and of outer space (interplanetary, intersolar) up to cosmological scales; experiments in material science, medicine, and biochemistry; and support of scientific spacecraft around the Earth and in the solar system. The technology base to support many scientific investigations would be greatly expanded. Incremental costs of scientific missions could, thereby, be greatly reduced, and mission capabilities vastly increased.

Important subsidiary technology decisions fall into place with a 1990 LTB goal.

An adjunct to the LTB goal would be a NASA *robotics-technology* program. Owing to the high radiation levels at GEO (as against LEO), any long-term presence of man there is inconceivable for the near future, say, the next 20 years. Shielding requirements to meet legal standards under today's health and occupational safety rules for a manned LTB at GEO would make the enterprise uneconomic. Rather, many of the support and maintenance operations (as well as update, retrofit, and disposal functions) will have to be accomplished by remote control—by unmanned, automated systems. An unmanned LTB in GEO will require great and exciting advances in robotics technology. The LTB technology goal will provide a focus to sustain advances in robotics, and it might well prove to be the only chance to have a national robotics program. It would create many and diverse applications elsewhere in the U.S. economy in decades to come.

With an unmanned LTB at GEO, the large structures for systems at GEO will have to be assembled, maintained, refurbished, and monitored elsewhere in space. The most benign and economic orbit, at LEO, will support well the LTB goal and allow rational planning of support systems and operations. Manned support of LTB missions can be executed from a *Space Operations Center* (SOC) in LEO, and SOC development should be a major NASA objective.

A SOC program in support of the LTB mission, it should be noted, can build importantly on the European Spacelab project and may provide a logical next step for *U.S.-European cooperation* in space for the next decade. No other program comes easily to mind for this purpose.

Moreover, the LEO-to-GEO transportation program today has no focus guiding decisions on the scale, capabilities, and technology requirements for an *Orbit Transfer Vehicle* (OTV) to round out the new Space Transportation System (STS). An LTB program will allow a cost-effective approach to this important, unresolved part of our STS development.

The analogy to past successful approaches along similar lines in the United States are several, among them the Highway Trust Fund (to achieve an integrated national highway program) and the World Bank (to bring about useful economic development projects in Third World countries). The analogy to the World Bank is the most appropriate and striking. Whereas it finances projects for economic development in poor nations (with appropriate transfer of technology), the role of the Space Bank would be to bring about the economic development of space and space technology: the buildup of a limited equity over time, say, $1 billion per year over 10 years; the leveraging of the funds of the Bank (by debt and equity participation) with funds of other institutions, industries, and countries; the funding of long-term, higher risk ventures as against what the market would accomplish anyhow; bridging the capital "poverty" of aerospace industry when compared to other sectors of U.S. industry; and fostering new projects and companies to overcome institutionalized lethargy. It would, in short, bring to bear the initiative and energy of a market economy, where central planning simply constitutes a bottleneck at taxpayers' expense.

The patents in aerospace technology held in NASA and elsewhere in the U.S. government should also be transferred to this Bank. It would be the Bank's job to bring about the best, earliest, and most profitable use of space technology in the market.

In summary, a vast exciting space-technology base has been created through NASA and other agencies and industry over the past two decades. To make maximum use of the opportunities in space, new goals, programs, and economic structures are needed:

- NASA should be given a challenging R&T goal for the 1980s. The need for a strong technological presence in geosynchronous orbit by 1990 fits this purpose. In the absence of such a goal, a NACA type of organization may be a more cost-effective use of taxpayers' money.
- Space-transportation services should be moved from NASA and other government agencies to a Space Transportation Company for operating and marketing *any* such services now performed by government.
- Space applications and technology utilization funding should be transferred from NASA to a Space Bank, to be funded over the next decade. This bank should become self-supporting by a certain date, say, 1990.

With such an "economic" structure in place, the future accomplishments of the U.S. space program will be many and varied—some totally unpredictable. The time to act is now.

Developing the Structure to Carry Man to the Planets

Leonard A. Harris
North American Rockwell Corporation

The present state of the structures art, with the progress one normally expects to see, will allow manned interplanetary missions in the 1980s. It should not be implied that no mission constraints will result or that no further structural research and development is required. This article discusses what must be done to make the vehicles feasible structurally, as well as to make them efficient and reliable.

A number of general technology advances can be expected. First, to establish subsystem performance requirements, design criteria are needed to provide an understanding of the assumptions, philosophy, and conditions which govern the structural design. They would also provide a basis for determining design requirements (loads, temperatures, stiffeners, etc., as well as the natural environment).

Since criteria have been imprecise in the past, they have stressed reliability. Likewise, because anticipated environments are inadequately known, arbitrary tests have been used. The result has been less-than-optimum efficiency.

During the 1970s, we can expect the structural engineer to design by synthesis, rather than by analysis, as he does now. Today, the designer conceives and draws the vehicle concept and, then, with others, analyzes it to see if it meets requirements (using standards of minimum weight or maximum payload). Because the time period between concept and final engineering evaluation is relatively long, only a few concepts can be considered. Generally speaking, this approach to design limits the potential benefits of the multifunctional use of materials—for example, the benefits derived when the structure is used as part of the thermal- and/or meteoroid-protection system, thereby doing double or triple duty.

In the present context, synthesis is the computerized technique which rapidly develops structural systems, each of which meets the specific mission requirements and criteria. Looking ahead, the computer displays vehicle or component geometry; the engineer makes changes with a light pen; and the computer displays back to him the effects of the changes.

Current experience with synthesis makes it clear that the lightest components do not necessarily produce the optimum vehicle. With increased use of synthesis, the word "optimum" will be redefined; it will relate more directly to performance, cost, and program schedule. Systematic change of components will be performed to determine the optimum performance of the *vehicle* or the system, not that of specific components.

Astronautics & Aeronautics, Vol. 7, No. 12, Dec. 1969, pp. 66–73.

The different character of a future comprehensive, manned planetary exploration program will further reduce the emphasis on minimum weight, maximum payload, and shortest design and fabrication schedule. Such a program could call for numerous launches to place more than a million pounds into Earth orbit each year. Although still striving to reduce weight, the designer will apply his effort selectively, cutting off pounds where they reduce cost most. Figure 1 illustrates how sharply the investment per pound in adding velocity to a payload rises at the higher velocities. These high velocities would apply to landers, vehicles returning from distant planets, and the Earth-entry module of a Mars mission. Cost spent reducing weight of these modules should be repaid by reducing launch vehicle costs and balancing the overall system to achieve minimum cost.

The program-cost sensitivities described by Fig. 2 pinpoint meteoroid shielding as a specific target for weight reduction. Structural research, therefore, might emphasize defining the meteoroid environment, improving analytical penetration mechanics using stochastic environmental models, formulating new models with multiple bumper sheets, and integrating thermal and meteoroid protection with the load-carrying structure for biggest results.

For a manned planetary mission, the need for accurate analysis is accentuated by the difficulties of effecting repairs in space and providing abort modes to accommodate structural failures. Failure analysis, therefore, should be extremely accurate, but this is currently not always the case. The theory and tests of shell buckling, for instance, still disagree, and current analysis procedures rely upon empirical factors. Consequently, as new structural concepts are introduced, additional tests must be made to verify theory or provide empirical design factors.

Because launch vehicles for manned planetary missions will probably have high slenderness ratios, the aeroelastic response of the launch vehicle and its effect on control systems will continue to be important. This will encourage

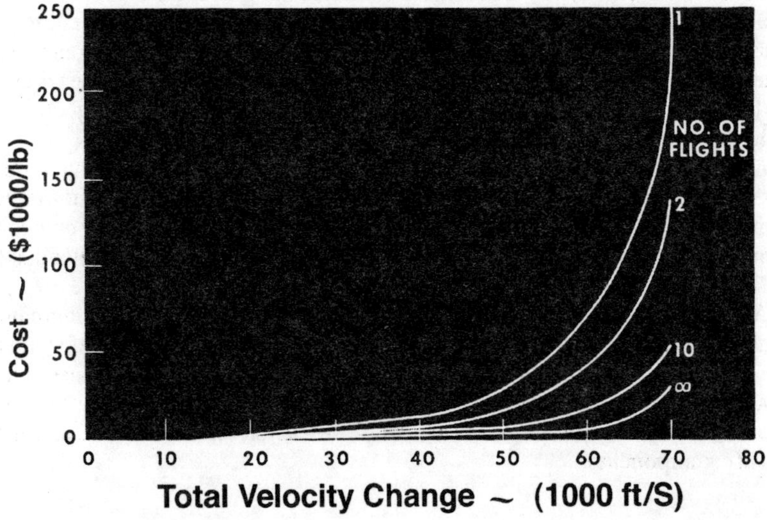

Fig. 1 Mars lander-mission transportation costs: nuclear modularized stages.

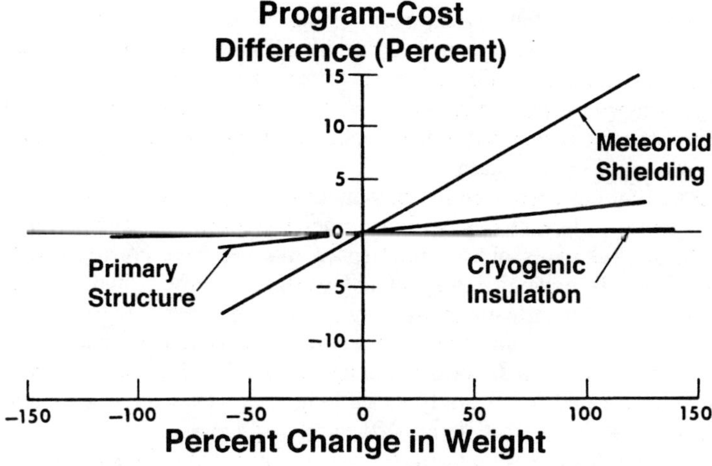

Fig. 2 Program-cost sensitivities to propulsion-module structural weights.

increased development of computer-analysis techniques, which can solve equations with thousands of unknowns.

Although rapid strides have been made in the linear analysis of structures, nonlinear analysis (including inelasticity) grows increasingly important. Historically, methods of stress and strain prediction have preceded failure analysis. Although solutions in the literature apply to numerous structures, materials, and environments, many include such simple assumptions that they must be used conservatively. Time effects compound this difficulty. Challenging problems are presented by quasi-static modes of failure (e.g., creep and failure from thermal stress), which are not only time dependent but require time-incremental analysis. Predicting failure of a fibrous composite (or any anisotropic) structure subjected to varying load and temperature distribution is a good example. The difficulty is increased when load and temperature act together over a long period.

Predicting the behavior of vehicle systems and preventing control-system instabilities is only possible with a solid knowledge of structural dynamics (frequency, mode shapes, damping, etc.). Advanced spacecraft will require three-dimensional analysis verified by test. The greater use of docked modules will increase reliance on test facilities to obtain the dynamic-response characteristics of mated vehicles. Models could reduce the high cost of full-scale testing and supply data earlier if structural-modeling techniques and scaling laws are improved. Simulation could provide information of the effects of construction types, structural joints, and thermal loads.

As some conditions of space and re-entry are difficult or impossible to reproduce on the ground, certain tests will have to be carried out in situ. An example is the particularly important test of 0-*g* dynamic responses of mated modules with or without fluid coupling. Some data could be obtained with existing Apollo CSM and Apollo Applications Program vehicles.

A new approach to the development testing of structures is in order. All too often, a basic concept is proven feasible in the laboratory, then dropped until needed on a production item. Development then takes place during the production schedule, compromising vehicle cost and performance. Development testing to bring theories and concepts to point of application should be funded and completed before production design starts.

Present structural-design philosophy includes the "factor-of-safety concept" against specific failure modes, such as general yielding or ultimate rupture. To satisfy the requirements of advanced space missions, the concepts of fracture mechanics must be expanded to evaluate the effects of flaw growth and criticality for both brittle and ductile materials (as well as composites) and to accommodate significant plastic behavior. For reasonably homogeneous materials, brittle-crack behavior, fatigue, and stress corrosion appear relatable. More-precise understanding must be pursued, particularly since long space exposure and high-energy propellants may change a material's fracture toughness.

From these general observations, we can now proceed to look at some specific effects of structure on space-vehicle systems.

Launch Vehicle Structure

Saturn V has shown us some of the problems of handling large, heavy propulsion stages.

But currently under study are nuclear stages about 30 ft in diameter and up to 150 ft long. Such vehicles would have to store liquid hydrogen for as long as 1400 days to reach Ganymede and return. Stages of this nature may have to be boosted individually and assembled in orbit. They must be extremely lightweight and will require a laminated, ultralight superinsulation on the exterior. The complete unacceptability of damage to this superinsulation will make the tanks even more delicate to handle.

Minimizing tank supports (to reduce heat leaks) and carrying an aerodynamic shroud (to protect a nuclear stage from aerodynamic heating during launch and from meteoroid damage during spaceflight) will further complicate handling.

Materials

The structural engineer always hopes for advances in materials to solve many of his problems. The strength-to-density ratios of commonly used structural materials (primarily aluminum, titanium, magnesium alloys, and steels) will undoubtedly improve, but, history says, at a moderate pace. For major advances in specific stiffness, the structural engineer must look to more unusual materials, such as beryllium and the advanced composites.

Beryllium's unusually high modulus-to-density ratio over its usable temperature range and its competitive strength-to-density ratio, as shown in Figs. 3 and 4, suit it for applications demanding stiffness or resistance to buckling. Its relatively high cost will limit its use to high-velocity modules. However, beryllium has not been widely used because it is brittle and costly. Low anisotropic ductility has proven its main drawback. Although efforts to increase the ductility of beryllium should continue, additional effort should be devoted to design concepts

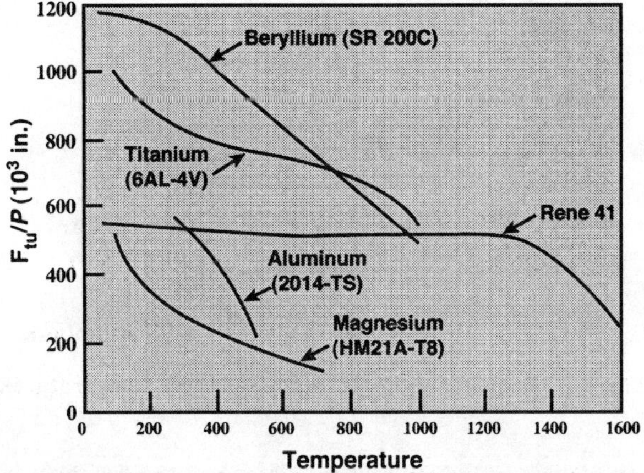

Fig. 3 Temperature effects on ultimate tensile-stress-to-density ratio.

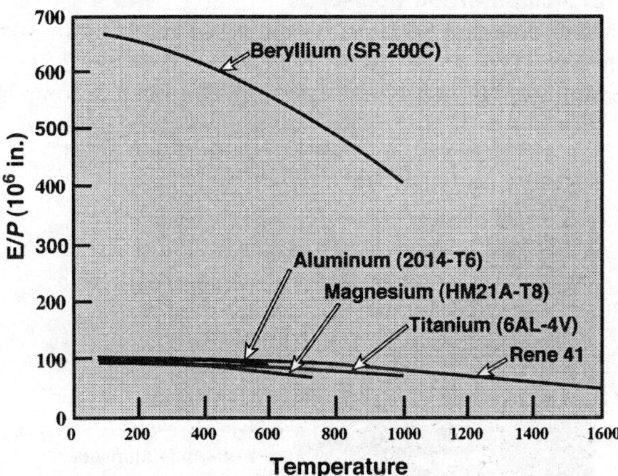

Fig. 4 Temperature effects on modulus-to-density ratio.

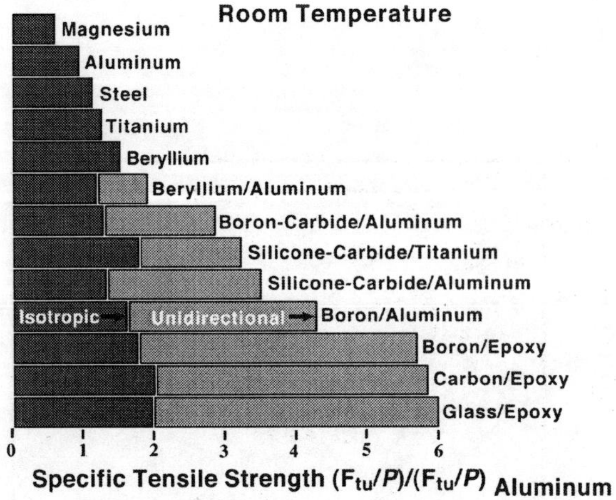

Fig. 5 Tensile-strength efficiency (at room temperature).

and fabrication techniques which take advantage of the desirable high stiffness/weight ratio within the limitations of its low ductility.

Fibrous composites offer the greatest design potential. The advanced composites cover a wide variety of combinations of high-strength and high-modulus filaments (boron, carbon, silicon carbide, beryllium) embedded in a load-transferring matrix (plastic, metallic, ceramic). They permit tailoring properties to meet design conditions through control of fiber and matrix combinations and the size, spacing, and orientation of the fibers.

Figures 5 and 6 compare specific strength and modulus of some composite materials with some typical structural materials. Glass/plastic-matrix materials will see increased use, but boron and carbon/plastic-matrix composites will pio-

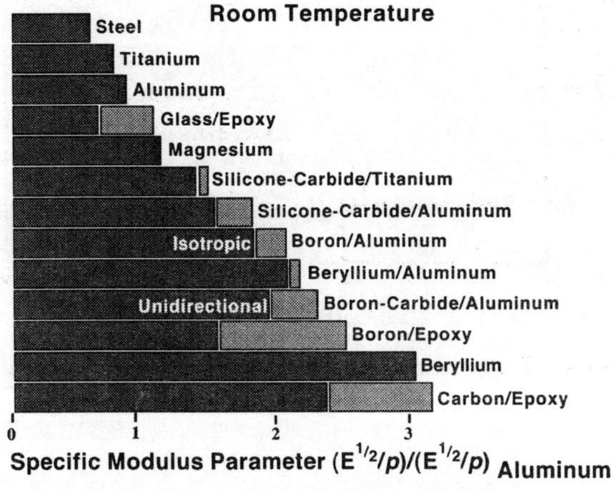

Fig. 6 Compressive-buckling efficiency.

neer the application of advanced composites. Metal-matrix systems are less well developed but will predominate in high-temperature uses.

If the new composite materials are to be ready for manned planetary missions, many problems must be solved. Orthotropic analysis methods, particularly discrete element techniques, are proving effective in design, and nonlinear analysis methods are now becoming available and will be valuable in assessing the behavior of joints. However, the field needs further development of techniques for optimizing the many material and geometric variables. Joint design continues to be a development requirement. Since composites have directional properties, multidirectional vehicle loads must be defined precisely to assure adequate strength for conditions not usually critical for isotropic materials. And the high cost of composite materials hampers wide use in flight vehicles. Those areas in the spacecraft where composites will be most cost-effective should be selected for initial application. Increased production will bring costs down for future applications.

In spite of their early state of advancement, the advanced composite materials have demonstrated performance that justifies their development for application in planetary mission vehicles, particularly in the higher velocity modules.

Cryogenic Insulation

Advances comparable to composites are not anticipated for the insulation surrounding cryogenic fuel tanks. Because of the present high efficiency of multilayered high-performance insulations (Fig. 7), we should not expect much improvement before the 1980s. There is, however, room for improvement in the design of cryogenic tank supports. As shown in Fig. 8, a large portion of the total heat enters through the supports. Improvements in both design concepts and materials (such as composites) will prove effective.

Heat flow can be impeded by reducing the temperature of the stage exterior surface. Here, coatings that lower the absorption/emission ratio show promise.

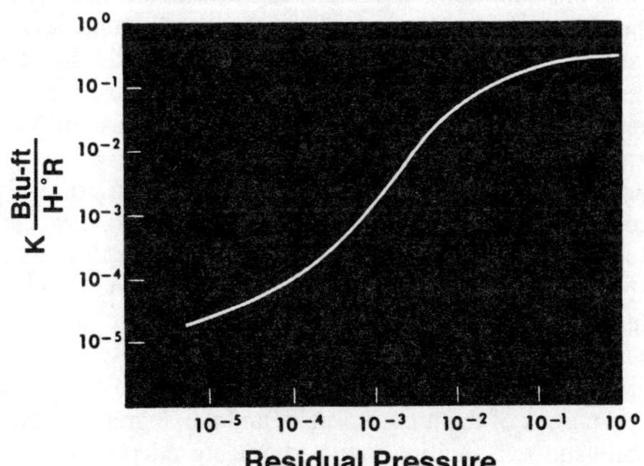

Fig. 7 Variation of superinsulation conductivity with pressure.

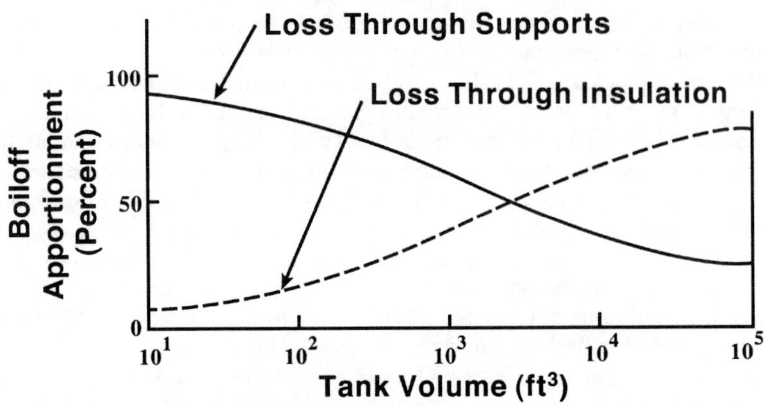

Fig. 8 Cause of boiloff losses for typical stage.

Current coatings have ratios in the range of 0.2–0.4 on a practical basis, resulting in surface temperatures of −150 and −80°F, respectively. Coatings under development have exhibited ratios as low as 0.05, which can result in a −250°F surface temperature under laboratory conditions. Unfortunately, the environment degrades the current best coatings, raising absorbtivity 30% in 1000 h or 70% in one year.

For planetary mission propulsion modules, development of the cryogenic insulation and meteoroid-shielding should proceed hand in hand (more about this later in the discussion of spacecraft).

System and Subsystem Testing

A launch vehicle experiences longitudinal loads from booster acceleration; lateral loads from wind shears and engine gimbaling; thermal loads from boundary-layer heating; and vibration loads from engine-exhaust noise, boundary-layer turbulence, and onboard machinery such as engines, pumps, and compressors. Methods of testing for the effects of combined loads need additional development, especially for severe static and thermal loads that must be precisely sequenced. The requirements for such testing will increase with the use of highly efficient, lightweight structural components.

Recent flight data indicate differences between design criteria and actual vibrations encountered. Where does testing fail? In the laboratory, electrodynamic shakers simulate the effects of local mechanical sources such as engines, pumps, and compressors. Proposed methods for simulating the random vibration caused by engine exhaust noise and boundary-layer turbulence use large reverberant- or progressive-wave acoustic test facilities. However, the response of structures to a fluctuating pressure field depends on the free-vibration characteristics of the structure, the properties of the pressure field, and the degree of coupling between the two. An acoustic test facility cannot duplicate all properties of the flight-environment pressure fields. Consequently, adjustments must be made to the spectral density of the pressure field in the acoustic facility (and/or the angle of wave incidence in a progressive-wave facility).

Underlying this problem, of course, is the need to predict the space environment accurately. Some spaceflight conditions, however, cannot be duplicated because of gravity, the atmosphere, and other hindrances. A review should be made of these concepts that cannot be realistically tested so that flight structure gets designed for spaceflight, not ground testing.

Many of the factors affecting launch vehicles also affect spacecraft design. In addition, some special factors, such as meteoroid protection, leakage control, and radiation, must be considered.

Meteoroid Protection

For planetary missions, as Fig. 9 shows, meteoroid protection for spacecraft and propulsion modules becomes a major weight saving target. First, there is insufficient information on the planetary-meteoroid environment. At the minimum, the analyst needs information on meteoroid population, mass, density, heliocentric velocity, and direction. Present techniques do not directly measure any of these values in either near-Earth or deep space. Satellites, telescopes, and radar have made numerous indirect measurements in near-Earth space. Some indirect measurements in deep space have been made by telescopes and by the Mariner probes (between Venus and Mars).

A recent and authoritative estimate of the near-Earth environment, issued by NASA to guide Apollo design, probably will not change drastically in the future. In deep space, estimates on the population of asteroidal meteoroids may be off by plus or minus some 2½ orders of magnitude. Measurements out to Saturn and in to Mercury by probes of the Pegasus type would provide a good start in correcting this situation. Considerable attention should be directed toward equip-

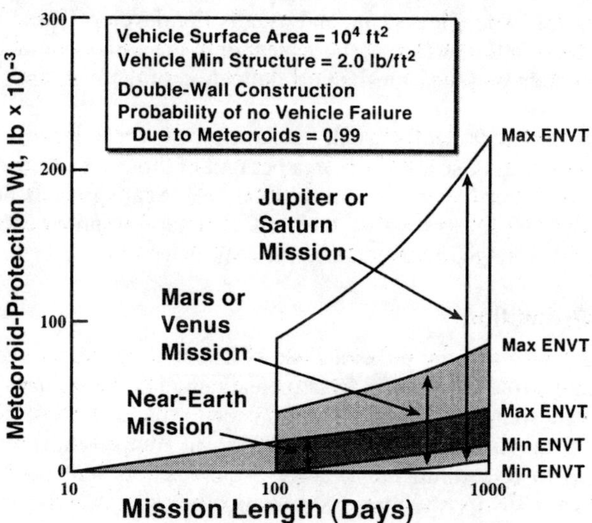

Fig. 9 Meteoroid-protection variations for interplanetary missions.

ping probes to make measurements directly. Such data will result in realistic structural weight.

Structural concepts should be developed to resist impact damage more efficiently. Tailored to mission and spacecraft, the concepts should integrate meteoroid protection, load-carrying functions, and thermal protection. Recent studies indicate that exterior structures should have at least two or three spaced sheets to minimize the weight for meteoroid protection.

Methods of analysis for particle impact should also be improved and verified by ground test with the highest velocity particle available. Detection and repair of punctures in manned spacecraft could simplify protection; present methods are inadequate.

Leakage

Manned spacecraft on long missions will tolerate considerably less leakage than Mercury, Gemini, and Apollo. Loss of 1.5 lb of 5-psi oxygen atmosphere per man-day has been permitted. This leakage during a four-man, 600-day mission necessitates 3600 lb of make-up oxygen. Allowing 2 lb per man-day for breathing, more than 40% of the oxygen carried would constitute the leakage fraction. If a regenerative system recovered 75% of metabolic oxygen, oxygen consumption would still be four times higher than if no leakage took place. A mixed-gas system operating at a higher pressure would accentuate the system effect of leakage.

Leakage through unsealed perforations in the shell could be eliminated before launch by sophisticated leak-detection techniques and repair. A 0.01-in.-diam pinhole leaks 1 lb of atmosphere per day; a pencil-size hole leaks 1000 lb.

Openings are necessary in the pressurized structure of the spacecraft for crew entrance or exit, electrical or fluid lines, mechanical actuating shafts, windows, etc. All have seals where leaks can, and usually do, develop. These leaks can be caused by defects in the surface of the sealant or in the seat, material degradation, insufficient contact pressure, mechanical deflection of the seat, and cold flow of the seal.

Current parametric tradeoffs usually assume an allowable leakage rate of X lb per sq ft of spacecraft or seat surface area per unit of time. A more useful measure may well be X lb per linear ft of mechanical seal. Sealing needs development, particularly through integration of function, structural requirements, materials, protection from damage, maintenance, and leak detection.

Radiation Protection

Vehicle structure and the judicious use of onboard structures and equipment may sufficiently protect personnel from typical radiation. Solar flares may require the addition of a somewhat heavier "storm cellar" shield to boost the equivalent shield from 5 to 10 g/cm^2 or more. Orbiting in trapped-radiation belts may necessitate up to five times the average 5-g/cm^2 equivalent shield for Earth. Neither Mars nor Venus has a trapped-radiation belt that is of any consequence, but Jupiter has an extensive one.

Of onboard radiation sources, isotopes usually emit alpha particles and few gamma rays. Low-power or low-thrust reactors, on the other hand, may require equivalent shields of 30–100 g/cm^2; high-thrust propulsion reactors, 200–400. But these radiation fields, unidirectional in terms of the crew, can be blocked with shadow shielding. Here again, onboard equipment can do much shielding.

Efforts are under way to evaluate an electric or magnetic field or a combination, such as a plasma, as an active shield. None of these is likely to be ready by the 1980s, but no missions now planned for then will need them.

To better define spacecraft, more work must be done on establishing allowable radiation limits for personnel, defining the natural environments, and developing techniques to make maximum use of onboard material for shielding.

Structural Analysis of Nonstructural Subsystems

A significantly high percentage of nonstructural subsystem qualification failures are structural failures. Examples, such as the failure of integrated circuit boards and valve seats and the unsatisfactory dynamic response of equipment, lines, and subassemblies, might be reduced if the parts were analyzed structurally. Such failures during qualification testing create serious schedule problems.

Attaching subsystem components to the basic vehicle structure changes local structural stiffness and mass characteristics, and this alters local structural load paths and vibration characteristics. Few precise methods exist to evaluate the effect of dynamic interaction on local structural loads and subsystem performance.

Many laboratory and flight-vibration test failures occur as a result of artificial laboratory test conditions. Examples of such conditions are unrealistic test-fixture design and criteria (load and frequency), as well as the method of inducing vibratory forces. These situations should be corrected and made consistent with the vehicle-design environment.

Entry Vehicle Structure

A manned spacecraft approaching a planet goes through three phases—high-speed entry, transition, and landing. An approach through a thin atmosphere, as on Mars, will require a much lower mass-to-drag ratio (the ballistic parameter) than those of current vehicles. Aeroshells developed for low structural mass have had high aerodynamic-drag coefficients. They either reduce the Mach number or increase the altitude for deployment of auxiliary drag devices and retropropulsion systems for landing.

Deployable-structure design lags that of fixed geometry. The readiness of deployable devices for mission application decreases markedly with increases in the Mach number at which they are to be deployed.

For the landing phase, a specialized structure must absorb impact forces and orient the payload for surface operations or launch for the return trip. Little attention has been given to developing landing configurations that would permit later travel over rough and unknown terrain.

Where will effort on structural technology pay off most in enhancing planetary mission objectives and reliability?

Table 1 presents an appraisal of heat shield technology. In spite of only fair simulation of the total flight environment in present ground test facilities, thermal protection during Earth entry at 35,000 fps has been verified by the success of the Apollo lunar flights. Such ablative composite heat shields, in the density

Table 1 Status of heat shield technology

	Earth entry		Mars entry
Factor	35,000 fps	50,000–55,000 fps	15,000–20,000 fps
Material	Charring composites (30–35 lb/ft^3)	Carbon composites (30–50 lb/ft^3)	Silicone composites (15–20 lb/ft^3)
Ground test simulation	Fair	Poor	Good
Ablation analysis	Fair–good	Unproven	Fair–good
Heat shield weight fraction	-10%	>10%	3–5%

range of 30–35 lb/ft^3, make up about 10% of the vehicle entry weight. The flight behavior of ballistic and semiballistic entry vehicles correlates well with predictions from analysis and ground tests. More development is necessary on thermal-protection systems for lifting vehicles because of their more complex flow fields. However, the Asset and Prime vehicles have performed successfully at Earth-orbital velocity. These vehicles subjected ceramic, refractory metal, superalloy, ablative heat shields to moderate fluxes for long entry times.

For speeds of 50,000–55,000 fps, typical of those reached on return to Earth from planetary missions, one can anticipate development of charring composites with higher carbon content for maximum energy dissipation per pound. Because present ground facilities cannot adequately simulate the high radiative-heating rates encountered, analytical methods should be improved to bolster confidence in the shields.

Test facilities that can handle the relatively mild heating of a Mars entry at 15,000–20,000 fps are available and have been used to provide a heat shield technology for that planet. Although the exact composition of the Martian atmosphere is not known, the thermochemical response of low-density silicone composites has been verified in a variety of stream chemistries and does not appear to be particularly sensitive to the ranges of compositions expected at Mars. Designers, therefore, can confidently predict heat shield weight fractions in the range of 3–5% of entry vehicle weight, depending on entry conditions and configurations. This range of uncertainty can be further narrowed for well-defined trajectories and entry conditions. The technology for other planets will probably fall out from work on Mars-Earth return.

Deployable devices for making the transition from high speed to a landing condition can yield a major vehicle improvement. For example, methods for deploying parachutes through a broader range of speed and dynamic pressure are needed. A completed NASA flight-test program showed that lightweight subsonic-type parachutes can be deployed successfully up to Mach 2 at low (10 psf) dynamic pressure. But, in other recent flight and wind-tunnel tests in this country, a variety of aeroelastic phenomena destroyed parachutes at higher speeds and dynamic pressures.

New decelerators, to develop lift as well as high drag, will evolve from attempts to circumvent such problems. Some now being considered are sketched in

Fig. 10. In addition to the conventional membrane-and-tension-string structure of parachutes, these concepts cover internal-pressure-stabilized structural components with or without internal shear-carrying membranes to stiffen them and fix external contours. They may be towed or closely coupled to produce either high drag or high drag plus some lift. Such surfaces are pneumatically or mechanically erected. Surfaces or entire vehicles constructed in this way introduce interactions between structure and thermal protection currently beyond the state of the art.

Structural analysis for all decelerators falls short, particularly during inflation when they take on "nonanalytic" shapes. Idealized inflation behavior can be formulated and bounding values of stress determined. At the same time, ad hoc wind-tunnel and flight tests will continue.

Commonly used fabrics absorb energy by stretching to withstand high shock loads. For this reason they will probably continue to be used even though advanced filament-reinforced membrane materials would be desirable for their high strength and temperature resistance.

The inflatable membrane structure is also a candidate for components of the landing gear and surface mobility systems. Spacecraft cannot land on water or be snatched from the air at target planets. Nor are they likely to roll to a good landing on wheels under winds on rough terrain. Further work is needed on legged landers to adapt them to vehicles with aerodynamic shapes.

Various impact devices have been proposed to fill this gap for severe landing conditions. Each embodies a special characteristic, such as low bearing pressure, resistance to tip-over, or ability to withstand omnidirectional or multiple impacts.

Packageable impact-attenuation and stabilization systems can be erected in the low-speed terminal phase before touchdown and can be installed on aerodynamic landing shapes. The emphasis here is on adaptability to the landing situation, rather than on energy-absorption efficiency as such, a factor which sometimes receives undue emphasis.

Ideally, touchdown and surface mobility should be integrated into a single system. Wheeled and tracked crawlers have so far been investigated separately. Floaters, hoppers, and flyers should also be studied in terms of available power,

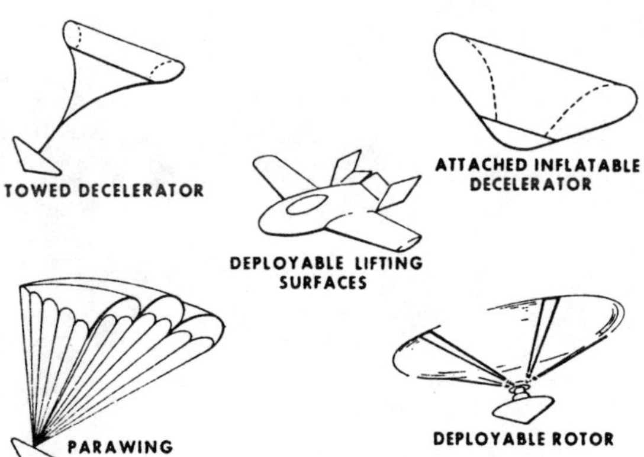

Fig. 10 Deployable surfaces for increased drag and lift.

modes of guidance, and distance to be traversed over rough terrain. The technological challenge here ranks with that of high-speed entry 10 years ago.

Concluding Remarks

In summary, the basic task of structural technology is to assist in the definition of mission plans and vehicle configurations. These grossly affect shape, heat shields, and other protective systems. The question is not *what is the most maneuverable shape or the most efficient heat shield or structure?* Rather, we must determine what combination of factors such as configuration, meteoroid shield, and structure best performs a mission and gives the most flexibility for planned and emergency mission changes. Greatest changes in structural concepts will come for entry and in the multifunctional use of materials.

Structures and Materials in the Long-Duration Manned Spacecraft

Hua Lin
Boeing Aerospace Division

Historically, structural design criteria have been developed through the accumulation of experience. As the experience with manned spacecraft is yet very limited, structural design is generally developed for each specific manned spacecraft system. Principal considerations include the mission requirements, natural and induced environments, and the major subsystem interfaces (including human factors). Meaningful design requirements can only be defined by simultaneous consideration of all pertinent factors.

Many of the structural and material considerations in spacecraft design are the same for manned and unmanned missions. Others impose much more stringent requirements in manned systems owing to the higher reliability level necessary for crew safety. A few, unique to manned spacecraft design, will be discussed briefly here, particularly with long missions in mind. All lead to ever-increasing demands for advanced materials, ingenious structural concepts, and better fabrication and simulation techniques.

Radiation Shield

The significant radiation environment in space may be categorized in three groups: solar radiation (solar flares and solar wind), galactic radiation, and trapped-radiation belts. These groups of radiation differ in intensity, energy spectrum, and direction.

In general, natural radiation in space does not impose any significant problem to unmanned spacecraft. Damage threshold of most structural materials is of the order of 10^8 to 10^{13} rads, and sensitive electronic equipment can function under a total dose of 10^3 to 10^4 rads. Man, however, has a much lower radiation tolerance of approximately 50 rads.[1] Radiation shielding is, thus, necessary.

Figure 1 shows radiation doses absorbed by aluminum as a function of thickness for a typical Mars mission (surface dose of 2.8×10^{10} rads). In Earth-Mars space, the galactic radiation, being relatively insignificant, is lumped into radiation due to solar wind. It can be seen that the solar wind gives rise to high near-surface doses but is not very penetrating—95% of particle energies being less than 5 kev. Radiation of this category is stopped by the normal spacecraft structure. It affects only surface materials, such as thermal coatings. On the other hand, radiation due to random solar events has a greater proportion of high-energy particles and is much more penetrating. A total additional shield thickness of

Astronautics & Aeronautics, Vol. 4, No. 3, March 1966, pp. 34–40.

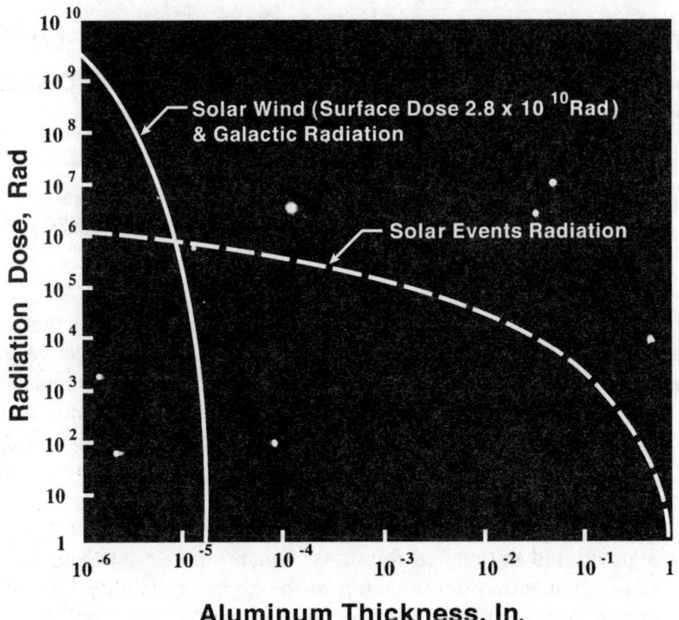

Fig. 1 Space radiation dose—15-month Mars mission.

approximately 1/2 in. of aluminum at 7 lb/ft^2 would be required to keep the interior safe for man. Such a shield completely over a large spacecraft would bring a prohibitive weight. The often-proposed solution would be a configuration maximizing the amount of shielding provided by the structure; more-sensitive elements would be placed in the center of the mass, and a small shelter would be provided for the crew to use during solar-storm activity. Another possible approach involves shielding only one wall of the spacecraft and orienting this wall to the sun during solar-storm events. Tradeoffs in shielding design must be made in terms of weight, trajectory, and mission duration.

In near-Earth missions, the significant source of radiation is that due to trapped protons in the Van Allen belts. Shielding requirements will involve a tradeoff with orbital altitude and inclination. Extended operations in the radiation belts will necessitate very heavy shielding.

Meteoroid Protection

Heretofore, space programs have been too limited in scope to have required additional weight solely for meteoroid protection. As manned spacecraft become larger and mission durations considerably longer; such a situation will be no longer true, and an adequate protection system will have to be provided at the least expense in precious weight.

Unfortunately, the meteoroid environment is not well established. Figure 2 shows the range of the cumulative number of meteoroids per unit exposure for a near-Earth environment. The broad band is drawn from composite data from

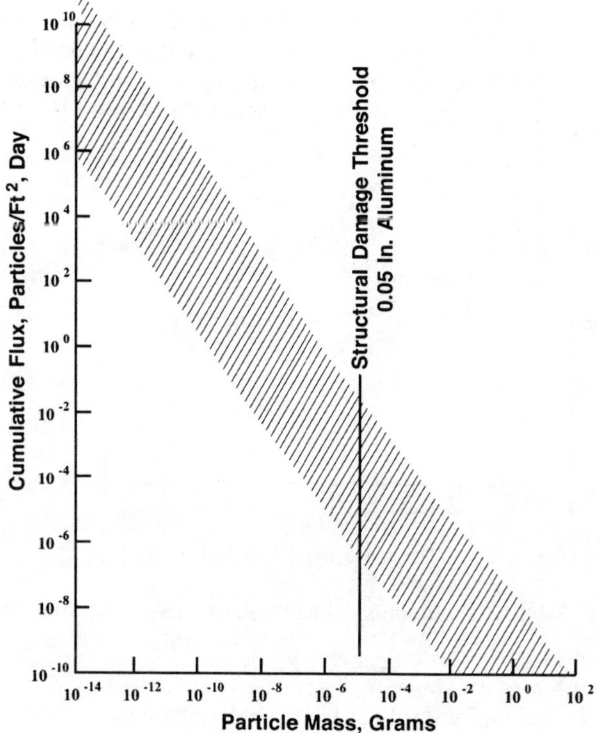

Fig. 2 Meteoroid environment.

meteoroid observations and from satellite-borne impact and penetration sensors. Corresponding information for other regions of the solar system is even less known. Additional effort in meteoroid data gathering and interpretation is urgently needed for future manned space missions.

The significance of meteoroids to manned spacecraft design depends greatly on the type of impact damage. It is obvious that very high velocity particles that can cause space cabin penetration and pressure-vessel failure deserve special consideration in long missions.[2]

The available information on spacecraft penetration is limited to the measurement of phenomena in the immediate vicinity of the penetrated wall. Data are needed for the effects on the entire space cabin due to shock, light, heat, and acoustic phenomena accompanying penetration. Full-scale cabin tests are necessary to permit simulation of realistic wall structure, cabin atmosphere, and geometric relations. Figure 3 gives an indication of the shield weight for a multilayer barrier structure. For the case shown, the basic structure may weigh approximately 1.0 lb/ft^2, but, for a 0.999 probability of no penetration, it may require about 3 lb/ft^2 for a three-sheet meteoroid barrier or 5 lb/ft^2 for a two-sheet barrier. The total protection-weight requirement increases with surface area exposed and with the duration of the mission.

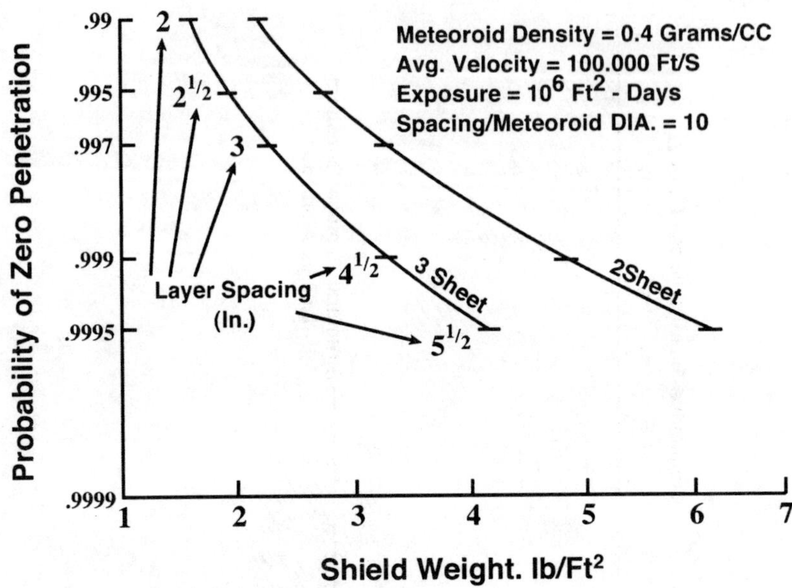

Fig. 3 Multilayer aluminum shield weight—15-month Mars mission.

A major portion of the structure of any spacecraft consists of pressure vessels—space cabins, fuel tanks, oxidizer tanks, and sundry gas bottles. For manned vehicles, crew safety rules out any meteoroid damage that may cause catastrophic fracture of a pressure vessel rendering the entire spacecraft unusable. Figure 4 shows some results of tests at Boeing depicting a relationship between the pressurization stress and the critical impact damage size that induces catastrophic fracture in 2014-T6 aluminum. The static failure curve was obtained by drilling holes and subjecting the specimen to a conventional tensile test. Allowable stresses are lowered drastically under impact conditions. Similar data are needed to determine this relation for all pressure-vessel materials, thicknesses, and operating temperatures. Moreover, materials and techniques must be developed to inhibit catastrophic growth of impact-induced cracks and to repair localized damage.

A sound theoretical treatment of the penetration of a hypervelocity projectile has yet to be developed. In recent years, however, empirical methods have been devised that permit a rational design of practical structural arrangements. The most serious limitation of the empirical methods has been the necessity to extrapolate laboratory data to meteoroid velocities (~100,000 fps). This limitation will be with us until some major improvements in laboratory launching techniques are developed. Indeed, the importance of obtaining higher-velocity data justifies considerable effort in developing new hypervelocity launching devices.

Cabin Sealing

The need to minimize leakage might be exemplified by a hypothetical extension of the Mercury spacecraft capability. The leakage through the seals on the Mercury

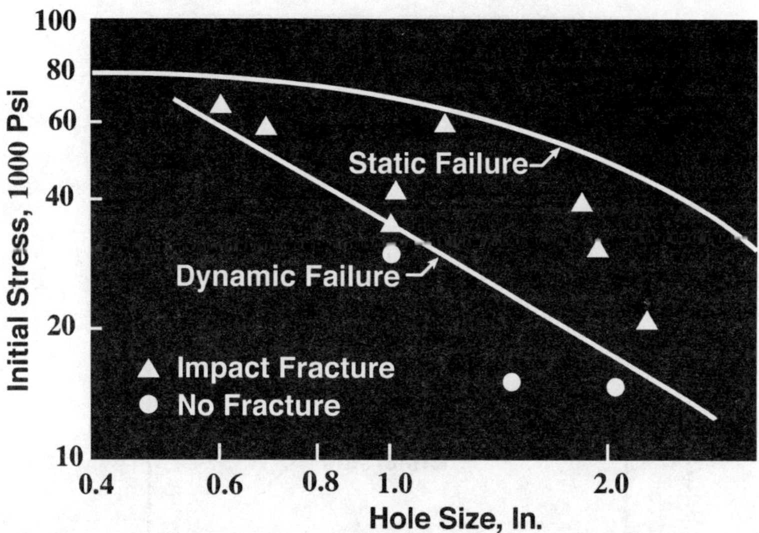

Fig. 4 Critical impact damage to 2014-T6 aluminum.

capsule has been reported as 2.24 lb of oxygen (at 5 psia) per day.[3] At this rate, 800 lb of stored atmosphere would be required for leakage alone for a one-year space trip, not considering the make-up required for dumping of airlocks or the manyfold increase in vehicle volume and complexity required to support a mission of this duration. Figure 5 gives some current and projected leak rates, indicating the marked improvement needed for some future manned systems.

Dynamic seals—seals in which there is linear or rotary movement with respect to the sealing surface—are undesirable because they have relatively high leakage rates. Static seals with a means of transmitting motion through the seal plane should be preferred. Other seal-design considerations are temperature and vacuum stability, radiation, internal atmospheric composition and pressure, number of cycles of opening and closing hatches, and rubbing of seal surface caused by structural bulging under pressure.

As cabin complexity and size increases, the sealing problem becomes more complex. While there are now many good designs and materials for seals, continuing work will be needed to determine the most efficient seal configuration and material combinations for use in window, hatch, and docking applications.

Thermal-Control Coating

Optimum thermal control of a spacecraft in terms of minimum weight and maximum reliability can be achieved by both active and passive methods. Active control may include heat exchangers, electric heaters, space radiators, etc. Passive control may utilize insulation, spacecraft attitude management, and thermal-control coatings.

For man to work efficiently in a ''shirtsleeve'' environment in space for an extended time, the temperature of his cabin must be kept within a much narrower

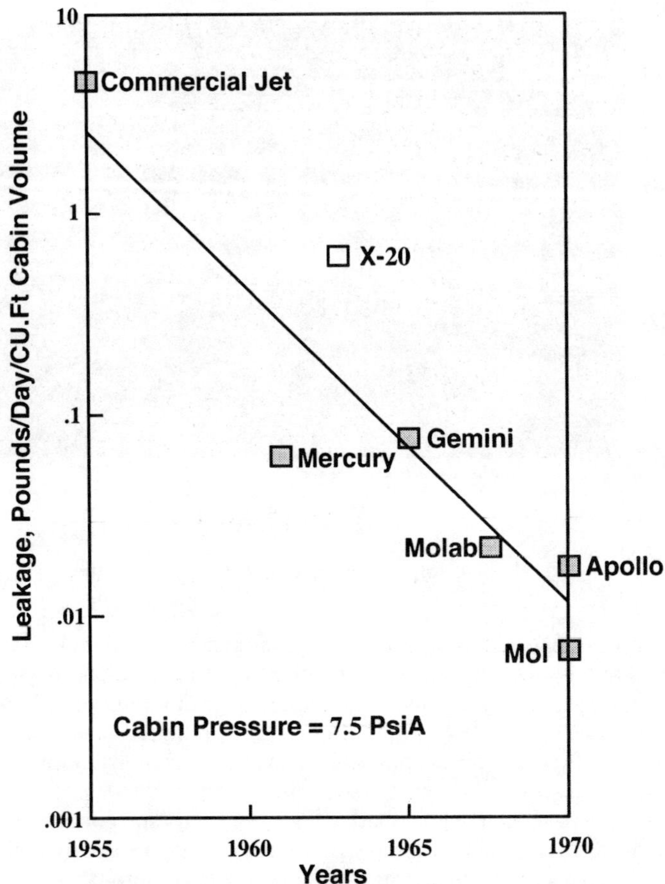

Fig. 5 Leakage requirements for manned missions.

limit than that inside an unmanned spacecraft. To achieve a temperature control between 55–85°F for a shirtsleeve space cabin,[1] with possibly large variations in power dissipation by the electronic and mechanical equipment, requires efficient and reliable thermal management and stable radiative properties of the coatings on the space radiators.

A coating gives thermal control by fixing a desired ratio of solar absorptivity α to emissivity ϵ for a given space environment. A smaller value of α/ϵ results in a cooler vehicle and vice versa. Higher individual values of α and ϵ also raise the rate at which thermal equilibrium is reached under a given set of conditions.

The thermal properties of many coating materials change under exposure to space environment. Figure 6 shows the effect of proton exposure on some coatings under vacuum conditions. Figure 7 shows the effect of ultraviolet exposure on one of these coatings and the effectiveness of one kind of filter coating.

It is apparent that coating properties may change significantly in the hostile environment of space. Many more tests are needed under combined environments

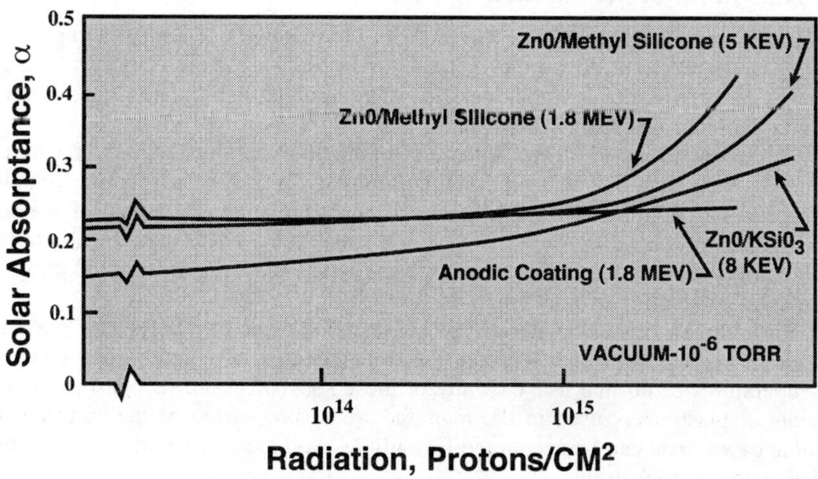

Fig. 6 Effect of proton exposure.

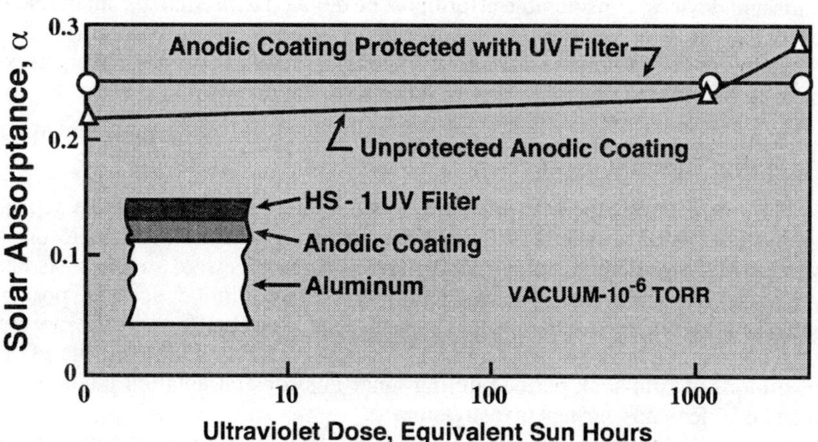

Fig. 7 Ultraviolet effect on an anodic coating.

of vacuum, ultraviolet, and proton exposures to develop coating-filter materials of stable properties for long-duration space missions.

Toxicity and Contamination

Gaseous contaminants are evolved from the materials of construction of the cabin and are produced by the metabolic processes of man. They can also be produced as byproducts of the operation of equipment. Particulate contamination can arise from mechanical abrasion.

The ultimate control of atmospheric contamination requires a thorough evaluation of each material and subsystem proposed for use in a specific vehicle. The evaluation consists of qualitative and quantitative determination of volatile substances produced by each material and subsystem. There must be a detailed accounting of the total amounts of each material used to permit a prediction of the total production of contaminants.

Virtually all organic materials of construction—such as polymers, plastics, elastomers, and adhesives—produce trace quantities of volatile substances at temperatures of normal usage. Many of these substances are recognized as solvents or plasticizers used in the manufacture or processing of the material. In some cases, they can be accounted for only as products of oxidation or thermal-decomposition reactions.

The electrical and electronic equipment used in the space cabin presents probably the greatest source of atmospheric contamination next to man. Such equipment often embodies a wide variety of organic materials, maintained in most cases at an elevated temperature, with consequent significant evolution of volatile, toxic substances. Due to complexity and multitude of material applications in electronic parts and devices, considerable effort must be devoted to minimize contamination by substitution or legislation of materials. For example, most room or low-temperature curing finishes, adhesives, potting compounds, and encapsulants will have to be omitted or replaced with more stable materials.

Re-entry Thermal Protection

The selection of a thermal-protection system for a re-entering vehicle depends upon the type of vehicle. For low-*L/D* re-entry, an ablation heat shield on the windward surface combined with radiative cooling of leeward surfaces has been proven out in the Mercury and Gemini spacecraft. Similar designs, possibly introducing combined ablation and transpiration concepts for specific portions of the structure, may be incorporated on future vehicles. However, the effects of prolonged cold-soak periods during lunar missions on ablation material performance demands further investigation.

Vehicles with higher *L/D* ratios (1–2) experience lower peak-heating rates but longer thermal-exposure periods (~30–60 min). This class, represented by Asset, M-2, HL-10, and X-20 vehicles, can be protected by either radiation shields, low-density ablators, transpiration cooling, or combinations of these. Work on the X-20 and the flight-test data from the Asset program have provided a background for the design of radiation-cooled structures. Considerable experience has been obtained in the use of ablation systems from NASA and DOD programs.

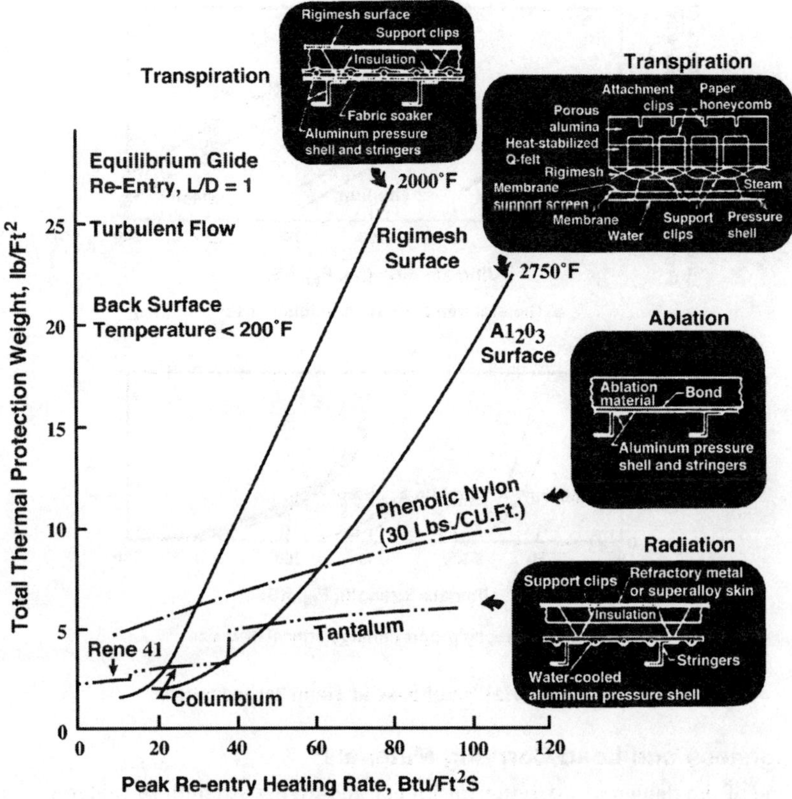

Fig. 8 Thermal-protection systems.

Research at Boeing on transpiration and permeable matrices has provided a basis for assessment of a transpiration-cooled design.[4,5]

Figure 8 compares weight of the various heat-protection concepts for orbital re-entry with $L/D = 1$. The curves are based on single usage. For multiple flights, the ablation-protection system would necessitate much more refurbishment.

Various developments have extended the potential refractory metals such as tantalum with aluminum-tin coating to temperatures above 3000°F for prolonged periods. It is expected that in the near future a capability of 3500°F and above will be achieved, which will make radiation cooling considerably more attractive. With the improvement in coating techniques, usage of permeable matrices can also be extended to higher temperatures. This makes transpiration cooling more competitive, especially for vehicles where refurbishment requirements and special considerations such as contour integrity and local hot spots, prevent the use of ablation protection. Much more effort will be needed to develop practical designs for a reliable thermal-protection system.

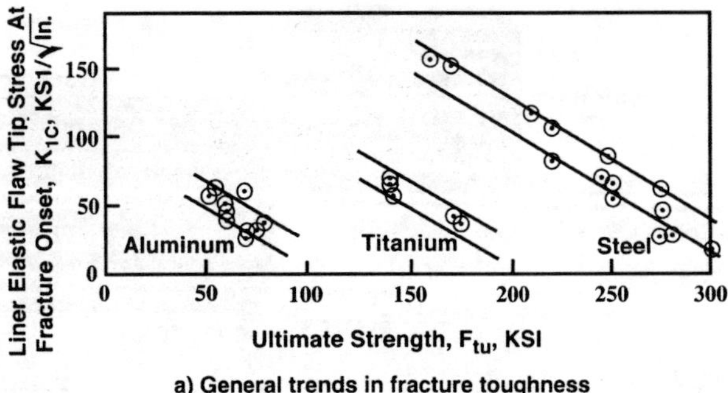

a) General trends in fracture toughness

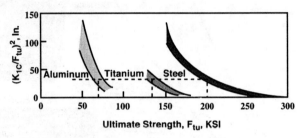

(K1C/Ftu)², in.
Aluminum
Titanium
Steel
Ultimate Strength, Ftu, KSI

b) Ordinate directly proportional to critical flaw size

Fig. 9 Material toughness at room temperature.

Toughness and Load-Carrying Materials

One of the dangers in striving for higher and higher strength of materials and, thus, lower weight in structures is that they generally become increasingly sensitive to small defects, which are either inherent in the material or are introduced during fabrication. The general trends in fracture toughness K_{IC} with increasing strength levels in aluminum, steel, and titanium alloys are illustrated by the test data shown in Fig. 9a. In Fig. 9b, the ordinate is directly proportional to the critical flaw size, thus placing the influence of increasing material strength levels into better perspective. For example, an increase in ultimate strength level from 50 to 60 ksi in aluminum shows a 16% drop in K_{IC}, but the critical flaw size is reduced by a factor of 2. A pressure vessel designed from a 200-ksi steel, a 135-ksi titanium, and a 70-ksi aluminum would all have approximately the same critical flaw size for a given operating factor of safety.

While such comparisons provide general guidance for initial material screening, it must be noted that material anisotropy, material forms, and applied stress cause differences in toughness throughout the structure. Consequently, critical flaw size in different portions of the structure will not be the same. In pressure vessels, weldments and base material loaded in the short transverse direction are particularly prone to low toughness, and when combined with pre-existing initial defects and high applied stresses, they are potentially dangerous.

From the point of view of crew safety, it is desirable that the material selected for a given structural application have large critical flaw size at the operational stress levels, low subcritical flaw growth rates, and low probability of flaw occurrence. Better nondestructive testing techniques must be developed to readily and positively identify flaws.

Concluding Remarks

The reliability that will be demanded of long-duration manned spacecraft confronts the designer with truly difficult problems. The recitations of them will no doubt fill books before long. Some of the structural and material considerations attributed to or accentuated by the presence of man have been briefly examined here. They underscore the need for research that will better establish the deep-space environment and its effect on materials and for much development work in creating new structural concepts and devising better fabrication and test techniques.

References

[1] Webb, P., (ed.), ''Bioastronautics Data Book,'' NASA SP-3006, 1964.

[2] Lundeberg, J. F., ''Meteoroid Design Criteria,'' SAE Preprint No. 650786, Oct. 1965.

[3] ''Environmental Control Systems and Selection for Manned Space Vehicles,'' ASD-TR-61-240, Part I, Dec. 1961.

[4] Evans, R. W., Crossland, F. J., and Baginski, W. A., ''Development of Practical Water/Steam Transpiration Cooled Systems,'' AIAA Preprint No. 65-290, July 1965.

[5] Koh, J. C. Y., del Casal, E. P., and Evans, R. W., ''Investigation of Fluid Flow and Heat Transfer in Porous Matrices for Transpiration Cooling,'' Final Rept., Contract AF 33(615)-1868, to be released.

Expandable Structures for Space

J. T. Harris and F. J. Stimler
Goodyear Aircraft Corporation

Expandable structures are seriously contemplated for space vehicles. The term "expandable structure" means a structure constructed of flexible materials which can be packaged into a small volume and, upon command, expanded into its useful shape by inflation.

The advantages of this structural concept exist in its unique capabilities for packaging, ease and reliability of deployment and erection, light weight, and structural recovery after overload.

Goodyear Aircraft has developed this concept into a large number of structural applications, including such flight articles as the nonrigid airship and the Goodyear Inflatoplane, as well as inflated radomes for large radar antennas.

Another development has been the stabilization balloon for high-altitude and high-speed application. This concept offers replacement of the parachute or other conventional drag bodies by providing positive and predictable characteristics in this regime of operation.

The outstanding advantages of this structural approach and the degree of success achieved in its application led naturally to its consideration for space application. When considered in the light of availability of payload and performance of present state-of-the-art boosters, the advantages of the packageability feature for minimum drag and lightweight structural feature for maximum range make this structural concept appear to have high potential for early success.

Expandable structures, fabricated from woven flexible materials, can now be constructed in any shape desired. Ordinarily an inflated structure takes the shape of a body of revolution. Modification of this shape was developed by introducing an internal member along the diameter of a cylinder and then, by shortening this member, creating a two-lobed cylinder. When this is repeated in each of the lobes thus formed, the ultimate result is essentially two flat fabric surfaces connected by a series of vertical elements which can be pressurized to act as a structure. This construction is known as "Airmat" and can be woven as one integral piece of cloth. The cross section of the Airmat can be varied by changing the gage blocks which regulate the depth of the Airmat on the loom. The original "Inflatoplane" wing was constructed of flat pieces of Airmat. However, a contoured NACA 0015 airfoil section is now maintained on the loom.

From a consideration of intended use and operating environment, the requirements of materials for space can be subdivided into three categories: 1) manned orbital operation, 2) unmanned orbital operation, and 3) re-entry. Manned orbital flight in space stations requires high-strength materials because of the structural

Astronautics, Vol. 6, No. 4, April 1961, pp. 30, 31, 92–94.

loads associated with a pressurized breathable atmosphere. Unmanned orbital operation requires lightweight, minimum-strength materials since the external applied loads will be small. Re-entry conditions dictate a material with the capability to resist the heat associated with this operation while retaining sufficient strength for satisfactory functional performance.

In view of these general requirements, Goodyear conducted, under NASA contract, design and stress analyses and development tests on cord-type materials for an interrupted-torus space station concept. Methods of fabrication and packaging were studied to simplify design and minimize weight. Butyl and neoprene elastomers used with nylon are presently being considered as adequate materials based on high-vacuum and ultraviolet test data and fabrication requirements.

A three-ply, nylon-neoprene cord fabric of 0.79-lb/ft^2 weight possessed strength characteristics above 2000 lb/in. A cord-type fabric was designed to develop full-strength characteristics of the yarns in the meridional direction where maximum stress occurs. The bias plies were used to supply structural rigidity in addition to improved strength for the fabric.

Continuous-torus and C-annular-torus configurations were generally 20–40 ft in diameter with about a 7- to 10-ft cross-section diameter. Working pressures of 7–10 psi were considered adequate. High-altitude balloon tests have been considered for the C-annular-configuration to investigate problem areas and conduct preliminary evaluations near operating conditions of a space station. Use of proven escape techniques and equipment and recovery of a space station upon completion of the tests make this a simple and inexpensive program. Artificial gravity for these concepts can be supplied by rotation about the torus axis.

Expandable structures have also been considered for ballistic and lifting re-entry vehicles because of the many advantages they offer. These structures of low density and light wing-loading make it possible to increase the re-entry corridor width, while their packageability makes it possible to utilize presently available boosters for model or full-scale tests. The lower re-entry temperatures achieved with vehicles of this type alleviate environmental and structural problems, thus orienting many design requirements within the state of the art.

In a typical lift-glide re-entry vehicle, parametric analysis was made for wing-loadings of 5 and 10 psf resulting in a maximum re-entry temperature of about 1500°F. The expandable structure consists of cylindrical and Airmat components supplemented by shear webs and cable attachments where necessary.

Unmanned Orbital Operation

A typical example of the application of expandable structures to unmanned orbital operations is the solar concentrator. Our company has pioneered the use of aluminized plastic film for an expandable parabolic reflector for spacepower generation. Methods of fabricating, packaging, deploying, and rigidizing reflectors have been considered for powers ranging up to 15 kW and for dish sizes of approximately 50 ft in diameter. A 10-ft-diam demonstration model has been constructed, under contract, for evaluation purposes. A clear plastic cover completely encloses the reflector. The shape of the reflector is attained by pressurizing the volume between the reflector and the hemispherical section. Rigidization is provided by releasing a lightweight foam between the reflector and the outer cover behind it. After sufficient time to insure permanent set of the foam, the transparent covering is cut from the periphery of the reflector and removed for efficient reflector operation. Two- and 10-ft models are presently being built of 1-mil Mylar to improve foaming techniques and fabrication methods. The concentrator efficiencies will be checked in a specially designed GAC test stand.

In the development of expandable structures for space application, Goodyear Aircraft has proceeded using the philosophy of extending the state of the art of known reliable materials to the new environment and conditions. This involves the application of the best qualities of existing materials after evaluation and testing in the space environment, rather than development of entirely new materials, and has accelerated performance and retained a confidence in reliability not otherwise possible.

Goodyear has had considerable experience, for instance, in the use of polymeric cloth and elastomer coatings. Extensive test and fabrication data are available for such materials as cotton-neoprene, nylon-neoprene, Fortisan-neoprene (Celanese Corporation), and Dacron-neoprene (DuPont). Data on the use of these materials with other elastomers such as butyl are also available. Expandable structures are fabricated with single-ply and multiple-ply fabrics, depending on the stiffness or strength requirements. Multiple-ply fabrics are cemented together at certain angles between the warp threads of successive layers to provide torsional stiffness or improved strength. Where strength is desired in only one direction, cord-type fabrics have been developed to optimize weight and volume. These fabrics consist of polymeric cords or filaments placed side-by-side and held together with the elastomer. The properties of the fabrics available for the applications mentioned previously are indicated in Table 1. These values are meant only to be representative, since in many cases several materials would offer suitable solutions to the problem, whereas material choice may have been dependent on availability of special consideration.

Generally speaking, the polymeric fabrics cannot withstand temperatures greater than 300–400°F. However, glass-fiber fabrics have been tested to temperatures of approximately 1000°F. It was found that the Sil-Temp cloth (Haveg Corporation), a newly developed material, had fairly good strength characteristics above 1000°F after a 2-h exposure. Additional developments in this polymeric fabric area could lead to a material suitable for applications in the 1000°F temperature range.

In defining a fabric for a specific application, the proper weave and yarn must be chosen to obtain acceptable permeability, coating adhesion, joining, flexibility,

Table 1 Characteristics of polymeric fabrics

Fabric type	Total weight, oz/yd	Working pressure	Tensile strength, lb/in.
Nonrigid airship	12–26	1–4 in. H_2O	50–550
Fabric radomes	37	13 in. H_2O	700
Inflatoplane	18	7 psi	450
Drag balloon[a]	4–6	1–2 psi	182
Space station	86	7 psi	2000

[a]Ballute.

Table 2 Weave types and ratings

Type of weave	Tear resistance	Coating support	Gas barrier
Plain (1 × 1)	3	3	1
Basket (2 × 2; 3 × 3)	1	1	3
Twill (2 × 2)	2	2	2

and structural characteristics, to name a few. Table 2 shows the type of weaves generally considered and their ratings for various characteristic properties (the lower the number, the better the value).

It is readily apparent that the twill weave is a good compromise for the properties considered. Flexibility of the cloth is dependent on whether the yarn consists of a single filament or twisted multiple filaments, in addition to the tightness and type of weave.

In the development of a fabric suitable for re-entry applications, superalloys and refractory metals were investigated with respect to strength, flexibility, strength-to-weight ratio, packageability, etc. It was also important to consider the material availability, workability, and growth potential, in addition to its high-temperature characteristics. For this reason Inconel X and Rene 41 were given the most consideration. Since for many representative applications a maximum temperature of 1500°F proved adequate, it was decided to select Rene 41 as most promising, as considerable wire-drawing experience was available and fabrication methods did not seem insurmountable. It is likely that metals such as Udimet 700 (Kelsey-Hayes Company) will replace Rene 41 in some applications as more fabrication experience is obtained.

Rene 41 wire of 0.0016-in. diameter has been woven into 100-by-200- and 200-by-200-count cloth (warp by fill count per in.) in plain, twill, and basket weaves. Representative re-entry-type tests were conducted on these materials with suitable surface coatings, but most data are available on the 200-by-200 plain weave, which was readily available early in the program. One other significant result is that the tear strength of woven cloth is greater than that of stainless steel shim stock of three times the cross-sectional area. The resulting fabric chosen for this re-entry drag body application was 200-by-200 plain-weave Rene 41 wire cloth of 0.0016-in.-diam yarn weighing about 13.60 oz/yd^2 with an included coating weight of 8.0 oz/yd^2 applied to one side.

For high-temperature applications, the coating of a metal cloth must provide protection from oxidation in addition to acting as an inflation-medium barrier. The coating material must have good adhesion to the basic material under static and dynamic conditions and good flexibility to facilitate packaging. Goodyear has conducted tests on several hundred coating materials suitable for both high- and medium-temperature fabrics.

Permeability tests were conducted for a specific heat cycle (T_{max} = 1500°F) at helium pressures of 0.5–1 psi. These exploratory tests indicated that Goodyear coatings CS105 and CS107 were the most promising of the two types tested. Basically these coatings consist of a mixture of silicone elastomer S2077 and a glass enamel. Controlled tests were conducted on CS105 coating to define its characteristic changes through a temperature range of −320–1800°F. The coating retained some degree of flexibility throughout the temperature range tested.

Similar tests of coatings on glass-fiber cloth showed that type-B silicone was very effective up to the 1000°F test.

Present fabrication techniques for metal-fabric structures consider resistance welding, high-temperature adhesives, ultrasonic welding, and brazing, to name a few. Joint efficiencies ranging from 70 to 90% have been obtained with small-diameter Rene 41 wire cloth during the preliminary development phases of recent programs.

Airmat itself has been constructed of such materials as nylon, Dacron, glass fibers, and metal wires for specific applications. The thickness has been varied from 0.4 to 15 in. for the polymers and up to 9 in. for the metal wires for constant cross sections. The maximum width of cloth on present looms is 54 in. However, future plans for a 120- to 240-in. size seem reasonable. The number of drop threads may vary from 8 to 90 per sq in., depending on operating pressures and surface smoothness, but 32 drop threads per square inch are generally used. The present plush looms have a maximum weaving distance of 3 in. between face cloths; however, greater depths in increments of loom capability are achieved through simple fabrication schemes.

The recommended operating pressures for Airmat vary from one-fifteenth to one-third of the burst pressure, depending on such requirements as time, weight, volume, cycling, etc. The maximum bending moment which can be resisted without wrinkling of Airmat is determined by the formula $M_{max} = \frac{1}{2} pt^2$ (lb-in./in. of width), where p is the operating pressure in psi and t is the Airmat thickness in inches. The thickness of the cover plies is determined by stresses due to pressure and buckling or twice the pressure stress, since these two stresses are equal at the wrinkling condition.

The total weights range from 20 to 90 oz/yd^2 with the elastomers and cover plies usually representing about 50–75% of this weight. A new development utilizing a film coating of Mylar (DuPont) and Videne (Goodyear Tire and Rubber) has cut the cover-ply weight to about 20% of the total Airmat weight. This should open new space-application areas.

Goodyear has been in the plastic-film field since the early 1930s, developing such products as Pliofilm, Vita-Film, vinyl film, Videne A, and Videne TC. The company has also developed film laminates—for example, laminates of Mylar to cloth or foams of different density and thickness as a means of improving general performance.

Films have been considered and used for such applications as high-altitude balloons, deceleration balloons, space antennas, re-entry bodies, solar collectors, and other space applications. Mylar, Videne, polypropylene, and polyethylene have been investigated for these applications. The properties of these materials are readily available in the company literature and will not be discussed here.

Considerable development work has also been done on polyurethane and polyether foams. The use of flexible and rigid foams is being considered as a rigidization method for expandable structures. Techniques have been developed to successfully deploy, pressurize, and foam-rigidize expandable structures in simulated space environments. Foam densities on the order of 0.10 lb/ft^3 have been achieved. A laminate material of Mylar film and lightweight foams is being considered as a means of rigidizing large lightweight space structures. Laminates of films and aluminum foil also seem suitable for these applications.

Verification of physical properties for expandable structures generally can be made using the same techniques and apparatus as that employed in rigid-structure testing. Since they are usually composites, however, this requires determination of the individual constituent properties as well as the final composite performance.

The bell jar and related equipment, shown in Fig. 1, are used at Goodyear to subject expandable materials to combined high-vacuum, ultraviolet-radiation, and elevated-temperature conditions. Short- and long-time tests have been conducted on elastomers, films, foams, and coated wire cloth to determine the effects of the simulated spatial environment on weight and performance characteristics.

The Pyrex bell jar is 18 in. in diameter and 30 in. high. It is possible to obtain pressures of approximately 5×10^{-7} mm of Hg (an altitude of approximately 150 mile) with the use of the liquid nitrogen trap. A GE quartz tube, mercury-vapor lamp is used as the ultraviolet source radiating in the wavelength band of from 2200 to 4000 A. The ultraviolet lamp is mounted vertically along the axis of the bell jar. Water-cooled coils are used to maintain specimen temperatures in the 55–300°F range. The specimens are arranged in the bell jar so that some are exposed to combined vacuum and ultraviolet light, while others are exposed to the vacuum only.

Tests on neoprene and butyl elastomers over an approximate 300-h period indicate that a rapid weight loss of about 5% was experienced only during the first 100 h where test temperatures were maintained at 155°F. This is apparently due to the early outgassing of the more volatile, low-molecular-weight constituents of the material. An increase in ultimate tensile strength (11–25%) and stiffness (10–35%) was noted for these test conditions.

Fig. 1 High-vacuum bell jar with ultraviolet source.

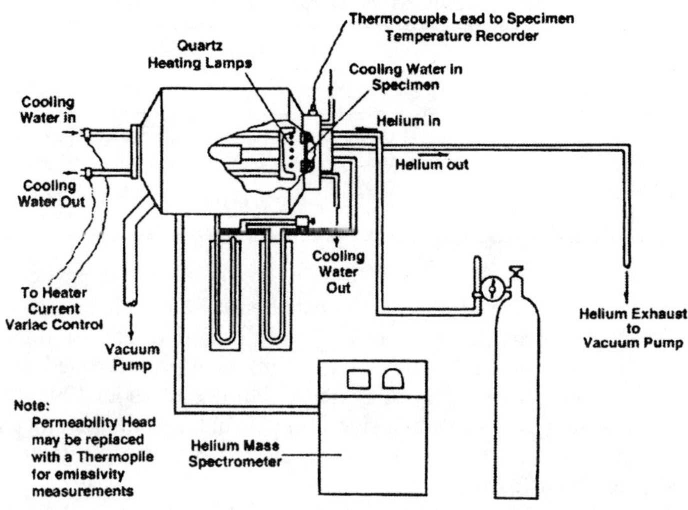

Fig. 2 Diagram of fabric-permeability apparatus.

High-vacuum (5×10^{-6} mm Hg) and ultraviolet tests were conducted for periods of 5 and 15 days on Rene 41 wire cloth coated with the Goodyear elastomer CS105 at the specimen temperature of 300°F. Except for a slight discoloration of the coating, due to the ultraviolet exposure, no deteriorative effects were noted.

The apparatus shown in Fig. 2 is used to determine fabric permeability under controlled conditions of temperature and pressure. The fabric sample is exposed to a controlled helium pressure on one side and a heated vacuum environment on the opposite side. The specimen temperature is controlled by the quartz heating lamps to simulate a representative temperature-time cycle. A constant pressure difference is maintained across the specimen, and the rate of leakage of the helium through the specimen is measured with a helium mass spectrometer specifically designed to analyze gas samples. Plots of helium concentration vs time permit determination of rate-of-change of helium concentration in the chambers as a measure of the fabric permeability.

Rene 41 wire cloth with filaments of 0.0016-in. diameter was tested for plain, twill, and basketweave conditions at a representative re-entry temperature cycle with a maximum temperature condition of about 1500°F and a time cycle of 5.5 min. The helium permeability ranges from 0.002 to 0.02 cu ft/sq ft/min for a pressure differential of 0.5 psi with the plain weave being the least permeable of the three types and the basketweave being the most permeable for similar coating conditions. All the specimens showed practically no helium leakage until the maximum temperature condition was attained. At maximum temperatures, the leakage rate increased and stayed practically constant for the remainder of the test.

Only enough elastomeric coating is applied to maintain a reasonable rate at operating pressures to minimize overall vehicle weight.

Table 3 Emissivities at various temperatures

Temp, °F	Emissivity	
	Coated surface	Opposite surface
500	0.63	0.72
800	0.82	—
1000	0.92	0.95
1200	0.92	0.95

The emissivity of Rene 41 wire cloth coated with CS105 elastomer was determined with the permeability apparatus, slightly modified. A 3-in.-diam-sample opening was used, and both sides of the sample were subjected to vacuum conditions of about 0.1 in. of Hg pressure (an altitude of about 120,000 ft). The quartz lamps were used to heat one side of the sample, while the other side was turned to a thermopile. A metal plate was placed between the lamps and the sample to provide diffused heating.

A comparison of the time rate-of-change of the thermopile voltage for the fabric sample and a standard black body (oxidized Inconel disk coated with camphor soot) at various temperatures resulted in the emissivities shown in Table 3.

The opposite surface possessed normal strikethrough characteristics for an 8-oz/yd^2 coating of elastomer.

In the development of the expandable structure concept for space application Goodyear Aircraft has demonstrated that polymeric fabrics, long useful for low-speed vehicles and ground applications, are now finding a place in high-speed and high-altitude applications and as space vehicle components.

It has also been demonstrated that application of lightweight low-density structures can limit re-entry temperature to 1500°F and make this operation feasible within the present state of materials development.

Materials and Structures for Space Stations

Robert S. Osborne, Clarence O. Keffer, and George Look
NASA Langley Research Center

A major concern in the development of manned orbiting space stations will be materials and structures, especially for the crew's quarters or cabin. Living quarters might be constructed of rigid modules or of a combination of rigid and inflatable sections or of inflatable sections made entirely of flexible materials. The state of the art of flexible materials and inflatable structures being many years behind that of rigid materials and structures, we concentrated effort on inflatables, as discussed here.

The quarters of a space station must have an exacting set of properties. Its structure and materials must be strong and light, able to withstand the space environment, nonhazardous to man, and, with inflatables, flexible and packageable. The listing here outlines specific properties and requirements. Of the loads mentioned, the largest one on the cabin wall will probably be imposed by internal pressure. The station would orbit below the Van Allen belts and so circumvent a major particle-radiation hazard.

Now, it is apparent from these requirements that no one material can be used to construct the cabin wall. A typical wall section will, therefore, probably be of the multilayer, or composite, type whether the wall be rigid or flexible. Going from the inside to the outside of the wall, there might be a coating for interior color and abrasion resistance, a coating for control of gas permeability, elements designed to take the structural loads, a layer of insulation, a micrometeoroid bumper, and, on the outside, coatings for temperature control.

Our early configuration studies indicated that the cabin might be toroidal. Several ways of constructing the load-carrying portion of inflatable toroids have thus been investigated. One of these employs a filament cage and bladder. A torus model, having an overall diameter of 24 ft and a cross-sectional diameter of 8 ft, has been built using this principle. The cage, designed to take the major pressure loads, is constructed of 80-mil dacron cords, or filaments, arranged meridionally around the torus. At the inner rim of the torus, the cords are wound around a 1-in.-diam cable clamped to a rigid 8-ft-diam central hub. An 8-mil-thick bladder made of butyl-impregnated nylon placed inside the cage contains the inflation gases and carries the relatively small local longitudinal loads. After initial inflation to shape during fabrication, the cords are combed parallel and stuck down to the bladder with polyurethane cement. This torus has an operating pressure of 7 psi, which corresponds to a wall stress of about 360 lb per linear in., and it has a design burst pressure of 35 psi, giving a safety factor of 5. It weighs 280 lb and has a volume of ≅2200 cu ft inflated.

Astronautics, Vol. 7, No. 9, Sept. 1962, pp. 36–39.

The composite material has a strength of over 1800 lb per linear in. and weighs only about 0.25 lb/sq ft. This material is foldable, and tests have indicated that the torus can be packaged around the hub so that it occupies only 2% of its inflated volume. Repeated folding and inflation tests have indicated no structural damage due to such packaging. In addition to structural-integrity, shape-stability, and packaging-and-deployment tests, the 24-ft model is being used to investigate arrangement of internal furnishings and controls, air circulation, and leak rates. Further testing will include deployment and leak-rate studies in a large vacuum chamber.

We have also investigated an isotensoid concept for structural skin. This involves filament winding—the winding geometry and cross-sectional shape of the structure being matched so that all primary loads are carried by filaments or cords having equal tension and no load is carried by the flexible binder or elastomer. We recently studied a 45-in.-outside-diam torus so constructed. It had 10-in.-diam cross sections and no central hub. It was made of multiple meridional windings and inner and outer equatorial reinforcing bands to take the overall longitudinal loads and to provide shape stability. The model was one-eighth scale but was designed for a pressure of eight times (56 psi) the operating pressure of 7 psi so that the design skin would have full-scale thickness and stresses. Design burst pressure was 280 psi. The materials used were 12-mil dacron cords and a polyurethane elastomer.

The resulting composite material had a tensile strength of over 1800 lb per linear in. (pli), or 56,000 psi based on the thickness of the material, which was only about 32 mil and weighed approximately 0.25 lb/sq ft. It was not readily foldable, however, probably because of the small amount of elastomer contained in the composite—only about 20%. Pressure tests indicated a failure of the structure at approximately twice the design operating pressure, or 40% of design burst pressure. The failure was not caused by breaking of the cords but rather occurred because of localized spreading apart of adjacent cords where they had no support perpendicular to their length except for the low-strength elastomer. This result was not unexpected, since the inner and outer equatorial bands do not provide resistance to local longitudinal loads. These deficiencies have been corrected in the design of a second 45-in.-diam torus. Double helix windings are utilized so that local longitudinal strength is provided anywhere on the torus, and foldability is improved by modifying the elastomer content.

A third way of constructing the structural skin employs the pattern-lay-up method, gores of a special three-ply fabric being joined to form the torus. Each ply is made of a flexible elastomer and flexible cord reinforcements, with all the filaments in a given ply being parallel to one another. The plies are arranged so that the cords of one are oriented meridionally on the torus, and the cords of the other two plies are oriented approximately 45 deg to either side of the meridional ply. This allows the fabric to handle the ≅2:1 stress ratio in the torus skin.

Samples of such composite materials have been constructed of a butyl-rubber elastomer and dacron cords. One of these fabrics is one-tenth of an inch thick and weighs about 0.6 lb/sq ft. It has a maximum tensile strength of 1800 lb per in., or 18,000 psi based on the 0.10-in. thickness. It is easily foldable, being near 65% elastomer.

Another fabric of comparable strength per inch has also been developed; it contains about 45% elastomer, is 0.07 in. thick, and weighs about 0.42 lb/sq ft. While its folding characteristics appear to be satisfactory, it is not as easy to fold as the 0.10-in. fabric.

One disadvantage of this type of construction is that gores must be put together in some manner so that the full strength of the basic material may be realized. But, as is well known, some elastomers such as butyl are hard to bond. We are presently working on this problem and are hopeful of achieving satisfactory results.

Many of the materials available for use in constructing a manned space station have been tested by manufacturers, and their results are available in current publications. The properties evaluated in their tests, however, are usually those of concern at the surface of the Earth. Little information is available on response to a hard vacuum, large temperature extreme, and to ultraviolet (UV) radiation—properties that determine the suitability of a material for space application.

It became apparent early in our work that pertinent materials must be exposed to a hard vacuum, UV radiation, and large temperature extremes over long periods and that degradation of properties due to this exposure must be carefully measured. We established a laboratory for such studies.

This laboratory contains three small high-vacuum systems. Two of the systems have 18-in.-diam by 30-in.-high chambers capable of simultaneously exposing several 10-by-1-in. samples. These have windows to admit UV light and liquid nitrogen cryopanels and heating coils to give temperatures from –300 to 300°F. They operate in the low 10^{-7} mm Hg range, as does the third.

The third system contains an automatic vacuum balance in a 6-by-10-in. chamber. This allows continuous monitoring of a sample's weight loss without removing it from the vacuum and, thus, gives an accurate indication of its rate of evaporation or decomposition in space. Our experience shows that removing a sample from vacuum and weighing it in air gives meaningless results because of indeterminate moisture pickup. This chamber also has a cryopanel, a heating coil, and a window for UV exposure.

Obtaining the desired UV exposure has been a problem. Conventional equipment, such as the mercury-arc source and quartz windows, only provides UV coverage down to about 2000 Å. But there are indications some materials may be affected by radiation near the Lyman-alpha line at 1216 Å. So our system reaches down to 1000 Å through use of a hydrogen-discharge-tube source and a lithium fluoride window specially designed to function with high pressure across it.

The laboratory also contains the usual equipment, such as an Instron tensile-testing machine for determining degradation of physical properties. A Dow-cell gas-transmission tester serves to measure permeability of wall materials to oxygen and nitrogen; sheet metal, such as aluminum and stainless steel, can be studied as well as composite fabrics. The Dow cells can easily measure gas-transmission rates at standard conditions as low as 0.001 cu ft/sq ft of surface per year.

It is vital to identify accurately gases that emit from material exposed to the space environment or that collect in a closed space station environment. This helps analyze breakdown mechanisms of materials such as polymers, and it aids in identifying toxic and noxious gases. To aid in identifying these gases, the laboratory includes a time-of-flight-type of mass spectrometer and a gas-liquid gas chromatograph. This equipment is used in conjunction with both the small vacuum chambers and a much larger one for evaluating life-support systems.

Our investigation of materials in the laboratory has emphasized long-term tests and has given some interesting results. For example, samples of three different filament-elastomer composite fabrics were exposed to a pressure of 4×10^{-7} mm Hg for 140 days at room temperature and without UV exposure. Breaking tensile strength and elongation of the samples were determined before and after exposures of 24, 61, and 140 days. A nylon-fairprene fabric which had an initial breaking strength of 335 lb per linear in. lost 5% of its strength the first 24 days in the vacuum and then retained an approximately constant strength for the rest of the test. A nylon-neoprene fabric having an initial breaking tensile strength of 282 pli and a dacron-silicone sample having a strength of 115 pli decreased in strength about 11% after the 140-day exposure. The important thing here was that the decrease occurred over the whole exposure period and that there were indications that continuing small decreases might be expected upon further exposure. Additional long-term tests are being run to verify these data. No trends were evident from the breaking elongation test results.

Data obtained in the vacuum-balance facility have indicated no serious weight losses for prospective flexible-wall materials. A 25-sq-in. sample of dacron-cloth fabric coated with a polyolefin elastomer was exposed to a pressure of 5×10^{-7} mm Hg at room temperature for 14 days. The sample lost only 0.07% of its weight, all of this loss occurring during the first 24 h. An 18-sq-in. sample of butyl-coated dacron fabric exposed to the same environment for 50 days lost 0.5% of its weight. This loss, however, occurred over a period of 30 days and shows the desirability, as indicated also for the strength-loss tests, of making sure that such exposures are continued for sizeable periods of time or until there is positive evidence that no further changes in properties are taking place.

It will be noted that no vacuum-exposure work on metals has been included in the program. This is because much is already known about the behavior of metals in the space environment and little is known about polymers. As indicated in the literature, there appears to be no problem concerning the use of most of the structural metals in space. Aluminum, steel, and titanium, for example, must be exposed to temperatures above 1200°F before they lose as much as 0.0004 in. of surface thickness per year. This would have no significance structurally. On the other hand, changes in the properties of surfaces could affect the absorptivity and emissivity of a material and so upset the thermal balance of a space station. But there is little evidence that the elements of the space environment being considered here have any appreciable effect on the mechanical properties of structural metals.

It is vital that a space station retain as much as possible of its initial supply of air. Resupply will be costly. So loss of air through walls, joints, seams, and feedthroughs must be cut to the minimum. Using the gas-transmission tester, several wall materials have been tested for leakage rates using a pressure difference of 15 psi across the sample. Results indicate that no measurable loss occurs through $\frac{1}{64}$-in. or thicker Type 2024 or 2014-T6 aluminum or Type 347 stainless steel. Also, resistance- and heliarc-welded seams tested in $\frac{1}{64}$-in. and $\frac{1}{32}$-in.

aluminum evidenced no gas transmission. Riveted joints, however, leak badly, and methods of sealing these joints using Saran-type films are being investigated.

Flexible filament-elastomer fabrics, on the other hand, have objectionably high leak rates. It was indicated, for example, that even using one of the most impermeable of the elastomers—butyl—that the leak rate through a 0.1-in.-thick wall would be on the order of 0.7 cu ft of room air per sq ft of surface area per year, at a pressure difference of 1 atm across the wall. This amounts to about 800 lb of air for a 150-ft torus with 10-ft cross section.

Experiments show, however, that a 1-mil-thick coating of a Saran-type film on the inside of the wall reduces this leakage to about 2 lb per year. A program is now under way to improve the bonding properties of Saran and to increase its flexibility in order to make it more compatible with heavy-fabric flexible wall construction. A second reason for using an impermeable film on the inside of the wall would be to keep the objectionable odors of some elastomers out of living quarters.

It appears that the problem of gases passing through the walls of a space station can be solved. But the prevention of leakage around hatchways, airlocks, actuators, feedthroughs, etc. is another matter. This is a difficult problem requiring a large effort. It is being investigated both within NASA and by companies under contract.

Micrometeoroid punctures pose one of the major problems with structures in space. It is not possible at present to simulate the velocities of micrometeoroids, and, therefore, we cannot tell exactly what will happen to materials exposed to them. However, some preliminary results are available at lower velocities. Aluminum spheres $\frac{1}{16}$ in. in diameter have been fired into samples of the 0.10-in. three-ply and 0.032-in. filament-wound dacron-elastomer materials previously discussed at velocities from 6000 to 13,000 fps. The resulting punctures had approximately the same diameter as the projectile, but the materials delaminated over 4 or 5 diameters. Firings into 2-in. cubes of these materials resulted in penetration-to-diameter ratios of about 3.5. This indicates that, for equal material weight, resistance to penetration is approximately the same as for aluminum, but a given fabric must be 2.5 times as thick as aluminum to have the same stopping power and would, therefore, be hard to fold.

The best solution to the micrometeoroid problem may be to use a bumper, a thin layer of material spaced on the order of an inch outboard of the structure. This approach is possible for inflatable or rigid structures.

Joints and seams constitute a problem area still requiring much work. Methods are required for putting together the 1800-lb/in. materials previously discussed with a reasonable factor of safety and with a minimum of additional bulk and weight. As an initial effort in this area, approximately 20 commercial cold-setting cements have been used to join 1-in.-wide strips of dacron and nylon fabrics employing neoprene, butyl, hypalon, and silicone elastomers. Simple lap joints having 1-in. overlap were used in these tests; the joined area equalled 1 sq in. The maximum joint strengths so far achieved have been 400 lb for neoprene, 275 lb for hypalon, 275 lb for butyl, and 45 lb for silicone. It is realized that higher strength can be achieved by increasing the length of the lap, but this first investigation was to determine the efficiency of existing commercial cements.

Future programs include studies of variations in type and length of overlap, vulcanizing, and mechanical seaming.

Our work in materials for manned orbiting space stations has only begun, and there is much more to be accomplished. We must study honeycomb structures, lubricants, rigid and flexible foams, and sealing methods. Advances must be made in filament-winding techniques. The application of glass filaments to flexible structures must be investigated. Finally, more information must be obtained on the effects of the space environment—especially micrometeoroids and particle and electromagnetic radiation—on prospective space station materials and built-up flexible and rigid wall structures.

Bibliography

''Elastomer and Cord Fabric Investigations, and Preliminary Design Studies Associated with an Expandable Orbital Space Station,'' Goodyear Aircraft Corporation, Contract NAS1-725, GER-9963, Aug. 26, 1960.

Jaffe, J. D., and Rittenhouse, J. B., ''Behavior of Materials in Space Environments,'' *ARS Journal,* Vol. 32, No. 3, March 1962.

Marketos, J. D., ''Optimum Weight of an Expandable Structure Having the Form of a Surface of Revolution—Applicable to Toroidal Space Stations,'' Society of Aeronautical Weight Engineers, Technical Paper No. 244, May 15, 1961.

''Preliminary Development and Design of an Expandable Torus Space Station,'' Goodyear Aircraft Corporation, Contract NAS1-620, GER-9766, April 22, 1960.

Schuerch, H., ''Space Structure Design with Composite Materials,'' Astro Research Corporation, ARS Preprint 1096-60, April 1980.

Schuerch, H. U., and Kyser, A. C., ''Isotensoid Torus Design,'' Final Rept., Astro Research Corporation, Contract NAS1-889, ARC-R-9, Nov. 9, 1960.

Space Platforms for Building Large Space Structures

Charles J. Goodwin
Grumman Aerospace Corporation

In the spring of 1974, thinking about how large space structures—particularly the very-low-density structure of the solar power station (SPS)—could be constructed, we foresaw the need for some sort of beam builder to operate in space. Work on this concept has proceeded, under contract to NASA Marshall, to the point where a ground demonstration machine that fabricates multibay, 1-m-deep, triangular section beams has undergone successful development trials.

In an early space demonstration being planned, the Shuttle would carry the beam builder into orbit on a sortie flight. There, a compound structure would be fabricated, assembled, tested, and left in orbit (Fig. 1). The beam builder would return to Earth with the Shuttle.

Naturally, a single Shuttle sortie flight imposes limits on the maximum size of structure that can be built. Furthermore, the beam builder, when developed, ought to remain in orbit between uses to avoid the cost of repeated launches. A space platform would answer these points.

This platform could have solar arrays, radiators, space-construction equipment, and docking provisions. It would be a free flyer, placed during the mid-1980s into a low-Earth orbit readily accessible to the Shuttle. Docked to the Shuttle, as shown in Fig. 1, the platform would enlarge the construction workspace and add power and cooling to handle more demanding payloads and extend the Shuttle's on-orbit duration. As a free flyer, it could support experiments, laboratories, observation instruments, and, in due course, habitation modules. At that point, it would become a full fledged space station.

For the most part, the platform will help fabricate and assemble structures and satellites. This contrasts to the repetitive series construction envisaged eventually for SPS production (Fig. 1).

Many of the platform functions, such as providing room for a construction workbench, an orbital depot, space manufacture, experimental laboratories, observation instruments, environmental exposure tests, logistics handling, docking, and eventually habitation modules, fall into the ''real estate'' category. The room needed for real estate functions suggests a platform quite a bit larger than the Shuttle cargo bay. Consider the list of equipment just to carry out construction workbench functions: a cherry picker or a crane as a general purpose mobility aid; berthing, storing, and tensioning equipment; a device for deploying thin films; a beam builder; outside lighting and sunshades; alignment, checkout, and test equipment; and mission-peculiar construction jigs. All but the last would

Astronautics & Aeronautics, Vol. 16, No. 10, Oct. 1978, pp. 44–47.

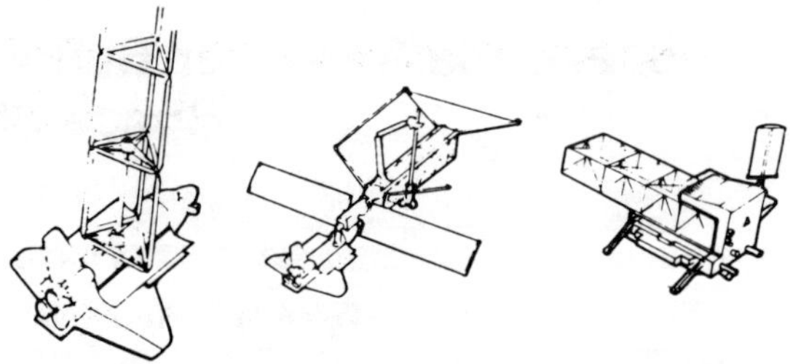

Fig. 1 Evolution of space platforms: left, early Shuttle sortie with beam builder; center, mid-1980s space platform to extend Shuttle's abilities; and right, later construction base in low-Earth orbit for building satellite power stations.

serve on several construction jobs. Because of the succession of one-of-a-kind structures to be built, this equipment will emphasize versatility rather than the extreme automation looked for in the full-size SPS production phase.

As the first step in determining the right size for a construction platform, we have drawn up a table (Table 1) of seven large space structures (LSS). These represent those required in the mid-1980s and appear too big for convenient construction direct from the Shuttle (the Shuttle cargo bay is 18 m long). Building some of these LSS breaks down into several distinct construction tasks. For instance, building the 100-m-diam radiometer divides into three: constructing a dish, constructing a tower, and final assembly. Assembling the seven LSS calls for completing 12 construction tasks.

The eventual orbits of the LSS present a problem. Four of the LSS must be transferred to geosynchronous orbit, and a low-inclination construction orbit would minimize the plane-change energy. The 100-m radiometer, on the other hand, must operate at an at least 55-deg inclination, and transfer from a low-inclination construction platform takes even more energy at low-altitude orbital velocities than from the ground. The best inclination for a low-altitude construction platform, therefore, depends on the mission and has yet to be determined.

For the cherry picker, we can come up with a more definite size. The analysis starts with four elements sketched in Fig. 2: a railed platform, a rail-mounted cherry picker that can run the length of the platform, a jig (represented by four sloping arms), and workpiece. At a reach of 70 m or more, the cherry picker can perform all tasks with fixed jigs. As the reach drops below 70 m, first one, then two, then three of the jigs have to provide a degree of freedom to the workpiece (sliding or rotating) to continue to permit all the tasks to be carried out. When the reach dips below 30 m, the number of feasible construction tasks falls off rapidly, even with the more elaborate jigs. For this group of LSS, the railed cherry picker should have a reach of 30 m.

For the method of construction proposed, a platform 65 m long would permit performing all 12 construction tasks if half the jigs gave a sliding or a rotating degree of freedom to the workpiece (Fig. 3). The frequency of the need for rotation proves the turntable a generic piece of equipment.

Table 1 Representative large space structures for construction system sizing

Description	Eventual orbit	Mass, kg × 10^3	Size, m
250-kW solar array to power a space station	29 deg, 400 km	2	50 × 50 
 2-MW satellite power development article module, 4 required	geosynch.	4	111 × 50
1-km mapping antenna	geosynch.	5	1000 × 3 
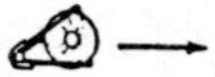 83-kW Brayton solar collector to power space station	28 deg, 400 km	10	30 diam × 40 long
Radiometer	55 deg, 400 km	75	100 diam × 200 long 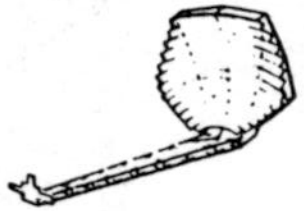
 Public service platform	geosynch.	28	60 diam × 70 long
600-kW satellite power development article	geosynch.	15	32 wide × 140 long 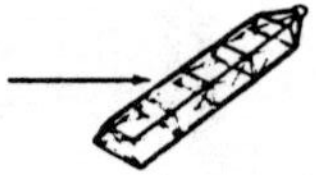

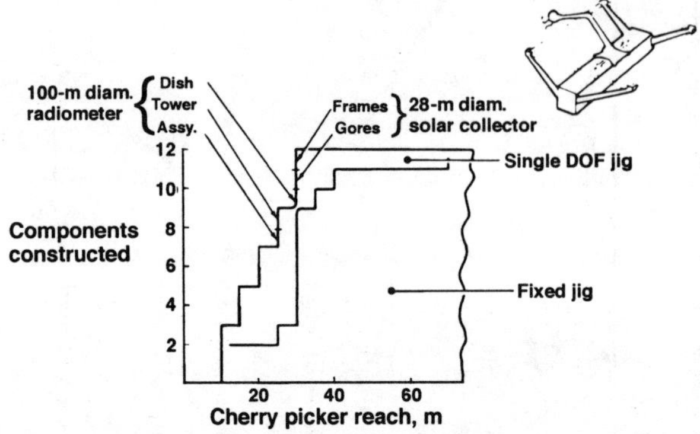

Fig. 2 Cherry picker on a railed platform capable of carrying 12 construction tasks represented by the seven large space structures in Table 1.

This combination of a 65-m-long railed platform and a travelling cherry picker with a 30-m reach represents one answer for this group of representative LSS. A reach of 30 m may approach the limits of feasibility for a highly articulated space-based cherry picker at the stiffness needed.

A different construction approach leads to another possible solution—a fixed-base cherry picker with a purely nominal size platform. In general, the smaller the platform, the larger the jigs, but this exchange does not influence the required cherry picker reach.

With suitable jigs, a cherry picker with a reach of 68 m will handle all 12 tasks (Fig. 4). Such a reach would probably make the cherry picker too flexible. If the construction sights are lowered by not building the 100-m radiometer, the reach shrinks to 37 m, but still significantly longer than the 30 m of the first approach. Bearing in mind also that there are many other potential demands for

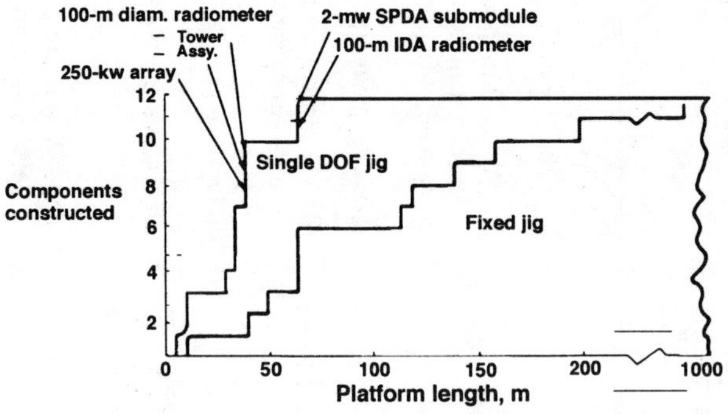

Fig. 3 Work platform 65 m long with a moving cherry picker having a 30-m reach needed to construct all 12 components of structures in Table 1.

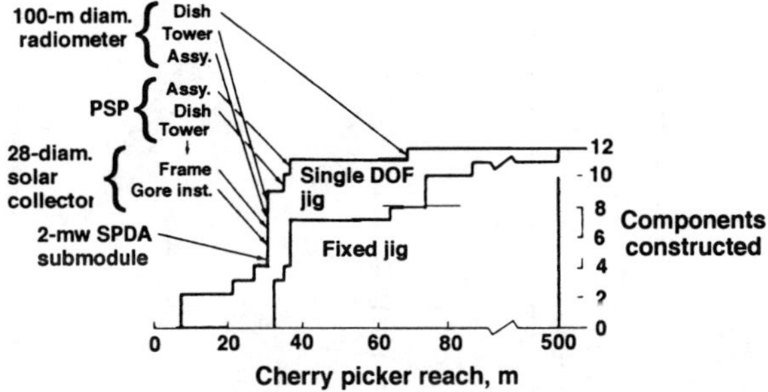

Fig. 4 Fixed cherry picker with a 68-m reach determined to be too flexible.

room on the platform over and above construction, the 65-m-long platform with the 30-m reach railed cherry picker appears the better direction to go.

With its approximate size established, consider the structure of the platform. Several structures have been proposed as building blocks from which to assemble LSS (Table 2). When deployed and assembled, they will be rather light, lightly loaded, spidery structures. Any of these or others that win the competition for sustained funding are likely to be too delicate by an order of magnitude for use in the platform itself. A special robust version of one of the LSS candidates or a Shuttle external tank fitted out before launch with strong points and rails and carried into orbit rather than dropped into the ocean appears better suited. The tank has the virtues of being remarkably strong and stiff and putting the bulk of the assembly work on the ground before launch. Because normal Shuttle operation boosts the external tank very nearly to orbital velocity, carrying it fully into orbit would cost only 2500 kg of orbiter payload. One external tank measures only 47 m long—short of our target. Joining two tanks end to end or using structure assembly jigs which extend beyond the ends of a single tank would cope with this problem.

From our studies emerges a picture of the space platform as a structure expanding the Shuttle's power and cooling capacity, extending its duration, and enlarging its construction capabilities. As a free flyer, the platform would serve many low-Earth-orbit missions. It could grow in an orderly way and would fit naturally and smoothly into a logical progression from Shuttle sortie to space platform to space station to SPS base.

In its construction role, in the 1980s it would handle many one-of-a-kind tasks which would demand versatile, rather than automated, equipment. Automation would come later with the SPS. We see a platform length of 50–70 m, and a railed cherry picker with a 30-m reach.

Thus, the major functions and the size of a platform are coming into focus. The optimum orbit inclination or inclinations, its required stiffness, and its structure remain to be settled.

Table 2 Typical large space structure construction element options

Element	Launch density, kg/m^3	Size limit		Intermediate joints	Space construction equipment
		Length	Depth		
Space-fabricated triangular beam	4–8,000	None	None	Yes	Beam builder
Space-fabricated geodetic beam	4–8,000	None	None	None devised	Beam builder
Dixie beam	100–200	1.5 bay	None	None devised	Dispenser, assembler
Unfolding beam	20–40	0.9 bay	None	Yes	None
Unfolding tetrahedron	25–50	None	0.5 bay	Needs none	None

Practical Design of Low-Cost Large Space Structures

John M. Hedgepeth
Astro Research Corporation

Martin M. Mikulas Jr.
NASA Langley Research Center

Richard H. MacNeal
MacNeal-Schwendler Corporation

Without a doubt, the 21st century will see numerous large space structural systems. But it will be the impending first decade of the Shuttle age that introduces us to large structures in space. The missions of these first-generation structural systems have been and are being studied; they do not yet present a clear picture. But the manner in which the first large space structures will be designed and the approaches that will be used in developing hardware for them are very predictable, as we hope to show in this article.

Over the past 20 years, our aerospace profession has evolved a highly effective way of developing space-systems hardware. The frequent failures of the early 1960s have faded to only dimly remembered nightmares—now, when the "go" button is pushed, success is expected. The proven techniques of space-system development entail high costs, but remembrance of those early, unfortunate days makes us reluctant to depart from our now-accustomed ways. A realistic examination of the prospective first generation of large space structures must keep that reluctance in mind.

The first generation of space structural systems will almost certainly be produced in very limited numbers. This means that the costs of development cannot be spread over a large production quantity. Consequently, the largest cost of a space structural system will continue to be the cost of design, analysis, test, and all the systems engineering and tons of paperwork necessary to give adequate assurance that the system will perform as intended. The next largest cost will be launch and space erection. The smallest, by far, will be the manufacturing costs of materials, hardware, fabrication, assembly, and inspection.

The engineer will, therefore, find only one practical way to design low-cost space structures: use approaches that reduce the cost of the design and development effort itself. Such approaches include:

- Design with criteria arrived at rationally.
- Design for simplicity, repeatability, and modularity.

Astronautics & Aeronautics, Vol. 16, No. 10, Oct. 1978, pp. 30–34.

- Assemble without adjustments.
- Design for testability on the ground.
- Achieve structural efficiency by configuration and material choice, rather than by squeezing down on the design margins.
- Prefabricate and preassemble before launch.

Criteria

The successful performance of any structure depends largely on the correctness of the loads and design criteria. Until now, most structures used in spaceflight have been designed primarily to withstand launch loads. A great deal of effort in the 1960s dealt with understanding these loads and developing criteria for design that would yield structures that could withstand the loads without being excessively heavy. The design process for launch structures has, therefore, matured on this solid foundation of rational design criteria.

So far, most "space" structures have been, in fact, "launch" structures, inasmuch as their primary design requirements have stemmed from the launch environment. But the large structures being contemplated will be deployed, erected, assembled, or fabricated in space. Such structures will be required to face the launch environment only in a packaged, predeployment state. Indeed, their primary design requirements will be derived from the spaceflight environment and will deal with phenomena as primary criteria which have been considered as only secondary in the past. The design of such genuine space structures will require a solid foundation of criteria similar to that created for the launch environment.

Prominent among the new criteria will be enough *stiffness* to avoid deleterious interaction with the control system. The control system acts essentially as a centering spring that overcomes a disturbing torque Q with an allowable angular displacement $\Delta\theta$. The effective stiffness of the control system takes the form

$$K = \frac{Q}{\Delta\theta}$$

The control frequency f_c may then be defined as

$$f_c = \frac{1}{2\pi}\sqrt{\frac{K}{I}}$$

where I is the mass moment of inertia of the spacecraft about the same axis as the applied torque.

For large structures in planetary orbit, gravity gradient exerts the dominant disturbing torque. A worst-case attitude assumed for a platform configuration necessitates the following control frequency for various orbits to achieve the desired pointing accuracy, as shown in Table 1.

Knowing the required control frequency, the engineer can then estimate the structural stiffness needed to avoid unstable interaction between structure and control system. Figure 1 gives sample results obtained by requiring gain stabilization. Some structural configurations, such as tension-stiffened ones, exhibit very low damping and, therefore, require a frequency ratio of 10 or more.

The foregoing very simple analysis serves to identify an upper limit on the structural-stiffness criterion. With advanced control techniques, it should be possi-

Table 1 Pointing accuracy

Orbit	Accuracy	f_c, Hz
Low (R = 6740 km)	0.1 deg	0.0053
Geosynchronous (R = 42,190 km)	0.1 deg	0.0053
Low	1 arcsec	0.10
Geosynchronous	1 arcsec	0.0066

ble to relax it. Even so, the structural frequencies needed to cope with control frequencies are much lower than those postulated in most current studies. Using too high a stiffness criterion will clearly cause excessive weight and cost.

Modularity

The high costs of development chargeable to a one-off production run can be reduced by insisting on simplicity of design detail even if that increases weight somewhat. Repeatability and modularity in the design can reduce cost even more than simplifying design detail, for they obviously save engineering and tooling costs. Repeatability also increases reliability since the design of each type of repeated part can be economically examined in enough depth to ensure its proper operation.

Assembly

Hardware experience on deployable booms has shown the advantage of designing modular structures for assembly without adjustment. That eliminates the costs of designing the adjusters, avoids the labor costs of training the adjusting personnel

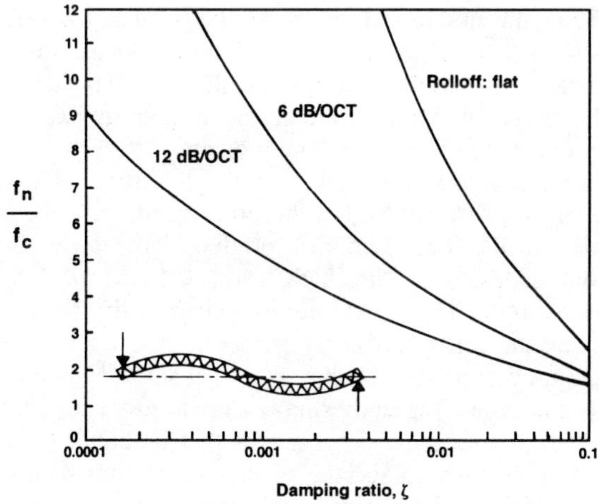

Fig. 1 Stiffness required for stable control.

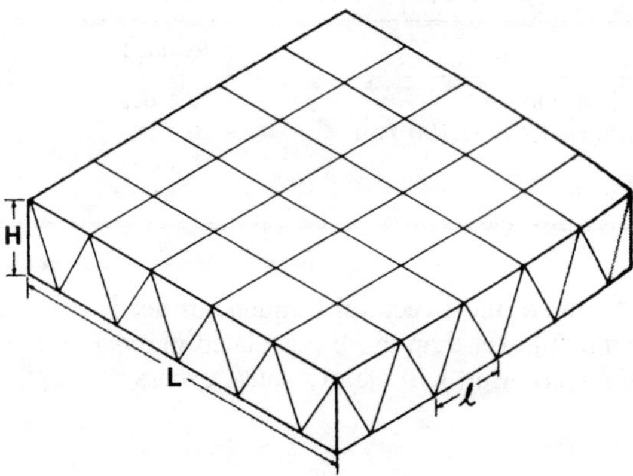

Fig. 2 Truss platform.

and inspecting their work, and reduces the costs of assembly fixtures. The savings in time and money greatly outweigh the additional expense of fabricating the details and subassemblies to the necessary close tolerances. Here again, such an approach greatly increases reliability because it eliminates human error that would have been entailed in the adjusting operation.

Repetitive elements needing no adjustment can yield very true structures. The Astromast deployable lattice column, for example, has all its longerons and diagonals made on the same tooling, and it has excellent straightness. So does the Extendable Support Structure for the Seasat Synthetic Aperture Radar. Assembled from repetitive elements bonded sequentially on constant-temperature tooling, it supports the 11-m-long antenna surface flat to $\pm$ 2 mm.

Figure 2 shows the theoretical limits on the accuracy achievable with such repetitive structures (truss platform shown). It relates the root-mean-square value of the lateral distortion σ_δ at the center of the platform to the random manufacturing error in the length of the truss elements—a quantity expressed in the form $\delta_{\Delta\ell/\ell}$, the rms value of the fractional length error.

If the truss platform has small depth, then length errors in the surface elements produce large surface distortions. On the other hand, with the truss very deep, the length errors in the long diagonals produce large local distortions. There exists an optimum design depth which, for equal fractional tolerances in the surface and diagonal elements, turns out to be half of the geometric mean of the overall size L and the cell size ℓ.

For the optimum design, the relative statistical surface distortion is half of the random length tolerance. Our studies have shown that, with care, a tolerance of one part in 100,000 is practical. This means that it should be possible to build kilometer-sized platforms to a surface accuracy of 5 mm without adjustment.

For a cell size of 10 m the analysis yields an optimum depth for a kilometer-sized platform of 50 m—considerably greater than would be initially supposed. The influence of deterministic effects such as thermal strains would be reduced at this depth. For a differential strain of 10^{-5}, for example, the maximum departure from a best-fit plane would be 12.5 mm and the average, 5 mm.

Testing

Ground testing forms a necessary part of a program development discipline. Ultimately, after enough experience with Shuttle space operations, such testing can be replaced by exposure to the actual space environment, but until then, we will insist on testing the structure on the ground before launch to ensure a successful mission. If the full structure cannot be tested, segments as large as possible will be. The cost of this testing depends on the design, and the design should be selected with that in mind. Incidentally, in many cases it will be the ground testing that causes the largest loads encountered by the structure.

Structural Efficiency

Spaceflight structures can be designed with a large degree of freedom. This allows the use of efficient structures often precluded by other environments. Thus, the structures can be made light in weight yet still provide generous margins of safety—margins very useful for the first generation of large structures because of the unfamiliarity of the design, assembly, and operational conditions.

Consider the 50-m-long diagonal elements needed for the platform truss shown in Fig. 2. One method for constructing such columns uses nested conical segments which are unpackaged and assembled in space. This approach, attractive because of its high packaging density, has been extensively studied.

Another approach has been considered. It involves an expandable truss as shown in Fig. 3. A short segment has been built and tested. The load-carrying element in this truss is the slender central tube. The lattice of longerons and

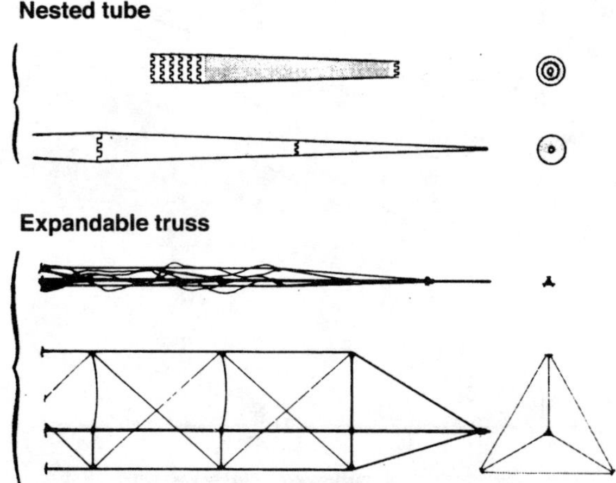

Fig. 3 Column types—nested tube, expandable truss.

diagonals makes up an outer truss which supports the central tube against lateral buckling by means of spokes. One of the spokes of each set is made somewhat lighter and longer than the other two. This lighter spoke then buckles, its Euler load supplying the necessary force to pretension the diagonals. The longerons are pretensioned when the truss is expanded (the spokes rotate and each longeron moves outward and longitudinally as a unit) by pulling at the end fitting, precompressing the central tube.

Figure 4 compares masses of the nested-tube and expandable-truss columns for graphite composite material (the superior material for space structures).

The structural efficiency parameters $M/l^{5/3}$ and P were found to be convenient for lightly loaded, minimum-gage columns. For the loading considered, tubular columns appear as a single line, independent of the column length. Other column concepts show some variation in the mass parameter with length, but the change is small.

The masses for the nested columns were taken from previous work (minimum wall thickness of 0.38 mm). The mass of the expandable truss is based on the center column and spokes having wall thickness of 0.7 mm.

It can be seen from Fig. 4 that the expandable truss has considerably less mass than the nested tubular column.

It is necessary to consider the required packaging constraints as well as the mass. Figure 5 compares the number of nested columns and expandable-truss columns that can be stowed in the Shuttle cargo bay. The parameters were chosen to be compatible with those in Fig. 4.

The nested column was assumed to be packaged with a density of 450 kg/m³ independent of length. A small sketch in the figure indicates the packaging configuration assumed. For the expandable-truss column, the diameters of the center column and spokes were obtained directly from the design procedure used. The other small sketch in the figure shows the resulting packaging configuration.

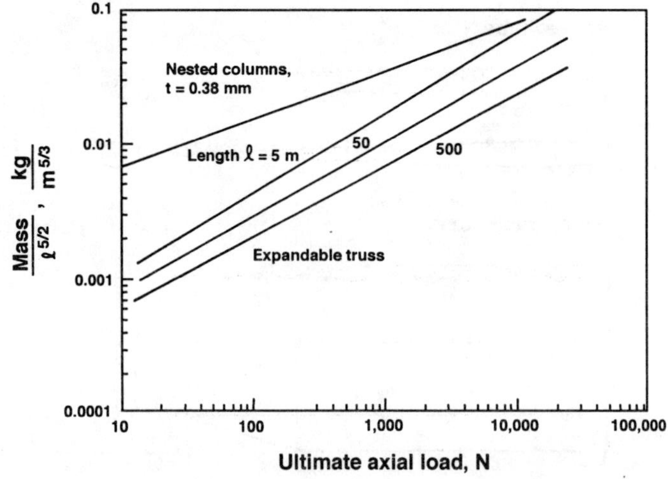

Fig. 4 Nested-tube vs expandable-truss columns in mass.

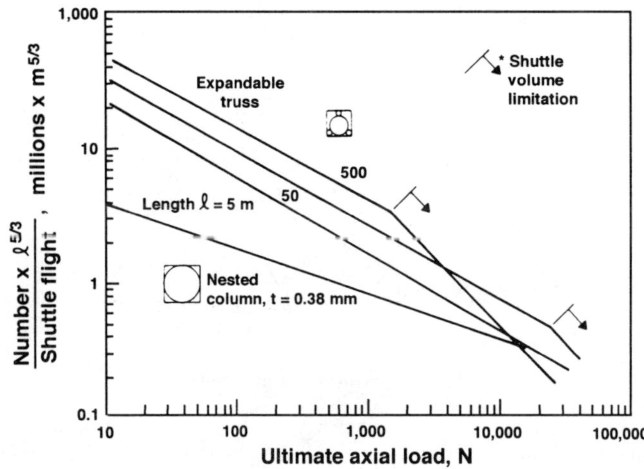

Fig. 5 Shuttle's capacity for expandable-truss vs nested columns.

Since the nested columns and expandable-truss columns package quite well, the Shuttle's mass-carrying ability limits the number of columns per flight in most cases. For high values of load or length, however, the number of expandable trusses per flight becomes limited by volume rather than mass, as indicated in the figure. For lower values of the design load, considerably more expandable truss columns than nested columns can be put on each flight of the Space Shuttle. This increase in number can be reflected directly in the amount of structure built with each flight.

The Seasat Extendable Support Structure already mentioned presents another example of taking advantage of configuration selection. The designer usually packages such long planar arrays in zigzag fashion and deploys them as a thin structure with, perhaps, scissor stiffening at the edge. In the Seasat case, the available package depth was only 22 cm, and extreme accuracy would have been required for fabrication and testing of the delicate structure. A deep truss entirely avoided these problems, and the robust structure has been fabricated and tested by straightforward techniques.

Such advantageous structural configurations tend to be complex, consisting of many elements and joints. Repeatability here avoids an otherwise severe reliability problem. Also, strong discipline must be used to keep the joints simple. Preferably all joints should be simple hinges (Fig. 6) having long hinge pins of small diameter to keep the torque low. Latches should be avoided. Joints should be given the necessary rigidity by closure springs with levers if the loads are large.

Following these rules has yielded good results. Several Astromasts have been flown without any hitches, including the magnetometer boom on the Voyager spacecraft. Another, the Seasat Extendable Support Structure, deployed smoothly and successfully, according to plan, after the launch in June.

Preassembly

The first-generation large space structures will probably involve some assembly. That will prove expensive; each second of the astronaut's time is very

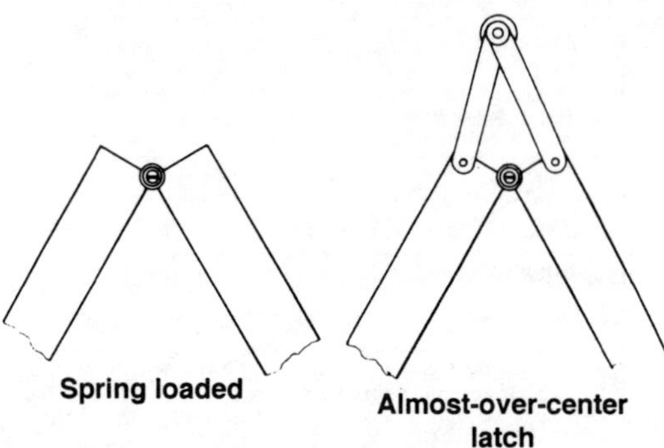

Fig. 6 Simple hinge joints.

valuable and automation is complex. Reducing cost will entail preassembly and easy deployment. This requires a great amount of ingenuity to keep an acceptable packaging density without introducing unacceptable complexity.

The Future

As the reader may surmise, the next decade will be one of great activity and learning for those of us working with space structural systems. Not only will we push outward the boundaries of structural and material technology, but also we will create a new discipline of hardware development. Growth in each of these areas will be necessary to build the foundation for the large structural systems of the 21st century.

Structures for Solar Power Satellites

Ralph H. Nansen and Harold Di Ramio
Boeing Aerospace Company

In a useful size a solar power satellite (SPS)—a large electrical power generating system in a geosynchronous orbit transmitting energy to Earth by microwave beam—has output at the ground interface between 5–10 gigawatts (GW). An SPS might use one of several methods of energy conversion—photovoltaic (solar cells at various concentration ratios), closed-cycle thermal engine, and thermionic—and these have been evaluated over the past several years. The currently preferred energy-conversion system, photovoltaic, uses annealable solar cells and no reflective concentrators, yielding a planar satellite as the baseline concept. Two antennas, one at either end of the satellite, together provide 5-GW useful output to two independent receiving stations on Earth.

Figure 1 shows the overall configuration of the baseline photovoltaic concept. Eight major modules made up of 32 bays each comprise the power-conversion area of the satellite. In overall planform, the satellite measures 21.28 by 5.30 by 0.47 km—truly a large structure.

The structural subsystem provides the backbone for the satellite. It employs 660-m-sq bays 470 m deep as repeating elements.

Design considerations for the structural system include pretensioning of the solar-cell blankets; modules strong enough to withstand being transferred assembled from low-Earth orbit to geosynchronous; and structure stiff enough for overall satellite control under the gravity gradients, thermal cycling, and transients that would be encountered in the geosynchronous environment.

Figure 2 shows a detail of a typical bay in the satellite and also depicts the method of supporting the solar-cell blankets within each bay. The solar-array panels are supported by a main web that attaches to a catenary at approximately 1-m intervals. In turn, the catenary attaches to the main structure at 20-m intersections. Spring-loaded piston cylinders exert a constant force on the solar-array support system to counteract thermal expansion and contraction. This arrangement also counteracts a movement of up to 2 m that could be caused by an acceleration of 10^{-4} g during the transfer to geosynchronous orbit.

The basic element of the 660-m bay, a triangular beam, also serves as the basic element throughout the satellite. Candidate chord concepts for the triangular beam element include tubular pin-ended members fabricated on the ground and open-chord and closed-chord sections automatically fabricated in orbit. The upper surface of the satellite experiences a maximum load of 3300 lb.

Figure 3 shows a 20-m triangular beam comprised of tapered tubes and notes construction features. The strut pattern for these beams provides an optimum

Astronautics & Aeronautics, Vol. 16, No. 10, Oct. 1978, pp. 55–59.

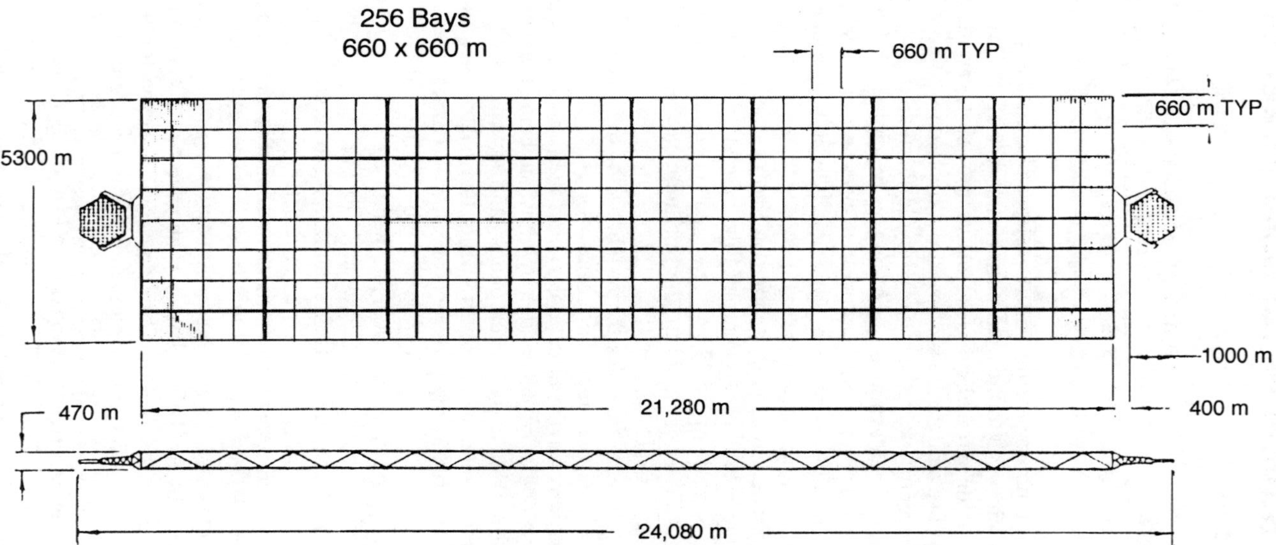

Fig. 1 Photovoltaic reference configuration with a total solar cell area of 97.34 km², total array area of 102.51 km², total satellite area of 112.78 km², and a minimum output to the sliprings of 16.43 GW.

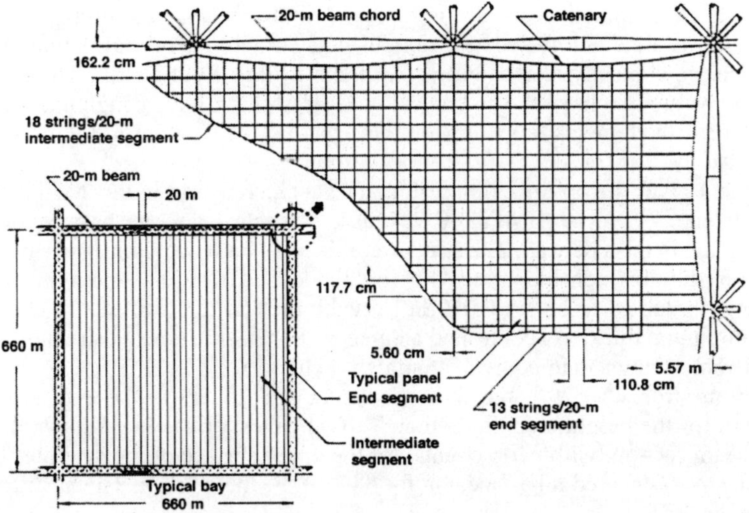

Fig. 2 Typical bay of the photovoltaic reference configuration made up of 32 20-m segments having a total of 566 strings, with each string having 541 panels.

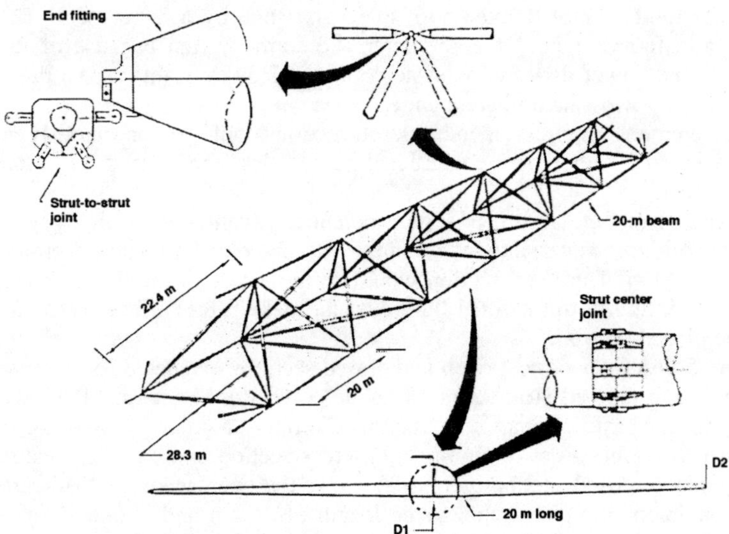

Fig. 3 Conical struts of 0.3-cm-thick graphite/epoxy.

weight arrangement for the beam and also allows beam intersections without changing the strut geometry within the beam. Each triangular beam comprises individual struts joined as shown in the Fig. 3 fitting detail. The fittings at each end of the strut aid automated assembly using a straight lateral translation. Two equal-length, conically shaped halves with end fittings form a strut. The center fitting joins the two halves. The conical or tapered shape allows high packaging density of strut elements fabricated on Earth and nested for transport to orbit for assembly.

Figure 4 gives the optimization results for the tapered strut in the 20-m beam. Graphite/epoxy material (HMF330C/934 resin) was selected for the basic laminate construction in the side wall. The struts were sized by both buckling and crippling (nonlinear-failure) considerations. For lightly loaded struts, a minimum gage thickness (assumed to be $t = 0.010$ in.) usually dictated the strut configuration. Since minimum thickness controlled a number of areas, the tapered struts proved considerably lighter than equivalent straight cylinders.

A 20-m strut, 45 cm diameter at the center and 15 cm at the end, proved optimum for the baseline configuration. Table 1 gives a mass breakdown of the 20-m beam section, with parts-counts, for the upper surface of the satellite. The graphite/epoxy tapered tubes account for 88% of the mass and metal end-fittings the remainder.

Using the 20-m beam with tapered struts as the basic element, the stiffness of the satellite was assessed at various aspect ratios. The parameters varied included external geometry, bay size, and structural depth. Table 2 gives results for four concepts evaluated for the first fundamental frequency of the truss and also for the truss with solar-cell blanket installed. Installing the solar cells decreases the basic first fundamental frequency of the truss alone by a factor of 3. The 4:1 aspect ratio (concept no. 2) is the reference configuration because of its low structural and power-distribution system masses. However, this aspect ratio also results in the lowest natural frequency.

Chord elements of either open or closed sections could be continuously manufactured in orbit rather than constructed as just described, using an automatic beam builder.

Automated manufacture offers such potential advantages as high packaging density in bringing material to orbit, continuous long elements, and fewer individual parts to ship. The open section would be the easier of the two to fabricate, but its susceptibility to torsional failure might make it less efficient structurally than the closed section.

Figure 5 describes typical open and closed sections evaluated for comparison with the 20-m tapered strut beam. Graphite/polysulfone material (P 1700) was selected for the basic laminate construction. Overall triangular 660-m beam length was used to establish the optimum beam cross section and chord geometry for both open and closed sections. Figure 6 shows the open section's load-carrying ability parametrically for a chord side length of 25 cm and thickness of 0.097 cm. Depth of the triangular beam was put at 5, 7.5, and 10 m. A 10-m-deep beam using 4-m batten spacing will carry the required load. Figure 7 illustrates the trend of the load-to-mass ratio efficiency for these sections using the selected chord geometry. A load-to-mass ratio of about 7.8 N/kg looks nearly optimum for the 10.0-m-deep beam.

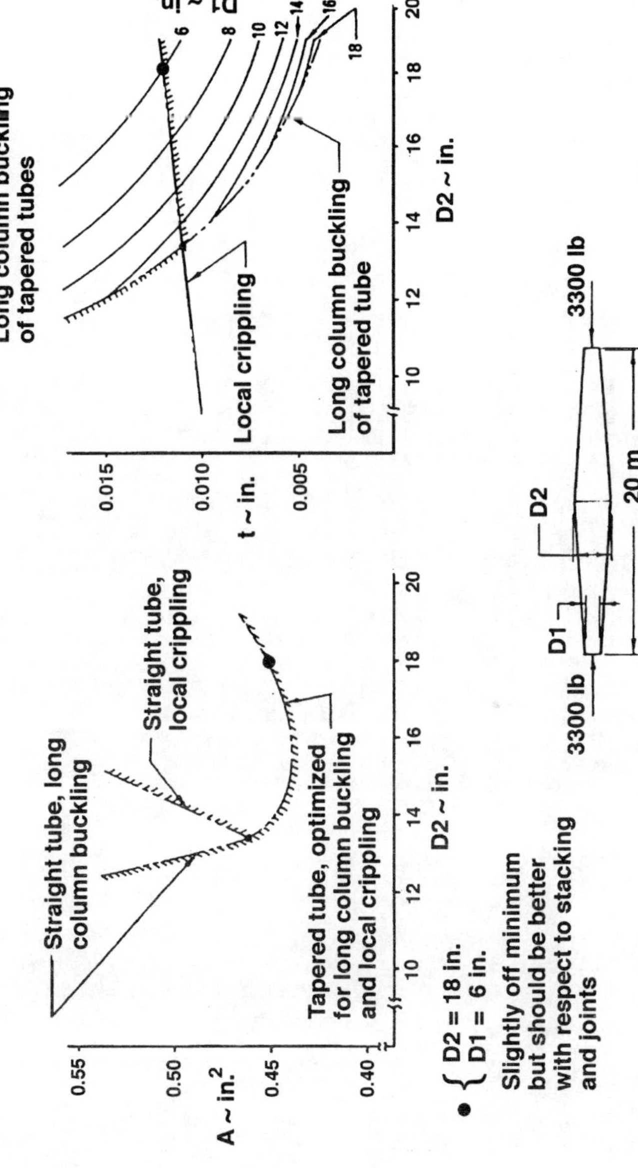

Fig. 4 Optimization of conical strut for $E = 20 \times 10^6$ psi.

Table 1 Mass of 20-m upper-surface beam section[a]

Item	Material	Principal dimensions	Mass/item, kg	No. of items	Mass per 20-m beam, kg
Tapered tubes					
Chord half section	G/E	D = 45.7 cm, d = 15.2 cm, L = 9.8 m, t = 0.30 mm	5.03	6	30.18
Side diagonal half section	G/E	D = 38.1 cm, d = 12.7 cm, L = 11.0 m, t = 0.25 mm	3.92	8	31.36
Base diagonal half section	G/E	D = 38.1 cm, d = 12.7 cm, L = 13.9 m, t = 0.25 mm	4.99	2	9.98
Base batten half section	G/E	D = 38.1 cm, d = 12.7 cm, L = 9.8 m, t = 0.25 mm	3.51	2	7.02
					Total 78.54
Tube center joints					
Chord center joint half section	Alum.	(Fits tube end with D = 45.7 cm)	0.38	6	2.28
Diagonal/batten center joint half section	Alum.	(Fits tube end with D = 38.1 cm)	0.31	12	3.72
					Total 6.00
Tube end fittings					
Chord end fitting	Alum.	(Fits tube end with d = 15.2 cm)	0.18	6	1.08
Diagonal/batten end fitting	Alum.	(Fits tube end with d = 12.7 cm)	0.15	12	1.80
Stud retention spring	Inconel		0.0009	36	0.03
Spring installation bolt	Steel		0.0004	72	0.03
Spring installation nut	Steel		0.0001	36	——
					Total 2.94
Strut interconnect fittings					
Apex fitting	Alum.	L = 5.7 cm, W = 5.7 cm, H = 4.2 cm	0.16	1	0.16
Base fitting	Alum.	L = 5.7 cm, W = 4.2 cm, H = 4.2 cm	0.14	2	0.28
Ball-end studs	Steel	D_{ball} = 0.95 cm, L = 3.4 cm	0.011	18	0.20
					Total 0.64

[a]G/E = graphite/epoxy. D = center diameter of tapered beam. d = end diameter. L = length. t = thickness of material. W = width. H = height. Total mass = 88.12 kg to give 4.41 kg/m.

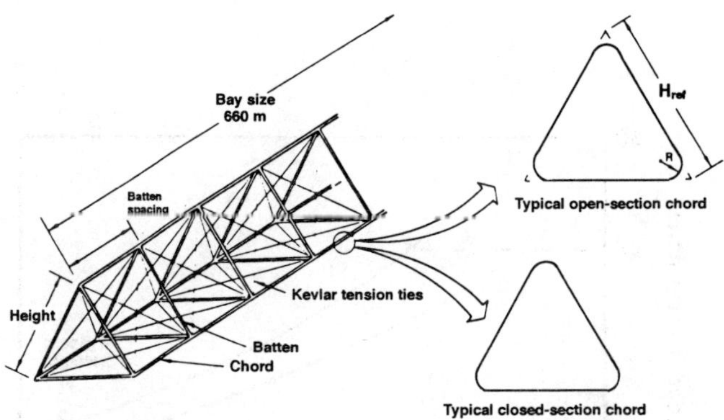

Fig. 5 Options for automatically fabricated continuous chord.

Table 2 Stiffness for structural concepts for SPS[a]

Concept	Width a, m	Length b, m	Bay size l, m	Depth h, m	Mass, kg	Area, m²	Truss f_1, Hz	Truss with solar cells f, Hz
1	10,240	10,240	640	406	4.16×10^6	1.05×10^8	0.0130	0.0035
2	5,120	20,480	640	406	4.19×10^6	1.05×10^8	0.0052	0.0014
3	6,240	16,640	520	566	5.81×10^6	1.04×10^8	0.0100	0.0032
4	6,080	18,240	380	683	10.30×10^6	1.11×10^8	0.0094	0.0037

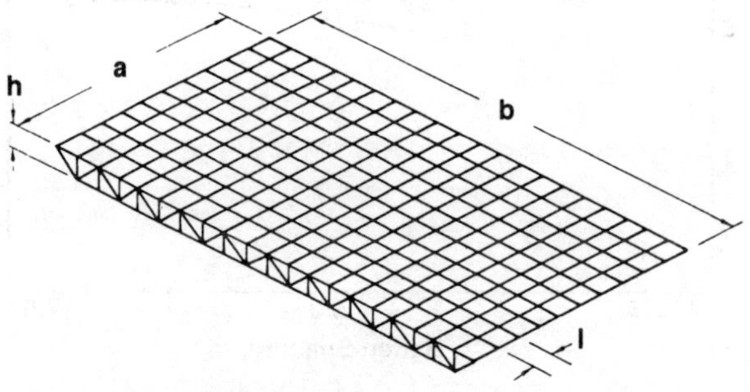

[a]For graphite composite 20-m beams having tapered struts with 12-in. max diameter and 0.009-in. wall thickness.

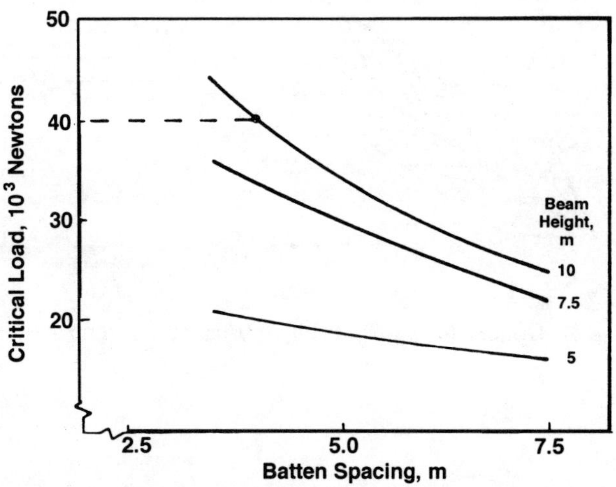

Fig. 6 Load-carrying ability of triangular beam with open chords.

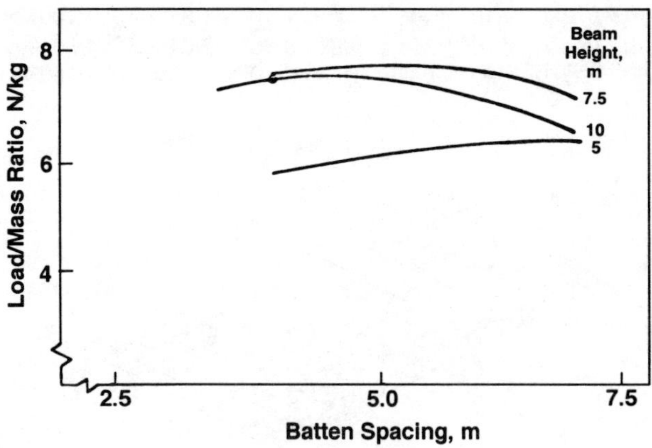

Fig. 7 Structural efficiency for triangular beams with open chords.

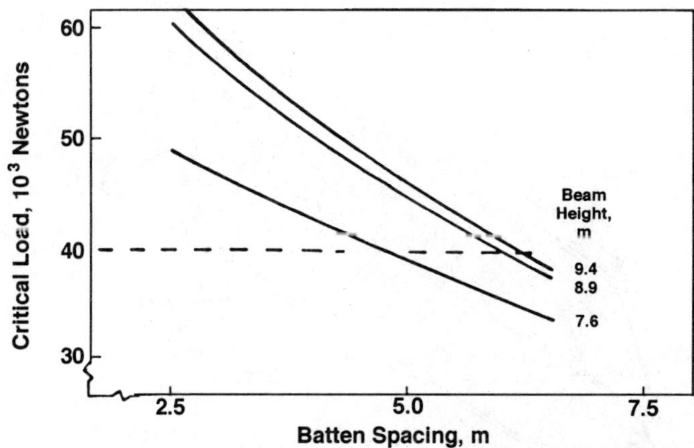

Fig. 8 Load-carrying ability of triangular beam with closed chords.

Similarly, the closed-chord sections were parametrically evaluated and load-carrying ability and mass ratios determined. Figure 8 depicts the load-carrying ability as a function of triangular beam height and batten spacing. An optimum structural efficiency of 10 N/kg is achievable using a closed-section chord element and a beam depth of 8.9 m with 6.0-m batten spacing (Fig. 9). Finally, Fig. 10 depicts the selected beam with closed section.

The relative structural mass of the satellite as a function of type of chord is shown in Table 3.

However, the structural mass proves only a small percentage of the total satellite mass for the structural concept using tapered strut members. This can be seen in the breakdown in Table 4.

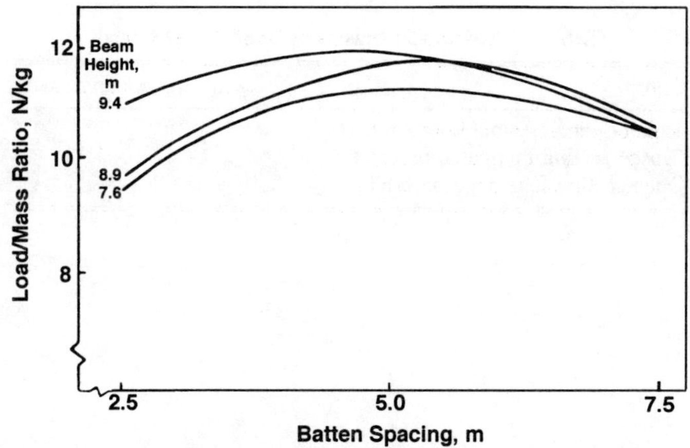

Fig. 9 Structural efficiency of triangular beams with closed chords.

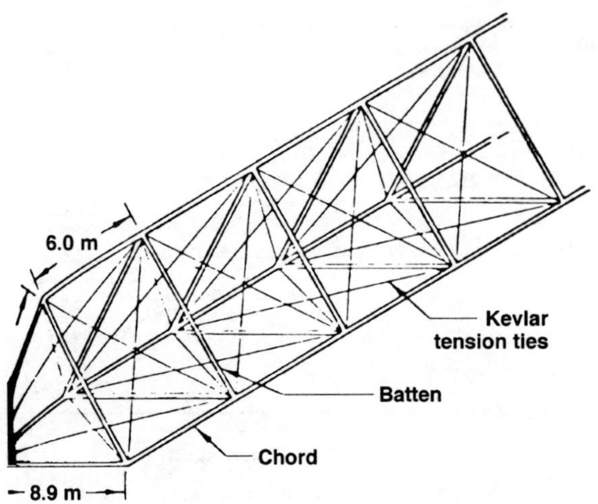

Fig. 10 Selected beam with closed chord.

Automated fabrication of closed- or open-chord sections would change satellite mass 1 and 2%, respectively. Such small changes for the automated-fabrication options should prove a minor factor in light of other programmatic options such as equipment requirements, on-orbit operations, and packaging—all largely yet to be assessed.

In short, all of the structural concepts discussed here appear to be feasible and potentially could be incorporated into an SPS design. The structural options might be compared qualitatively as follows.

The tapered tube has the advantage of less mass per SPS (but the difference runs probably less than 1000 tons, or about 1%), higher manufacturing rate, easier structural-integrity verification, and more structural-design flexibility.

Table 3 Structural mass as a function of chord

Chord	Relative mass
Tapered tube assembled on orbit	1.0
Closed section fabricated on orbit	1.2
Open section fabricated on orbit	1.4

Table 4 SPS baseline satellite mass

System	Mass, 10^3 kg	Percent of total
Structure	5,385	7
Energy conversion	43,750	57
Power distribution	2,398	3
Power transmission	25,212	33
Other	249	0
Total	76,994	

The continuous chord has the advantages of higher packaging density if manufactured in space (either open or closed chord adequate), less joint slop (analysis indicates not a problem), and lower machine complexity. In overall construction operations, number of parts, and cost, the two approaches show no significant difference. The use of a beam machine may present a problem of maintaining alignment precision for the tapered tube and beam straightness for the continuous chord.

Either approach will work. Technology-verification work will decide the issue.

An Entree for Large Space Antennas

Robert V. Powell and Albert R. Hibbs
Jet Propulsion Laboratory, California Institute of Technology

Imagine it is 1987. You are driving through the boondocks of the Northwest, and you need to be available to your office 1000 miles away. You proceed on your way with confidence, knowing that you can be reached, far from any urban area, on your mobile communications transceiver. The message alert sounds, and you are in contact with the home office, barely perceiving the fraction-of-a-second delay between transmission and response. By now you have become used to rural mobile communications relayed across the country by an acre-sized antenna in geostationary orbit.

Back now in 1977, we can see such a service ahead of us, for the era of using large antennas in space is dawning. With the transportation of the Shuttle soon to be available, the increasing availability and excellent properties of composite materials, and new powerful tools in structural analysis, we will be able to put up space objects whose dimensions measure in the hundreds of meters, with surface accuracies in the tens of microns and near-zero temperature coefficients. Clearly, one of the early useful forms for such large space structures will be space antennas. New approaches to antenna system analysis and design[1] as well as developments in related disciplines (e.g., pointing and control) will aid exploitation of large structures in space.

Applications will not be limited to rural mobile communications or even to communications in general but will include exciting new methods for space exploration and observations of Earth. Radio astronomy will benefit from the opening of a new spectral region in the submillimeter microwave band. High-resolution mapping of the brightness distribution of celestial radio sources will be aided by very long baseline interferometry (VLBI) with at least one end of the baseline provided by a large space antenna in low-Earth orbit. Earth observations will benefit from high-resolution radiometric observations of sea state, soil moisture, and other parameters directly affecting our economic well-being. Video data rates from the far outer planets will be made possible by an orbiting deep-space relay station that exploits the 0-*g* environment for higher antenna performance than possible on the ground.

Applications in communications, Earth observations, and radio astronomy, which are of particular interest to NASA, provide technical arguments for reflectors. Other applications, such as those requiring extreme beam agility, provide a complementary technical case for the phased-array or the lens antenna designs.

Astronautics & Aeronautics, Vol. 15, No. 12, Dec. 1977, pp. 58–64.

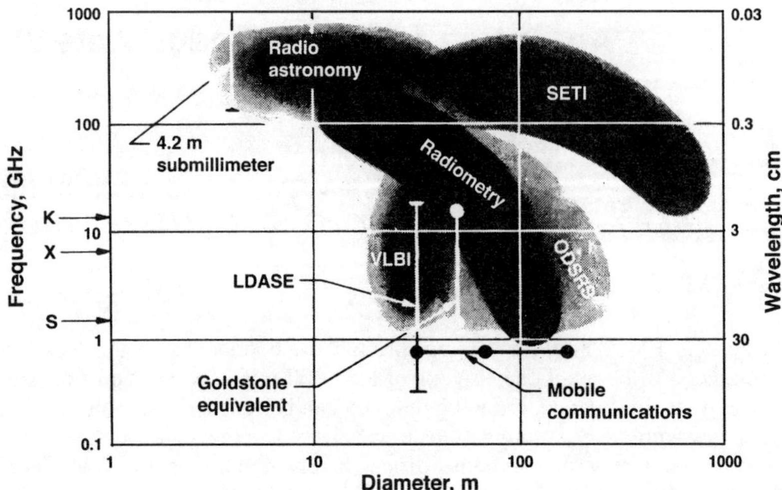

Fig. 1 Large space antenna application regimes. Shaded areas indicate regions of aperture size and frequency where large space antennas will likely have their most important application. Definitions: SETI (Search for Extraterrestrial Intelligence); VLBI (very long baseline interferometry); ODSRS (orbiting deep-space relay station); LDASE (large deployable antenna Shuttle experiment). The graph shows a candidate ODSRS design with performance comparable to the current Goldstone 64-m antenna plotted as a Goldstone equivalent. It shows a range of antenna sizes appropriate to a number of mobile communications satellite (Mobcomsat) designs plotted for a specific mobile communications frequency allocation especially attractive for satellite communications.

We believe, however, that the large reflectors will continue to offer an elegant means to use multiple beams and efficient multispectral sensors and may be the only practical way to achieve very high gain antennas in space.

We will discuss at greater length a number of applications of such large reflectors. We believe we have conceived a sensible and promising method for taking timely and cost-effective action on some of the early steps in a large space antenna development program.

Regions of Applications

It is convenient to characterize large space antennas in terms of two parameters: frequency and aperture size. Figure 1 illustrates in a general way the frequency and aperture regime pertaining to many of the applications considered here. The boundaries on the figure are not intended to exclude other applications nor to claim, for example, that radio astronomy would never be carried out with a 100-m dish at 10 GHz. They are intended, rather, to show where we believe certain applications will make *early demands* for large space antenna technology.

Potential Applications

Some highly promising candidates for potentially useful applications were identified in a study.

The mobile communications satellite (Mobcomsat) represents a probable early application for large space antennas. Studies made by a number of NASA centers (primarily Lewis Research Center and Goddard Space Flight Center) of the possible applications and cost benefits of such satellites led to a conclusion that a market exists for a commercial service that complements the planned urban mobile communications network for rural or remote users.

A possible Mobcomsat configuration would provide a national voice network for hundreds of thousands of users in remote areas using handheld or vehicle-mounted transceivers with whip antennas. It would use a 75-m-diam antenna with a beamwidth of 0.3 deg to blanket the continental United States with approximately 200 beams. Hawaii and Alaska would be covered by separate beams. The multibeam configuration permits low-cost ground terminals and spectrum conservation through frequency reuse. Demand switching provides full interconnection between all users. Beam isolation is ensured by placing adjacent beams on different frequencies. Studies of a number of possible frequencies and antenna sizes have shown that, above a few hundred thousand users, the cost becomes ground-terminal dominated, resulting in a projected cost per user of between $50 and $150 per month. This cost includes the satellite and ground-terminal procurement and maintenance.

An orbiting deep-space relay station (ODSRS) could, in a single antenna, perform most of the duties of the current three-station Deep Space Network (DSN).[2] Navigation benefits, in the form of simultaneous two-station tracking, would encourage retaining at least one ground-based tracking antenna. Performance improvements would accrue from the higher gain achievable in a gravity-free environment as well as from the use of the higher frequencies precluded from use by ground antennas because of atmospheric attenuation. For example, assuming a fixed spacecraft aperture, 50 kb/s from Pluto would require a 200-m ODSRS diameter at Ku-band. Such a system would also accommodate 5 Mb/s from Jupiter. Use of the higher frequency, however, would complicate the spacecraft pointing.

One concept assumes a rigid structure assembled in low-Earth orbit and subsequently transported to geostationary orbit as an integral assembly. Thrusts for the orbit transfer would be held to below 0.2 g. The erectable concept for ODSRS would yield a performance improvement over the current ground antennas. It is interesting to note that an ODSRS performance equivalent to the current Goldstone performance at X-band seems to be achievable with a 45-m deployable mesh antenna. Such an antenna would be limited, however, in its growth potential. The fact that the performance of a 45-m antenna in space equals Goldstone's suggests a low-cost way to evaluate subsystems for a future ODSRS.

The opening of a new spectral region from 1 to 0.1 mm (in the so-called submillimeter range) presents an exciting prospect for radio astronomy. Opening of a new portion of the spectrum in the past has, without exception, led to exciting discoveries. Observations above 300 GHz using a space-based antenna would especially interest radio astronomers because such observations cannot be made readily from the ground through the atmosphere. It should also be noted that the Boltzmann distribution generally peaks in the submillimeter range, making it a good portion of the spectrum in which to observe line spectra.

A space antenna operating in this spectral region would also aid observations of interstellar clouds, extragalactic radio sources, and cosmic background. Apertures of 10–30 m and rms surface accuracies of 10–50 microns in conjunction with submillimeter-wave receiver development will make that prospect realizable. In the case of a 10-m antenna with an expected surface accuracy of 50 microns being evaluated on the Shuttle, application as an observatory would dictate, of course, a free flyer.

The possibility of resolving unambiguously the brightness distribution of celestial radio sources to 10^{-3} arcsec presents an equally exciting prospect for radio astronomy. Such mapping would permit investigating such radio sources as quasars, galactic nuclei, interstellar masers, and pulsars. Right now interest runs high in component pairs in quasar-like objects with apparent recessional velocities greater than the speed of light.

To do high-resolution mapping unambiguously requires many baselines. A large-aperture antenna in low-Earth orbit in conjunction with a ground-based antenna readily provides the large number of baselines required.

Many uses have been identified for high-resolution Earth-looking radiometry, one of the most intriguing being the possibility of remotely measuring soil moisture to assist in planning agricultural uses and predicting yield. Other exciting applications include high-resolution observation of sea state in terms of water temperature, wind velocity, foam coverage, and salinity; snow cover and ice, including navigable waterways; the structure of hurricanes, other clouds, and precipitation rates and distributions. A 100-m aperture in a 400-km orbit would yield, for instance, a resolution on the order of 1 km at 1.4 GHz.

High-resolution radiometry complements broad-swath mappers such as Seasat and Nimbus (both high resolution and broad swath are difficult to achieve in the same sensor).[3] Large-aperture radiometers in geostationary orbit could continuously observe a selected region with Seasat-like resolution.

We are beginning to get a clear picture of cost-benefit payoffs in Earth observations. We believe that picture will show a requirement for the high resolution offered by large apertures in space.

A configuration for an erectable 100-m-aperture antenna for Earth-looking radiometry can handle several frequencies. To gain the high-beam efficiencies required for radiometry, the system would use only a segment of the larger parabolic surface to permit the feed to be offset and outside the beam of the antenna. Unambiguous signatures and the observation of different phenomena demand simultaneous observations at a number of frequencies and two polarizations.

The fact that multispectral observations require the same footprint independent of frequency makes the antenna, in practice, a constant-footprint instrument. Illuminating smaller portions of the total aperture at higher frequencies yields a footprint independent of frequency. This leads to an alternative concept, a hybrid design with a mesh peripheral section surrounding a precisely surfaced center. This concept may permit at least partial self-deployment.

Technology

The industry has made much progress on the technology of large antennas over the past several years.[4]

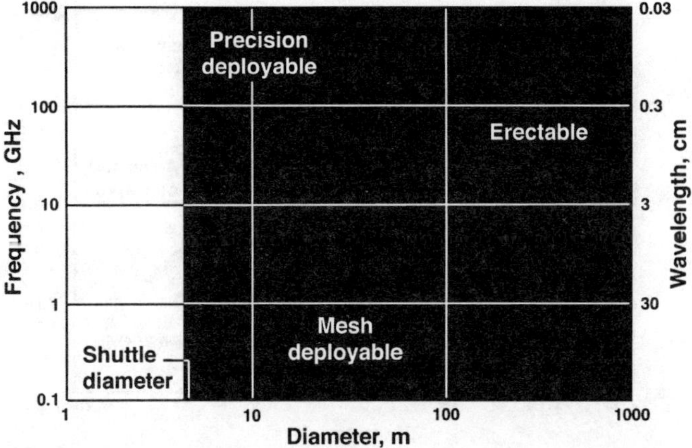

Fig. 2 Appropriate regions for application of deployable and erectable large space antennas.

Figure 2 indicates the status of the technology. It illustrates regions where types of antenna should be applicable. As a point of definition, we exclude antennas that can be accommodated in the Shuttle diameter (less than 4.2 m).

There are two broad categories: deployable and erectable. The deployable we subdivide into two regions: precision and mesh. The erectable category will ultimately encompass an antenna type sometimes called manufacturable.

A deployable fits either the high-frequency or the large-diameter region, but not both in the same antenna. We recognize the boundaries between these regions are somewhat arbitrary. In this article, we concentrate on the deployables because we believe that, in terms of technology readiness, they permit early application of large space antennas.

A survey of industry work reveals the mesh deployable technology to be more mature (Fig. 3). A 9-m wrapped-rib, mesh deployable antenna has been successfully demonstrated, in fact, on NASA's ATS-6 satellite. Concepts for large self-deployables that have been developed to the point of detail design include the Lockheed Missiles and Space Company's wrapped-rib mesh antenna, Harris Corporation's radial-rib dual-mesh antenna, General Dynamics' parabolic erectable-truss antenna, and TRW's radial-rib antenna.

Many other promising concepts and preliminary designs for high-performance self-deployable large space antennas are being pursued. They include variations on the Maypole, a bicycle-wheel-like structure with a rigid deployable rim and tension cables terminating at the ends of a deployable column. The basic structure then supports a mesh reflector shaped by mesh ribs supported by additional cables. Methods used to shape the mesh define design variations of this concept. Also being considered are advanced concepts that replace mesh with thin concentric conical rings extending the frequency range.

More-advanced precision deployable antennas would employ rigid panels interconnected by hinges, foldable to fit within the Shuttle, and capable of self-deployment. TRW's Advanced Sunflower, a preliminary design for an 8-m

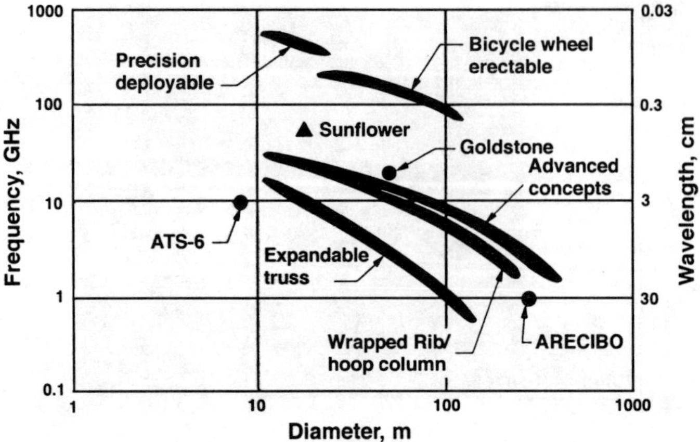

Fig. 3 Large antenna performance—current and projected. Noted for reference: the general size and typical use frequency of certain existing and projected antennas and the ground-based Goldstone and ARECIBO systems.

antenna, can operate to frequencies well over 60 GHz; it indicates the industry's growing abilities. A ground antenna developed at CalTech by Robert Leighton presents an exciting example of what might be done in space. By innovative fabrication and careful attention to detail, Dr. Leighton has been able to fabricate a 10-m parabolic antenna from hexagonal segments in such a way that the surface and the supporting structure can be disassembled for transport and reassembled yet maintain a best rms fit to a parabola of *50 microns*. This suggests the feasibility of a Shuttle-launched precision deployable antenna of comparable performance in space.

A Developmental Scenario

Arriving in the era of large space antennas must begin with a first step more modest than a 75-m Mobcomsat or a 200-m ODSRS. Having reviewed a number of experiment options, we conclude that a deployable antenna with 30-m diameter and 1-mm surface accuracy makes a logical first step.[5] Flight experience with an antenna system of this size in space would permit validating performance prediction models—an especially important engineering exercise now because these large space structures cannot be tested on the ground.

Confirming the validity of a model would permit extrapolating from 30 m to much larger antennas using the same design concept. The investment in the 30-m project would not only reduce risk for applications requiring larger antennas but also would immediately make available a deployable-refurlable reusable antenna suitable for Shuttle-based experiments.

Antennas of this type have been under consideration for a Shuttle demonstration for several years. We believe it is now time to undertake a serious design effort leading to a flight project—a large deployable antenna Shuttle experiment (LDASE). Using a 30-m-diam mesh deployable-refurlable antenna, in an opening series of experiments, the LDASE would be attached to the Shuttle—probably to a pallet of the Spacelab. Its surface accuracy would support operations to perhaps 15 GHz. In the initial experiment, perhaps in 1982 or 1983, LDASE would be given a measuring device to map its surface very accurately under various thermal and dynamic loadings. An RF beacon in orbit would make possible gain-pattern measurements. At the conclusion of some five days of experimenting, the antenna would be refurled and returned to the ground for retrofitting of follow-on experiments.

After the Shuttle-based validation experiments, the LDASE antenna, suitably modified, could be placed in orbit as a free flyer—e.g., as one member of a VLBI pair. Alternatively, it could be linked with an appropriate spacecraft and transported in the furled state to geostationary orbit and then deployed on station as, for example, a precursor to a Mobcomsat.

Such an LDASE seems to us a crucial first step in large space antenna development. Figure 4 describes one scenario for this development. LDASE, as the scenario indicates, could produce a large number of derivatives.

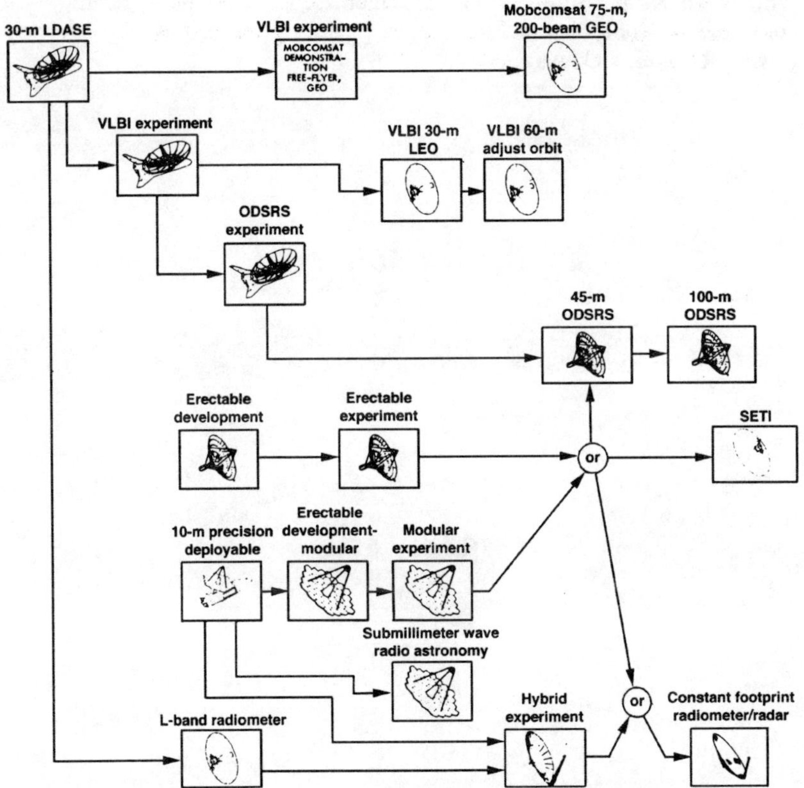

Fig. 4 Large space antenna development scenario. Note the large number of derivatives from the initial LDASE experiments.

We see the LDASE as the keystone not only for large mesh deployables, but also for precision deployable antennas, and since it provides a vehicle for early operational experiments with large space antennas, we also think it would advance erectable and manufacturable space-structures technology.

In short, we see the age of the large space antenna ready to begin, and we believe that LDASE represents its threshold.

References

[1] Bathker, D., (ed.), ''Microwave Performance Characterization of Large Space Antennas,'' JPL Publication 77-21, May 15, 1977.

[2] Renzetti, N. A., ''Network Function and Facilities,'' *The Deep Space Network Progress Report,* Document 42-40, May and June 1977, p. 1.

[3] Gloerson, P., and Borath, F., ''Scanning Multi-Channel Microwave Radiometer for Nimbus-G and Seasat-A,'' *IEEE Journal of Oceanic Engineering,* Vol. OE-2, No. 2, April 1977.

[4] Freeland, R. E., ''Industry Capability in Large Space Antenna Structures,'' in preparation.

[5] Freeland, R. E., Smith, J. G., Springett, J. C., and Woo, K. E., ''Large Deployable Antenna Shuttle Experiment,'' paper presented at 21st Meeting of the AAS, Denver, CO, Aug. 26–28, 1975.

Magellan: Aerobraking at Venus

Henry H. Curtis
Martin Marietta

On August 3, Magellan became the first planetary mission ever to apply aerobraking as a means of significantly altering both its orbit and mission objectives. Aerobraking uses the spacecraft's structure as a decelerator during its flight through an atmosphere. Although Magellan was not designed for atmospheric flight, it was flown through the atmosphere of Venus 730 times to halve its orbit period, eliminating the need for large propulsive burns. The technique's demonstration sets a precedent for other spacecraft and also multiplies Magellan's science returns. NASA has approved funding for a fifth cycle of Venus exploration to collect a high-resolution map of the planet's polar gravity field from a near-circular orbit—a capability that was not anticipated until two years ago.

Magellan had been in a 3-h 15-min elliptical orbit measuring 8470 km at apoapsis (the point farthest from the planet) and 172 km at periapsis (closest approach). From that orbit, the spacecraft had just completed the secondary mission objective of measuring the equatorial gravitational field. To achieve similar resolution at the north and south poles, the orbit was reduced from 8470 km apoapsis to 542 km.

Making the orbit near circular required a velocity reduction of 1221 m/s, which would have used 900 kg of liquid propellant. However, only 95 kg was available. Aerobraking was the only way to drastically modify the orbit. Magellan had no heat shield; so the multiple-pass method was used, whereby atmospheric drag on the spacecraft, over multiple passes through the upper atmosphere, slowly reduces the apoapsis altitude. This method generates lower temperatures and smaller aerodynamic forces than does the single-pass method.

As Magellan neared periapsis from the north pole and encountered the upper atmosphere, forces and temperatures began to increase. As it flew through periapsis, it experienced the greatest pressure and temperatures, factors that decreased as the spacecraft rose over the south pole. Magellan probably flew through the atmosphere in the free molecular flow regime. If so, no shock waves or turbulence around the back sides of surfaces occurred. There may be some evidence that transitional flow occurred. Only 9 N of force acted on the entire spacecraft.

After each pass through the atmosphere, the apoapsis altitude was diminished by 11 km. Over 70 days, the orbit period was reduced by 105 to 94 min. While apoapsis altitude decreased, so did the periapsis altitude. The duration of flight through the atmosphere grew geometrically from 4 to 12.5 min in 70 days. Atmospheric drag and increasing gravitational forces could quickly have caused the spacecraft to fall deeper into the atmosphere and burn up. To prevent this

Aerospace America, Vol. 32, No. 1, Jan. 1994, pp. 32–36, 41.

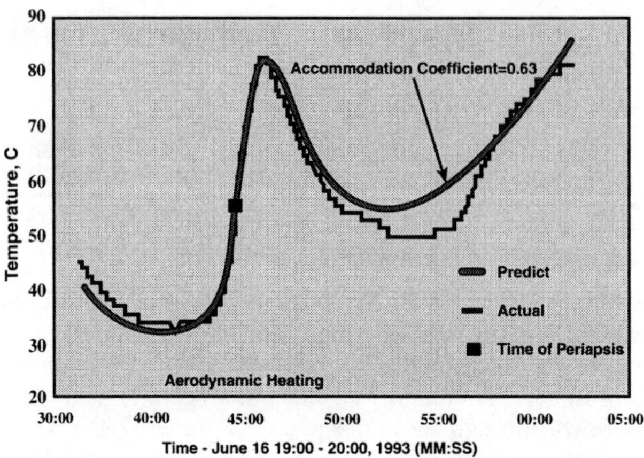

Fig. 1 Predictions of Magellan's solar panel temperatures during atmospheric flight closely matched actual readings.

effect, Magellan's 0.9-N thrusters were fired continuously at the apoapsis orbit location, raising periapsis altitude and preventing destruction of the spacecraft. Six different magnitudes and directions of these firings, called orbital trim maneuvers (OTMs), were available onboard the spacecraft to control the periapsis altitude.

A Risky Operation

Executing aerobraking with minimal analysis and a small operations staff was risky, but Magellan had already achieved all its primary and secondary mission objectives. To achieve a near-circular orbit in 80 days, the spacecraft was aerobraked aggressively. The Magellan team decided to accept the risk that a major spacecraft anomaly or component failure could mean spacecraft destruction.

The payoff, of course, was significant. In addition to a global gravity mapping opportunity, aerobraking was a boon to atmospheric science. The atmospheric density and structure of the Venus atmosphere were measured at altitudes never before investigated. Magellan measured the atmosphere as many as 18 times more often per day than had the previous spacecraft, Pioneer Venus orbiter. Also, Magellan aerobraking occurred at a different time in the 11-year solar cycle than had earlier measurements.

The upper atmosphere of Venus consists primarily of atomic oxygen and carbon dioxide. Very little was known about their interaction with the optical solar reflectors, thermal blankets, and white paint covering Magellan's exterior. Insignificant damage was expected from these molecules, based in part on results of a Magellan materials experiment on the Shuttle in August 1992. Surprisingly, only 63% of the expected solar panel heating occurred (Fig. 1). We do not know why, but the reduced heating enabled Magellan to fly lower in the atmosphere and shrink the orbit in only 70 days of aerobraking, compared with a predicted duration of 80 days.

Earlier data collected by Pioneer indicated that the atmosphere contracts on the shadow side of Venus. Since Magellan's orbit moved into the "night side" during aerobraking, periapsis altitude had to be lowered to maintain the desired amount of atmospheric drag. Another unique feature of the Venus environment is the existence of a periodic wave in the atmosphere. Every four Earth days, the density of the atmosphere rises and falls at a given altitude. This variability could not be anticipated; so the spacecraft was flown across the top of the waves to avoid overheating or overtorquing. However, these waves were not the only unpredictable forces acting on the spacecraft. Gravitational forces had not been measured over the poles to the desired resolution—but that was the reason for aerobraking.

Atmospheric flight drove several operational requirements, including a special entry attitude (for aerodynamic stability) that did not allow real-time telemetry to be received, a need to record data during the drag pass, and a need to play back these data at a faster transmission rate immediately after the drag pass. Also, the atmospheric drag caused the orbit period to shrink with each orbit, requiring the placement of orbital activities to be adjusted lest they drift out of position.

This was accomplished by rewriting the onboard command software to automatically reduce the time between periapsis and apoapsis events. The real rate of orbit shrinkage often did not match the commanded rate. Thus, the spacecraft could have turned to the entry attitude too early or too late. Command timing corrections, sent to the spacecraft as often as every 12 h, remedied this when the atmosphere became unpredictable.

The critical aerodynamic data were stored onboard during the drag pass. Immediately after this pass, the high gain antenna (HGA) was pointed toward Earth to report the stored spacecraft telemetry. The data were quickly checked to determine if an emergency exit was required. As the orbit period approached 94 min, the total time needed for communicating with the spacecraft was reduced to 12 min/orbit. The risk of mission failure demanded that controllers be able to command propulsive maneuvers rapidly to raise periapsis altitude. The onboard commands were arranged so that OTMs could be executed after receipt of a single ground command. In addition, commands were stored onboard to perform a 12-min emergency OTM burn to raise periapsis roughly 13 km and halt aerobraking.

Hiding from the Sun

Also conducted during the orbits were "star calibrations" to maintain gyroscope calibration and rotations to keep the spacecraft cool. Magellan, a three-axis stabilized craft, uses gyroscopes as references to maintain continuous control of the spacecraft body coordinate system relative to a celestial reference. Because Magellan orbited Venus in a solar radiation environment equal to two Earth suns, it was heavily insulated with multilayer blankets and solar reflectors. To reduce heating, it was necessary to spend half of each orbit in an attitude where the HGA hid most of the spacecraft from the sun.

Fortuitously, this "hide" attitude also allowed transmission of low-rate telemetry to Earth through the medium gain antenna (MGA). As the orbit shrank, MGA

Although Magellan was not designed for atmospheric flight, it was flown through the atmosphere of Venus 730 times to halve its orbit period, eliminating the need for large propulsive burns.

downlink time decreased to zero. Apoapsis orbital activities alternated between a star calibration and an OTM opportunity. Orbital activities remained the same throughout aerobraking, but the star calibration position was changed several times to maintain star visibility.

While Magellan's hardware could not be altered for aerobraking, its computers could be reprogrammed to perform the necessary tasks. This capability was used extensively to make aerobraking a reality. Alterations were made in the onboard attitude-control, fault-protection, and command programs. The attitude-control system parameters were also changed significantly.

Most of the time Magellan's orientation was manipulated by reaction wheels, which can produce a maximum torque of 0.18 Nm. Aerodynamic torques during the drag pass easily exceeded the capabilities of these wheels. Thus, 0.9-N hydrazine thrusters, producing a torque of 0.8 Nm, were used only during the drag pass to maintain attitude control. The spacecraft was aerodynamically stable, with the parabolic HGA trailing. Flight through the atmosphere with the HGA trailing resulted in a slow pitching oscillation and some roll about the velocity vector.

To conserve propellant the control system gains were redesigned to allow for as much as 14 deg of angular rotation about a body axis before a corrective thruster pulse was fired. Maximum angular rates were 0.63 deg/s before a thruster was actuated. After exiting the atmosphere, the thrusters fired to reduce angular position and rate errors to levels at which the reaction wheels could control the spacecraft. Attitude errors and rate errors were stored onboard during each drag pass and transmitted immediately after each one.

To maximize aerodynamic drag the solar panels were rotated before the drag pass so that they presented a large flat area to the flow field. The panels' active surfaces faced away from the flow. During this period, the onboard batteries supplied sufficient power. After the drag pass, the panels were turned to track the sun. The period of the final near-circular orbit was selected to enable full recharging of the onboard batteries with each orbit. This limited the final orbit to no less than 94 min.

Heat Poses Threat of Melting

With respect to damage from aerodynamic effects, the major concerns were overheating the HGA and melting the solder connecting the solar cells. The HGA could deform at 180°C; the solder was expected to melt at 170°C. Periapsis altitude was controlled throughout aerobraking to keep the solar panels below 160°C and the HGA below 170°C. Unfortunately, the temperature sensor on the HGA had failed earlier in the mission. These flight-allowable temperatures significantly exceeded those previously permitted and represented a newly acceptable level of risk.

The temperature of the solar panels during the drag pass was stored onboard and telemetered immediately after each pass. Because of the entry configuration of the spacecraft and the protective thermal blankets, aerodynamic heating did not significantly raise internal spacecraft temperature.

Magellan can no longer play back data stored on the tape recorders; so the command-and-data subsystem computer program was altered to allow storage of spacecraft telemetry in redundant memory. This form of storage, created just for aerobraking, was also used before and after the process to substitute for the lost tape-recording capability.

The aerobraking mission was designed to a maximum dynamic pressure of 0.4 Pa. Initially, periapsis altitude was maintained within an imaginary "corridor" defined by a maximum and a minimum desired dynamic pressure. The maximum desired dynamic pressure at the start of aerobraking was 0.32 Pa. Later in the process, it was changed to 0.20 Pa to reflect extra margin for the variability of the night-side atmosphere. An OTM was performed to lower the dynamic pressure back into the corridor when Magellan was expected to reach the maximum dynamic-pressure limit. The maximum dynamic-pressure limit of the corridor became the OTM trigger mechanism. The minimum limit ensured that aerobraking progressed as rapidly as possible. During aerobraking, the focus shifted to keeping dynamic pressure below the maximum desired limit to slow down orbit decay.

The maximum dynamic-pressure limit can be defined in periapsis altitude terms: the lower the altitude, the higher the dynamic pressure. The limit envisioned in periapsis altitude is not fixed over time. This is due to longer-term trends in atmospheric density with local solar time. Aerobraking began at 11:00 a.m. and ended at 6 p.m., both Venus local solar time. The rate of periapsis altitude decay was determined by the amount of drag during each pass and the local gravitational field. The history of the actual altitude shows abrupt changes caused by periapsis raise and lower OTMs (corridor OTMs). A typical corridor OTM was a 25.3-s burn of eight 0.9-N thrusters. This OTM nominally changed periapsis altitude by 0.8 km.

A Three-Phase Process

Aerobraking was broken down into three phases: an incremental lowering of periapsis into the atmosphere (walk-in phase), a sustained period of orbit decay

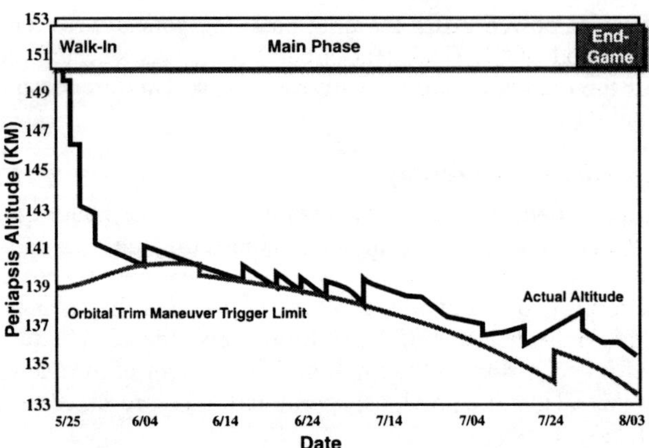

Fig. 2 Aerobraking comprised three phases: incremental lowering of periapsis into the atmosphere (walk-in), a sustained period of orbit decay at a relatively constant periapsis altitude (main), and a final rapid increase in forces and heating (end-game).

at a relatively constant periapsis altitude (main phase), and a final rapid increase in forces and heating with frequent OTMs until a last-minute propulsive exit out of the atmosphere (end-game phase) as shown in Fig. 2. The walk-in phase was designed to calibrate analytical models and allow escape if the atmospheric, heating, or control models were wrong.

Beginning on May 25, four OTMs were performed to reduce periapsis altitude 30 km into the aerobraking corridor. During this five-day period, it was observed that the most recent Venus atmospheric model was predicting densities roughly 1.7 times greater than observed. The walk-in phase was completed two days later than planned because of the conservative atmospheric model.

During the main phase, May 29 through July 27, nine OTMs were performed. Six raised periapsis and three lowered it. The down maneuvers were necessary to maintain the desired rate of orbit decay as dusk approached on Venus. The corridor OTM trigger limit was lowered to 0.20 Pa on July 26.

As aerobraking progressed, the variation of atmospheric density (including the periodic wave phenomenon) was approximately 5% instead of the expected 12%. Moreover, the temperature of the solar panels was reaching only 81°C at the OTM trigger point instead of 120°C. Because of the lower atmospheric variability and lower temperatures, the OTM trigger dynamic pressure was increased to 0.35 Pa on June 11. This change allowed a faster rate of apoapsis decay and reduced the aerobraking mission phase by six days.

Actual dynamic pressures did exceed the desired limits several times prior to OTMs. This was expected, and margins had been allocated so that no OTMs were executed on an emergency basis. In one case, the atmosphere expanded, increasing dynamic pressures 30% in 36 h. A scheduled OTM was moved 10 orbits earlier in response.

Falling Propellant Consumption

Typically, the thrusters fired 2500 pulses, each lasting 11 ms, in association with each drag pass. Most firings took place outside the atmosphere during the pre- and postdrag pass reconfigurations of the control system. Aerodynamic angle of attack averaged +0.5 deg, and oscillations of ±14 deg occurred. Two to three complete oscillations were observed at a rate of approximately 0.7 oscillation/min. Based on propellant consumption for the first 17 days, it was apparent that too much time had been set aside for the reconfiguration of control; so the spacecraft was wasting propellant. The time on thruster control was reduced by 5 min on June 12, causing average propellant consumption to fall from 45 to 20 g with each pass.

Midway through the main phase, the rate of orbit shrinkage was slowed to keep the apparent size of the planet small relative to the star field. Selection of OTM magnitude was optimized so that the maximum rate of apoapsis decay was achieved while maintaining more than 102 min of orbit period on July 27. This was necessary to maintain star visibility during a 5-min interval of the orbit.

Challenges at End-Game

During the end-game phase, which began July 27, the local gravity field caused a rise in periapsis altitude, while the atmosphere dropped away. The combined effect was a continuous decrease in dynamic pressure; frequent up-OTMs were not necessary as feared. The end-game phase lasted seven days, ending with the second exit OTM on the 69th day of aerobraking. Two exit OTMs were performed on consecutive orbits by using the emergency OTM commands stored onboard. These thruster burns raised the periapsis altitude by 24.9 to 161 km, reducing dynamic pressure to pre-aerobraking levels.

After several ground-commanded star calibrations were made to regain pointing performance, three exit OTMs were executed to place Magellan in the final near-circular orbit. By August 5, the final orbit had been achieved. It was 94.5 min in duration at 541-km apoapsis by 197-km periapsis.

The end-game was challenging. Several processes began to break down. As the orbit period decreased, a larger proportion of the orbit was spent rotating the spacecraft. Magellan was not designed for nearly continuous rotations; so the onboard calculation of gyroscope drift became inaccurate. Aerobraking could not have continued for another 48 h before the antenna pointing would have been lost. In addition, predictions of orbit shrinkage and command timing were diverging from the measured values.

With three command timing uplinks a day, tolerances were barely maintained. The Magellan team worked hard; for the previous 70 days, they had averaged 3.3 transmissions a day (one a day was the usual average). At 6:18 a.m. MDT on August 3, the telemetry consoles in Denver and Pasadena showed the successful completion of the first exit OTM. Aerobraking had ended.

Magellan had been designed and built to fly an elliptical orbit in a vacuum. Now the orbit is near circular, and the spacecraft has been successfully flown many times through Venus's upper atmosphere. The craft was not damaged—some surfaces even indicate they may have been cleaned by the atmosphere. Only 38 kg of propellant was used for all aerobraking activities, instead of 900 kg.

Other missions can now contemplate use of multiple-pass aerobraking using the techniques and operations processes demonstrated by Magellan to reduce launch weight and cost. The atmospheric measurements will be used to better define Venus atmospheric models for future missions. At press time, 6% of the gravity-mapping orbits had been finished; by October 1994, a high-resolution global gravity map of Venus should be a reality.

Aerobraking and Aerocapture for Planetary Missions

James R. French and Manuel I. Cruz
Jet Propulsion Laboratory, California Institute of Technology

Every planetary exploration program manager dreams of a propulsion system that allows his spacecraft to enter orbit about the target planet without expending any of the precious mass launched from Earth. Needless to say, we cannot deliver that kind of performance. However, by clever use of natural resources—the planetary atmospheres in this case—we can come close.

Discussion of modifying the orbits of spacecraft by lift and drag generated in planetary atmospheres appears in some of the earliest literature on spaceflight, both serious and fictional. Since the beginning of the space age, aerodynamic forces have been used extensively for landing either on Earth or elsewhere. But no one has used aerodynamic forces to capture a vehicle into a closed orbit and/ or to modify an orbit. Generally, inadequate navigation or inadequate knowledge of the atmospheres has prevented it. These shortcomings are being overcome, while pressures for more delivered payload are increasing; so the time has come to consider using aerodynamic insertion and orbit modification for planetary exploration in the 1980s and 1990s.

The first, and simplest, technique called aerobraking uses drag during successive passes through the upper atmosphere to circularize a highly elliptical orbit. Strictly adhering to successive passes and the upper atmosphere keeps operations and technology simple and mass low. The technique calls for relatively high orbit periapsis and relatively low amount of energy removed per pass (Fig. 1).

The second technique, aerocapture, transfers a vehicle into a closed stable orbit from a hyperbolic flyby trajectory. Since taking out all the energy in one pass calls for precise control, this technique requires a higher level of technology. On the other hand, it offers substantially higher performance gains and applies to more targets than aerobraking.

Aerobraking requires first inserting the vehicle into a stable orbit by some means, presumably rocket propulsion. For Venus, Earth, and Mars, which are the most probable targets for an aerobraking mission, the best orbits fall between roughly 12 and 48 h. Below 12 h, the advantage of using aerobraking diminishes rapidly, and, above 48 h, circularization times become long, and the saving in propulsion diminishes. In some cases, stability of the orbit might also be a factor. Sensitivity to perturbations might cause escape or unacceptable variations in periapsis altitude.

Astronautics & Aeronautics, Vol. 18, No. 2, Feb. 1980, pp. 48–55, 71.

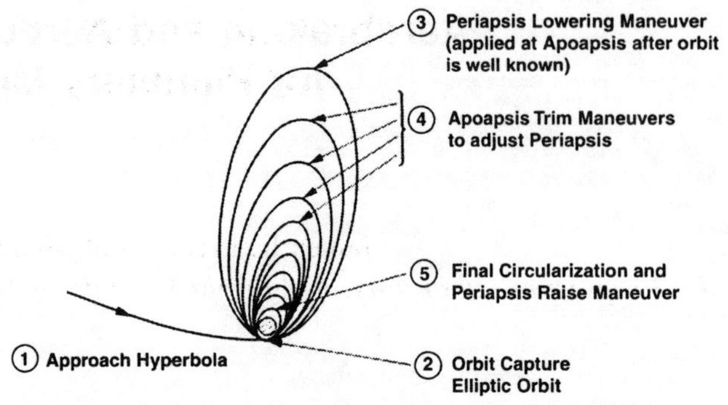

Fig. 1 Typical aerobraking mission profile.

To see early use, aerobraking must demand no major technology advances and have minimum impact upon the primary data-taking mission of the spacecraft and upon the overall spacecraft design. The spacecraft must, thus, make its atmospheric passes at a relatively high altitude where density is low. Ideally, the vehicle should remain in free molecular flow. This allows using a vehicle with a front shield but without a closed afterbody, since, in free molecular flow, the wake closes by molecular diffusion, which occurs too slowly to pose a threat to the spacecraft.

A path that would keep circularization times reasonably short, but still compatible with material limits, would place the vehicle not truly in the free molecular regime during braking. Fortunately, the vehicle would not swing over to continuum flow either, but into the transitional region between the two regimes. This holds true regardless of the target planet, since the governing factor is density. The altitudes differ for each target. In this transitional region, the flow tends to expand around the shield and try to close the wake more rapidly than in true free molecular flow. As a result, the spacecraft may need some local protection for components close to the edge of the shield. Determining whether the wake will close in transitional flow calls for a Monte Carlo analysis, which has not yet been done. Preliminary estimates based upon rocket nozzle expansion data imply that objects along a 10 deg cone angle from the shield edge will experience heating equal to one sun at Venus. Most items will remain little affected during the brief braking pass of 15–20 min, but delicate hardware such as high gain antennas or solar arrays may need local thermal protection.

Typical aerobraking missions to Mars, Venus, and Earth call for relatively large drag areas for reasonable circularization times (Table 1). Structures of such size could not fit in the Shuttle or any other launch vehicle. They will have to be deployable or erectable. Present technology can support building deployable antennas up to 15 m in diameter that can take the aerobraking of loads less than 0.1 g. The brake would probably have a high-temperature metal foil in place of the usual woven Inconel or molybdenum wire mesh, although the mesh might do perfectly well depending upon porosity. The latter case would lead to a dual-function drag brake/antenna, a perfectly reasonable concept usually viewed with horror by conservative project management. A drag brake serving just one function could accept much poorer surface accuracy than an antenna, so would cost substantially less to fabricate.

For a Venus Orbiter Imaging Radar (VOIR) spacecraft, the aerobraking system (Fig. 2) would replace one solid-propellant motor used for orbit circularization on the all-propulsive version (Fig. 3). Aerobraking reduces launch mass (Fig. 4) but increases final mass in orbit because of the nonjettisonable additional equipment required.

In shaping the "badminton bird" configuration of the VOIR aerobrake, the designer strives principally for static aerodynamic stability. Although static stability does not appear absolutely essential to successful aerobraking, it holds down demands for active attitude control. If the design shown enters the atmosphere at a modest angle of attack, it would seek a zero angle of attack. With no active control, the vehicle would oscillate about ± 1 deg in angle of attack. This motion would brake acceptably well but would allow the vehicle to leave the atmosphere with some rate in pitch. Momentum wheels already planned for the vehicle would hold the vehicle to zero angle of attack throughout the pass. The propulsive version could get by with a low-capacity wheel in one axis, but the aerobraking version would need a larger unit.

Aerobraking could find use on a Mars mission (Fig. 5), following up Viking with a rover or sample return. The vehicle would insert itself into an elliptical orbit using the propulsion system. From this orbit, the lander within the large bioshield would descend to the surface. The bioshield and support structure would be jettisoned, leaving a clean forward shield. The orbiter would then aerobrake into a low circular orbit from which it would conduct geoscience measurements to support a rover or rendezvous with an ascent vehicle from the lander for return to Earth. A Mars sample return may apply aerobraking at Earth as well, since mass limitations may preclude carrying enough propellant to Mars and back to enter an orbit from which the Shuttle could retrieve the spacecraft. In both instances aerobraking would save propulsion mass.

Can we control and predict the orbit well enough to insure the vehicle's not entering permanently or experiencing excessive heat? Analysis says we can.

Table 1 Typical aerobraking parameters

Planet	Venus	Earth	Mars
Vehicle mass, kg	1,630	1,630	1,630
Drag brake diameter, m	9	9	9
Initial orbit period, h	12	12	12
Periapsis altitude, km	132	86	123
Initial apoapsis altitude, km	37,218	18,247	40,033
Periapsis velocity, m/s	9,594	4,604	10,363
Shield temp, °K	664	571	592
Final periapsis altitude, km	691	556	692
Number of orbits	244	53	344
Elapsed time, days	44	10	61
ΔV to circularize, m/s (at 300 km)	205	142	225

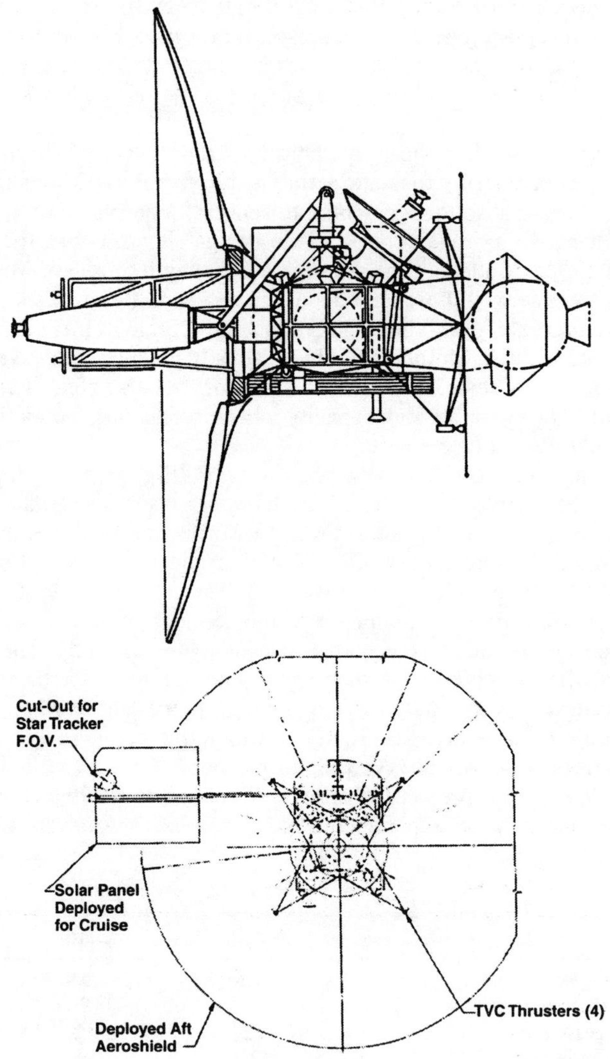

Fig. 2 Aerobrake configuration of Venus Orbiter Imaging Radar (VOIR) spacecraft.

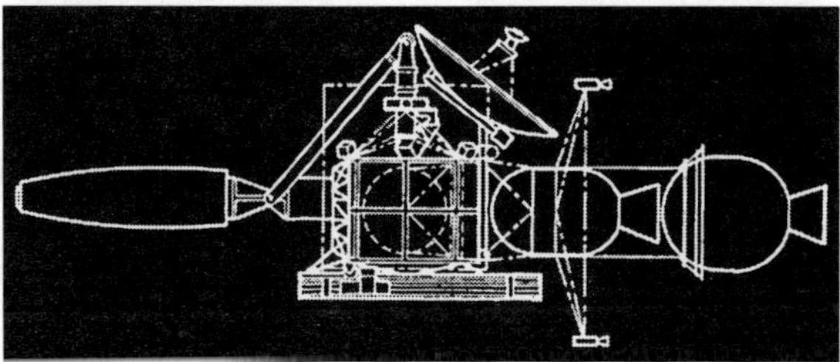

Fig. 3 Propulsive configuration of VOIR spacecraft.

During the early phases of the mission when the orbit would be quite elliptical, the intervals between periapsis passes would give adequate time for determining orbit. Toward completion, orbital periods would drop into the 1.6–2.0-h range and make things more difficult. However, analysis of the Venus mission shows periapsis could be predicted to an accuracy of ± 0.7 km for five orbits ahead—much better than the 2–3 km needed. In any case, onboard instrumentation would make the primary check by measuring acceleration. Excessive drag deceleration and heating on one pass would signal that periapsis was too low or that the atmospheric density profile was changing. The command computer system would raise the periapsis by firing a thruster at the next apoapsis. Sufficient margin designed into the vehicle and orbit would insure that no worst-case combination of periapsis drift and atmospheric variation would cause damage to the vehicle in one orbit.

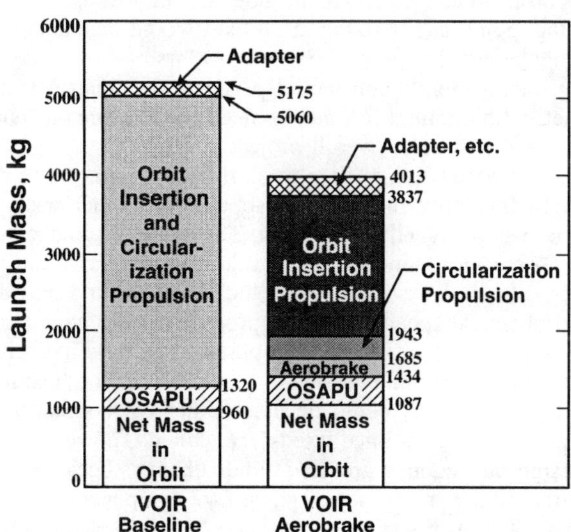

Fig. 4 VOIR launch mass savings using aerobrake. OSAPU: Orbit Stabilization and Attitude Propulsion Unit.

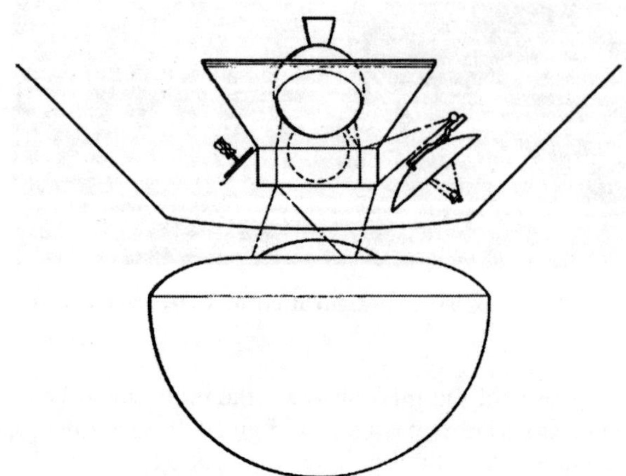

Fig. 5 Aerobraking Mars orbiter/lander.

To begin aerobraking, a small apoapsis propulsion maneuver would lower periapsis into the atmosphere after radio occultation data had revealed the atmospheric profile. Raising periapsis could stop braking at any time for tracking, troubleshooting, or terminating aerobraking. When to terminate depends upon the acceptable degree of risk. As the orbit nears a circle and control becomes more and more difficult, prudence dictates terminating early enough to avoid any hazard. For the Venus mission, the decision has been made to terminate when apoapsis drops to about 700 km altitude with the periapsis at 120–130 km. Starting from that point, circularizing at 300 km would take about 250 m/s ΔV. Similar philosophies would apply at Earth and Mars.

Aerobraking, then, offers a substantial performance improvement for use at terrestrial planets with minimal risk and no need for advanced technology. Costs appear very modest relative to overall project costs.

Aerocapture uses aerodynamic drag to shift to a closed planetary orbit from a hyperbolic flyby trajectory and, in doing so, eliminates fuel-costly retropropulsion. It advances a step beyond aerobraking. The vehicle would vary its in-plane lift-to-drag (L/D) ratio to maintain constant drag in the continuum flow regime until the proper velocity is reached. Then, the vehicle would pull up and climb above the atmosphere. At apoapsis, a small propulsive burn would raise periapsis out of the atmosphere. Aerocapture would place the vehicle in orbit at the small cost of the aeroshell system and propulsion mass for orbit adjustment.

Changing the in-plane L/D could be done either by modulating pitch or by rolling an aerodynamically trimmed fixed-L/D vehicle. The latter method simplifies vehicle design but requires greater attitude changes because bank reversals must equalize the effects of the out-of-plane L/D component.

Aerocapture applies at almost any atmosphere-bearing celestial body. Performance advantages begin to diminish at planets with gravity wells greater than at Uranus—Jupiter and Saturn—due to the higher entry speeds and greater aerothermal-protection mass required. At some planets with large gravity wells, this problem can be circumvented by performing aerocapture at atmosphere-bearing satellites such as Saturn's Titan.

A nominal aerocapture mission has two critical phases: approach navigation and atmospheric flight. Approach trajectory correction maneuvers (TCMs) outside the atmosphere can be controlled by commands computed on Earth. Once the vehicle enters the atmosphere, trajectory maneuvers must be computed and commanded onboard. Then, Guidance Navigation and Control (GN&C) works on a closed loop using adaptive guidance algorithms to correct the atmospheric flight trajectory to meet its end boundary condition.

The complete sequence for an aerocapture vehicle could go as follows (Fig. 6). A NASA twin-stage inertial upper stage (IUS) would launch the vehicle from the Shuttle. During cruise, the vehicle would orient itself approximately normal to the ecliptic plane. This attitude would allow communicating to Earth by a leeward side skin-mounted phased-array antenna (data-rate requirements set by the approach navigation system) and facilitate attitude control using externally mounted sun and star sensors. In this attitude, the vehicle could readily carry out approach navigation—maintain celestial lock, obtain images of planets and/or satellites against the star background, and transmit the radio and optical data to Earth for computation of the approach TCMs.

Long-duration approach navigation would determine the capacity of the power supply. Most missions would easily meet navigation power demands, since the spacecraft science would require radioisotope thermoelectric generators (RTGs). Missions depending on solar panels could mount extra solar arrays on the aeroshell skin or expendable appendages.

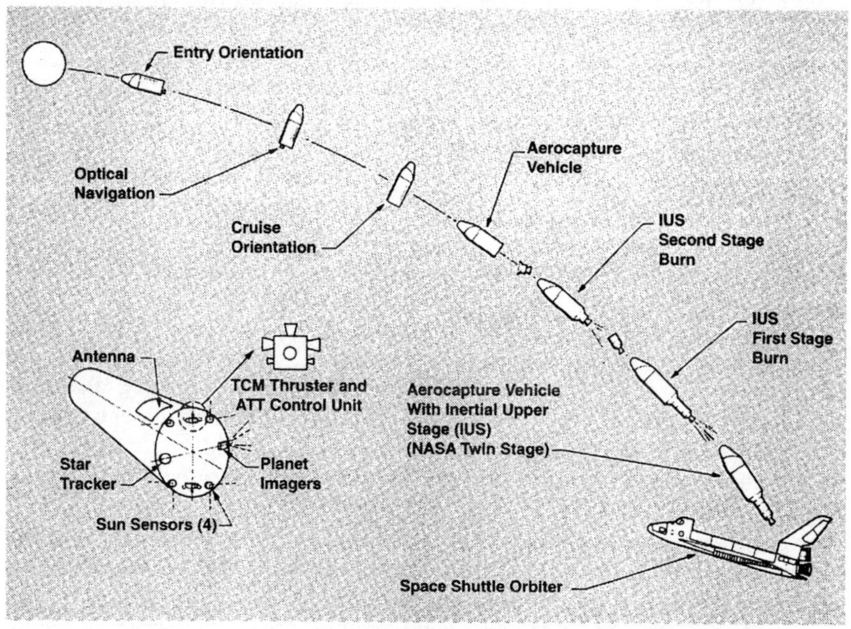

Fig. 6 Interplanetary cruise portion of aerocapture mission.

Approach navigation would size the expandable cruise antenna and/or solar arrays since the aerocapture vehicle must be navigated to a better approach accuracy than an all-propulsive mission. Approach navigation would not trigger a major technology push or hardware development but would call for processing more data closer to encounter than an all-propulsive mission. Current optical navigation or an equally accurate navigation scheme such as advanced radiometrics could carry out approach navigation which would include possible TCMs at E − 1 day. The data-transmission rate needed for optical navigation would require more antenna gain than radiometric navigation and would lead to a trade between antenna size and transponder power. At the inner planets, antenna size and transponder power would not have to be extraordinarily great, even though the entry corridors are tight. Entry corridors narrow to the order of ±1 deg at Mars and ±1.2 deg at Venus and Earth. These required entry corridors result primarily from the shallow atmospheres (low scale height of 5–6 km) of the inner planets and their daily variations in height. Telecommunications does not drive design for the inner planets because of the short distances back to Earth. At the outer planets, distances would be greater but data rates lower since entry navigation would not have to be as precise. The entry corridors at Titan would range from ±8 deg at an entry speed of 6 km/s to ±12 deg at an entry speed of 12 km/s. The wider corridors result from atmosphere density scale heights at the outer planets and satellites about 10 times greater than for the inner planets because of lighter component gases.

The density scale height is inversely proportional to the product of the gas molecular weight and planet surface gravity. Therefore, the difference in density scale height between the inner and outer planets is primarily attributed to the gas components. At Titan, on the other hand, it is a combination of the gas components and low gravity.

With such a large entry corridor to aim at, the most difficult job for the navigation system would become adjusting entry speed to reduce aeroshell mass. At the lower entry speeds for Titan, steady-state loads would not exceed those at Mars—3.0 Earth g in peak aerodynamic load and 60 w/cm^2 peak convective stagnation heating rate on an equivalent 1-m radius sphere. At high entry speeds for Titan, entering the atmosphere at the bottom of the corridor could result in Venus-type aerocapture steady-state loads of 40 Earth g in peak aerodynamic load and 500 w/cm^2 peak convective stagnation heating rate. These peak loads would occur at the beginning of an entry at a steeper flight path than nominal. The drag would eventually stabilize at a constant 1.5 Earth g for an entry speed of 6 km/s or 3.0 Earth g for an entry speed of 12 km/s. A GN&C algorithm would keep the drag constant; so drag would depend primarily on entry speed. Shrinking the entry corridor by an order of magnitude would reduce entry loads by the same ratio. This would reduce stiffening of the aeroshell and thermal-protection mass.

Expendable power sources and antenna to meet peak navigation demands may impose integration problems. Using closed-loop autonomous approach navigation may avoid those problems, lighten the aeroshell at the outer planets, and improve navigation accuracy. Systems under consideration include 1) Star Tracker Economic Long Life Attitude Reference System and 2) Space Sextant. The higher cost of these advanced interplanetary navigation systems must be traded against the extra systems engineering and integration complexity of expendable add-ons.

Three factors will determine the aerocapture vehicle aerodynamic shape and performance characteristics: accuracy, volumetric efficiency, and vehicle-design mass.

Accuracy sets *L/D*. During the initial entry, the vehicle must have sufficient *L/D* for control to keep drag constant without crashing or skipping out. *L/D* in turn defines the entry corridor, and, to a certain degree, higher *L/D* vehicles widen entry corridors.

During constant drag, very little lift would be required to maintain the vehicle in equilibrium with gravity. The constant drag phase would place the vehicle into a known trajectory, so would have the effect of filtering out previous uncertainties. Since the rate of velocity depletion becomes known during constant drag, the GN&C system can readily compute a new reference trajectory to meet its required end boundary conditions. Then, during the exit maneuver, most of the trajectory control errors would occur. Shortening this period would limit the effects of these errors. This could be done by having the vehicle pull up to its highest *L/D* and climb out of the atmosphere at its fastest rate.

Aerocapture accuracy improves with increasing *L/D* until it reaches a maximum around *L/D* = 1.5 (Fig. 7). The figure of merit is a root sum squared of output errors associated with input errors dealing with vehicle-design uncertainties, density scale height uncertainties, trajectory control and knowledge uncertainties, and uncertainties in the measurements of state-of-the-art inertial sensors.

Of the aerodynamic shapes covering the 0–3.0 range of hypersonic *L/D* in Fig. 8, the very high *L/D* entry geometries tend to lose volumetric efficiency and significantly increase the ballistic coefficient (W/C_DA).

A factor of 2 increase in W/C_DA would require the vehicle to drop to a starting altitude where the density was greater by that factor. Thermal and structural loading would increase. Also, the very high *L/D* vehicles have sharp leading edges which cause thermal problems.

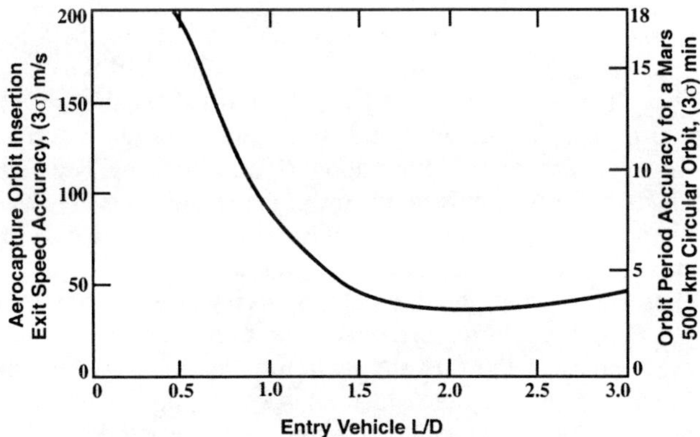

Fig. 7 How entry vehicle *L/D* affects aerocapture accuracy. For an entry mass of 5000 kg.

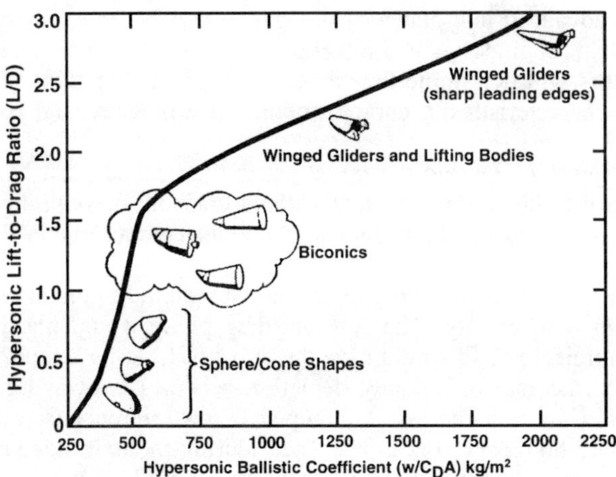

Fig. 8 Performance of aerocapture shapes. For a 5000-kg mass and 14-m^2 cross-sectional area.

Biconic shapes have the best combination of high L/D for insertion accuracy, low hypersonic ballistic coefficient, and volumetric efficiency. They also have a great deal of flight experience both as passive and actively maneuvering flight vehicles. Biconic shapes would fit well within the Shuttle bay along with the IUS and possibly a solar electric propulsion stage (SEPS). Since the spacecraft designed for missions in the 1980s and 1990s will be sized and designed for the Shuttle bay, payload integration to the aerocapture shell should pose no problem.

A Mars sample return (MSR) mission would fit into an aerocapture vehicle as shown in Fig. 9. This mission presents the most difficulties because it includes orbiters, landers, rovers, and ascent and rendezvous vehicles. It would require an aerocapture vehicle with the maximum Shuttle/IUS allowable payload length of 10.3 m.

A Saturn orbiter dual probe (SO2P) mission would fit inside a much shorter 6.6-m aerocapture vehicle (Fig. 10) to make room for a SEPS as well as IUS in the Shuttle bay. This mission would have the most difficult trajectory phasing since it would have to gather and retransmit data from the Saturn probe relay link before proceeding to aerocapture about Saturn by using Titan's atmosphere on its outgoing trajectory. Even with its low gain expendable antennas, the vehicle could service the relay link without difficulty since the data rates (~100 bps) fall far below those required for approach navigation (~3 kbps).

A VOIR spacecraft would be the simplest to integrate within an aerocapture vehicle (Fig. 11) and would require only the 6.6-m version.

For the MSR mission, the forward section of the aerocapture vehicle would serve as an entry system for the lander after aerocapture. This forward section would be approximately 6.6 m long with about the same aerodynamic characteristics as the complete aerocapture vehicle. Thus, a generic system of aerocapture vehicles would cover foreseeable planetary exploration. Very ambitious missions would use the long vehicle and less ambitious ones, the chopped-off version.

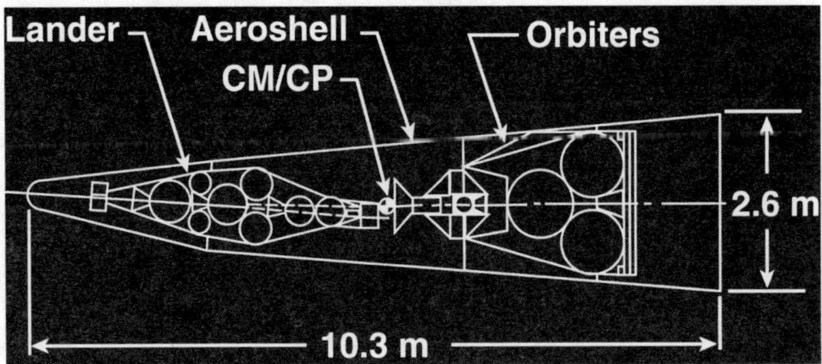

Fig. 9 Mars sample return spacecraft in aerocapture vehicle. Aerocapture vehicle L/D=0.47, C_D=0.47. Masses in kg as follows: orbiters, 1935; lander, rover, ascent vehicles, 1750; aeroshell for entry and re-entry, 330; orbit trim, reaction control system propulsion, inerts, 700. Total: 4715.

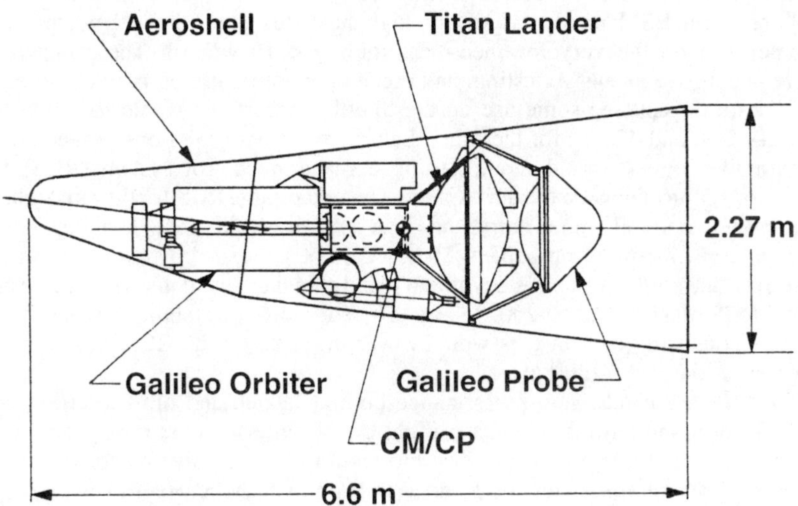

Fig. 10 Saturn orbiter two-probe spacecraft in aerocapture vehicle.

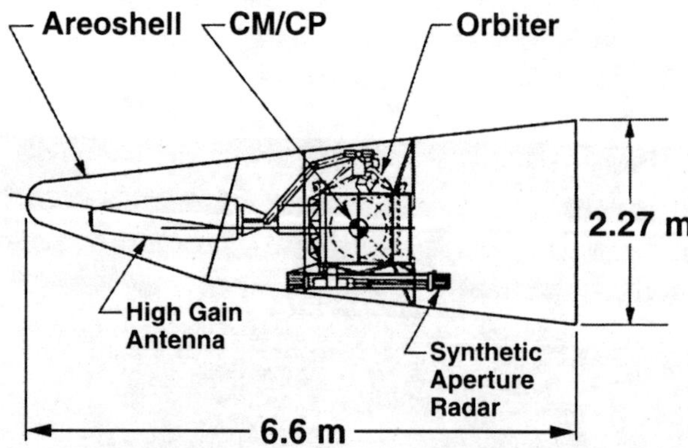

Fig. 11 VOIR spacecraft in aerocapture vehicle. Aerocapture L/D = 1.40, C_D = 0.45. Masses in kg as follows: orbiter, 1500; aeroshell, 420; orbit trim, reaction control system propulsion, inerts, 1100. Total: 3020.

The aerocapture vehicle thermal-protection system or aeroshell would be relatively light and would use current technology in its structural and aerothermodynamic design—an aluminum honeycomb ring-stiffened shell with low-density ablators distributed over it. Elastomeric silicone materials (ESMs), silicone ablative materials, and RTV-coated Nomex felt have been recommended for the ablators with ESM on the windward high heat-flux side ($\dot{q}$ > 80 w/cm^2) and Nomex felt on the very low heat-flux side ($\dot{q}$ < 10 w/cm^2). These materials have undergone extensive testing, have seen operational use on military re-entry vehicles, and will see some use on the Shuttle. Aeroshells would have masses of 330, 200, and 420 kg for the Mars, Saturn, and Venus missions, respectively.

Attitude control would require using a reaction control system (RCS) for trajectory control outside the atmosphere. Using the same RCS for the atmosphere roll maneuvers and initial entry and final exit three-axis rate damping would burn a large amount of propellant. Most would be expended by correction in the roll axis due to the vehicle's asymmetry and the short available moment arms. Total RCS mass would be 200–250 kg. The addition of a stability augmentation system consisting of trim tabs would lower the mass of the RCS plus stability augmentation to 80–120 kg.

Aerocapture would gain performance both at launch and orbit insertion and significantly shorten interplanetary flight cruise time for missions to the outer planets. Normally, interplanetary trajectories for all-propulsive planetary orbiter and/or lander missions must be planned to keep hyperbolic excess velocity (V_∞) to a minimum at both launch and arrival. The aerocapture vehicle would be insensitive to small variations in entry speed (on the order of hundreds of meters per second) since the thermal-protection system is not sensitive to such changes. Therefore, aerocapture vehicle interplanetary trajectories could be optimized for minimum launch energy alone for the inner planets.

At the outer planets, high approach V_∞ does not significantly increase the throw weight for an all-propulsive mission since multiple encounters at the satellites would pump down the large initial orbit period (~200 days). If the cruise flight time to an outer planet became prohibitive (nominally 8.2 years to Saturn and 13 years to Uranus), a SEPS could be used to power faster trajectories (higher approach V_∞) which could reduce flight times by as much as a factor of 2. Slowing down the additional velocity produced by SEPS at approach would demand additional fuel on an all-propulsive mission and reduce the in-orbit payload. The aerocapture vehicle could tolerate the higher approach and entry speeds without significantly greater aeroshell system mass (from about 220–300 kg initial mass for a Saturn Titan-assisted aerocapture mission).

Aerocapture could deliver not only larger throw weights for VOIR, MSR, and SO2P (Figs. 12–14) but also could place in orbit more than twice as much mass as an all-propulsive spacecraft. Missions like VOIR could carry out more science, such as returning an atmospheric sample from the high Venusian clouds. Aerocapture could significantly shorten the flight to Saturn for SO2P without greatly reducing its mission mass margin by using SEPS in a solar electric Earth gravity-assisted (SEEGA) trajectory (Fig. 13). Aerocapture would do even more for a mission to orbit Titan. A Titan orbiter mission using anything other than aerocapture is presently impossible.

For manned planetary missions, aerocapture would serve twice—once at the planet under investigation and once on return to Earth. For example, a manned Mars mission could be done with a Shuttle and aerocapture tanker vehicles.

These performance results show aerocapture-subsystems cost should be low compared to cost of the overall mission. Also, on outer planets missions, saving the operations cost of long-duration all-propulsive missions would greatly offset

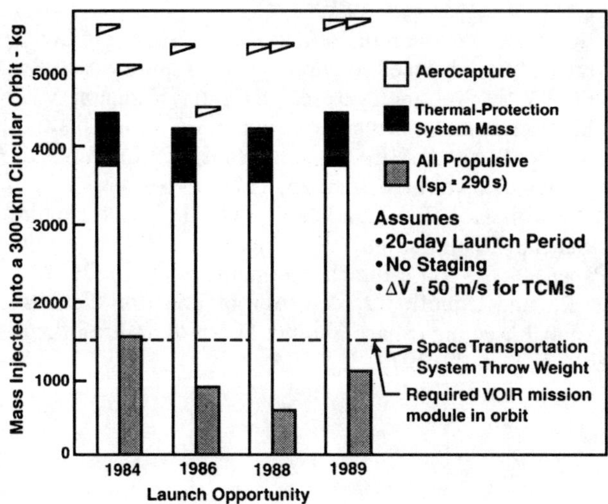

Fig. 12 Venus aerocapture performance. For launch by Shuttle and twin stage IUS.

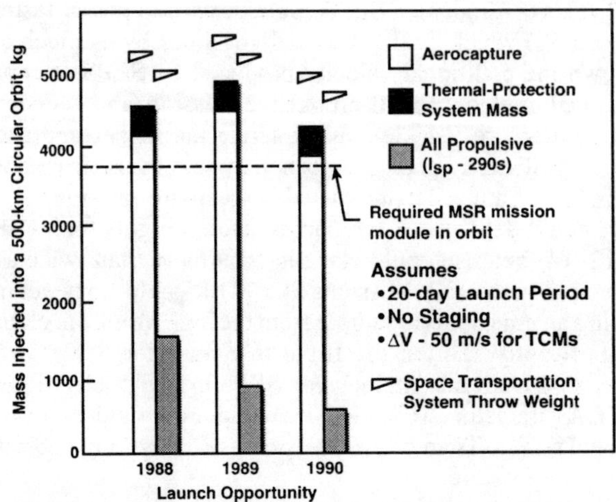

Fig. 13 Mars aerocapture performance. For launch by Shuttle and twin stage IUS.

the expense of an aerocapture vehicle. And, in many cases, aerocapture would make a mission possible.

Bibliography

Cruz, M. I., ''The Aerocapture Vehicle Mission Design Concept,'' NASA/AIAA Conference on Advanced Technology for Future Space Systems, AIAA Paper No. 79-0893, NASA Langley Research Center, Hampton, VA, May 8, 1979.

Cruz, M. I., ''Aerocapture Vehicle Mission Design Concepts for the Inner and Outer Planets,'' AAS/AIAA Astrodynamics Specialist Conference, AIAA Paper No. 79-115, Provincetown, MA, June 25, 1979.

Cruz, M. I., Armento, R. F., and Giles, W. H., ''Aerocapture—A System Design for Planetary Exploration,'' 30th International Astronautical Federation Congress, Paper No. 79-160, Munich, Germany, Sept. 16, 1979.

French, J. R., and Uphoff, C. W., ''Aerobraking for Planetary Missions,'' AAS 1979 Annual Meeting (Space Shuttle: Dawn of an Era), Paper No. 79-286, Los Angeles, CA, Oct. 29, 1979.

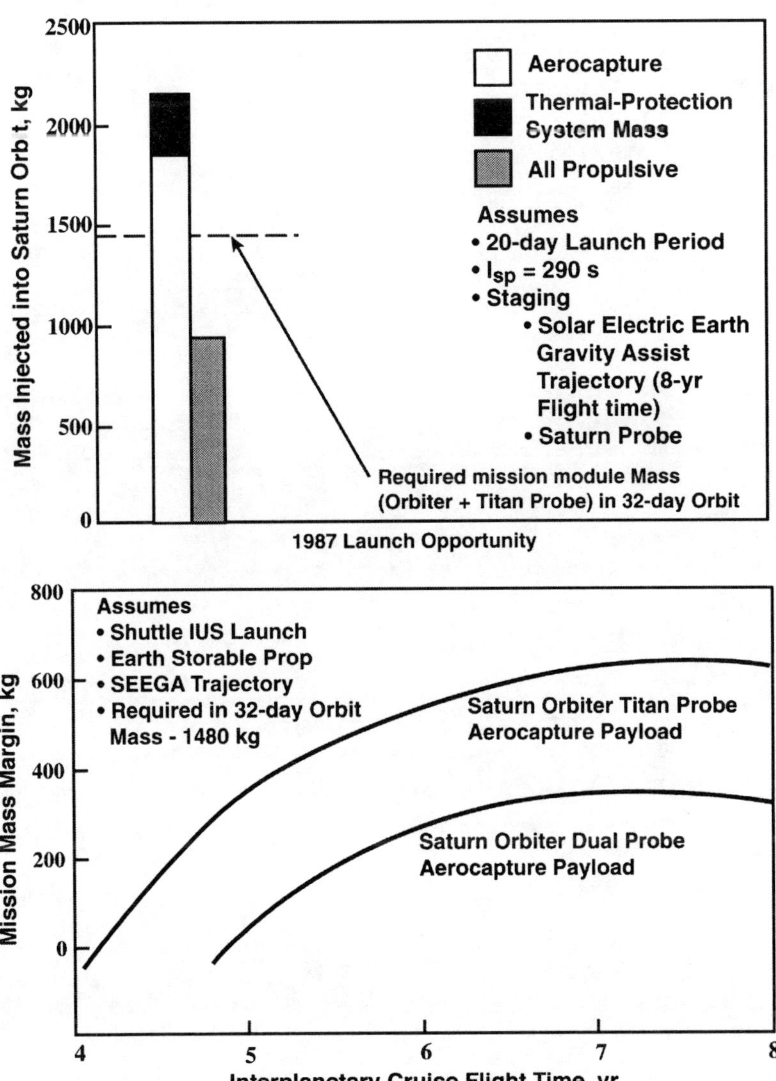

Fig. 14 Saturn/Titan aerocapture performance. For launch by Shuttle and twin stage IUS and a solar electric Earth gravity-assisted trajectory powered by a solar electric propulsion stage.

A New Cryogenic Storage System for Spacecraft

Norman Meyers, Robert J. Dannenmueller, and Ivan G. Hughes
McDonnell Douglas Astronautics Company

Long-term space stations with a number of men in the crew will present the vehicle designer with many technical problems. Few areas will be more critical than efficient cryogenic storage systems of high reliability, minimum weight, and minimum cost, for near-term missions will span years rather than the current weeks or months of operations.

The conventional vacuum-jacketed concept—the early approach to spacecraft cryogenic storage systems—offers advantages primarily during ground hold. Thermal efficiency restricts its application, however, to missions of relatively short duration. Several attempts have been made to improve the performance of vacuum-jacketed systems, but the requirements for minimum weight and overall system volume dictate a minimum vacuum annulus, which, when coupled with the necessity for multilayer radiation shielding and/or vapor-cooled radiation shields, creates difficult vacuum pumping problems and increases manufacturing costs considerably. Various techniques such as ion pumps and "getters" have been applied to the problem of obtaining the necessary annulus pressures, but results have been questionable.

With these problems and concerns in mind, some time ago we reviewed and analyzed all parameters affecting the performance of cryogenic storage systems in the ground-hold, launch, and space environments and concluded that a single-wall system could equal or exceed the performance of the vacuum-jacketed concept. This article describes the recent successful development of such a single-wall cryogenic storage system.

The potential of this system prompted design, fabrication, and test, under simulated prelaunch and orbital environments, of a 24- by 20-in.-diam cylindrical model to verify the predicted thermal behavior of a system. Prior investigations had disclosed that, for the single-wall concept, two distinct types of insulations are required: one tailored specifically for ground hold and the other for the space environment. In addition, it was concluded that a ground cooling coil located at the tank wall could maintain the stored fluid indefinitely. The test results, using liquid oxygen (LOX) as the stored fluid and liquid nitrogen (LN_2) as the ground coolant, compared favorably with a state-of-the-art vacuum-jacketed system. This encouraged design and development of a prototype single-wall system of flight weight.

Astronautics & Aeronautics, Vol. 7, No. 9, Sept. 1969, pp. 62–67.

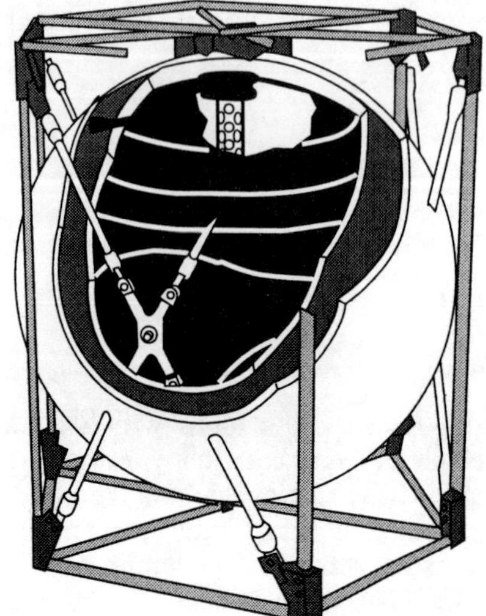

Fig. 1 Single-wall cryogenic storage system.

The development plan followed advanced requirements already specified for a vacuum-jacketed cryogenic storage system, which was designed to fit the physical and functional requirements of the Apollo Service Module (SM) Sector 1. Figure 1 depicts a prototype system. It consists of a 1000-psia pressure vessel of Inconel 718 suspended by glass-fiber tension members within a truss assembly, a combination ground-cooling/regenerative-cooling coil located on the vessel wall, a mass-quantity probe, and a two-component insulation system. Inconel 718, selected for its exceptional mechanical properties, does not necessarily represent the ideal vessel material in terms of strength-to-weight ratio. On the other hand, an unusual forming operation made it possible to manufacture the vessel at a fraction of the cost normally incurred for high-pressure vessels.

Closed-cell polyurethane foam was chosen for the ground-hold insulation because of its low weight, thermal efficiency, and fabricability. Combined with the ground refrigerant, it permits a reliable, predictable assembly. The second phase of the insulation system consisted of 100 layers of perforated, crinkled, aluminized polyester film. Special precautions must be taken with this material to realize its high performance potential. The packing density of the reflective films should be in the range of 50–70 layers/in., and the perforations must be random to avoid direct radiation leakage through the assembly. Similar precautions govern the overlap between adjacent sheets of the reflective film. The perforation percentage depends upon a tradeoff between the degradation of insulation from radiation leakage and the necessity for allowing the gases trapped between the insulation layers to vent rapidly during launch. Moreover, since this class of insulation is effective only at very low pressures, the pressure within the layers should reach 10^{-5} to 10^{-6} torr as rapidly as possible.

The entire vessel assembly is suspended by 12 glass-fiber tension members within a tubular, six-sided aluminum truss—a configuration fitting the Apollo SM compartment but readily adaptable to other vehicles. An extensive development program determined the optimum tension-member length and thermal-conductivity/area-ratio configuration. The rods selected, wrapped with radiation shielding, contribute a total heat flux of only approximately 1 Btu/h into the system.

The vapor barrier to prevent moisture absorption in the insulation system during ground hold consists of a polyethylene membrane. The area within this membrane is purged with a noncondensable gas during ground hold and is vented during ascent.

For high-demand expulsion rates, the system can carry an electric heater/blower assembly mounted externally, as shown in Fig. 2, or internally mounted in the conventional manner. Static destratifiers can be used internally in place of a blower. Other subsystems include a mass gage, indicating fluid density, and a fill-and-discharge circuit that permits either natural or forced expulsion.

A single stainless steel tube bonded to the exterior surface of the pressure vessel with aluminized polyurethane resin serves the dual ground- and regenerative-cooling system. To insure the most favorable heat-transfer characteristics during both phases of cooling, the tube has been carefully routed to intercept the major sources of heat input. During ground hold, a cryogen is circulated through the tubing to eliminate fluid loss entirely, and, depending on the cooling time and system pressures used, the ground cooling can also subcool fluid in the tank.

During orbital storage, especially for long missions, the regeneration function of this tubing becomes very significant in terms of overall system thermal efficiency. The fluid is withdrawn by isenthalpic expansion from the 1000 psia operating pressure in the tank through the regeneration valve and expansion

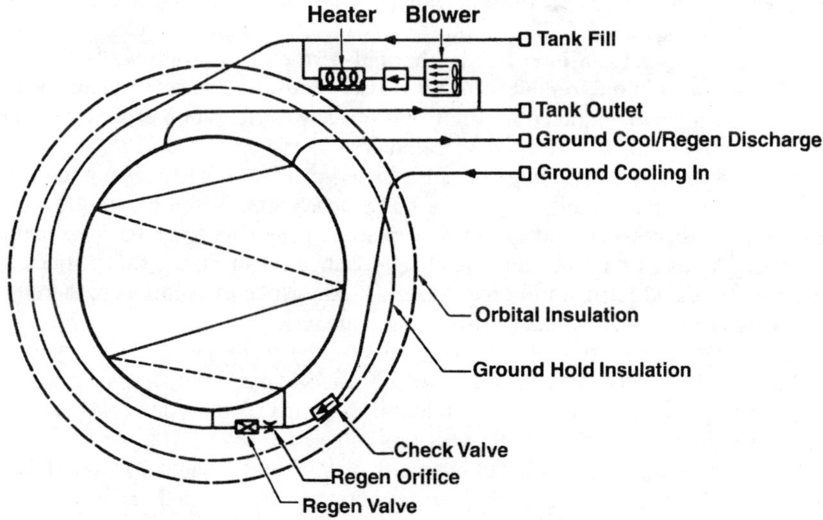

Fig. 2 Schematic of a single-wall cryogenic storage system with an electric heater/blower assembly.

Table 1 Original system design objectives

	Oxygen	Nitrogen	Hydrogen
Size sphere, in.	39	39	39
Material – pressure vessel	Inco 718	Inco 718	Optional
Capacity – lb	1200	850	75
Ullage – %	4.43	4.35	2.97
Vibrational loading – three axis, *g*	10	10	10
Dry system weight – lb (w/o truss)	247	247	<247
Maximum operating pressure, psia	1000	1000	1000
Proof pressure, psia	1670	1670	1670
Design burst pressure, psia	2200	2200	2200
Heater capacity at 28 VDC, W	400	400	400
Ground-hold heat leak, Btu/h	500	525	600
Ground coolant flow rate, lb/h	9	9	75 He 30 SH_2
Orbital heat leak, Btu/h	15	15	17
Ground-hold time without cooling – to 1000 psia, h	18	13	3.2
Orbital hold time – 14.7 psia to 1000 psia – no heaters, h	754	492	145
Orbital hold time – 14.7 psia to 1000 psia with heaters fully energized, h	3.6	2.5	1.0

orifice to the tubing, which is at a relatively low pressure. The refrigeration capacity of this expanded fluid is nominally 40–50 Btu/lb for LN_2 storage.

The single-wall system without regenerative cooling extends orbital-hold times well beyond the ability of present vapor-cooling systems. The addition of the regeneration valve and Joule-Thompson expansion system introduces very little weight penality yet can increase the thermal performance 25–40%.

Development of the Joule-Thompson orifice hardware and performance verification testing was completed in October 1968. The refrigeration process obtained from the expansion was verified as isenthalpic over a temperature range of 160–410 R—the maximum coverage for this test set-up. (This range does not represent a limitation of the expansion process.) The nearly 100% isenthalpic efficiency of the expansion technique, coupled with mass-flow data from the verification testing, has confirmed the analytical approach used in sizing the orifices for specific flow and refrigeration requirements. An orifice assembly can, therefore, be tailored for a specific set of mission parameters.

Table 1 gives the original design objectives—basically those of the cryogenic storage system selected as a baseline for performance comparison, but more conservative from the strength standpoint, because of the comprehensive test sequence planned for this single-wall prototype.

Thermal testing was conducted in an 8-ft-diam altitude chamber at the McDonnell Douglas Astronautics Propulsion Laboratory. Results indicated a thermal performance, both ground and orbital in the nonregeneration mode, equalling or bettering predicted values. The orbital heat-leak rate was less than 9.0 Btu/h, and, during ground hold, there was no loss of fluid.

Thermodynamic testing of a cryogenic storage vessel with highly efficient insulation proves challenging because the vessel must have constant internal pressure for results to be valid. Tests involve maintaining a constant absolute pressure inside the vessel and measuring the mass-flow rate of the vent gas as an indication of the heat flux through the insulation. Any reduction in control pressure lowers the energy-storage capacity of the fluid; to re-establish an equilibrium condition, the excess energy is dissipated by a more rapid vent rate, even though the heat flux remains essentially constant. The reverse is true for a pressure rise. Either occurrence falsifies the heat-leak record.

Using the 39-in. prototype tank as an example, with 850 lb of liquid nitrogen onboard and testing at near atmospheric pressure in the tank, a 1-psi fluctuation in pressure causes a 300-Btu change in potential energy. Commercially available pressure controllers are, at best, limited to ±0.01-psi fluctuation, which introduces a ±3-Btu/h discrepancy, or a maximum possible error of 6 Btu/h. Since the total heat flux into the tank in this example is approximately 9 Btu/h, a 6/9 (66.7%) error can cause an instantaneous reading. The test period must, therefore, be increased until the instantaneous 6-Btu/h error becomes an acceptable portion of the accumulated quantity of heat that has entered the fluid. For instance, a 100-h test on the test tank would bring a total heat flux of approximately 9 Btu/h × 100 h, or 900 Btu. Thus, maximum measurement error would be 6/900 Btu, or 0.667%.

As another complicating aspect of current pressure controllers, the pressure-control abilities claimed for them assume a constant temperature environment. Because these devices generally use a reference pressure volume, a fluctuation of 1°F in the sample case can cause a pressure fluctuation of 1/530 × 14.7 = 0.0277 psi, with an accompanying 8.3-Btu fluctuation in apparent heat absorption.

So an extremely sensitive pressure controller and a precisely controlled temperature environment are necessary to minimize test duration. This problem becomes even more acute with increasing tank size.

Additional nonregeneration subcritical testing, completed in February 1969 at the NASA Manned Space Center in Houston, yielded the following excellent results: equilibrium steady-state orbital heat leak of 8.56 Btu/h, equivalent to an orbital boiloff rate of 0.100 lb/h, and overall thermal conductance of 0.0255 Btu/h R. Orbital time accumulated at equilibrium; subcritical operation during this test phase totaled some 300 h. Based on this information, we predict the nonregeneration supercritical, equilibrium heat leak to be 6.50 Btu/h.

All test data to date, both here and at NASA Marshall, confirm the single-wall concept as a prime candidate for cryogenic storage systems on current and projected long-duration spacecraft. Figures 3–5 show predicted performance based on experimental results to date, and Figs. 6 and 7 present a conservative indication of system performance for a 60-in. tank applicable to a long-duration multiman space station.

The following brief discussion of the analysis will serve to underscore the simplicity of this single-wall concept.

Thermal performance was estimated by summing the radiative and conductive heat fluxes from the thermal environments of a typical Earth-orbital flight at 140-n-mile altitude. Assuming an α/ϵ ratio of 1.0, the equilibrium temperature for a typical spacecraft is approximately 520 R. The pressure at 140-n-mile

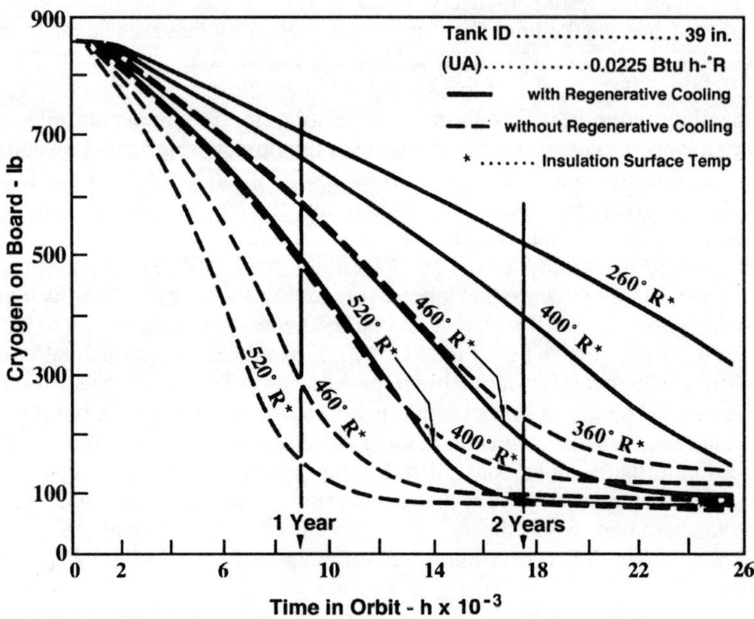

Fig. 3 Liquid nitrogen mass history with and without regenerative cooling.

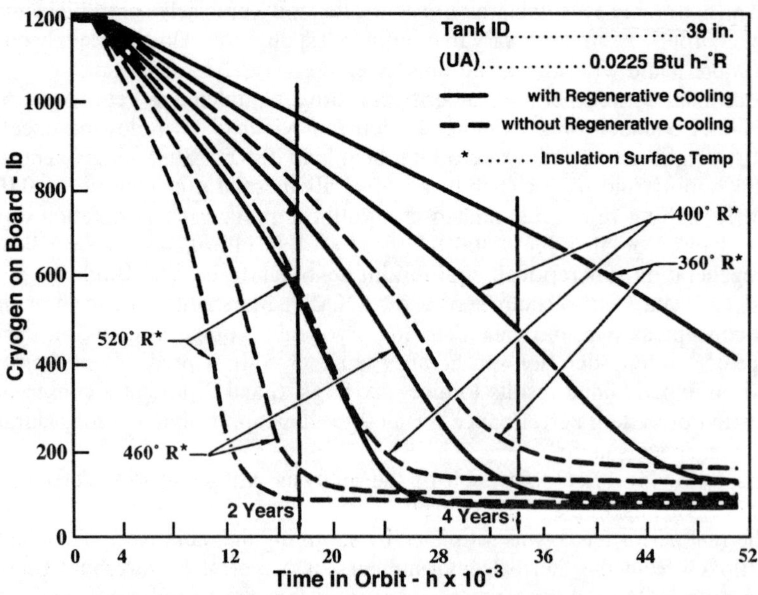

Fig. 4 Liquid oxygen mass history with and without regenerative cooling.

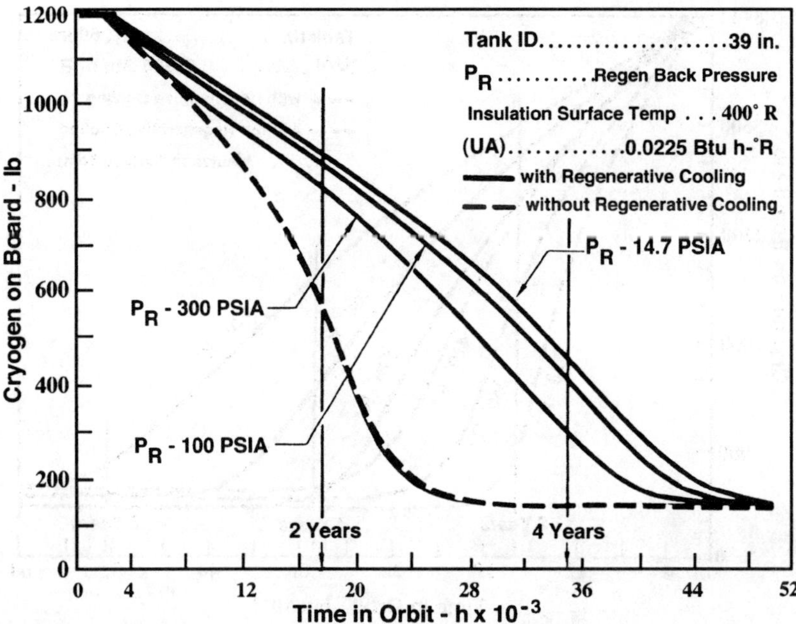

Fig. 5 Effects of regenerative back pressure on orbital oxygen storage.

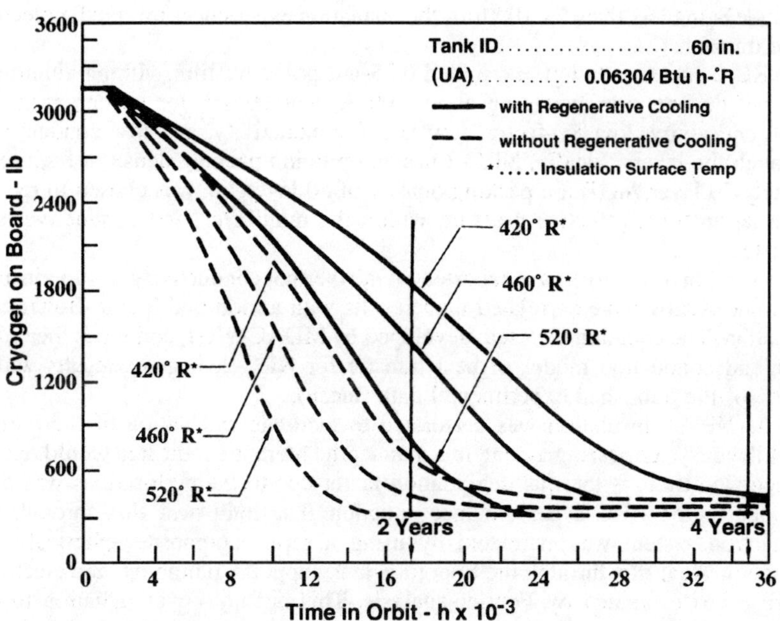

Fig. 6 Liquid nitrogen mass history.

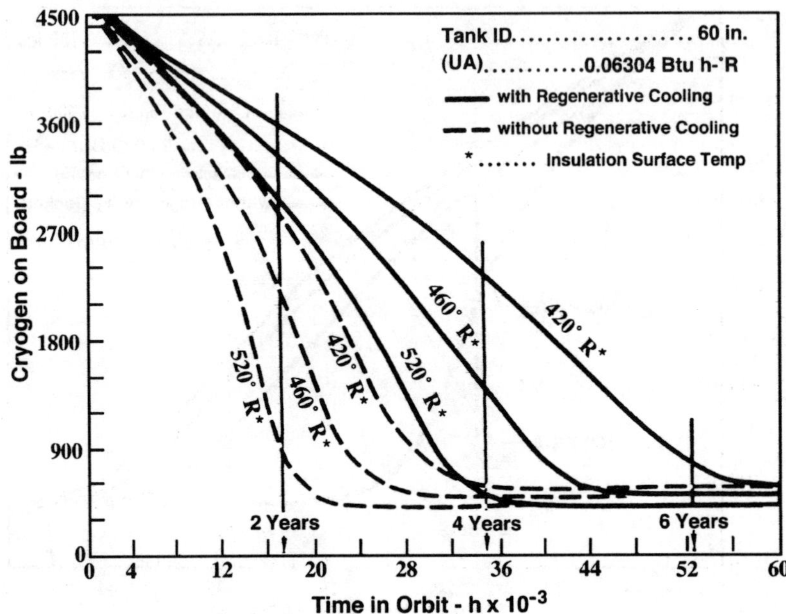

Fig. 7 Liquid oxygen mass history.

altitude being less than 1×10^{-6} torr, the insulation experiences minimal molecular heat transfer.

NRC-2 superinsulation, a crinkled 0.25-mil polyester film with an aluminum deposit on one side averaging about 300 Å, was chosen for its demonstrated high emissivity, low $K\rho$-factor, low lateral conductivity, and low conductivity through the layers. Ideally, NRC-2 has an optimum packing density of approximately 70 layers/in. But a packing density of 50 layers/in. was chosen to reduce evacuation time, effective pressure within the multilayer, and conductive heat transfer.

Several investigators have reported on the thermal conductivity of superinsulation but usually have correlated their results with a modified Stefan-Boltzmann equation. The equation we used, developed by MDAC-West, combines the radiation and conduction modes of heat transfer for NRC-2. It gives results within ±6% of the published experimental data (mean).

The NRC-2 insulation was perforated to facilitate evacuation of the purged multilayer system during ascent, in a manner the literature indicated would reduce evacuation time. A thermal-degradation factor due to the perforations was then applied to the NRC-2 performance equation. The total heat flux through the insulation system was estimated by using a series composite spherical-shell equation. Heat flux through the fiberglass tank supports, plumbing, and electrical wiring was estimated by Fourier analysis. This did not cover radiation to the outside of the supports and plumbing since a few layers of multilayer insulation minimized this flux. The internal radiation in the supports and plumbing was also neglected because of the high *L/D* ratios and the curved nature of the plumbing.

Electrical wiring for the tank heaters and the mass-quantity probe exits the tank through the vent line. This tends to minimize the heat flux during nonregenerative venting. However, during periods of increasing pressure and regenerative cooling, this flux must be accounted for because energy will either be conducted into the bulk fluid or will lower the quantity of the fluid before expansion, thus degrading its refrigerating capacity. Insulation degradation due to penetrations was minimized by attaching the tank supports tangent to the tank wall and by routing the fill and discharge lines gradually through the insulation system.

In terms of the typical temperature-entropy diagram shown in Fig. 8, the thermodynamic processes of the bulk fluid and the expanded fluid become readily apparent. The tank is filled at 1 (saturated liquid). However, since there is a small ullage percentage, the average specific volume is represented as 1′. As energy leaks into the system, the temperature increases (1′ to 2), which causes the pressure to increase. The slope of the temperature-pressure relationship depends on the initial percentage of ullage because this determines the primary mode of pressure increase. The time required to reach the discharge pressure is approximated through the following relationship:

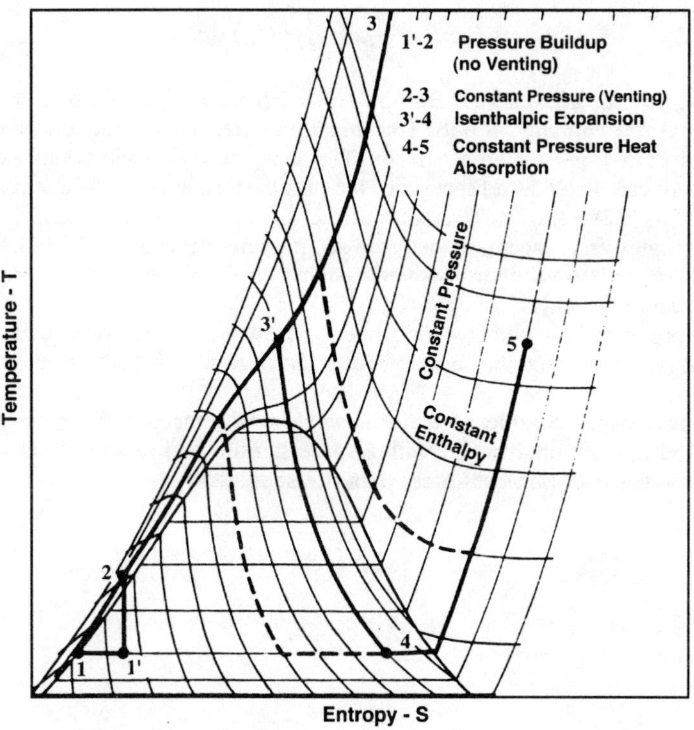

Fig. 8 Thermodynamic process with regenerative cooling.

$$t = \frac{M\,(u_1 - u_2) + m\bar{C}_p\,(T_1 - T_2)}{\bar{Q}}$$

where t = discharge time, M = mass onboard, u_1 = internal energy at 1 (N.B.P.), u_2 = internal energy at 2 (vent pressure), $\bar{Q}$ = average total heat flux into system, m = mass of tank and equipment affected by temperature increase, $\bar{C}_p$ = average specific heat of tank and equipment, and $T_{1,2}$ = temperature at 1 or 2.

The process from 2 to 3 has constant pressure. Regenerative expansion takes place at any point (3′ to 4) to some reduced pressure (4). The process now is one of constant-pressure heat absorption as the fluid is circulated about the tank until it reaches the temperature of the shell (5). At this point, the fluid is routed to the supply system. The change in enthalpy from 4 to 5 represents the refrigeration ability of the expelled fluid.

The mass-depletion curves for cryogens in an orbital environment were calculated assuming the insulation scheme to be in equilibrium and the normal boiling point liquid at $t = 0$ h. As the pressure increases to 1000 psia, no discharge of fluid occurs, and the process is one of constant volume. The specific energy absorption E_{sp} of the cryogen is $E_{sp} = u_2 - u_1$.

After the pressure reaches 1000 psia and fluid discharge is initiated, the process has constant pressure. The specific energy absorption into the tank is calculated by balancing the total energy of the system by the expression:

$$E_{sp} = u_2 \frac{\rho_1}{\rho_2} u_1 \frac{1}{\rho_2} \int_{\rho_1}^{\rho_2} h(\rho)\,\mathrm{d}\rho$$

where E_{sp} = specific energy, Btu/lb; u = internal energy, Btu/lb; ρ = density, lb/ft^3; and h = enthalpy, Btu/lb. This equation is readily adaptable to numerical analysis or a finite difference form. The mass-depletion plots and expulsion rates were calculated as a function of the outside-insulation surface temperature, assumed as the average temperature of the surrounding environment.

The single-wall concept, so analyzed and properly designed and fabricated, has proven superior to vacuum-packeted systems in performance, cost, complexity, weight, and reliability.

The program now aims at two major objectives: to improve performance without reducing reliability and to refine fabrication techniques to reduce time and cost.

We have every confidence that this single-wall concept will be successfully employed in large, multiman orbiting space stations and in unmanned satellites using cryogens for purposes other than life support.

Applications of Materials to Solid-Rocket Nozzles

Edward W. Ungar
Battelle Memorial Institute

Solid-propellant rocket engines offer distinct advantages over liquid-propellant engines, particularly simplicity and constant readiness. However, one disadvantage is that the solid rocket's fuel cannot be used directly to cool the exhaust nozzle, as is done commonly in liquid engines. The solid-rocket nozzle must either be capable of operation in the environment produced by the combustion product or some type of a built-in thermal-protection system must be used.

The conventional bell nozzle is the type which is currently used most widely. It is a fixed-geometry device which has a fixed expansion ratio and, thus, with a fixed combustion pressure, has the optimum shape at only one altitude. Advanced nozzle concepts, such as the plug nozzle[1-3] and the expansion-deflection nozzle,[4] which are being studied, offer the advantage of altitude compensation.

The design of minimum-weight nozzles for solid rockets has become increasingly difficult because of increases in burning times and the introduction of propellants with higher and higher performances. Burning time is a critical factor because the nozzle must maintain its integrity while exposed to exhaust products as long as an engine operates. Propellants currently being developed produce exhaust products with higher temperatures and greater chemical activity than those used previously. In addition, these products of combustion often carry a high concentration of particles in suspension.

In view of these trends, our purposes here are to identify problems involved in both the selection of materials and the development of nozzles for use in solid-propellant rocket engines and also to examine approaches being taken to solve these problems.

Structural Design Concepts

The environment produced by the combustion of a solid propellant imposes restrictions on the selection of materials for use in uncooled exhaust nozzles. These restrictions are in addition to those normally associated with selecting materials for flight vehicles. Let us approach the materials-selection problem by a brief description of the hypothetically ideal material.

An ideal material must operate for extended periods essentially at equilibrium with the hot combustion products. The material must, therefore, 1) be sufficiently refractory to withstand the maximum combustion temperatures contemplated while retaining a reasonable strength, 2) be chemically compatible with combus-

Astronautics, Vol. 6, No. 4, April 1961, pp. 24–26, 83–86.

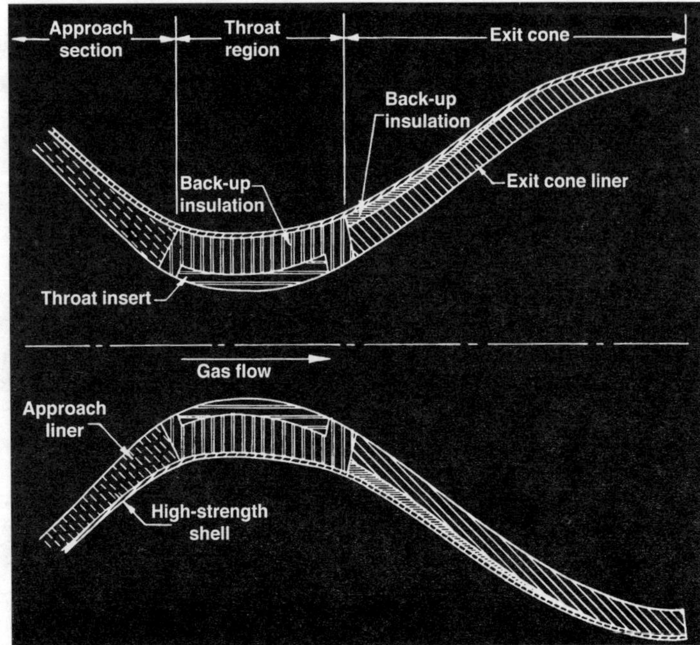

Fig. 1 Modified hot-structure-type of nozzle.

tion products, 3) be resistant to both thermal and mechanical shock, 4) be sufficiently tough to withstand mechanical abrasion resulting from high-velocity particle impacts, 5) have a high strength-weight ratio, 6) be ductile, 7) be readily available and present no serious fabrication difficulties, and 8) have reasonable cost.

Unfortunately, this ideal material does not exist and probably will not be available in the foreseeable future. Therefore, schemes are required which allow the use of materials of lesser potential. These schemes involve the use of several materials in a composite structure that utilizes the desirable properties of each.[5] The effects produced by undesirable properties (such as high density) must be minimized in the design of the composite structure. A composite nozzle structure is illustrated in Fig. 1.

Hot Structures

A ''hot'' structure is one which operates continuously with exposed surfaces so hot as to be almost adiabatic. Materials for the exposed element of a hot structure must possess at least the first three properties of the ideal material. The supporting materials of the composite structure must compensate for deficiencies in the other properties listed. Low thermal conductivity, thus, becomes desirable to prevent overheating of the less refractory components in the composite.

There are advantages in the use of hot structures and also certain difficulties involved in attaining such a system in practice. The principal advantages are reproducibility of engine performance and possibility of a minimum-weight design.

Heat Sink Structures

The heat sink concept is based on the assumption that a structure which would ultimately fail or melt in a severe thermal environment may perform satisfactorily while it is being heated. Thus, for a finite time after introduction into the environment, the structure is satisfactory, but failure by melting or loss of strength will result from exposure for longer periods of time.

To extend the time of satisfactory operation, the material must be capable of rapidly transferring this heat impulse into the interior of the structure. That is, the material must possess a high thermal conductivity. When the mass of the structure must be minimized for a particular total energy absorption, the product of specific heat and available temperature range as a solid must be large. When rate of absorbing heat becomes the dominant requirement, the thermal diffusivity is the more important material property.

In practice, composite nozzle structures can be used that employ highly refractory materials on the exposed surface but also have an appreciable heat sink capacity in the underlying material. The exposed surface layer is, thus, cooled by the underlying materials sufficiently to delay surface melting or appreciable loss in strength beyond the required operating time. These structures might be termed modified hot structures.

Ablative Structures

Ablation systems have been studied extensively in connection with the reentry problem. The ablation process consists of the formation of relatively cool gas and/or liquid melt-layer by decomposition, melting, sublimation, or melting with subsequent boiling. A good general discussion of this subject, pertinent to nozzle design, was presented by Adams.[6]

The evolution of a gas may come about through thermal decomposition of a plastic material, boiling, or sublimation. Usually, the temperature of the evolved gas is considerably lower than the mainstream gas temperature. The gas then cools the boundary layer, resulting in a net decrease in the heat flux to the solid body. The evolution of gas also thickens the boundary layer, which also tends to decrease the heat flux into the solid body by decreasing the heat-transfer coefficient. Thus, a material which generates a large volume of gas is desirable.

The effectiveness of the gas as a coolant depends to a large extent on the amount of energy it absorbs while being heated as a gas. The total heat-absorbing capacity also depends on the energy absorbed by the gas during its formation. This energy of formation is all the heat which goes into latent heat for melting and subsequent boiling or sublimation, plus the heat of formation if the gas is formed from a chemical decomposition process.

A liquid melt-layer results when the surface material melts to form a highly viscous liquid. The thermal protection resulting from the presence of the liquid arises from the energy absorbed by the material during the melting process and from the insulating effect of the melt layer. The temperature of the solid-liquid interface is fixed by the melting process, although the temperature at the outer edge of the melt is generally higher. Energy is absorbed by the liquid when its temperature is raised above the melting point. If the temperature at the outer edge of the melt is high enough for boiling to occur, then the additional beneficial thermal effects resulting from boiling are obtained.

Effective Heat of Ablation

The effective heat of ablation is defined as the ratio of the heat-transfer rate to a nonablating surface to the rate at which material is lost from the surface on a mass basis. Because the protection offered by the ablation process depends on interactions between ablation products and the hot gas, the effective heat of ablation is a function of gas temperature and composition.

However, the point of primary interest in this discussion is that the heat flux is a predominant factor influencing the performance of an ablative material. Therefore, the chamber pressure can be more important than the flame temperature. This importance of heat flux is similar to the case of a hot structure that has not yet attained thermal equilibrium.

Environmental Criteria

The interior surface of a rocket nozzle is exposed to a high-temperature, chemically reacting, high-velocity, abrasive flow of combustion products that is intimately related to the properties of the propellant employed. The environmental factors which affect nozzle performance may be divided into three groups, namely, thermal factors, chemical factors, and physical factors. Complex interrelationships among these factors are generally involved in the deterioration of a nozzle surface.

Table 1 summarizes some of the salient features of the nozzle environment. The Materials Advisory Board has estimated that nozzles will have to handle combustion temperatures of about 8300°F in the period 1970–75.[7]

The heat flux to a nozzle throat depends on both the chamber pressure and the enthalpy of the combustion products. In addition, condensed exhaust products affect the heat-transfer processes, primarily by forming an insulating layer between the structure and the gas stream. Nongaseous exhaust products result from the addition of metals to fuels. Numerous metal additives are being considered for future propellants.[8]

Thermal conditions vary significantly along the length of a nozzle. Figure 2 shows the manner in which heat flux varies as a function of axial distance in a nozzle.

The chemistry of the expanding gas in a rocket nozzle is not completely understood, owing to the complex composition of solid propellants and the lack of high-temperature kinetic data. The gas consists of a large number of compounds, and the combustion process may still be going on during expansion in the nozzle. Furthermore, the equilibrium composition is changing as the static temperature and pressure decrease. Then, too, the extent to which the exhaust products approach equilibrium at every place during expansion is not known because neither the reactions occurring nor the kinetics of these reactions are known accurately.

A number of studies of the flow of reacting mixtures are in progress. In addition to the difficulties associated with defining the gas composition, there is a further

Table 1 Nozzle environments and philosophies

Time period	1960–65	1965–70	1970–75
Representative flame temp, °F	6300	7000	8300
Estimated heat flux to throat, Btu/sq-in.-s (at 600 psi)	~30	~35	~60
Chemistry of exhaust products	Dissociated, generally reducing	Dissociated, generally reducing	Highly dissociated, fluorine may be present
Nongaseous exhaust species	Primarily aluminum oxide	Depending on propellant: oxides of aluminum lithium beryllium magnesium	Depending on propellant: oxides of lithium beryllium magnesium
Available basic materials for equilibrium operation	Tungsten, some carbides	Some carbides, nitrides, and borides	None
Pertinent design philosophy for throat region in composite nozzle structure	Modified hot structure 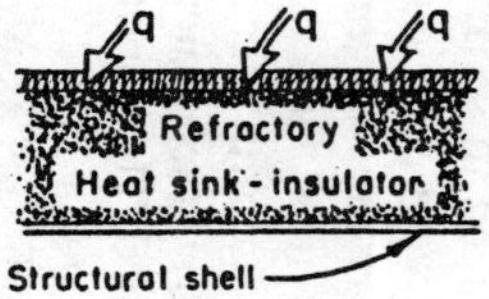 q q q Refractory Heat sink - insulator Structural shell	Modified hot structure 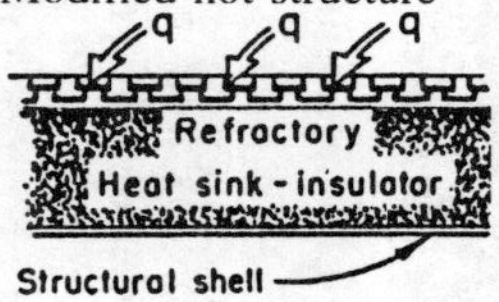 q q q Refractory Heat sink - insulator Structural shell	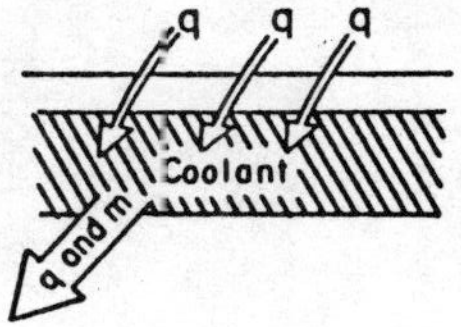 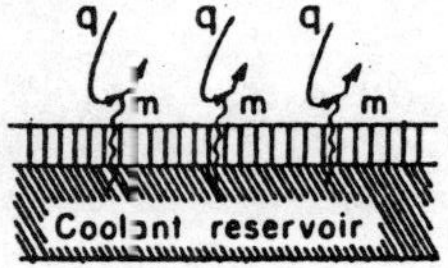 Cooled structure q q q Coolant q and m q q q m m m Coolant reservoir

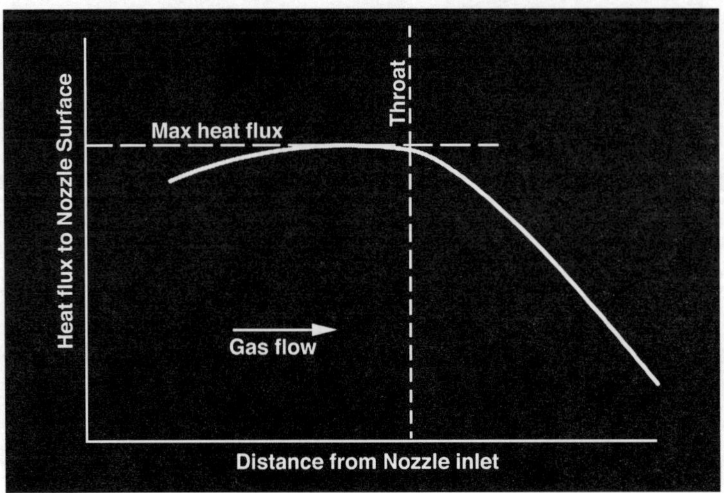

Fig. 2 Heat-flux distribution in a nozzle.

difficulty because the gas may react with the interior surface of the nozzle. Reactions with the interior surface of the nozzle are of primary interest in this discussion.

Laboratory studies of surface reactions influencing nozzle performance are hindered by the complexity of the mixture and the fact that, even if a particular species does not appear as a reactant, it can act as a catalyst. There is reason to believe that the presence of a molten oxide layer on a surface affects the chemical reactions. It would tend to separate gas species from the nozzle surface, thus reducing gas-solid reactions, if the molten layer is sufficiently dense. The oxide may also act as a catalyst, which may be undesirable. It is difficult, if not impossible, at the current state of the art to define all of the specific reactions which can produce harmful effects on nozzle performance. Consequently, the compatibility of each surface material with each chemically active exhaust product species must be studied. The possibility of catalytic action on each of the reactions should not be neglected.

The exposure conditions associated with future propellants will undoubtedly be more severe than those now encountered. These new fuels will introduce higher heat fluxes, associated with higher flame temperatures, and also greater chemical reactivity in the exhaust gas. For example, the introduction of fluorine into solid propellants, as discussed by Farber,[9] will both raise flame temperatures and introduce additional highly reactive chemical species into the combustion products.

Materials Development, Utilization

In a composite nozzle structure, several classes of materials are generally used, each material performing the function for which it is best suited. The choice of material depends largely on the tasks which must be performed: load carrying, heat absorption, insulation, erosion, resistance, chemical protection, ablative cooling, etc. Only after the task or tasks have been identified can the best class or specific material be identified.

Graphite

Graphite is a "workhorse" material in high-temperature applications. It sublimes at a temperature of about 6700°F, which is higher than most materials. It is not a very strong material at low temperatures, but its strength increases with temperature up to about 4500°F. At high temperatures, graphite is considered a strong material. Graphite has a tendency to oxidize exothermically at relatively low temperatures. It has found wide use as a nozzle-insert material when the exhaust products are generally reducing. The brittleness and high thermal-expansion coefficient of graphite lead to design difficulties when it is employed in a composite structure. Some of the design problems encountered with graphite are discussed by Krochmal and Anthony.[10] The high sublimation temperature of graphite and its relatively high specific heat and high latent heat of sublimation make it attractive as a heat sink material. In addition to its thermal properties, graphite has a relatively low density (2.25 g/cc), which tends to make it attractive for flight applications.

A vapor-deposition process of fabricating graphite has led to the development of a graphite coating with anisotropic thermal properties. This development—pyrolitic graphite—may eventually have an important effect on nozzle design in the near future. The thermal conductivity of pyro-graphite along the surface as formed is an order of magnitude greater than its thermal conductivity in the normal direction. The high-temperature region can be limited to a thin layer near the surface, reducing insulation requirements. It is possible that conduction of heat along the surface can appreciably reduce heating below the surface at a region of high heat flux by providing a heat-leakage path to a region of lower heat flux.

Due to its anisotropic thermal expansion, there are still many problems concerning the use of pyro-graphite. Also, many early rocket-firing tests were hampered by inconsistent coatings. When one designs with pyro-graphite, its anisotropic thermal-expansion properties must be considered. As experience in the use of the material is gained, it will probably find many applications in rocket nozzles.

Refractory Metals

A few important metals that are among the most refractory materials available are tungsten, tantalum, and molybdenum. The refractory metals have high tensile strength and good resistance to thermal shock when compared with most ceramic materials. The high density of these metals limits their use in nozzles to specialized applications, such as throat inserts. As with other metals, the refractory metals suffer from a loss in strength at temperatures approaching melting point, and this limitation becomes important because throat-insert materials must operate at near-melting temperatures. Structural and fabrication problems sometimes become severe—particularly the brittleness of molybdenum, the difficulties encountered in fabricating large parts of tungsten, and the poor thermal-shock resistance of tungsten. Alloys of the refractory metals have been used to obtain a reasonable compromise between ease of fabrication and high-temperature properties.

Tungsten, with a melting point of 6160°F, is widely used in the high-temperature region of nozzle structures. The high density of tungsten and difficulties in fabrication have led to its use only in critical nozzle regions, such as the throat. Dense, forged sintered tungsten is available. Thin, hot-spun tungsten nozzle forms have also been fabricated.[5] The current need for tungsten parts has stimulated development of forming methods. At present, high-temperature physical and thermodynamic property data are limited. Additional data are needed above 3000°F so that this material can be utilized to its full potential in nozzle designs.

Ceramics

The metal oxides have high melting temperatures, although not quite as high as tungsten. The metal oxides are brittle and more susceptible to thermal shock than are most metals. The densities of the ceramic oxides are considerably less than those of the pure refractory metals.

Metallic compounds, such as the carbides, nitrides, and borides, are similar in many respects to the oxides, with the exception of their chemical stability. The metallic compounds are sometimes referred to as the hard metals because of their high conductivities and metallic luster in addition to hardness. Whereas the oxides are usually stable in an oxidizing environment and unstable in a reducing environment, the opposite is usually true for the carbides, nitrides, and borides.

Melting Points

The melting points of the metallic compounds are not necessarily as high as either of their constituents. For example, tungsten carbide has a melting point of about 4710°F, whereas pure tungsten melts at about 6160°F, and graphite sublimes at 6700°F. This effect of depressed melting point is one that is commonly encountered. On the other hand, hafnium carbide and tantalum carbide both melt at temperatures higher than 7000°F. However, the former is extremely expensive, and the latter has a high density and an unusually strong tendency to oxidize.[11] A number of carbides and nitrides offer the possibility of operation at temperatures of the order of 7000°F. The promising materials are generally highly susceptible to thermal- and mechanical-shock failure. Their limitations can probably be overcome by the development of composite materials.

The same specialized applications which have led to the use of the composite nozzle structure have also led to the current interest in composite materials for applications within the structure. The composite-material concept permits materials to be designed which satisfy particular needs. A composite material may incorporate a ductile material and a refractory brittle material to achieve a combination that will withstand both the thermal shock and the high temperatures associated with a rocket-engine firing. The Avcoite family of materials, described by Scala,[12] is typical of this type of composite. These composites incorporate a brittle refractory in a metal honeycomb. It is believed that the composite-material concept will be needed in order to utilize the high melting temperatures of some carbides and nitrides.

A second form of composite material is being applied in ablation systems. The performance of an ablation material depends to a large degree on the cooling capacity of the ablation products, rather than on the high-temperature properties of the solid material. Hence, the desirable ablation materials are those which produce large quantities of gas, preferably with a low molecular weight, and/or viscous melt-layers. Therefore, the most commonly used ablation materials are composites of glass-like materials and plastics resins.

Common resins are the phenolics, melamines, silicones, epoxies, and polyesters. The glass is generally used in fiber form, and various glassy materials have been used in ablating composites. They improve the strength characteristics of the composite material. A large number of material combinations are now available under a number of trade names. The recent introduction of graphite cloth on the market has lead to the consideration of impregnated graphite cloth as a nozzle material.

Values of the effective heat of ablation, ranging from about 1000 to about 6000 Btu/lb, are now attainable. The reason that ablating materials are not applicable to the high heat-flux, long-time exposure obtained in the throat of a rocket engine can best be illustrated by a simple example. Consider a material with an effective heat of ablation of 2000 Btu/lb and a density of 100 lb/cu ft. This material may be considered to be representative of the general class of ablating reinforced plastics. It is also reasonable to assume a heat flux in a nozzle-throat region of the order of 20 Btu/sq in./s. Thus, a linear erosion rate of about 0.17 in./s would be obtained with this material, if it were used at the nozzle throat. If the burning time was of relatively long duration or the nozzle diameter small, the total throat erosion would be intolerable. On the other hand, if a short-burning-time application is considered, then the application of an ablating material may be permissible in large nozzles.

Reinforced plastic composites usually have a low thermal conductivity besides a low density, which make them useful backup insulation in hot structures. Further, when used in ablative structures, the reinforced plastic can provide both the ablative surface and the thermal insulation required to protect the structural shell. Thus, refractory materials may not be required for the shell in either case, and steel and titanium become logical choice because of their high strength-to-weight ratios.

Future Development Requirements

The problems involved in the development of uncooled nozzles are intimately related to the nature of the exhaust products. The velocity increment delivered by a rocket stage increases approximately linearly with propellant specific impulse and as a logarithm of $1/(1 - m_f)$, where m_f is the propellant mass fraction (propellant weight/total weight). Some loss in mass fraction may be acceptable to take advantage of improved propellants. But there are limits to the mass-fraction losses that can be accepted without compromising performance to the point where new propellants are not worthwhile. Therefore, continued nozzle improvements are warranted.

The present nozzle-design philosophy is based primarily on the modified hot-structure concept. Nozzle requirements during 1965–70 will require either that the exposed element of the hot structure be more refractory than the refractory metals or that the structure provide more heat sink than is now provided. Advances in the area of composite materials containing extremely refractory ceramics will determine which philosophy will be adopted. An alternative philosophy, namely, that of the cooled structure, might be utilized in the 1965–70 period and will probably be required in the 1970–75 period. The cooled structure can use either a passive coolant on the back surface of the exposed material or an active coolant which is injected through a porous nozzle surface.

Advances in materials and nozzle-design technology will be necessary to use to advantage advanced propellants. There is need for additional research on exhaust product environments and expansion processes, on composite materials which utilize constituents more refractory than tungsten, and on improved ablation materials. Moreover, we need improved high-temperature property data for the materials now available to permit their more effective utilization.

References

[1] Berman, K., ''The Plug Nozzle: A New Approach to Engine Design,'' *Astronautics,* Vol. 5, No. 4, April 1960, p. 22.

[2] Berman, K., and Neuffer, B., ''Plug Nozzle Flexibility,'' *Astronautics,* Vol. 5, No. 9, Sept. 1966, p. 30.

[3] Chamay, A., et al., ''An Experimental Evaluation of Plug-Cluster Nozzles,'' ARS Preprint No. 1608-61, Feb. 1961.

[4] Rau, G. V. R., ''The E-D Nozzle,'' *Astronautics,* Vol. 5, No. 8, Aug. 1960, p. 34.

[5] Levy, A. V., and Liggett, H., ''Composite Ceramic-Metal Systems for 3000–6000° Service,'' ARS Preprint No. 1572-60, Dec. 1960.

[6] Adams, M. C., ''Recent Advances in Ablation,'' *ARS Journal,* Vol. 29, No. 9, Sept. 1959, p. 625.

[7] ''Report on Materials for Large Solid-Propellant Rocket Motors,'' Materials Advisory Board, National Academy of Sciences—National Research Council, Rept. MAB-150-M, June 12, 1959.

[8] Blackmal, A. W., and Kuebl, D. K., ''The Use of Binary Light Metal Mixtures and Alloys as Additives for Solid Propellants,'' ARS Preprint No. 1595-61, Feb. 1981.

[9] Farber, M., ''Fluorine Solid Propellants,'' *Astronautics,* Vol. 5, No. 8, Aug. 1960, p. 34.

[10] Krochmal, J. J., and Anthony, F. M., ''Application and Processing of Non-Metallics (Graphite) to Leading Edges of Hypersonic Re-Entry Vehicles,'' Symposium on Processing Materials for Re-entry Structures, WADC-TR 60-58, 297, May 1960.

[11] Porter, H. B., ''Rocket Refractories,'' NAVORD Rept. 4893, NOTS 1191, Aug. 1955.

[12] Scala, E. E., ''Composite Materials,'' ARS Preprint No. 1098-60, April 1960.

Index